Lecture Notes in Computer Science 11454

Commenced Publication in 1973
Founding and Former Series Editors:
Gerhard Goos, Juris Hartmanis, and Jan van Leeuwen

More information about this series at http://www.springer.com/series/7407

Paul Kaufmann · Pedro A. Castillo (Eds.)

Applications of Evolutionary Computation

22nd International Conference, EvoApplications 2019
Held as Part of EvoStar 2019
Leipzig, Germany, April 24–26, 2019
Proceedings

 Springer

Editors
Paul Kaufmann ⓘ
University of Mainz
Mainz, Germany

Pedro A. Castillo ⓘ
University of Granada
Granada, Spain

ISSN 0302-9743   ISSN 1611-3349   (electronic)
Lecture Notes in Computer Science
ISBN 978-3-030-16691-5   ISBN 978-3-030-16692-2   (eBook)
https://doi.org/10.1007/978-3-030-16692-2

Library of Congress Control Number: 2019936010

LNCS Sublibrary: SL1 – Theoretical Computer Science and General Issues

This Springer imprint is published by the registered company Springer Nature Switzerland AG
The registered company address is: Gewerbestrasse 11, 6330 Cham, Switzerland

Preface

This volume contains the proceedings of Applications of Evolutionary Computation, the 22nd International Conference, EvoApplications 2019, held as part of EvoStar 2019, in Leipzig, Germany, April 24–26, 2019.

EvoStar, or Evo*, is the leading event on bio-inspired computation in Europe. EvoAPPS, as it is familiarly called, aims to show the applications of research in this field, ranging from proofs of concept to industrial case studies. At the same time, under the Evo* umbrella, EuroGP focused on the technique of genetic programming, EvoCOP targeted evolutionary computation in combinatorial optimization, and EvoMUSART was dedicated to evolved and bio-inspired music, sound, art and design. The proceedings for all of these co-located events are available in the LNCS series.

This volume combines research from the following domains: engineering and real-world applications, games, image and signal processing, vision and pattern recognition, life sciences, networks, neuroevolution, numerical optimization, and robotics.

This year, we received 66 high-quality submissions, most of them well suited to fit in more than one domain. We selected 20 papers for full oral presentation, while a further 24 works were presented in short oral presentations and as posters. All contributions, regardless of the presentation format, appear as full papers in this volume (LNCS 11454).

Many people contributed to this edition: we express our gratitude to the authors for submitting their works, and to the members of the Program Committee for devoting such a big effort to review papers pressed by our tight schedule.

The papers were submitted, reviewed, and selected through the MyReview conference management system, and we are grateful to Marc Schoenauer (Inria, Saclay-Île-de-France, France) for providing, hosting, and managing the platform.

We would also like to thank the local organizing team led by Hendrik Richter from the Leipzig University of Applied Sciences, Germany, for providing such an enticing venue and arranging an array of additional activities for delegates. Our appreciation also goes to the Leipzig University of Applied Sciences for the patronage provided in support of the event.

We would like to acknowledge Pablo García Sánchez (University of Cádiz, Spain) for his continued support in maintaining the Evo* website and handling publicity.

We credit the invited keynote speakers, Risto Miikkulainen (University of Texas, USA) and Manja Marz (Friedrich Schiller University Jena, Germany), for their fascinating and inspiring presentations.

We would like to express our gratitude to the Steering Committee of EvoApplications for helping with the organization of EvoAPPS: Stefano Cagnoni, Anna I. Esparcia-Alcázar, Mario Giacobinni, Antonio Mora, Günther Raidl, Franz Rothlauf, Kevin Sim, and Giovanni Squillero.

We are grateful to the support provided by SPECIES, the Society for the Promotion of Evolutionary Computation in Europe and Its Surroundings, and its individual members Marc Schoenauer (President), Anna I. Esparcia-Alcázar, (Secretary and Vice-President), Wolfgang Banzhaf (Treasurer), for the coordination and financial administration.

And last but not least, we express our continued appreciation to Anna I. Esparcia-Alcázar, from Universitat Politècnica de València, Spain whose considerable efforts in managing and coordinating Evo* helped toward building our unique, vibrant, and friendly atmosphere.

March 2019

Paul Kaufmann
Pedro A. Castillo
Jaume Bacardit
Carlos Cotta
Gusz Eiben
Francisco Fernández
James Foster
Kyrre Glette
Emma Hart
Giovanni Iacca
Juanlu Jiménez-Laredo
Oliver Kramer
J. J. Merelo
Julian Miller
Monica Mordonini
Trung Thanh Nguyen
Sebastian Risi
Günter Rudolph
Sara Silva
Stephen Smith

Organization

EvoApplications Coordinator

Paul Kaufmann   Mainz University, Germany

EvoApplications Publication Chair

Pedro A. Castillo   Universidad de Granada, Spain

Local Chair

Hendrik Richter   University of Leipzig, Germany

Publicity Chair

Pablo García Sánchez   University of Cádiz, Spain

Engineering and Real-World Applications Chairs

Sara Silva   LASIGE, University of Lisbon, Portugal
Emma Hart   Napier University, Edinburgh, UK

Games Chairs

Julian Togelius   Tandon School of Engineering, New York University, USA
Alberto Tonda   Université Paris-Saclay, France

Image and Signal Processing Chairs

Stephen L. Smith   University of York, Heslington York, UK
Monica Mordonini   Universitá di Parma, Italy

Life Sciences Chairs

James Foster   University of Idaho, USA
Jaume Bacardit   Newcastle University, UK

Networks and Distributed Systems Chairs

Juan Julián Merelo-Guervós   Universidad de Granada, Spain
Juan L. Jiménez-Laredo   RI2C/LITIS, Université du Havre Normandie, France

Neuroevolution and Data Analytics Chairs

Julian Francis Miller   University of York, UK
Sebastian Risi   IT University of Copenhagen, Denmark

Numerical Optimization: Theory, Benchmarks and Applications Chairs

Günter Rudolph   University of Dortmund, Germany
Oliver Kramer   University of Oldenburg, Germany

Robotics Chairs

Agoston E. Eiben   Vrije Universiteit Amsterdam, The Netherlands
Kyrre Glette   University of Oslo, Norway

General Chairs

Carlos Cotta   Universidad de Málaga, Spain
Giovanni Iacca   University of Trento, Italy
Trung Thanh Nguyen   Liverpool John Moores University, UK
Francisco Fernández   Universidad de Extremadura, Spain
  de Vega

EvoApps Steering Committee

Stefano Cagnoni   University of Parma, Italy
Anna I. Esparcia   Universitat Politècnica de València, Spain
Mario Giacobinni   Università degli Studi di Torino, Italy
Antonio M. Mora   Universidad de Granada, Spain
Günther Raidl   Technische Universität Wien, Austria
Franz Rothlauf   Mainz University, Germany
Kevin Sim   Edinburgh Napier University, UK
Giovanni Squillero   Politecnico di Torino, Italy
Cecilia di Chio   University of Southampton, UK
  (Honorary Member)

Program Committee

Ahmed Hallawa   RWTH, Germany [General]
Alex Freitas   University of Kent, UK [Life Sciences]
Anca Andreica   Babes-Bolyai University, Romania [General]
Anders Christensen   University Institute of Lisbon, Portugal
                         [Neuroevolution and Data Analytics]
Andres Faina   IT University of Copenhagen, Denmark [Robotics]

Andrew Turner	Simomics Ltd., York, UK [Neuroevolution and Data Analytics]
Anil Yaman	Eindhoven University of Technology, The Netherlands [Neuroevolution and Data Analytics; General]
Anna I. Esparcia	Universitat Politècnica de València, Spain [Engineering and Real-World Applications]
Anthony Brabazon	University College Dublin, Ireland [Engineering and Real-World Applications]
Anthony Clark	Michigan State University, USA [Robotics]
Antonio Della Cioppa	University of Salerno, Italy [Image and Signal Processing]
Antonio Fernández Ares	Universidad de Granada, Spain [Games]
Antonio Gonzales Pardo	Basque Center for Applied Mathematics, Spain [Games]
Antonio J. Fernández Leiva	Universidad de Málaga, Spain [Games]
Antonios Liapis	University of Malta, Malta [Games]
Bernabé Dorronsoro	Universidad de Cádiz, Spain [Networks and Distributed Systems]
Bill Langdon	University College London, UK [Networks and Distributed Systems; Numerical Optimization]
Carlos Cotta	Universidad de Málaga, Spain [Engineering and Real-World Applications; General]
Carlos M. Fernandes	Instituto Superior Técnico, Universidade de Lisboa, Portugal [General]
Cedric Buche	ENIB, France [Games]
Charly Lersteau	Liverpool John Moores University, UK [General]
Clara Pizzuti	ICAR-CNR, Italy [Life Sciences]
Claudio Rossi	Universidad Politecnica de Madrid, Spain [Robotics]
Daniel Hernández	Instituto Tecnológico Nacional, México [Engineering and Real-World Applications; General]
David Camacho	Universidad Autónoma de Madrid, Spain [Games]
David Pelta	Universidad de Granada, Spain [Numerical Optimization]
Doina Bucur	University of Twente, The Netherlands [General; Networks and Distributed Systems]
Dominik Sobania	Mainz University, Germany [General]
Edward Keedwell	University of Exeter, UK [Image and Signal Processing]
Emma Hart	Edinburgh Napier University, UK [Engineering and Real-World Applications]
Ernesto Tarantino	Italian National Research Council, Italy [Engineering and Real-World Applications]
Evelyne Lutton	INRA, France [Image and Signal Processing]
Fabio Caraffini	De Montfort University, UK [General; Life Sciences]
Federico Liberatore	Universidad Carlos III, Spain [Games]

Ferrante Neri	De Monfort University, UK [Engineering and Real-World Applications; Numerical Optimization]
Francesco Fontanella	Università degli studi di Cassino, Italy [Engineering and Real-World Applications; Image and Signal Processing]
Francisco Chávez de la O	University of Extremadura, Spain [General]
Francisco Fernández de Vega	Universidad de Extremadura, Spain [Engineering and Real-World Applications]
Francisco Luna	Universidad de Málaga, Spain [Networks and Distributed Systems]
Frank Veenstra	Edinburgh Napier University, UK [General; Robotics]
Frank W. Moore	University of Alaska Anchorage, USA [Engineering and Real-World Applications]
Frederic Guinand	Université du Havre Normandie, France [General; Networks and Distributed Systems; Life Sciences]
Gabriel Luque	Universidad de Málaga, Spain [General]
Gareth Howells	University of Kent, UK [Image and Signal Processing]
Geoff Nitschke	University of Cape Town, South Africa [Neuroevolution and Data Analytics; Robotics]
Giovanni Iacca	University of Trento, Italy [Life Sciences]
Giovanni Squillero	Politecnico di Torino, Italy [Engineering and Real-World Applications]
Gregoire Danoy	University of Luxembourg, Luxembourg [Networks and Distributed Systems]
Guenter Rudolph	University of Dortmund, Germany [Numerical Optimization]
Gul Muhammad Khan	University of Engineering and Technology, Pakistan [Neuroevolution and Data Analytics]
Gustavo Olague	CICESE, México [General; Image and Signal Processing; Networks and Distributed Systems; Neuroevolution and Data Analytics]
Heiko Hamann	University of Lübeck, Germany [Engineering and Real-World Applications; Neuroevolution and Data Analytics; Robotics]
Ignacio Arnaldo	MIT, USA [Networks and Distributed Systems]
Igor Deplano	Liverpool John Moores University, UK [General]
Ivanoe De Falco	ICAR- CNR, Italy [Image and Signal Processing]
Iwona Karcz-Duleba	University of Wroclaw, Poland [Numerical Optimization]
James Foster	University of Idaho, USA [General]
János Botzheim	Budapest University of Technology, Hungary [Engineering and Real-World Applications; General]
Jared Moore	Grand Valley State University, USA [Engineering and Real-World Applications]

Jean-Baptiste Mouret   Institut des Systémes Intelligents et de Robotique,
                           France [Neuroevolution and Data Analytics;
                           Robotics]
Jean-Marc Montanier   Norwegian University of Science and Technology,
                           Norway [Engineering and Real-World Applications;
                           Neuroevolution and Data Analytics]
Jhon E. Amaya   Universidad Nacional Experimental del Táchira,
                           Venezuela [General; Neuroevolution and Data
                           Analytics]
Joel Lehman   Uber AI Labs, USA [Neuroevolution and Data
                           Analytics]
Johan Hagelback   Blekinge Tekniska Hogskola, Sweden [Games]
Joost Huizinga   Uber AI Labs, USA [Robotics]
José Carlos Bregieiro   Polytechnic Institute of Leiria, Portugal [General;
   Ribeiro   Networks and Distributed Systems]
Jose Ignacio Hidalgo   Universidad Complutense de Madrid, Spain [Networks
                           and Distributed Systems]
José Manuel Colmenar   Universidad Rey Juan Carlos, Spain [Networks
                           and Distributed Systems]
Jose Santos Reyes   Universidad de A Coruña, Spain [Robotics]
Joshua Auerbach   Champlain College, USA [General; Neuroevolution
                           and Data Analytics]
Juan Julián Merelo-Guervós Universidad de Granada, Spain [Networks
                           and Distributed Systems]
Juan L. Jiménez-Laredo   RI2C/LITIS, Université du Havre Normandie, France
                           [Networks and Distributed Systems]
Julian Francis Miller   University of York [Image and Signal Processing]
Kai Ellefsen   University of Oslo, Norway [Neuroevolution and Data
                           Analytics; Robotics]
Lucia Ballerini   University of Edinburgh, UK [Image and Signal
                           Processing]
Maizura Mokhtar   Heriot-Watt University, USA [Engineering
                           and Real-World Applications]
Malcolm Heywood   Dalhousie University, Canada [Engineering
                           and Real-World Applications]
Marc Ebner   Ernst Moritz Arndt Universität Greifswald, Germany
                           [Image and Signal Processing]
María Arsuaga-Ríos   CERN, Switzerland [Networks and Distributed
                           Systems; Neuroevolution and Data Analytics]
Mario Cococcioni   NATO Undersea Research Centre, Italy [Engineering
                           and Real-World Applications]
Mario Giacobini   University of Torino, Italy [Engineering
                           and Real-World Applications; General;
                           Life Sciences]
Markus Wagner   University of Adelaide, Australia [Engineering
                           and Real-World Applications]

Mengjie Zhang   University of Wellington, New Zealand [Image
                             and Signal Processing]
Michael Guckert   University of Lille, France [Engineering
                             and Real-World Applications; General]
Michael Lones   Heriot-Watt University, UK [Image and Signal
                             Processing]
Mihai Polceanu   ENIB, France [Games]
Mohammed Salem   University of Mascara, Algeria [Engineering
                             and Real-World Applications; Games]
Moshe Sipper   Ben-Gurion University of the Negev, Israel [General;
                             Games]
Nadarajen Veerapen   Technische Hochschule Mittelhessen, Germany
                             [General; Engineering and Real-World
                             Applications]
Neil Urquhart   Edinburgh Napier University, UK [Engineering
                             and Real-World Applications]
Nelishia Pillay   University of Pretoria, South Africa [Engineering
                             and Real-World Applications]
Nicolas Bredeche   Sorbonne Université, France [Robotics]
Oliver Kramer   University of Oldenburg, Germany [Image and Signal
                             Processing; Numerical Optimization]
Pablo García-Sánchez   Universidad de Cádiz, Spain [Games; Networks
                             and Distributed Systems]
Pablo Mesejo Santiago   Universidad de Granada, Spain [Engineering
                             and Real-World Applications; Image and Signal
                             Processing]
Paul Kaufmann   Mainz University, Germany [Networks and Distributed
                             Systems]
Penousal Machado   University of Coimbra, Portugal [Engineering
                             and Real-World Applications; Image and Signal
                             Processing; Neuroevolution and Data Analytics]
Petr Posik   Czech Technical University, Czech Republic
                             [Numerical Optimization]
Philip Bontrager   New York University, USA [Neuroevolution and Data
                             Analytics]
Rafael Nogueras   Universidad de Málaga, Spain [General; Networks
                             and Distributed Systems; Neuroevolution and Data
                             Analytics]
Rafael Villanueva   Universitat Politècnica de València, Spain [Networks
                             and Distributed Systems]
Ran Wang   Liverpool John Moores University, UK [General]
Raúl Lara-Cabrera   Universidad Autónoma de Madrid, Spain [Games;
                             General; Neuroevolution and Data Analytics]
Renato Tinos   University of Sao Paulo, Brazil [Engineering
                             and Real-World Applications]

Contents

General

Image and Signal Processing

Life Sciences

Networks and Distributed Systems

Engineering and Real-World Applications

A Comparison of Different Many-Objective Optimization Algorithms for Energy System Optimization

Tobias Rodemann[✉]

Honda Research Institute Europe,
Carl-Legien-Strasse 30, 63073 Offenbach/Main, Germany
tobias.rodemann@honda-ri.de

Abstract. The usage of renewable energy sources, storage devices, and flexible loads has the potential to greatly improve the overall efficiency of a building complex or factory. However, one needs to consider a multitude of upgrade options and several performance criteria. We therefore formulated this task as a many-objective optimization problem with 10 design parameters and 5 objectives (investment cost, yearly energy costs, CO_2 emissions, system resilience, and battery lifetime). Our target was to investigate the variations in the outputs of different optimization algorithms. For this we tested several many-objective optimization algorithms in terms of their hypervolume performance and the practical relevance of their results. We found substantial performance variations between the algorithms, both regarding hypervolume and in the basic distribution of solutions in objective space. Also the concept of desirabilities was employed to better visualize and assess the quality of solutions found.

Keywords: Many-objective optimization · Energy management · Desirabilities

1 Introduction

Many-objective optimization algorithms are a hot topic in the field of evolutionary computation, due to both increasing interest from the application side as well as a number of unsolved issues. In this work we will evaluate the performance of several well-known many-objective optimization (MAO) algorithms on a challenging real-world problem: finding the optimal configuration of the energy system of a heterogeneous business building complex. The usage of local energy production and storage facilities has become increasingly interesting both in terms of energy costs and CO_2 emissions. Facility management is therefore looking how to optimally invest in extensions to the current building energy system. Example extensions are large-scale Photo Voltaic (PV) systems or battery storage capacity. There is typically a large number of options for adding new modules or optimizing the usage of existing ones. With increasing system

© Springer Nature Switzerland AG 2019
P. Kaufmann and P. A. Castillo (Eds.): EvoApplications 2019, LNCS 11454, pp. 3–18, 2019.
https://doi.org/10.1007/978-3-030-16692-2_1

complexity the number of objectives to be considered increases, resulting in a difficult optimization problem for the decision makers, especially considering the high investment costs. One has to consider that different modules might be linked, for example optimal battery and PV system size. A serious problem for both optimization and analysis of solutions is the wide range of objective values, which can vary over many orders of magnitude. In order to deal with the latter issue we propose to use desirability functions as discussed in [1].

There is a number of recent works targeting many-objective optimization of energy management systems [2,3]. For single objective optimization of energy systems, methods from the field of Linear Programming (LP) are commonly used since they are comparatively fast and provide performance guarantees [4,5]. However, LP approaches are based on simplified system models where many aspects are difficult to consider such as battery aging, temperature-dependent efficiencies, or dynamic user behavior. For this reason we employ a very detailed building simulation based on the Modelica standard [6,7] that can model most real-world effects. The simulator is treated as a black box by the optimization, meaning that we do not make any assumptions about the structure of the problem. The ability of evolutionary algorithms to handle these types of problems is one of their greatest assets. Please consult Fig. 1 for a view of the simulation environment with our building model and Fig. 2 for a sketch of the information flow from optimizer to simulator and back.

We will first introduce the specific application example. Afterwards we are going to sketch the optimization algorithms used in this paper and the experimental setting. We will then show results of different MAO algorithms using the hypervolume indicator. Furthermore, we will compare results found by the optimization with manually selected baseline results. Finally, we summarize our results and present some future work.

Note that our target is not a ranking of optimizers, but a better understanding if, from a practical point of view, the use of different optimizers for this specific problem is necessary.

An initial study on this application was already published earlier [8]. In the present work we compare a much larger variety of optimization algorithms since prior experiments suggested that some optimizers struggled with this application. We furthermore used findings from this earlier work to adapt the optimization problem (different parameter ranges), added new parameters and a new objective to include battery aging.

In order to capture essential seasonal effects one would normally consider at least one full year but to be able to perform more test runs we reduced the simulated time from one year to a single month. This could in practice lead to sub-optimal performance of the selected configuration, but allows us to thoroughly compare different optimizers. In the real application case, one would obviously revert to the simulation of at least one year using one of the better performing algorithms as identified by this study.

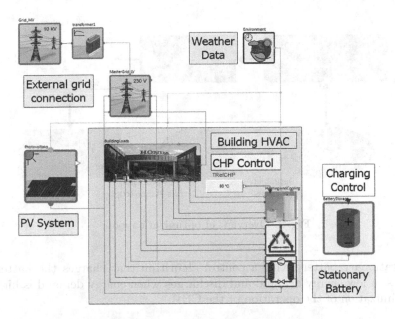

Fig. 1. System view of building simulator. The model contains a detailed simulation of the building's Heating-Ventilation-A/C system, including heat storage and co-generator (CHP), plus modules for Photo Voltaics (PV), grid connection, and a stationary battery including corresponding controller.

1.1 Overview

The target of the optimization is to find optimal configurations for the energy layout of a medium-sized research campus with about 200 employees. With configuration we mean a combination of different modules like PV system, battery or heat storage, and settings of a controller.

The building has a rather conventional load profile with peak loads above 500 kW mostly around noon, due to HVAC (heating/ventilation/air - conditioning) demands and a baseload of about 200 kW. Total annual energy costs for gas and electricity are in the range of several 100,000 Euros. The campus is already equipped with a co-generator for heat and power (CHP) that provides 200 kW electric and 300 kW thermal output. Further improvement options have been identified and are described below.

1.2 Investment Options

There are several promising areas for investment that define the search space for the configuration optimization presented here:

1. A large-scale PV system on the building roof or car-port
2. An extension of the internal heat storage

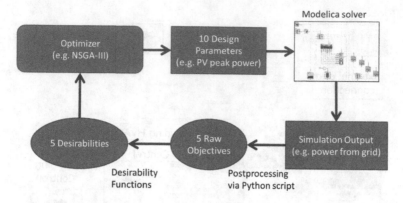

Fig. 2. Overview of optimization loop.

3. A stationary battery with a control algorithm that charges the battery in times of low energy demand and discharges when energy demand is high
4. Optimization of the operation of the CHP

1.3 Design Parameters

Based on the above defined investment options we decided to consider in total 10 independent parameters:

1. Inclination angle α_{PV}, orientation angle β_{PV}, and peak output power $PPeak_{PV}$ (in kW) of PV system.
2. Capacity E_{Batt} (in kWh), and linked to that maximum charging/discharging power P_{batt}^{max} (in kW), of a stationary battery storage system, where we assume a maximum charging power of 1 C (i.e. battery fully charged or discharged within one hour).
3. Min and max battery state-of-charge (SOC) level SOC_{min} and SOC_{max}.
4. Charging u_{Batt}^{charge} and discharging $u_{Batt}^{discharge}$ threshold for the battery controller. The battery will discharge when current load exceeds $u_{Batt}^{discharge}$, and correspondingly charge when demand is below u_{Batt}^{charge}. The battery will charge or discharge to keep the load below the given charging threshold level (as long as the battery is within its SOC limits and P_{batt}^{max} allows). The charging controller is visualized in Fig. 3.
5. Total volume of heat storage in m^3, V_{stor}. From this value we also derive the diameter of the heat storage d_{stor} as $d_{stor} = \sqrt{V_{stor}/3}$.
6. Operation threshold of the CHP system. The generator will only turn on if ambient temperature is below this level. This was implemented to avoid too frequent on/off switches (resulting in high maintenance costs) and is realized by using an upper (u_{CHP}^{high}) and lower (u_{CHP}^{low}) temperature threshold with a difference of $1.0°$ as $u_{CHP}^{low} = u_{CHP}^{high} - 1.0$. The CHP is turned off when ambient temperature exceeds u_{CHP}^{high} and restarted when it falls below u_{CHP}^{low}.

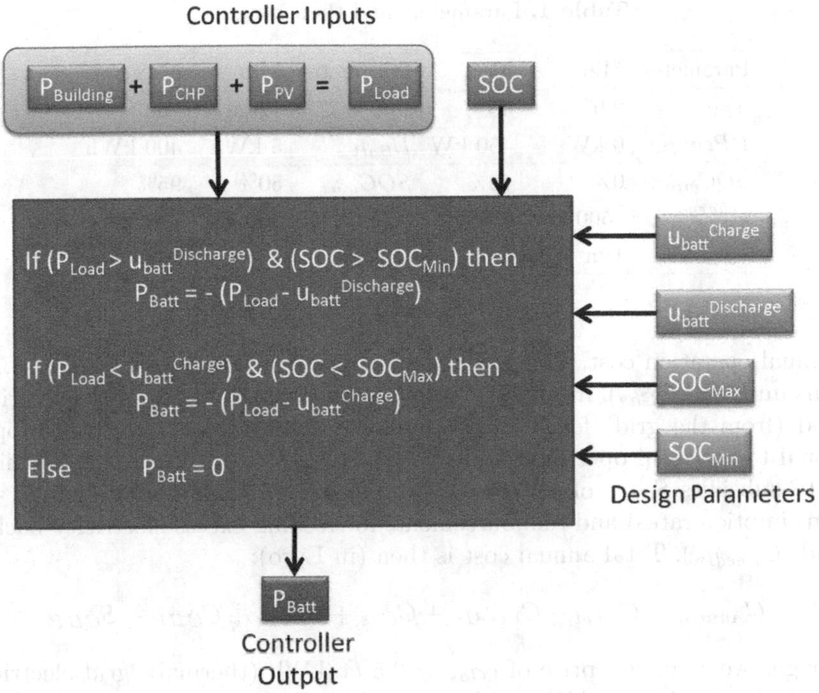

Controller Inputs

Fig. 3. Controller logics for stationary battery. The main controller inputs are the state-of-charge (SOC) of the stationary battery and the sum of building energy consumption and energy produced by CHP and PV (P_{CHP} and P_{PV}) (note that produced energy is by convention negative). The controller output is the charging power of the battery P_{Batt}.

Please consult Table 1 for a list of minimum and maximum values for all design parameters. Those ranges have been defined in cooperation with a building energy specialist considering actual physical constraints (e.g. available space) of the building.

1.4 Objectives

The quality of an investment solution will depend on a number of factors. We have chosen the following five objectives which cover the main factors in the respective domains. Values are computed based on outputs of the simulator.

1. Initial investment cost. We limit ourselves to the main (hardware) purchasing costs: PV system (as a function of peak power) $C_{PV} = 1000$ Euro $\cdot PPeak_{PV}$ (in kW), battery (total capacity) $C_{Batt} = 250$ Euro $\cdot E_{Batt}$ (in kWh), and finally heat storage (volume) $C_{HeatStor} = 700$ Euro $\cdot V_{stor}$ (in m^3). Total investment cost is the sum of all module costs (in Euro):

$$C_{Invest} = C_{PV} + C_{Batt} + C_{HeatStor}. \tag{1}$$

Table 1. Parameters and their ranges.

Parameter	Min	Max	Parameter	Min	Max
α_{PV}	0.0°	45°	β_{PV}	120.0°	250°
$PPeak_{PV}$	0 kW	450 kW	E_{Batt}	5 kWh	400 kWh
SOC_{min}	0%	40%	SOC_{max}	50%	95%
u_{Batt}^{charge}	−500 kW	0 kW	$u_{Batt}^{discharge}$	300 kW	700 kW
V_{stor}	1 m³	5 m³	u_{CHP}^{high}	10° C	25° C

2. Annual operation cost. This is the sum of grid electricity cost (C_{GridE}), gas consumption (C_{Gas}) from CHP and conventional boilers, peak electricity load (from the grid) fee, CHP maintenance cost C_{CHP} (which is proportional to the total operation hours h_{CHP}: $C_{CHP} = 4.3$ Euro $\cdot h_{CHP}$) minus CHP subsidies S_{CHP} of 4.35 ct per produced kWh (assuming very high self-consumption rates) and reimbursements for feeding excess electricity back to grid (C_{FeedIn}). Total annual cost is then (in Euro):

$$C_{Annual} = C_{Grid} - C_{FeedIn} + C_{Gas} + C_{Peak} + C_{CHP} - S_{CHP}. \qquad (2)$$

For gas we assume a price of $c_{Gas} = 2.5$ ct/kWh (thermal), grid electricity costs of $c_{El} = 13.0$ ct/kWh, and a feed-in tariff of $c_{FeedIn} = 7$ ct/kWh. The peak load cost is 76 Euros for each kW at the highest energy load (averaged over a 15 min interval) within the whole calendar year. All consumption values are provided by the simulation tool.

3. Annual CO_2 emissions. Reducing CO_2 emissions is a high-priority task. We compute the combined CO_2 from purchasing grid electricity $emis_{grid}$ (at 500 g/ kWh, approximated data for Germany) and gas $emis_{gas}$ (for CHP and boilers, at 185 g/kWh thermal), based on simulated electricity and gas consumption. Total CO_2 emission is then:

$$emis_{total} = emis_{grid} + emis_{gas}. \qquad (3)$$

4. Resilience. The availability of local energy production and electric storage capacity would allow a system operation even in case of severe grid malfunctions. This ability is termed resilience and in our specific case refers to the duration the company could operate if no grid power is available. Specifically we are using the minimum of battery charging level $BatLevel$ (in kWh) divided by current grid load E_{Load} (in kW), over the complete simulation period. This represents a worst case scenario (grid failure occurs at worst possible time). Obviously, one could also consider other formulations, like mean instead of min, or using a fixed "emergency" power demand. The resulting resilience (in seconds) in our case was computed as:

$$resi = \min \left(\frac{BatLevel}{E_{Load}} \right) \cdot 3600 \qquad (4)$$

5. Battery lifetime. We are using a coarse model of battery aging based on calendaric and cyclic aging. Please note that the quantitative results should be viewed with care since battery aging strongly depends on the underlying battery chemistry. In this work a generic Li-Ion battery was used. The simulation tool will compute the final battery State-Of-Health (SOH_{1m}) after a one month of operation, which represents the remaining battery capacity (relative to the initial capacity). Battery lifetime (in years) will then be computed as the number of years before state of health falls to $\theta_{SOH} = 90\%$:

$$\text{lifetime} = \frac{1}{12} \frac{\log(\theta_{SOH})}{\log(SOH_{1m})} \qquad (5)$$

The first three objectives (investment cost, annual cost, CO_2 emissions) are minimized, while resilience and battery lifetime are maximized. Results for yearly costs and emissions are approximated by simply scaling up all values from the monthly simulation by a factor 12.

1.5 Desirabilities

With 5 different objectives that vary within different ranges, visualization of solutions becomes a challenge. In order to avoid various scaling operations at different stages of the optimization process, we decided to employ the concept of desirability functions [1,9]. The desirability $d(Y)$ for an objective Y is given by:

$$d(Y) = \exp(-\exp(-(b_0 + b_1 Y))) \qquad (6)$$

The closer the desirability score to one, the more satisfying the quality of the objective value. For controlling the shape of the desirability function, we define two pairs of values to compute the parameters b_0 and b_1. These are two example objective values ($Y^{(1)}$ and $Y^{(2)}$) and corresponding desirability scores ($d^{(1)}$ and $d^{(2)}$), for details see e.g. [1].

$$b_0 = -\log(-\log(d^{(1)})) - b_1 Y^{(1)} \qquad (7)$$

$$b_1 = \frac{-\log(-\log(d^{(2)})) + \log(-\log(d^{(1)}))}{Y^{(2)} - Y^{(1)}} \qquad (8)$$

Specific values for (d, Y) pairs are given in Table 2. In practice one would first discuss with the actual decision maker about her preferences and then adapt the desirability functions accordingly. Here the selection was based on the baseline configurations described in the next section. To formulate the optimization as a minimization problem we internally used $1 - d(Y)$ as objective values, i.e. 0.0 represents the optimal value and 1.0 the worst possible solution, which is additionally used as the reference point for the hypervolume computation with an $\epsilon = 10^{-5}$ added.

Table 2. Control points for desirability function for different objectives.

Objective	$d^{(1)}$	$Y^{(1)}$	$d^{(2)}$	$Y^{(2)}$
C_{Invest} (Euro)	0.9	100,000	0.1	600,000
C_{Annual} (Euro)	0.9	350,000	0.1	400,000
$emis_{total}$ (t)	0.9	2,100	0.1	2,200
$resi$ (s)	0.1	1	0.9	900
$Lifetime$ (y)	0.1	10	0.9	30

1.6   Simulation System

Since we want to investigate a variety of different aspects of the energy system we need a simulation tool that can handle various physical domains (electricity, heat, cold, e-mobility among others). We therefore decided to use the Modelica [6] simulation language that is based on physical equations and well suited for simulation of complex systems including non-linear effects. The model was built based on an analysis of the real building and smart meter measurements of energy consumption over several years. The simulator provides all required output information like energy consumption or battery charge levels. A typical simulation now runs for approximately 10 s, but some configurations require far more than this. We therefore decided to stop any simulation that runs longer than a threshold value of 20 s and assign worst possible values for all objectives. See [8] for a discussion of this issue.

2   Optimization Task

Our application instance is a many-objective optimization problem (MaOP) with 10 parameters and 5 objectives. Prior work (e.g. [10]) investigated several algorithms for the optimization of a hybrid car controller and found that they exhibit different performance qualities. We now want to compare several optimization algorithms of different types on our test application. Due to the long simulation (fitness computation) times, we performed all tests on a computing cluster, allowing us to run all solutions of a single generation in parallel. For all algorithms we use a population size of $\mu = 35$ individuals. The total number of objective function calls was limited at 5250. All experiments were repeated 10 times with different random number seeds. Design parameters were hard limited to the range specified above. The software we employ for the optimizers is PlatEMO V1.5 [11]. All hyperparameters were chosen as the default values from the above software, since an extensive hyperparameter scan (as for example via irace [12]) is way too time-consuming. Please also consult [13] for some thoughts on selecting the proper optimizer for compute-intensive application problems.

We used the following algorithms to cover a broad range of approaches, which span the range from 2001–2016 as the year of publication: NSGA-II [14], NSGA-III [15], RVEA [16], IBEA [17], SPEA [18], KnEA [19], GDE3 [20], MOPSO [21], PICEAg [22], TwoArch2 [23].

Note that there are 77 algorithms within PlatEMO V1.5, so that any selection is to a substantial degree arbitrary. It is not our target to benchmark these algorithms but to evaluate if there is a practical difference in performance between different optimizers. As far as possible, evaluation of solutions was parallelized. As this is not trivially possible for steady-state algorithms (like [24, 25]), those were not tested. Average run-times for all algorithms (a single (out of 10) runs for one specific optimizer) are around two hours but would be around 24 h for steady state algorithms.

3 Results

Due to the large run-times we only used default settings for all optimizers and kept population sizes for all optimizers at the same value (35 individuals). Since we do not know the true Pareto front we are using the hypervolume as the main performance indicator.

3.1 Baseline Solutions

For a realistic performance assessment in addition to hypervolume values we compare results for our optimized configurations with some manually chosen baseline configurations. The results (objectives) of these basic configurations were also used to determine reasonable values for the parameters of the desirability function. Please note that actual results for the explicit simulation of a complete year are substantially different (mostly better) compared to the values shown here, which were extrapolated from a single month.

Current Configuration. The baseline configuration of the building has a 10 kW peak PV system, no stationary battery, a moderate heat storage of 1.3 m^3 and a CHP threshold setting of 17 °C. The corresponding parameter values and the resulting objectives are found in Table 3 under the row *Current*. Investment cost is minimal but yearly costs and CO_2 emissions are high, and the system does not have any resilience capacity.

Moderate Expansion. A moderate expansion configuration is simulated by setting PV size, battery capacity, and heat storage volume to intermediate values ((Max-Min)/2). The battery is used at a range from 20–80% capacity, it is charged when energy production exceeds demand ($u_{Batt}^{charge} = 0$ kW) and discharges when energy demand is above 600 kW ($u_{Batt}^{discharge} = 600$ kW). Inclination and orientation angle of the PV system and CHP threshold are the same as for the current configuration. Parameters and objectives can be found in row *Moderate*. We see that with an investment of 280,000 Euros, a reduction of annual costs by around 9,500 Euro, and CO_2 emissions by 34 tons/year is possible compared to the current solution. In addition the system now has a resilience, being able to operate on its own for 450 s in case of a grid failure. With the current desirability settings this value is considered sufficient.

Full Set-Up. The third configuration we analyze represents a maximum investment with the largest possible size of PV system, battery, and heat storage. The remaining parameters are kept as before, except for battery SOC settings which we set to the range 5–95% and a more ambitious battery discharge threshold of 500 kW. It is interesting to see that the effects are almost linear (roughly double investment costs and double savings on annual costs and emissions). This indicates that the larger PV system can be fully used.

Table 3. Parameters (*top*) and objective values (*bottom*) of different alternative configurations. For all configurations $\alpha_{PV} = 35°$, $\beta_{PV} = 180°$, $u_{high} = 17\,°C$ is used. Desirabilities are shown in brackets.

Parameter	$PPeak_{PV}$	E_{Batt}	u_{Batt}^{charge}	$u_{Batt}^{discharge}$	SOC_{min}	SOC_{max}	V_{stor}
Current	10 kWp	0	—	—	—	—	$1.3\ m^3$
Moderate	230 kWp	200 kWh	0 kW	600 kW	20 %	80 %	$3\ m^3$
Full	450 kWp	400 kWh	0 kW	500 kW	5%	95%	$5\ m^3$

Objective	C_{Invest}	C_{Annual}	$emis_{total}$	$resi$	lifetime
Current	0.0 (1.0)	382285 € (0.46)	2202 t (0.08)	0.0 s (0.0)	—
Moderate	281185 € (0.72)	372735 € (0.65)	2168 t (0.42)	449 s (0.61)	26.6 y (0.75)
Full	550665 € (0.18)	363517 € (0.78)	2140 t (0.69)	1223 s (0.96)	26.5 y (0.73)

3.2 Optimizer Results

Now we take a look at the results of the various optimization runs. Our main quality indicator is the maximum hypervolume of the parent population in the final generation. All hypervolume values are computed in desirability space. For a better comparison of hypervolume values we present final hypervolume results in a boxplot, see Fig. 4.

Our findings show that there is a substantial variation in both the mean HV values of different optimizers as well as variations over different runs of the same optimizer. The MOPSO method obviously struggles substantially with a lowest HV of 0.06. This might be due to sub-optimal hyperparameter settings, an implementation problem or some feature of our application problem. The remaining optimizers all stay in a range of HV values of [0.20, 0.29]. Still this means that some runs produced 50% better hypervolume values than other runs. Another interesting observation is that we can't see a direct trend for increasing performance over time of publication. The oldest (SPEA2) and the newest (RVEA) algorithm are very similar in their HV values.

Results from the optimization runs could be used in a number of ways, some of which will be shown exemplary below. First we plot desirabilities for all final solutions from all runs in a parallel coordinate plot (Fig. 5). This type of representation allows us to understand the relations between the objectives for all

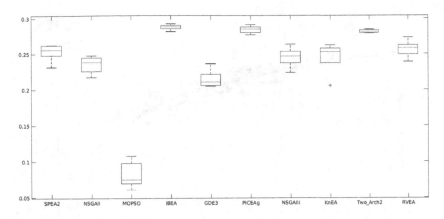

Fig. 4. Boxplot of hypervolume results for all runs and optimizers. Note that optimizers are sorted by year of publication - from oldest (2001, left) to newest (2016, right).

populations. We see a large spread of solutions in desirability space. A clear correlation is that investment cost is inversely proportional to annual cost and emissions, while the latter two are strongly correlated. But for example for resilience and battery lifetime the relations are more diverse. A very interesting finding is shown in Fig. 6, where we plot desirability values for objective 2 (annual cost) vs objective 1 (investment cost) for all runs of 5 optimizers. Each symbol represents one final solution. Individual runs show patterns different in detail but overall rather similar. We can see that solutions from different optimizers can cover different parts of objective space. In our example solutions from IBEA and RVEA strongly overlap, while MOPSO, SPEA, and especially PICEAg occupy different areas (note that some of the shown solutions might be dominated).

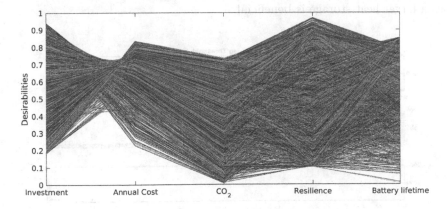

Fig. 5. Parallel coordinate plot of desirabilities

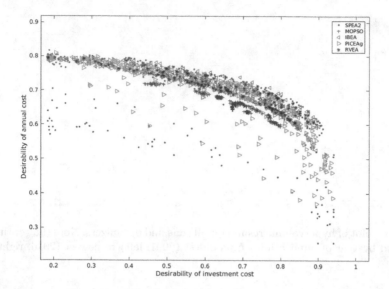

Fig. 6. Desirabilities for emissions vs. initial investment for different optimizers

To better understand the solutions we also studied the ranges of found parameters. Due to space limitations we can only show three exemplary parameter histograms (Fig. 7). We see that almost all solutions select a charging threshold of 0, which means that the battery is charged as soon as surplus power is available (*top*). This makes sense as feeding energy back to the grid is less desirable than self-consumption. In the middle panel we see a large variety in discharging thresholds which is probably due to a low impact of the battery for the overall energy flows (the battery is mainly used for resilience and peak shaving, but the latter is a very tricky business). The bottom row finally demonstrates that in all cases a larger heat storage is beneficial.

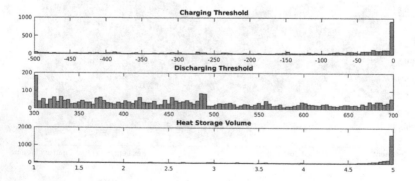

Fig. 7. Histograms of parameter values of all final solutions from all runs. *Top*: battery charging threshold u_{Batt}^{Charge}, *middle* battery discharging threshold $U_{Batt}^{discharge}$, *bottom* heat storage volume V_{stor}

Table 4. Best and worst objective values (desirabilities in brackets) for all found solutions, please compare these values to results from Table 3.

Objective	Max. desirability	Min. desirability
C_{Invest}	11,600 Euros (0.94)	550,000 Euros (0.18)
C_{Annual}	359,000 Euros (0.84)	393,000 Euros (0.22)
$emis_{total}$	2134 t (0.74)	2225 t (0.01)
$resi$	1252 s (0.97)	1.7 s (0.10)
$lifetime$	28.7 a (0.85)	17.4 a (0.01)

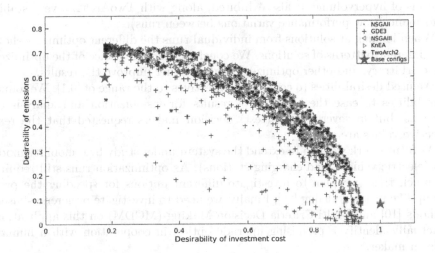

Fig. 8. Results (yearly emissions vs. investment cost) for 5 selected optimizer (all runs) plus initial configurations.

A look at the minimum and maximum desirability values for all objectives over all final solutions from all runs (Table 4) shows that we can find a large variety of solutions with higher peak desirabilities (except for investment cost) compared to the three baseline configurations (see Table 3).

3.3 Comparison to Baseline Solutions

In Fig. 8 we show for 5 selected optimizers and all runs the desirabilities for objective 3 vs objective 1. We also added the three reference solutions. It is obvious that coverage in desirability space is not the same for all optimizers. In relation to the baseline solutions we see that most found solutions are clear improvements and both middle and full extension baseline configuration are clearly dominated by the majority of solutions (this is also true for other objectives).

4   Summary and Outlook

For the application of building energy system optimization we compared 10 different many-objective optimization algorithms using the hypervolume indicator. We found that different optimizers exhibit substantial differences in their performance as indicated by the hypervolume of the final population. Interestingly, we don't see a clear trend for increasing performance with more recent algorithms compared to older work. This means that at least when using the default hyperparameters we don't see any clear benefits from using more recently published algorithms compared to older ones. Among the 10 test optimizers, and without any tuning of hyperparameters, IBEA was the best performing algorithm in terms of hypervolume. It also exhibited, along with TwoArch2, a very stable performance (low performance variations between runs).

When looking at solutions from individual runs the different optimizers show very different patterns of solutions. We conclude that the choice of the optimizer is not arbitrary, and other optimizers might further improve the results.

We used desirabilities to confine all objectives to the range of [0,1]. We found desirabilities to ease the handling of results for visualization and analysis of solutions, but in several cases (test) decision makers requested that the real objective values are shown.

As future work we plan to extend the system under study by e-mobility modules (electric vehicles and charging stations). As optimization runs still require too much time, we plan to investigate different options for speeding the process up like meta-modeling [26]. Finally, we need to investigate preference-based methods [10] and Multi-Criteria Decision Making (MCDM) on this application to actually identify a promising upgrade option in cooperation with a human decision maker.

References

1. Wagner, T., Trautmann, H.: Integration of preferences in hypervolume-based multiobjective evolutionary algorithms by means of desirability functions. IEEE Trans. Evol. Comput. **14**(5), 688–701 (2010)
2. Yang, R., Wang, L.: Multi-objective optimization for decision-making of energy and comfort management in building automation and control. Sustain. Cities Soc. **2**(1), 1–7 (2012). http://www.sciencedirect.com/science/article/pii/S221067071100059X
3. Fadaee, M., Radzi, M.: Multi-objective optimization of a stand-alone hybrid renewable energy system by using evolutionary algorithms: a review. Renew. Sustain. Energy Rev. **16**(5), 3364–3369 (2012). http://www.sciencedirect.com/science/article/pii/S1364032112001669
4. Khodr, H.M., Vale, Z.A., Ramos, C., Soares, J.P., Morais, H., Kádár, P.: Optimal methodology for renewable energy dispatching in islanded operation. In: IEEE PES T D 2010, pp. 1–7 (2010)
5. Naharudinsyah, I., Limmer, S.: Optimal charging of electric vehicles with trading on the intraday electricity market. Energies **11**(6), 1416 (2018)

6. Fritzson, P., Bunus, P.: Modelica – a general object-oriented language for continuous and discrete-event system modeling. In: Proceedings of the 35th Annual Simulation Symposium, pp. 14–18 (2002)
7. Unger, R., Mikoleit, B., Schwan, T., Bäker, B., Kehrer, C., Rodemann, T.: Green building - modeling renewable building energy systems with emobility using Modelica. In: Proceedings of Modelica 2012 Conference. Modelica Association, Munich, Germany (2012)
8. Rodemann, T.: A many-objective configuration optimization for building energy management. In: Proceedings of IEEE WCCI (CEC) (2018)
9. Ogino, Y., Iida, R., Rodemann, T.: Using desirability functions for many-objective optimization of a hybrid car controller. In: GECCO 2017 Conference Companion (2017)
10. Cheng, R., Rodemann, T., Fischer, M., Olhofer, M., Jin, Y.: Evolutionary many-objective optimization of hybrid electric vehicle control: from general optimization to preference articulation. IEEE Trans. Emerg. Top. Comput. Intell. **1**(2), 97–111 (2017)
11. Tian, Y., Cheng, R., Zhang, X., Jin, Y.: PlatEMO: a MATLAB platform for evolutionary multi-objective optimization. CoRR abs/1701.00879 (2017). http://arxiv.org/abs/1701.00879
12. López-Ibáñez, M., Dubois-Lacoste, J., Cáceres, L.P., Birattari, M., Stützle, T.: The irace package: iterated racing for automatic algorithm configuration. Oper. Res. Perspect. **3**, 43–58 (2016)
13. Rodemann, T.: Industrial portfolio management for many-objective optimization algorithms. In: Proceedings of IEEE WCCI 2018 (CEC) (2018)
14. Deb, K., Pratap, A., Agarwal, S., Meyarivan, T.: A fast and elitist multiobjective genetic algorithm: NSGA-II. IEEE Trans. Evol. Comput. **6**, 182–197 (2002)
15. Deb, K., Jain, H.: An evolutionary many-objective optimization algorithm using reference-point-based nondominated sorting approach, part I: solving problems with box constraints. IEEE Trans. Evol. Comput. **18**(4), 577–601 (2014)
16. Cheng, R., Jin, Y., Olhofer, M., Sendhoff, B.: A reference vector guided evolutionary algorithm for many-objective optimization. IEEE Trans. Evol. Comput. **20**(5), 773–791 (2016)
17. Zitzler, E., Künzli, S.: Indicator-based selection in multiobjective search. In: Yao, X., et al. (eds.) PPSN 2004. LNCS, vol. 3242, pp. 832–842. Springer, Heidelberg (2004). https://doi.org/10.1007/978-3-540-30217-9_84
18. Zitzler, E., Laumanns, M., Thiele, L.: SPEA2: improving the strength pareto evolutionary algorithm for multiobjective optimization. In: Proceedings of Evolutionary Methods for Design, Optimisation and Control with Application to Industrial Problems, pp. 95–100 (2001)
19. Zhang, X., Tian, Y., Jin, Y.: A knee point driven evolutionary algorithm for many-objective optimization. IEEE Trans. Evol. Comput. **19**(6), 761–776 (2015)
20. Kukkonen, S., Lampine, J.: GDE3: the third evolution step of generalized differential evolution. In: Proceedings of the 2005 IEEE Congress on Evolutionary Computation, pp. 443–450 (2005)
21. Coello, C.C., Lechuga, M.S.: MOPSO: a proposal for multiple objective particle swarm optimization. In: Proceedings of the 2002 IEEE Congress on Evolutionary Computation, pp. 1051–1056 (2002)
22. Wang, R., Purshouse, R.C., Fleming, P.J.: Preference-inspired coevolutionary algorithms for many-objective optimization. IEEE Trans. Evol. Comput. **17**(4), 474–494 (2013)

23. Wang, H., Jiao, L., Yao, X.: Two Arch2: an improved two-archive algorithm for many-objective optimization. IEEE Trans. Evol. Comput. **19**, 524–541 (2015)
24. Zhang, Q., Li, H.: MOEA/D: a multiobjective evolutionary algorithm based on decomposition. IEEE Trans. Evol. Comput. **11**(6), 712–731 (2007)
25. Bader, J., Zitzler, E.: HypE: an algorithm for fast hypervolume-based many-objective optimization. Evol. Comput. **19**(1), 45–76 (2011)
26. Chugh, T., Jin, Y., Miettinen, K., Hakanen, J., Sindhya, K.: A surrogate-assisted reference vector guided evolutionary algorithm for computationally expensive many-objective optimization. IEEE Trans. Evol. Comput. **22**(1), 129–142 (2018)

Design of Powered Floor Systems for Mobile Robots with Differential Evolution

Eric Medvet[⊠], Stefano Seriani, Alberto Bartoli, and Paolo Gallina

Department of Engineering and Architecture, University of Trieste, Trieste, Italy
emedvet@units.it

Abstract. Mobile robots depend on power for performing their task. Powered floor systems, i.e., surfaces with conductive strips alternatively connected to the two poles of a power source, are a practical and effective way for supplying power to robots without interruptions, by means of sliding contacts. Deciding where to place the sliding contacts so as to guarantee that a robot is actually powered irrespective of its position and orientation is a difficult task. We here propose a solution based on Differential Evolution: we formally define problem-specific constraints and objectives and we use them for driving the evolutionary search. We validate experimentally our proposed solution by applying it to three real robots and by studying the impact of the main problem parameters on the effectiveness of the evolved designs for the sliding contacts. The experimental results suggest that our solution may be useful in practice for assisting the design of powered floor systems.

Keywords: Multi-objective optimization · Automatic design · Swarm Robotics · Evolutionary Robotics

1 Introduction and Related Work

Mobile robots are playing a role of increasing importance in current society. They are and will be aiding humans in performing tasks which may be dangerous, fatiguing, or boring due to repetitiveness. Examples range from persons transportation and goods delivering to surveillance.

One significant challenge in the field of mobile robots is how to provide power to the robots. The most common solution is to use batteries, which have, however, the apparent limitation that have to be recharged, causing a stop in the robot operations. A viable alternative consists in delivering power to the robot by electrifying the environment where it moves: this solution builds on the long-established experience on supplying powers to mobile machines, the most notable examples being in transportation systems, e.g., trains [1] and trams [2]. Another large family of solutions is based on wireless power delivery, often realized by means of resonating coils [3,4].

© Springer Nature Switzerland AG 2019
P. Kaufmann and P. A. Castillo (Eds.): EvoApplications 2019, LNCS 11454, pp. 19–32, 2019.
https://doi.org/10.1007/978-3-030-16692-2_2

In this paper, we focus on powered floors: the surface on which the robot moves is covered with conductive strips, interleaved by narrow non-conductive strips, alternatively connected to positive and negative poles of a power source; the robots receive the power from the strips using a number of sliding contacts positioned on the bottom of their bodies. This solution fits well the scenario of lab experimentation with Swarm Robotics [5] or Evolutionary Robotics [6]. In this settings, small prototypical robots have to run for long time, without obstacles to their movements (as, e.g., wires for power supply), in a physically constrained environment which has to be easily observable and accessible to researcher, for experiment "debugging" purposes. The latter condition may be not met if, e.g., robots get the power from the floor and the ceiling. Indeed, powered floors have already been used in similar settings [7] and also inspired the design of robotic platforms tailored purposely to experimentation [8].

Designing a working powered floor system (i.e., robots and floor) requires to decide the width of the strips and the positions of the sliding contacts on the robot. While the first task is not subjected to many constraints, for the positions of the contacts one has to take in to account the shape of the robot and the presence of moving parts, parts which should not be covered (e.g., sensors), or parts which are too far or too close to the floor. Moreover, the fundamental requirement is that the contacts should be positioned in a way that guarantees that at least one contact is on a positive strip and one on a negative strip for any rotation and position of the robot. A fixed contacts design which aims at satisfying this condition, consisting of four contacts, has been already proposed in the 50 s by Claude Shannon [9, p. 678], but it might be unsuitable for some robots due to the constraints described above.

We here propose a method, based on Differential Evolution (DE) [10], for finding automatically the positions of the sliding contacts, given a description of the region of the robot which is suitable for hosting them, which guarantee that the robot can receive power from the floor in any rotation and position. DE is an Evolutionary Algorithm (EA) for continuous optimization which can be used for solving single- or multi-objective problems. We formally define the condition for the robot to be always powered, given the positions of its contacts, and we introduce, based on this definition, three quantitative objectives suitable for driving the search with DE. We experimentally assess the ability of our proposed solution to find the contact positions for three real robotic platforms (Thymio II, mBot, and Elisa-3) and investigate about which combination of the objectives is more effective and efficient in solving this task. Then, we explore the design space of the powered floor system problem: we focus on the two most relevant problem parameters, the width of the conductive strips and the maximum number of sliding contacts. The experimental evaluation suggests that our proposed solution based on DE is capable of designing automatically the salient part of a powered floor system, i.e., the positions of the contacts, in many different settings. The experiments also show that the quantitative objectives that we adopted for driving the search are indeed effective.

2   Scenario: Mobile Robots on Powered Floor

We consider the scenario in which a mobile robotic platform moves on a powered floor.

The *powered floor* is composed of strips of conductive material interleaved by strips of non-conductive, insulating material (see Fig. 1b). The conductive strips are alternatively connected to the positive and negative poles of a constant power source. For the purpose of this work and without loss of generality we assume that the strips length is infinite, there are infinite strips, all the conductive strips have the same width, and all the non-conductive strips have the same width.

The mobile robotic platform (the *robot*) is equipped with an *array of sliding contacts*, whose positions are fixed with respect to the geometry of the robot. For the purpose of this work we assume that each of the contacts is always in physical contact with the floor and is point-like (i.e., its size is zero along every axis). The robot may move, by translating and rotating, on the powered floor without any constraint.

The robot actually gets the power from the powered floor if there is at least one contact on a positive conductive strip and at least one contact on a negative conductive strip: if this condition is met, we say that the robot is in a *powered condition* and we call the contacts which enable this condition the *powered contacts*. We remark that there are no constraints on which contact is on which polarity: a simple rectifier circuit can be used to obtain a constant power source out of the powered contacts. The design of that circuit is orthogonal to this work.

The goal of the present study is to propose a method that, given (a) a maximum number n of sliding contacts and (b) a description of a *feasible region* for the position of the sliding contacts with respect to the geometry of the robot, finds a *design of the array* of up to n sliding contacts which (a) are in the feasible region and (b) allow the robot to be in a powered condition for any possible position, i.e., any possible combination of translation and rotation. We call a *powered array* an array of contacts which satisfies the two previous conditions, i.e., an array which guarantees that the robot is always in the powered condition.

2.1   Formalization

Without loss of generality, we assume that the strips are (all) parallel with the y-axis. This implies that the only relevant dimension is along the x-axis.

We denote by w the width of the conductive strips and by v the width of the non-conductive strips. We assume that the first strip starts at $x = 0$, ends at $x = w$, and is connected to the positive pole.

We denote by x the position of the a reference point of the robot, by ω the rotation of the robot, and by $A = \{(r_1, \phi_1), \ldots, (r_n, \phi_n)\}$ the array, i.e., the list of the n positions (r_i, ϕ_i), expressed in polar coordinates, of each ith contact with respect to the reference point of the robot. Figure 1a shows a schematic representation of the considered scenario and the corresponding notation; Fig. 1b shows a prototype of the powered floor and a robot.

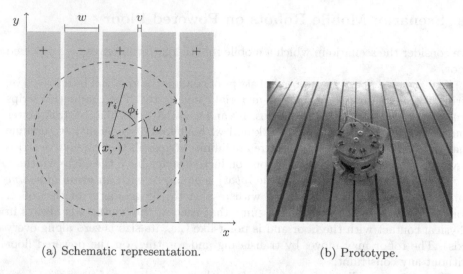

(a) Schematic representation.   (b) Prototype.

Fig. 1. A schematic representation of the considered scenario (a) and the picture of a prototype of the system (b). In the former, the conductive strips are colored in gray, whereas the non-conductive strips are not colored; the robot reference point is in $(x, \cdot)$ (the y coordinate is irrelevant) and its rotation is ω; there is a single sliding contact (r_i, ϕ_i).

We assume that the description of the feasible region for the contacts is provided as a function $v : [0, +\infty] \times [-\pi, \pi[\mapsto \{0, 1\}$. A position (r, ϕ), expressed in polar coordinates, is in the feasible region if and only if $v(r, \phi) = 1$.

A sliding contact (r_i, ϕ_i) is over a positive strip if and only if $\exists k \in \mathbb{N} :$ $2k(w + v) \leq x + r_i \cos(\phi_i + \omega) \leq (2k + 1)(w + v) - v$. The same condition can also be expressed in terms of remainder of the division, which allows to remove k: $(x + r_i \cos(\phi_i + \omega)) \bmod (2w + 2v) \leq w$. We define a function f_c^+ which is 1 if the condition is met and 0 otherwise:

$$f_c^+(r, \phi, x, \omega) = \begin{cases} 1 & \text{if } (x + r_i \cos(\phi_i + \omega)) \bmod (2w + 2v) \leq w \\ 0 & \text{otherwise} \end{cases} \tag{1}$$

Similarly, we define the function f_c^- for negative strips:

$$f_c^-(r, \phi, x, \omega) = \begin{cases} 1 & \text{if } w + v \leq (x + r_i \cos(\phi_i + \omega)) \bmod (2w + 2v) \leq 2w + v \\ 0 & \text{otherwise} \end{cases}$$

$$\tag{2}$$

We can hence write the powered condition of a robot in position x and rotation ω as:

$$f^+ (A, x, \omega) = \sum_{i=1}^{i=n} f_c^+ (r_i, \phi_i, x, \omega) v(r_i, \phi_i) \geq 1 \tag{3}$$

$$f^- (A, x, \omega) = \sum_{i=1}^{i=n} f_c^- (r_i, \phi_i, x, \omega) v(r_i, \phi_i) \geq 1 \tag{4}$$

where the factor $v(r_i, \phi_i)$ means that only contacts which are in the feasible region can be taken into account. The powered condition can also be written as:

$$f^{+-} (A, x, \omega) = \min \left(f^+ (A, x, \omega), f^- (A, x, \omega) \right) \geq 1 \tag{5}$$

where f^{+-} is the number of powered contacts with the robot in position x and rotation ω. Finally, the condition that the robot is in a powered condition in any position x and rotation ω, i.e., the condition for A being a powered array, can be written as:

$$f (A) = \min_{\substack{x \in [0, 2(w+v)[\\ \omega \in [0, 2\pi[}} f^{+-} (A, x, \omega) \geq 1 \tag{6}$$

The problem may hence be tackled by maximizing $f (A)$. Note that, beyond guaranteeing that an array is powered if $f (A) \geq 1$, the chosen f is such that the larger its value, whose upper bound is $\lfloor \frac{n}{2} \rfloor$, the better the positions of the contacts of the array, since there are more powered contacts.

2.2   Secondary Objectives

It can be seen that, due to the presence of min and mod operators and of the binary co-domain of the function v, f may be extremely non-smooth. Figure 2 shows how f^{+-} varies with the position x (left) and rotation ω (right) of a robot with 5 sliding contacts equally spaced on a circle of radius $r_0 = 20\,\mathrm{mm}$ centered in the reference point—i.e., $(r_i, \phi_i) = (r_0, (i-1)\frac{2\pi}{5})$ with $i \in \{1, \ldots, 5\}$—which moves on a powered floor with $w = 9\,\mathrm{mm}$ and $v = 1\,\mathrm{mm}$: for this specific contacts array A, $f (A)$ is 0, which means that there are positions and rotations for which the powered condition is not met.

The non-smoothness of f may negatively affect the effectiveness and efficiency of the search for a powered array. In order to address this problem, we consider a function $\hat{f}$ which is the average value of f^{+-}, instead of the min value, as in Eq. 6:

$$\hat{f} (A) = \frac{1}{4\pi(w+v)} \iint_{\substack{x \in [0, 2(w+v)[\\ \omega \in [0, 2\pi[}} f^{+-} (A, x, \omega) \, \mathrm{d}x \, \mathrm{d}\omega \tag{7}$$

Instead of measuring the number of powered contacts in the worst condition as f, $\hat{f}$ measures the average number of powered contacts across all the possible conditions. It is worth to note, however, that while $f \geq 1$ implies the the robot

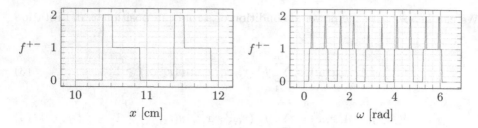

Fig. 2. Values of the function $f^{+-}(A, x, \omega)$ for a simple array of 5 contacts $\left(r_0, (i-1)\frac{2\pi}{5}\right)$ (equally spaced on a circle of radius $r_0 = 20\,\text{mm}$ centered in the robot reference point) for different positions (x, left) and rotations (ω, right) of the robot, with $w = 9\,\text{mm}$ and $v = 1\,\text{mm}$.

is always (i.e., for any x and ω) in a powered condition, $\hat{f} \geq 1$ does not. That is, $\hat{f}(A) \geq 1$ does not guarantee that an array A is a powered array. In the next sections, we will show how we used f and $\hat{f}$ *together* to drive the search for a powered array.

The function f does not explicitly take into account the relative positions of the sliding contacts. In particular, it does not penalize designs of the contacts array where the sliding contacts are too close to each other: those design might be harder to be realized in practice. To take into account this aspect, we introduce a third objective function d which measures the average distance of the contacts to their closest contact:

$$d(A) = \frac{1}{n} \sum_{i \in V} \min_{j \in V, j \neq i} \sqrt{r_i^2 + r_j^2 - 2r_i r_j \cos(\phi_i - \phi_j)} \tag{8}$$

where $V = \{1 \leq i \leq n : v(r_i, \phi_i) = 1\}$ is the set of the indexes of the contacts which are in the feasible region—i.e., as for f and $\hat{f}$, sliding contacts which are not in the feasible region are not taken into account by d.

2.3   Optimization

We resort to Differential Evolution (DE) [10] for solving the problem of the design of contacts array. DE is an EA which can be used for optimization in continuous (real-valued) search space both for single- and multi-objective problems and hence fits the scenario of this study. There exist many variants of DE which are commonly identified with the DE/a/b/c naming scheme [11]: in this work, we used the DE/rand/1, i.e., a variant without the crossover operator which selects the individual to be mutated randomly.

In brief, DE/rand/1 evolves a population of n_{pop} vectors $\boldsymbol{x} \in R^{n_s}$ as shown in Algorithm 1. The population is first initialized (lines 1–6) by randomly setting solutions elements. Then, the following steps are repeated until a termination criterion is met: (i) three different individuals $\boldsymbol{x}, \boldsymbol{y}, \boldsymbol{z}$ are randomly selected in the population (with uniform probability); (ii) a new solution $\boldsymbol{x'}$ is built in which each element x'_j is set either to x_j (with a probability $1 - c_r$) or to a linear combination $x_j + d_w(y_j - z_j)$ of the corresponding elements of the selected individuals (with a probability c_r); (iii) if $\boldsymbol{x}'$ is fitter than $\boldsymbol{x}$, then $\boldsymbol{x}'$ replaces $\boldsymbol{x}$ in the population, otherwise it is discarded. A usual termination criterion consists in having performed n_{ev} iterations, which correspond to n_{ev} fitness evaluations.

Parameters : search space size n_s; population size n_{pop}; number of evaluations n_{ev}; crossover rate c_r; differential weight d_w

1   $P \leftarrow \emptyset$
2   **foreach** $i \in \{1, \ldots, n_{pop}\}$ **do**
3     $\boldsymbol{x} \leftarrow 0$
4   **foreach** $j \in \{1, \ldots, n_s\}$ **do**
5       $x_j \leftarrow U(0, 1)$
6   **end**
7     $P \leftarrow P \cup \{\boldsymbol{x}\}$
8   **end**
9   **foreach** $i \in \{1, \ldots, n_{ev}\}$ **do**
10    $\boldsymbol{x}, \boldsymbol{y}, \boldsymbol{z} \leftarrow \mathrm{PickRandomly}(P, 3)$
11    $\boldsymbol{x}' \leftarrow \boldsymbol{x}$
12   **foreach** $j \in \{1, \ldots, n_s\}$ **do**
13   **if** $U(0, 1) < c_r$ **then**
14       $x'_j \leftarrow x_j + d_w(y_j - z_j)$
15   **end**
16   **end**
17   **if** $\boldsymbol{x}' \succ \boldsymbol{x}$ **then**
18     $P \leftarrow P \setminus \{\boldsymbol{x}\}$
19     $P \leftarrow P \cup \{\boldsymbol{x}'\}$
20   **end**
21   **end**

Algorithm 1. Differential Evolution in the DE/rand/1 variant.

In a multi-objective problem, the comparison between two individuals $\boldsymbol{x}, \boldsymbol{x}'$ for determining the fittest one can be done in several ways. In this work, we use lexicographical ordering. Let $\boldsymbol{f}(\boldsymbol{x}) \in \mathbb{R}^m$ be the m-dimensional fitness vector of the solution $\boldsymbol{x}$ and let assume, without loss of generality, that we want to maximize fitness objectives. We say that $\boldsymbol{x}'$ is fitter than $\boldsymbol{x}$, denoted by $\boldsymbol{x}' \succ \boldsymbol{x}$, if and only if $\exists k \leq m : (\forall i < k : f_i(\boldsymbol{x}') = f_i(\boldsymbol{x})) \wedge (f_k(\boldsymbol{x}') > f_k(\boldsymbol{x}))$.

In this study, we explored different options for driving the search, i.e., different $\boldsymbol{f}$ based on f, $\hat{f}$, and d functions. Concerning f and $\hat{f}$, the actual global minimum or average value of $f^{+-}(A, x, \omega)$ for $x \in [0, 2(w + v)[$ and $\omega \in [0, 2\pi[$ (see Eqs. 6 and 7) cannot, in general, be obtained. Instead, a numerical approximation can be computed. In this work, we compute the values of f^{+-} for the n^2_{points} (x, ω) pairs resulting from evenly sampling the two corresponding domains and take the minimum (for f) or average (for $\hat{f}$) of those values.

3   Experiments and Results

We aimed at gaining insights about our proposed solution from the point of view of the evolutionary optimization and of the considered application. In particular, we aimed at answering the following research questions:

RQ1: are the three objectives f, $\hat{f}$, and d, possibly combined, suitable for driving the search for a powered array?

RQ2: how do the main problem parameters (maximum number n of contacts and strips width w) affect the search for a contact array which always meets the powered condition?

3.1   Robots

We applied our proposed method for the automatic design of contacts arrays to a set of problems inspired by the geometry of three real robotic platforms: Thymio II[1], mBot[2], and Elisa-3[3]. The three robots, shown in Fig. 3, have similar size and were designed to ease learning and experimenting, in particular for children. They are also suitable (and have been actually used [12–14]) for experimenting with Evolutionary Robotics [6].

Fig. 3. Thymio II (left), mBot (center), and Elisa-3 (right) robots.

---

[1] https://www.thymio.org/en:thymio.
[2] https://www.makeblock.com/steam-kits/mbot.
[3] http://www.gctronic.com/doc/index.php/Elisa-3.

The feasible regions for the three robots are determined by the respective shapes and positions of the wheels and other parts which are designed to be in contact with the floor:

$$v(r, \phi) = \begin{cases} 1 & \text{if } -50\,\text{mm} \le r\sin\phi \le 0 \wedge |r\cos\phi| \le 75\,\text{mm} \\ 1 & 0 \le r\sin\phi \le 30\,\text{mm} \wedge |r\cos\phi| \le 110\,\text{mm} \\ 0 & \text{otherwise} \end{cases} \quad \text{for Thymio II}$$

(9)

$$v(r, \phi) = \begin{cases} 1 & \text{if } |r\cos\phi| \le 45\,\text{mm} \wedge |r\sin\phi| \le 45\,\text{mm} \\ 0 & \text{otherwise} \end{cases} \quad \text{for mBot}$$

(10)

$$v(r, \phi) = \begin{cases} 1 & \text{if } 25\,\text{mm} \le r \le 30\,\text{mm} \\ 0 & \text{otherwise} \end{cases} \quad \text{for Elisa-3}$$

(11)

A graphical representation of the feasible regions is shown in Fig. 4.

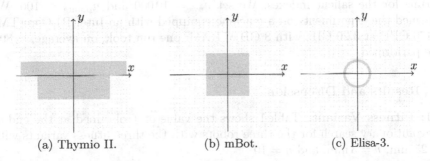

(a) Thymio II.   (b) mBot.   (c) Elisa-3.

Fig. 4. Graphical representation of the feasible regions for the three robots, using Cartesian coordinates x, y, rather than polar coordinates r, ϕ: regions whose points are such that $v(r, \phi) = 1$, with $r = \sqrt{x^2 + y^2}$ and $\phi = \tan^{-1}\frac{y}{x}$, are plotted in green. The scale is the same for the three plots.

3.2   Parameters and Procedure

Setting appropriate values for the salient DE parameters (population size n_{pop}, crossover rate c_r, and differential weight d_w) is not trivial [15]. Based on some preliminary experiments and previous knowledge, we set $n_{\text{pop}} = 100$, $c_r = 0.8$, and $d_w = 0.5$. We verified that small variations of those parameters do not substantially alter the qualitative findings of the experimental evaluation.

Concerning the representation, we encoded the position of the contacts in the array by means of a function $e : (\mathbb{R}^+ \times [0, 2\pi])^n \mapsto \mathbb{R}^{2n}$ as follows:

$$x = e(A) = \left(\frac{r_1}{r^*}, \frac{\phi_1}{2\pi}, \dots, \frac{r_n}{r^*}, \frac{\phi_n}{2\pi} \right)$$

(12)

where n is the maximum number of sliding contacts and r^* is the maximum value of r for which $v(r, \phi) = 1$, i.e., the distance of the farthest point of the feasible region from the reference point. This way, the initialization procedure of the population, which sets the elements of $\boldsymbol{x}$ by sampling $U(0, 1)$ (see Algorithm 1) is appropriate for the domains of r_i, ϕ_i.

Concerning the fitness $\boldsymbol{f}$ of the individuals, we experimented with three different options:

$$\boldsymbol{f}(\boldsymbol{x}) = \boldsymbol{f}_\mathrm{M}(\boldsymbol{x}) = (f(e^{-1}(\boldsymbol{x})))$$
$$\boldsymbol{f}(\boldsymbol{x}) = \boldsymbol{f}_\mathrm{MD}(\boldsymbol{x}) = (f(e^{-1}(\boldsymbol{x})), d(e^{-1}(\boldsymbol{x})))$$
$$\boldsymbol{f}(\boldsymbol{x}) = \boldsymbol{f}_\mathrm{MAD}(\boldsymbol{x}) = (f(e^{-1}(\boldsymbol{x})), \hat{f}(e^{-1}(\boldsymbol{x})), d(e^{-1}(\boldsymbol{x})))$$

where e^{-1} is the inverse of the function e. The M option corresponds to driving the search using only the function f, i.e., a single-objective optimization, whereas MD and MAD use also d and $\hat{f}$.

For each experiment, we performed 30 independent runs by varying the initial random seed: we present mean μ and standard deviation σ computed across the runs for the salient indexes. We set $n_\mathrm{ev} = 10\,000$ and $n_\mathrm{points} = 100$. We performed the experiments on a machine equipped with an Intel(R) Core(TM) i5-3470 CPU at 3.20 GHz with 8 GB of RAM: one run took, on average, $\approx 80\,\mathrm{s}$ to be performed.

3.3   Results and Discussion

RQ1: Fitness Variant. Table 1 shows the value of f obtained at the end of the evolutionary search for the three robots with the three fitness variants with $w = 25\,\mathrm{mm}$, $v = 3\,\mathrm{mm}$, and $n = 10$.

Table 1. Final value of f for the three robots and the three search variants with $w = 25\,\mathrm{mm}$, $v = 3\,\mathrm{mm}$, and $n = 10$.

Robot	MAD		MD		M	
	μ	σ	μ	σ	μ	σ
Elisa-3	2	0	1.48	0.51	1.76	0.44
mBot	2.03	0.19	2	0	2	0
Thymio II	2	0	0.24	0.44	0.83	0.38

It can be seen from the figures in Table 1 that there are differences among the robots and among the fitness variants. With the simplest fitness variant M, our proposed solution is able to obtain, on average, a final f of 1.76 for Elisa-3, 2 for mBot, and 0.83 for Thymio II. We recall that an array of contacts is a powered array (i.e., with that array the robot is in a powered condition in any position and rotation) when $f \geq 1$ (see Eq. 6): hence, our proposed solution with the M variant is able to design a powered array for two of the three robots. For the Thymio II robot, the M variant is not able to design a powered array, neither is the MD variant. With the MAD variant, instead, our solution always designs a powered array, i.e., for each robot (note also the σ values in Table 1). We hypothesize that the reason for which the search struggles for the Thymio II robot is because its feasible region has a more complex shape than those of the other robots.

By analyzing the raw results, we verified that MAD outperforms M and MD because, when the evolution stagnates on a value of f, improvements in the value of $\hat{f}$ allow to gradually improve the array design. In other words, thanks to its better smoothness, $\hat{f}$ can drive the search when f cannot. The d function—which, we recall, measures the average distance among contacts in the feasible region (see Eq. 8)—does not seem to be able to drive the search better than f: the MD variant, in facts, obtains worse results than the M variant. This difference between $\hat{f}$ and d is not surprising, however. On one hand, we introduced $\hat{f}$ with the precise goal of mitigating the poor smoothness of f, hence as a helper for driving the search together with f. On the other hand, d represents a different goal, i.e., designing a contact array which, by avoiding contacts which are too close, is easier to be realized. From a different point of view, f and $\hat{f}$ are not competitive objectives, whereas f and d are: in the extreme case, d pushes the search towards removing contacts (i.e., moving them away from the feasible region), because this increases the average distance; a counter-effect of removing a contact is that f may become lower. Note, however, that since we use lexicographical ordering of objectives for designs comparison (see Sect. 2.3), a change in a design which negatively affects f will never be kept during the search, regardless of the fact that it improves d.

Figure 5 shows the average values of f during the search for the three robots (plots) and the three variants (line colors).

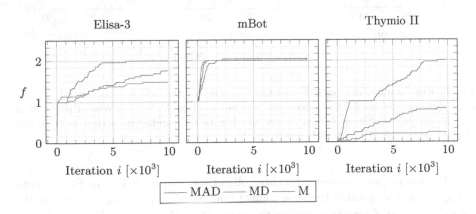

Fig. 5. Value of f during the evolution for the three robots (plots) with the three fitness variants (line color) with $w = 25\,\mathrm{mm}$, $v = 3\,\mathrm{mm}$, and $n = 10$.

The figure makes apparent the fact that, for two on three robots (Elisa-3 and Thymio II), the MAD variant is more effective and more efficient than M and MD; that is, the final value of f is greater and it is reached in fewer iterations. For the mBot the differences among M, MD, and MAD variants are negligible: we motivate this finding with the fact that the feasible region for mBot is easier than for Elisa-3 and Thymio II. In particular, it is more regular than the one of the latter and much larger than the one of the former.

RQ2: Problem Parameters. We considered only the best performing variant of our proposed solution (MAD) and performed a set of experiments by varying the value for $w \in \{15, 20, 25, 30, 35\}$mm and for $n \in \{4, 5, 6, 8, 10, 12, 15\}$, i.e., the width of the conductive strips and the maximum number of contacts in the array, respectively. Figure 6 presents the average final values of the three objectives $f, \hat{f}, d$ for each robot (color line) and each one of the two parameters (w and n, row of plots)—as for the previous experimental campaign, we performed 30 runs for each combination of parameters.

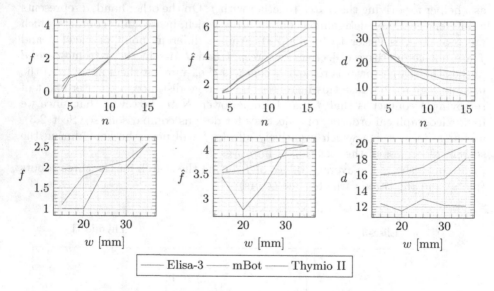

Fig. 6. Final values, averaged across the 30 runs, for the salient functions f, $\hat{f}$, and d for different values for the maximum number n of contacts (top row of plots, with $w = 25\,\mathrm{mm}$) and different values for the width w of the conductive strips (bottom row of plots, with $n = 10$) and with $v = 3\,\mathrm{mm}$.

The dependency between the three objectives and n, visible in Fig. 6, looks as expected. The greater the (maximum) number of contacts n, the greater the values of f and $\hat{f}$ and the smaller the value of d. Concerning $f, \hat{f}$, we observe that increasing n results, obviously, in more contacts in the array, and hence an opportunity for more powered contacts. Concerning the average distance among contacts d, the finding is sound: if more contacts are spread on the same feasible region, they are in general closer. Interestingly, the two leftmost plots in the top row of plots of Fig. 6 also constitute an evidence of the better smoothness of $\hat{f}$ with respect to f: the lines for the former are essentially straight lines, whereas in the lines for f sort of steps can be spotted in the plot.

From the bottom row of plots of Fig. 6 it can be seen that the relation between w and the objectives is more complex than the one between n and the objectives. For example, the final $\hat{f}$ for the Elisa-3 robot has a minimum for $w = 20\,\text{mm}$, i.e., greater $\hat{f}$ values are obtained for both $w = 15\,\text{mm}$ and $w = 25\,\text{mm}$. More in general, it can be seen from these three plots that the relation between the objective and the parameter w varies across objectives and across robots. From a very high level point of view, these plots suggest that the larger the conductive strips, the better: we motivate this finding by observing that increasing w while keeping v (width of non-conductive strips) constant corresponds, basically, to increasing the ratio of the floor area which is actually powered. Concerning the average contacts distance d, Fig. 6 suggests that the wider the conductive strips, the larger the distance among contacts.

4 Conclusions and Future Work

In this paper, we focused on powered floor systems for delivering power to mobile robots. We considered the problem of automatically finding the positions of the sliding contacts which guarantee that the robot actually receives power from the floor in any position and rotation, the positions being constrained in a given feasible region defined depending on the robot shape and equipment. We introduced a formal formulation for the problem including three objectives to be maximized and constraint to be respected. We tackled the resulting multi-objective optimization problem with Differential Evolution (DE).

We experimentally verified that our proposed solution was indeed able to design effective arrays of sliding contacts for three real robots (Thymio II, mBot, Elisa-3). We investigated about the ability of the three objectives to drive the search with DE. We also experimentally explored the impact of the two most important problem parameters (the width of the conductive strips and the maximum number of sliding contacts) on the effectiveness of the contact arrays designed automatically with DE.

The experimental results suggest that our solution may be useful in practice for assisting the design of powered floor systems. Nevertheless, we think that our work might be extended to take into account finer details of the considered scenario as, for example: (i) a more realistic model for the contact-strip conductivity (which is here binary); (ii) the preference for arrays where contacts on positive and negative strips are, in general, balanced; (iii) the possibility of imposing symmetries in the array.

References

1. Shing, A., Wong, P.: Wear of pantograph collector strips. Proc. Inst. Mech. Eng. Part F: J. Rail Rapid Transit **222**(2), 169–176 (2008)
2. Pastena, L.: A catenary-free electrification for urban transport: an overview of the tramwave system. IEEE Electrification Mag. **2**(3), 16–21 (2014)
3. Wang, J., Hu, M., Cai, C., Lin, Z., Li, L., Fang, Z.: Optimization design of wireless charging system for autonomous robots based on magnetic resonance coupling. AIP Adv. **8**(5), 055004 (2018)
4. Yang, M., Yang, G., Li, E., Liang, Z., Lin, H.: Modeling and analysis of wireless power transmission system for inspection robot. In: 2013 IEEE International Symposium on Industrial Electronics (ISIE), pp. 1–5. IEEE (2013)
5. Brambilla, M., Ferrante, E., Birattari, M., Dorigo, M.: Swarm robotics: a review from the swarm engineering perspective. Swarm Intell. **7**(1), 1–41 (2013)
6. Nolfi, S., Bongard, J., Husbands, P., Floreano, D.: Evolutionary robotics. In: Siciliano, B., Khatib, O. (eds.) Springer Handbook of Robotics, pp. 2035–2068. Springer, Cham (2016). https://doi.org/10.1007/978-3-319-32552-1_76
7. Watson, R.A., Ficici, S.G., Pollack, J.B.: Embodied evolution: distributing an evolutionary algorithm in a population of robots. Robot. Auton. Syst. **39**(1), 1–18 (2002)
8. Klingner, J., Kanakia, A., Farrow, N., Reishus, D., Correll, N.: A stick-slip omnidirectional powertrain for low-cost swarm robotics: mechanism, calibration, and control. In: 2014 IEEE/RSJ International Conference on Intelligent Robots and Systems (IROS 2014), pp. 846–851. IEEE (2014)
9. Sloane, N.J., Wyner, A.D.: Claude Elwood Shannon: Collected Papers. IEEE press, Piscataway (1993)
10. Storn, R., Price, K.: Differential evolution-a simple and efficient heuristic for global optimization over continuous spaces. J. Glob. Optim. **11**(4), 341–359 (1997)
11. Das, S., Suganthan, P.N.: Differential evolution: a survey of the state-of-the-art. IEEE Trans. Evol. Comput. **15**(1), 4–31 (2011)
12. Heinerman, J., Rango, M., Eiben, A.: Evolution, individual learning, and social learning in a swarm of real robots. In: 2015 IEEE Symposium Series on Computational Intelligence, pp. 1055–1062. IEEE (2015)
13. Heinerman, J., Zonta, A., Haasdijk, E., Eiben, A.E.: On-line evolution of foraging behaviour in a population of real robots. In: Squillero, G., Burelli, P. (eds.) EvoApplications 2016. LNCS, vol. 9598, pp. 198–212. Springer, Cham (2016). https://doi.org/10.1007/978-3-319-31153-1_14
14. Silva, F., Correia, L., Christensen, A.L.: Evolutionary online behaviour learning and adaptation in real robots. Roy. Soc. Open Sci. **4**(7), 160938 (2017)
15. Brest, J., Greiner, S., Boskovic, B., Mernik, M., Zumer, V.: Self-adapting control parameters in differential evolution: a comparative study on numerical benchmark problems. IEEE Trans. Evol. Comput. **10**(6), 646–657 (2006)

Solving the Multi-objective Flexible Job-Shop Scheduling Problem with Alternative Recipes for a Chemical Production Process

Piotr Dziurzanski[1]([✉]), Shuai Zhao[1], Jerry Swan[1], Leandro Soares Indrusiak[1], Sebastian Scholze[2], and Karl Krone[3]

[1] Department of Computer Science, University of York, Deramore Lane, Heslington, York YO10 5GH, UK
piotr.dziurzanski@york.ac.uk
[2] Institut fur Angewandte Systemtechnik Bremen GmbH, Wiener Strasse 1, 28359 Bremen, Germany
[3] OAS AG, Caroline-Herschel-Strasse 1, 28359 Bremen, Germany

Abstract. This paper considers a new variant of a multi-objective flexible job-shop scheduling problem, featuring multisubset selection of manufactured recipes. We propose a novel associated chromosome encoding and customise the classic MOEA/D multi-objective genetic algorithm with new genetic operators. The applicability of the proposed approach is evaluated experimentally and showed to outperform typical multi-objective genetic algorithms. The problem variant is motivated by real-world manufacturing in a chemical plant and is applicable to other plants that manufacture goods using alternative recipes.

Keywords: Multi-objective job-shop scheduling ·
Process manufacturing optimisation ·
Multi-objective genetic algorithms

1 Introduction

Manufacturing process scheduling is arguably one of the most widely studied optimisation problems [1]. In this problem, manufacturing jobs are assigned to machines at particular times in order to optimise certain key objectives, such as makespan or total workload of machines. Numerous versions of this problem have been proposed, starting from an original Job-shop Scheduling Problem (JSP) coined by R.L. Graham in 1996, and including flexible JSP (FJSP), where operations can be processed on any compatible resource [2]. However, the classic JSP and its popular extensions are limited to certain classes of rather artificial problems and do not scale well to the problem sizes found in industry [3]. Due to the NP-hard nature of the problem, exact solutions are not generally possible for real-world problems. The multi-objective nature of these problems further

© Springer Nature Switzerland AG 2019
P. Kaufmann and P. A. Castillo (Eds.): EvoApplications 2019, LNCS 11454, pp. 33–48, 2019.
https://doi.org/10.1007/978-3-030-16692-2_3

exacerbates the difficulty of obtaining good solutions. A variety of metaheuristics have been applied to the gamut of JSP variants. Amongst these, multi-objective genetic algorithms (GAs) have been applied particularly successfully, as surveyed in [4].

The real-world scenario motivating the research described in this paper is related to a manufacturing process for mixing/dispersion of powdery, liquid and paste components, following a stored recipe. The main optimisation objective of this case study is to increase production line utilisation and, consequently, to decrease the makespan of batch production. The recipes can be executed on different compatible resources. Various recipes can be used to produce the same commodity. Consequently, the decision problem includes the selection of the multisubset (i.e. a combination with repetitions) of the recipes and their allocation to compatible resources, such that the appropriate amount of goods are produced with the minimal surplus in the shortest possible time. As such, the problem resembles, to a certain degree, FJSP with process plan flexibility (FJSP-PPF) [5] or the earlier-formulated "JSP with alternative process plans" [6]. However, none of the papers known to the authors considers process planning by recipe multisubset selection to satisfy both the criteria of the shortest makespan and the minimal surplus of the ordered commodities.

The main contribution of this paper is the formulation of a new variant of a multi-objective FJSP in which a number of commodities can be produced with a set of recipes. A single commodity can be manufactured with a few different recipes whose executions produce different amounts of the commodity and which have different resource compatibility and manufacturing time. The objectives are to minimise the makespan and produce the commodities in the amounts as close to the ordered ones as possible, i.e., to minimise the discrepancies between the ordered quantities and the manufactured ones for each resource. A chromosome encoding for the described problem has been proposed and the classic MOEA/D multi-objective genetic algorithm has been tuned with customised problem-specific genetic operators: mutation and elitism. The applicability of the proposed approach is evaluated experimentally and showed to outperform the classic multi-objective genetic algorithms.

The rest of this paper is organised as follows. After the brief survey of related works in Sect. 2, system model and problem formulation are presented in Sect. 3. The proposed approach for the targeted manufacturing scheduling problem is provided in Sect. 4. The motivating real-world use case is outlined in Sect. 5, followed by experimental results and conclusion in Sects. 6 and 7, respectively.

2   Related Work

One of the first applications of a multi-objective genetic algorithm to manufacturing scheduling was described in [7]. The authors of that paper aimed to find a set of nondominated solutions with respect to the minimal makespan, total flow-time and the maximum tardiness. The fitness value of an individual has been computed as a weighted sum of these three criteria, but the weight values were randomly specified whenever a pair of the parent solutions were selected. This led to the creation of the solution space where each point was generated using a different weight vector. These solutions were then improved by local search. The problem solved by this algorithm is a classic JSP, where the sizes of the assumed plant and taskset were limited. Hence, the settings can be viewed as rather abstract, and the manufacturing scheduling was used just for illustrating potential applicability of the algorithm, rather for direct real-world applicability. In contrast, the problem considered in this paper describes a real-world scenario that is viewed as a challenge by a business partner. A more recent multi-objective genetic algorithm MOAE/D [8], used in this paper as the baseline, has employed some ideas from [7], such as generating various solutions from objective weighted sums.

Several real-world scheduling problems have been deeply researched, typically being solved by customised multi-objective GAs. For example, in [9], a real-world manufacturing problem originating from a steel tube production has been described by extending the classic FJSP and solved using a multi-objective GA with two objectives, namely reduction of the idle time on machines and waiting time of orders. The authors of that paper stressed that it was virtually impossible to apply the earlier research works on JSP in practice as they were based on overly simplified models and assumptions. In that paper, the production routes depend on the orders and a certain production stage could be processed on various homogeneous machines. The model proposed in that paper can be used in numerous job production problems, but is inappropriate in the case of batch manufacturing. In particular, it does not consider recipe selection or minimisation of the commodity surplus, which is addressed by the model proposed in this paper. Readers can refer to the survey presented in [10] to appreciate the complexity of the batch manufacturing in general. The factory model introduced in this paper is capable of describing the majority of the features from the general batch scheduling classification presented in [10], including the "sequence-depending setup", in which sequences of two manufacturing jobs scheduled to be processed subsequently by the same machine can require a time gap of a certain length between them (corresponding to e.g. cleaning the machine in a physical plant).

An interesting real-world problem related to textile batch dyeing scheduling has been described in [11]. Similarly to the problem described in this paper, both the temporal features and the weight of the products are considered. In the textile dying industry, cloths of the same colour can be batched together as long as their total weight does not surpass the capacity of the manufacturing resource. However, for the problem addressed in this paper, the resources are capable of producing only an exact weight of a given commodity, not lower or higher, and the total amount of a manufactured commodity is only influenced with the selection of the recipes multisubset to be executed. Instead of a batching heuristics, a method for recipe multisubset selection that optimises a set of criteria would be desirable.

A number of multi-objective GAs applied to manufacture scheduling problems has been surveyed in [4]. That survey covered assorted types of scheduling problems, including JSP, FJSP, dispatching in a flexible manufacturing system (FMS) and integrated process planning and scheduling (IPPS). The problem described in this paper follows certain realistic assumptions from those problems, such as the presence of alternative machines with different efficiency from FJSP or storage facilitation from FMS. The production planning and scheduling are performed simultaneously as in IPPS. However, none of the reviewed papers allowed selection of a multisubset of recipes for producing the same type of a commodity. Similarly, none of those papers addressed the problem of minimising the surplus of the produced commodities. Both these features are essential to the problem analysed in this paper and they therefore feature in the proposed solution. This objective is also not mentioned in survey [12], which addressed the variability of the objective functions used for multi-objective FJPs. The objectives enumerated in that survey were related to various features of the production process, instead of the amount of the produced commodities.

From this literature survey, it may be concluded that to date there is no proposed FJSP variant that is compatible with the considered real-world scenario. Consequently, a new customisation of FJSP is needed, together with an algorithm capable of solving this problem on a practical scale. Both are presented in the following sections.

3   System Model and Problem Formulation

The problem considered in this paper is an extended version of the classic FJSP, in which each taskset Γ includes a set of independent recipes γ_j, $j = 1, \ldots, n$. Recipe γ_j produces u_j units of certain commodity δ_l, $l = 1, \ldots, r$ and can be executed by one of resources defined by a set Λ_j, including at least one resource $\pi_i \in \Pi$, $i = 1, \ldots, m$. A recipe γ_j needs $t_{i,j}$ time units while executed on resource π_i.

The plant is supposed to satisfy order O, comprised of o_l units of commodities δ_l. The difference between the actually produced amount of commodity δ_l, θ_l and the ordered amount of commodity δ_l , o_l, is referred to as surplus and computed by:

$$\sigma_l = \theta_l - o_l. \tag{1}$$

Several instances of a single recipe γ_j can be scheduled to produce a sufficient amount of goods. These instances are later referred to as $\gamma_{j,k}$, $k = 1, \ldots, \mu_j$, where μ_j denotes the minimal number of recipe instances that satisfy the ordered amount of δ_l and is defined later in this paper.

Certain sequences of recipe instances γ_{j_1,k_1} and γ_{j_2,k_2}, which manufacture different commodities, can require a time gap of a certain length between them if scheduled to be processed subsequently by the same resources (it corresponds to e.g. cleaning the machine in a physical plant). Thus, the instance ordering can influence the makespan. This ordering is controlled with priority $p_{j,k} \in \mathbb{N}_0$ of a recipe instance $\gamma_{j,k}$. Priorities are ordered decreasingly, so from two recipe instances scheduled to a single resource, the one with a lower value of priority will be executed earlier.

Given a set of recipes Γ, a set of resources Π and an order O, the problem is to assign resources and priorities to a multisubset of recipes from Γ so that the total processing time (makespan) is minimised and the amount of each manufactured commodity is higher or equal to the order, $\theta_l \geq o_l$, but the surpluses of each commodity, σ_l, are minimised.

4   Proposed Approach

As summarised by Michalewicz [13], to apply a genetic algorithm to solve a particular problem, each of solution representation, fitness function and evolutionary operators need to be specified. These aspects of the proposed algorithm are focused on in this section.

Let us consider recipe γ_j producing u_j units of a certain commodity δ_l. To determine the upper bound on the number of this recipe instances in the recipe multisubset to be allocated to resources, the lowest number of the recipe execution leading to producing sufficient units o_l of an ordered commodity θ_l needs to be determined. This value can be computed using equation

$$\mu_j = \left\lceil \frac{o_l}{u_j} \right\rceil. \tag{2}$$

Consequently, the cardinality of the multisubset is upperbounded with

$$\eta = \sum_{j=1}^{n} \mu_j. \tag{3}$$

The solution to the problem can be then described with a chromosome of length 2η (η for resource allocation and η for priorities), following the encoding proposed in the following subsection. Hence the genes can be addressed as $\tau_1, \ldots, \tau_\eta$, where τ_1 and τ_η correspond to recipe instances $\gamma_{1,1}$ and γ_{n,μ_n}, respectively.

As the considered real-world scenario includes several objectives aiming at minimising the makespan and the surplus of each commodity, the multi-objective genetic algorithm techniques briefly described later in this section needs to be applied.

4.1   Genetic Representation of Metrics

In genetic algorithms, candidate solutions are treated as individuals. During the optimisation process, these individuals are evolved using a set of bio-inspired operators, described briefly in SubSect. 4.2. In this section, individuals' encoding that facilitates the manufacturing process optimisation and reconfiguration are proposed.

Since in the considered problem each metric assumes a value from a certain, predefined domain, so-called value encoding of chromosomes needs to be applied. This encoding, in contrast to e.g. the traditional binary encoding, allows each gene to directly correspond with a certain value of one variable of the optimisation problem and assume values from the domain of that variable only. For example, a gene representing a certain recipe instance allocation can assume only values corresponding to the compatible resources. To produce a required amount of the ordered commodities, a certain multisubset of recipe set Γ needs to be applied rather than all recipes from this set. The maximal number of recipes that needs to be considered is upperbounded to a certain value η, computed with Eq. (3).

The role of the GA is to allocate the recipe instances to resources and schedule them in time. The encoding has hence to embrace both the spatial and temporal scheduling. Consequently, in the proposed encoding a chromosome contains genes of two types, as shown in Fig. 1. For η recipes (i.e., τ_1 to τ_η) that need to be scheduled, the number of genes is thus equal to 2η. The odd η genes (R_x in the figure) indicate the target resource for η recipe instances, $G_{2x+1} \in \{\emptyset, \pi_1, \dots, \pi_m\}$, where symbol $\emptyset$ denotes the situation that certain recipe instance has not been scheduled for execution. The remaining η genes (ξ_x in the figure) specify the priorities of the recipe instances, $G_{2x} \in \mathbb{N}$, where $x = 1, \dots, \eta$. The priorities are sorted in descending order, i.e. priority 0 is the highest. The aim of introducing priorities is to determine the processing orders of recipes allocated to the same resource and thus to determine the temporal scheduling. This ordering does not change the amount of produced commodities but can influence the makespan due to the sequence-dependent setups discussed earlier. The value of the solution represented by such chromosome is then evaluated using a plant model based on interval algebra described in [14]. The details of the applied plant modelling are out of the scope of this paper.

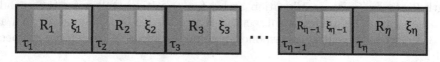

Fig. 1. Genes in a chromosome for manufacturing processes with alternative recipes

4.2   Evolution-Inspired Operators

In a typical GA, evolutionary operators (e.g., crossover and mutation) are applied to a set of individuals for advancing offspring with better solution quality. For the considered optimisation problem, a number of customised genetic operators needs to be proposed. The influence of the operators described below is experimentally evaluated in Sect. 6.

The mutation operator is customised for the studied optimisation problem in the following way. Instead of assigning random configurations (i.e., priorities and allocations) to recipe instances, the proposed operator mutates allocations of recipe instances via two approaches. Depending on the current allocation value of a given recipe instance, the first approach switches its allocation either to a randomly chosen resource that is compatible with the recipe (if this recipe instance is not allocated) or to *no allocation* (i.e., reject for production, in the case that the recipe instance has a valid allocation)[1]. The second approach mutates the allocation of an allocated recipe instance to another compatible resource (if possible). If the recipe instance is not allocated, no action is performed under the second approach. A mutation factor $F \in [0.0, 1.0]$ is introduced to specify the approach to be applied for each allocation mutation. As for priorities, the proposed operator simply assigns random values (but within the predefined priority range) to recipe instances that are chosen to be mutated.

As shown in Sect. 6, compared to a traditional mutation operator, the proposed problem-specific mutation improves the quality of generated solutions and helps to expand the search range of the multi-objective genetic algorithm.

Elitism is often applied by GAs to guarantee the solution quality of each generated population, where a limited number of best solutions are passed to the next generation directly without any operation. In this work, the elitism mechanism is modified to serve the purpose of minimising the sum of discrepancy scores of all commodities. At the end of each generation, solutions that contain the least discrepancy score of each commodity are selected and are used to form an individual that contains the best manufacturing configuration (in terms of the discrepancy scores) being found for each commodity. This individual is then added to the population (if possible, by replacing a randomly chosen individual with a lower fitness) and will involve into future evolution.

Such a simple but effective mechanism can arguably improve the solution quality (in terms of the total discrepancy scores) produced by the proposed multi-objective GA. In particular, solutions with a minimised sum of discrepancy scores for all commodities can be obtained among all solutions generated by the GA. This feature is highly desirable as the cost of storing over-produced commodities can be effectively reduced.

4.3   Customisation of MOEA/D

The problem analysed in this paper is characterised with multi-objective criteria, since not only does the makespan need to be minimised, but also the amount of manufactured commodities should be as close to the ordered amounts as possible to minimise the storage costs. The diversity of these criteria makes it difficult to convert such multi-objective optimisation problem into a single-objective weighted sum of these objective values. Depending on the current

---

[1] Note, due to the applied value encoding, a recipe instance can either be allocated to a resource among its feasible allocations or not be allocated at all. A recipe that is not allocated will be not scheduled for manufacturing.

situation, some solutions with a low weighted sum of objectives may not be acceptable due to, e.g., insufficient storage space for a certain commodity. An end-user should be then informed about a wide set of Pareto-optimal solutions to select the final solution based on his/her knowledge of the problem. The set of the alternative solutions presented to the end-user should be then diverse and, favourably, distributed over the entire Pareto front. This expectation is in line with the properties of the MOEA/D algorithm proposed by Zhang and Li in [8] with Tchebycheff Approach [15] adopted for multi-objective decomposition. With the Tchebycheff Approach, minimising a typical optimisation problem $F(x) = (f_1(x), \ldots, f_m(x))^T$ can be decomposed to the following:

$$\texttt{minimise}\, g^{te}(x|\lambda, z^*) = \max_{1 \leq i \leq m} \{\lambda_i(f_i(x) - z_i^*)\}, \tag{4}$$

where m is the number of objectives in the targeted optimisation problem F(x), $f_i(x)$ gives the value of objective i based on solution x, $x \in \Omega$ indicates a given solution in the decision space (i.e., Ω), λ is a set of uniformly distributed weight vectors, z^* indicates a set of reference points and $z_i^* = min\{f_i(x)|x \in \Omega\}$ gives the reference point of objective i.

The applied multi-objective optimisation algorithm is outlined in Algorithm 1, named MOEA/D-RS, i.e. MOEA/D for recipe scheduling problems. It follows the basic MOEA/D principles but is integrated with the proposed evolution-inspired operators for scheduling recipe in the context of manufacture.

The applied multi-objective optimisation algorithm treats differently the initial generation (lines 1–5) and the remaining generations (lines 6–14), as detailed below. For the initial population, MOEA/D calculates the neighbours (i.e., $B(i)$ for individual i) of each individual based on the Euclidean distance of their associated weight vectors (i.e., λ^i for individual i) (line 4). The set of ideal points (i.e., z) is initialised as the best objective values found in the initial population (line 5).

For the following populations, in each generation (line 6), MOEA/D iterates each individual in the current population (line 7) and selects individuals from the neighbours of the currently-examined individual (line 4) for creating new offspring. The rationale behind neighbourhood selection is the optimal solution of $g^{te}(x|\lambda^i, z^*)$ should be close to that of $g^{te}(x|\lambda^j, z^*)$ if the Euclidean distance between λ^i and λ^j is low, which indicates any g^{te}'s with a weight vector close to λ^i can help optimising $g^{te}(x|\lambda^i, z^*)$ [8]. With parents selected, new individuals are then generated by the proposed genetic operators described in Sect. 4.2.

Once a new individual (say individual y) is generated, values of its key objectives are calculated based on the fitness function (line 9). The ideal points' set z is updated if y contains a better value for any objective (line 10). This new individual will replace any given individual j in the neighbourhood of the currently-examined individual if $g^{te}(y|\lambda^j, z) \leq g^{te}(x^j|\lambda^j, z)$ (line 11). A set of recipe instance allocation and scheduling solutions, EP, is then updated (if necessary) based on the procedure given in line 12. At the end of each generation, we integrate the proposed elitism algorithm into the original MOEA/D and generate an elite individual as described in Sect. 4.2. Accordingly, z and EP are updated (if necessary) based on the objective values of the elite (line 13).

Algorithm 1. Pseudo-code of MOEA/D-RS algorithm

---

inputs   : Resource set Π; Chromosome size 2η; Neighbourhood size T;
            Population size N; Uniform spread of N weight vectors λ^1, λ^2,..., λ^N;
outputs   : EP (a set of recipe instance allocation and scheduling solutions);

1 Set $EP = \emptyset$
2 Generate N random individuals with recipe instance allocations (or recipe
   instance rejection) and priorities as the initial population;
3 Evaluate the key objective values of each individual in the initial population;
4 Compute the Euclidean distances between any two weight vectors and find T
   closest weight vectors to each weight vector. For each $i = 1, \ldots, N$, set
   $B(i) = \{i_1, ..., i_T\}$;
5 Initialise ideal points $z = (z_1, z_2, ..., z_m)$ based on the objective values obtained
   from all individuals of the initial population;
6 **while** *not termination condition* **do**
7 | **for** *i=1,...,N* **do**
8 | | Randomly select two neighbours from $B(i)$, generate a new individual y
   | | via genetic operators proposed in Subsection 4.2 to the selected
   | | neighbours.
9 | | Evaluate the key objective values of y;
10 | | For each $j = 1, \ldots, m$, if $z_j > f_j(y)$, then set $z_j = f_j(y)$;
11 | | For each $j \in B(i)$, set $x^j = y$ if $g^{te}(y|\lambda^j, z) < g^{te}(x^j|\lambda^j, z)$;
12 | | Remove all individuals in EP that are dominated by y and add y to EP
   | | if no individuals dominate y.
   | **end**
13 | Generate an elite individual employing the operator described in
   | Section 4.2, evaluate its objectives' values, and add it to the current
   | population (if eligible), update z and EP.
   end
14 **return** EP;

Once the `termination condition` is met, the algorithm is finished with EP
returned as the Pareto Front (line 14).

5   Real-World Scenario

The considered real-world scenario is based on the process manufacturing of
mixing/dispersion of powdery, liquid and paste recipe components, following a
stored recipe. The main optimisation objective of this case study is to increase
production line utilisation and, consequently, to decrease the makespan of batch
production. The optimisation process can be viewed as scheduling operations,
as described by recipes, both spatially (i.e. to a particular production line) and
temporarily (i.e. to a particular time slot). Depending on the selected production
line, the time to produce a product may vary significantly, which influences the
percentage of manufacturing time that is truly productive, known as Overall
Equipment Effectiveness (OEE). Further impact on the OEE is due to the size
of batches to be produced.

In the considered scenario, the recipes for each batch produce a certain amount of commodity. Consequently, to satisfy a (daily) order for a certain commodity, one or more recipes for producing such commodity have to be selected and scheduled to resources. However, the sum of the commodity amount produced by any selection of recipes may be different from the daily order amount for that commodity. If a certain commodity cannot be produced in the required amount, some commodity surplus is expected. As the total amount of the produced commodity cannot be higher than the available storage space and the surplus storage can be expensive, additional optimisation objectives can be defined: not only the makespan, but also the surpluses of each produced commodities have to be minimised. This observation leads to the conclusion that multi-objective optimisation techniques, as described earlier in this paper, can be applied.

The example plant consists of a set of mixers, Π. There are five identical 5 tonne mixers, named Mixer 1-5 (π_1, π_2, π_3, π_4, π_5, respectively), and two identical 10 tonne mixers, named Mixer 7 (π_6) and Mixer 8 (π_7). There are two special 10 tonne mixers: Mixer 9 (π_8) and Mixer 10 (π_9). Four types of white paints can be produced in the factory and each mixer can be used to produce any commodity. However, the amount of paint produced during one manufacturing process and processing time vary depending on the mixer type and paint type. For each combination of mixer type and paint type, there is a unique recipe, summarised in Table 1 (the paint type names are in German). The storage tanks, connected with the mixers via pipelines, limit the amount of the paints that can be produced as they have limited capacity and each tank can store only one type of paint. In case two recipes producing a different paint type are executed by the same mixer in sequence, a short sequence-dependent setup interval of the length provided by the business partner is enforced.

6   Experimental Results

Based on the real-world case study described in Sect. 5, this section aims at the considered manufacture scheduling problem and presents experiments investigating (i) the optimisation results of NSGA-II [16], original MOEA/D [8] (referred to as MOEA/D hereafter) and the proposed problem-specific algorithm based on MOEA/D (i.e., MOEA/D-RS); (ii) the efficiency of the evolutionary operators proposed in Sect. 4.2; and (iii) the scalability of the proposed multi-objective genetic algorithm for optimisation problems in manufacture scheduling.

The factory setting is given in Table 1. Based on the amount to produce for each commodity, we first generate sufficient number of recipe instances for each recipe type. The number of instances for a given recipe γ_j can be determined using Eq. (2). For instance, a requirement of 45 tonnes of "Std Weiss" would lead to 5 instances of γ_1 and 4 instances of γ_2, γ_3 and γ_4, respectively. Each recipe instance is then assigned with random scheduling parameters (i.e., a random allocation from its compatible resources and a priority) and will be then added to the initial population. To provide fair comparison, general GA parameters applied in all tested algorithms include: PopulationSize = 100, MaxGeneration = 100,

Table 1. Example recipe characteristics for a certain factory

Paint - commodity δ_l	Recipe γ_j	Compatible resources $\pi_i \in \Lambda_j$	Amount of produced commodity u_j	Time $t_{i,j}$ of executing recipe γ_j
Std Weiss δ_1	γ_1	$\{\pi_1, \pi_2, \pi_3, \pi_4, \pi_5\}$	5 t	90 min.
	γ_2	$\{\pi_6, \pi_7\}$	10 t	60 min.
	γ_3	$\{\pi_8, \pi_9\}$	10 t	45 min.
	γ_4	$\{\pi_8, \pi_9\}$	10 t	45 min.
Weiss Matt δ_2	γ_5	$\{\pi_1, \pi_2, \pi_3, \pi_4, \pi_5\}$	5 t	90 min.
	γ_6	$\{\pi_6, \pi_7\}$	10 t	60 min.
	γ_7	$\{\pi_8, \pi_9\}$	10 t	45 min.
	γ_8	$\{\pi_8, \pi_9\}$	10 t	45 min.
Super Weiss δ_3	γ_9	$\{\pi_1, \pi_2, \pi_3, \pi_4, \pi_5\}$	4 t	120 min.
	γ_{10}	$\{\pi_6, \pi_7\}$	8 t	90 min.
	γ_{11}	$\{\pi_8, \pi_9\}$	8 t	60 min.
	γ_{12}	$\{\pi_8, \pi_9\}$	8 t	60 min.
Weiss Basis δ_4	γ_{13}	$\{\pi_1, \pi_2, \pi_3, \pi_4, \pi_5\}$	6 t	60 min.
	γ_{14}	$\{\pi_6, \pi_7\}$	12 t	45 min.
	γ_{15}	$\{\pi_8, \pi_9\}$	12 t	30 min.
	γ_{16}	$\{\pi_8, \pi_9\}$	12 t	30 min.

`CrossoverRate = 1.0`, `MutationRate = 0.8`. One-point crossover operator is applied. The factor that controls the mutation approach (i.e., F) in MOEA/D-RS is set to 0.3 across the evaluation.

Figure 2 presents the optimisation results of three multi-objective GAs for the considered optimisation problem with input values of $o_1 = 45$, $o_2 = 40$, $o_3 = 30$, $o_4 = 20$ (in tonnes). The makespan obtained by each GA is presented as a box with a blue frame and is associated with the blue Y-axis on the left-hand side of the figure while the discrepancy scores (boxes with black frames) are associated with the Y-axis on the right-hand side of the figure. The navy box gives the sum of discrepancy scores of all commodities for each solution. As shown in the figure, the optimisation results given by NSGA-II and MOEA/D are similar, but are both outperformed by the proposed problem-specific GA in terms of the minimal value obtained for each objective. The Diversity Comparison Indicator (DCI) [17], a quality indicator commonly applied for assessing the diversity of Pareto front approximations in many-objective optimisation, is applied to the obtained Pareto Fronts and returns $(0, 0, 1)^2$, which indicates that at least one

---

[2] In general, each numerical value in a tuple obtained with DCI corresponds to a certain front quality in relation to the remaining fronts under comparison. These values are upperbonded with 1 and a higher value denotes a better relative front quality.

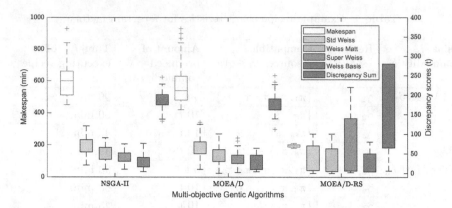

Fig. 2. Optimisation results NSGA-II, MOEA/D and MOEA/D-RS for $\delta_1 = 45t$, $\delta_2 = 40t$, $\delta_3 = 30t$, $\delta_4 = 20t$

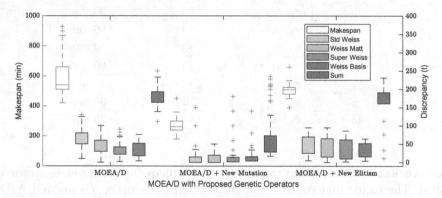

Fig. 3. Influence of the proposed evolutionary operators for an example order

solution returned by MOEA/D-RS strictly dominates[3] any solution obtained by either NSGA-II or MOEA/D. Such observation shows that the generic multi-objective GAs may not be suitable for the studied problem, and certain problem specific multi-objective evolutionary algorithms (e.g., the one proposed in this paper) are desirable. The experiment has been repeated 5 times for 5 slightly different recipe characteristics and ordered amounts of commodities. In all the conducted experiments, the same value of DCI, i.e. $(0, 0, 1)$, has been obtained. These conclusions were confirmed by Sign Test with the probability exceeding 99.99%.

Figure 3 investigates the efficiency of evolutionary operators proposed in Sect. 4.2, where each operator is integrated into MOEA/D alternatively and is compared with the original MOEA/D. As shown in the figure, MOEA/D with the proposed mutation operator applied demonstrates an overall better optimisation

---

[3] For two solutions p and q, p strictly dominates q if p has a better value than q for any objective.

results for each objective compared to MOEA/D. A tuple returned by DCI for the two obtained fronts is equal to $(0, 1)$, which means that at least one solution generated by the customised algorithm strictly dominates any solution generated by the original MOEA/D. The efficiency of the elitism operator is demonstrated by the results of the sum of discrepancy scores for all commodities (i.e., the navy box). With this operator, the applied algorithm can have solutions that contain the least amount of over-produced commodities (i.e., the sum of discrepancy scores), and hence, can greatly reduce the storage cost required by factories. As reported by the DCI test, the values of the original MOEA/D and MOEA/D with elitism operator are 0.318 and 0.682, respectively.

As confirmed by the experiment given in Fig. 2, by integrating all these operators into one algorithm, MOEA/D-RS demonstrates the best performance among all tested multi-objective algorithms under the studied problem scenario. The above experiments have been repeated for 5 slightly different recipe characteristics and ordered amounts of commodities. Similar results have been obtained with average DPI values $(0, 0, 1)$ for (NSGA-II, MOEA/D, MOEA/D-RS), $(0, 1)$ for (MOEA/D, MOEA/D + Customised Mutation), and $(0.389, 0.611)$ for (MOEA/D, MOEA/D + Customised Elitism) throughout the evaluation.

The scalability of the proposed problem-specific evolutionary algorithm is investigated with optimisation results of "makespan" and "sum of discrepancy scores" presented in Figs. 4 and 5, respectively. This experiment is performed based on scaling both the size of the factory (i.e., number of mixers) and the amount of commodities required for production. A scale factor $i = 1, \ldots, 10$ is introduced to control the size of factory and production, where the number of resources $NoR = 10 \times i$ and $o_1 = 45 \times i$, $o_2 = 40 \times i$, $o_3 = 30 \times i$, $o_4 = 20 \times i$ (in tonnes). Accordingly, the set of compatible resources for each recipe type is modified to cope with the increased number of resources while scaling the factory size, where recipes $\{\gamma_1, \gamma_5, \gamma_9, \gamma_{13}\}$, $\{\gamma_2, \gamma_6, \gamma_{10}, \gamma_{14}\}$ and $\{\gamma_3, \gamma_4, \gamma_7, \gamma_8, \gamma_{11}, \gamma_{12}, \gamma_{15}, \gamma_{16}\}$ are compatible with resources $\{\pi_1, \ldots, \pi_{NoR \times 0.5}\}$, $\{\pi_{NoR \times 0.5+1}, \ldots, \pi_{NoR \times 0.7}\}$ and $\{\pi_{NoR \times 0.7+1}, \ldots, \pi_{NoR \times 0.9}\}$, respectively. The rest of resources that are not associated with any recipe (i.e., $\{\pi_{NoR \times 0.9+1}, \ldots, \pi_{NoR \times 1}\}$) are defined as not applicable for producing the given commodities, and hence, will not be considered in manufacturing.

As given in Fig. 4, optimisation results of makespan obtained by MOEA/D demonstrate an observable increasing trend while incrementing scale factor i. In contrast, MOEA/D-RS has a lower increasing rate and outperforms MOEA/D with each i. Similar observations are obtained in Fig. 5, where MOEA/D-RS again outperforms MOEA/D with each i. However, the increasing rate of results obtained by MOEA/D has a much higher increasing rate, where the minimal amount of over-produced commodities with $i = 10$ reaches 3651 tonnes. However, the results obtained by MOEA/D-RS remains better and effectively limit the amount of over-produced commodities to 289 tonnes with $i = 10$. This experiment shows that under the considered manufacture optimisation problem, MOEA/D-RS yields a more diverse front (in terms of minimising objectives

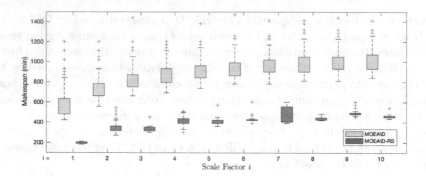

Fig. 4. Makespan optimisation results of original MOEA/D and MOEA/D-RS by scaling an example scenario

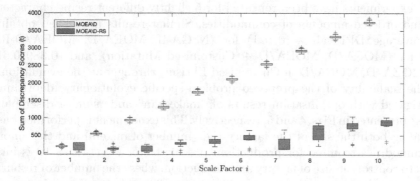

Fig. 5. Sum of discrepancy scores optimisation results of original MOEA/D and MOEA/D-RS by scaling an example scenario

"makespan" and "sum of discrepancy scores") than that of MOEA/D, and confirms the efficiency of MOEA/D with enlarged size of the studied problem.

In addition, the performance of the proposed multi-objective optimisation algorithm is not achieved via sacrificing its run-time efficiency, as shown in Fig. 6. The execution times of both MOEA/D and the proposed MOEA/D-RS algorithm demonstrate an increasing trend with the increment of the factory size and production requirement. With $i \leq 8$, both algorithms demonstrate similar execution times yet MOEA/D-RS can provide results outperforming that of MOEA/D (see Figs. 4 and 5). With $i > 8$, MOEA/D-RS has slightly higher execution time, yet the difference between the optimisation results by two algorithms is further enlarged, especially the sum of discrepancy scores given in Fig. 5 for $i = \{9, 10\}$.

Summarising the experiments above, we conclude that compared to generic evolutionary operators, the proposed problem-specific operators have better efficiency for recipe scheduling problem in manufacturing. With the proposed operators, the modified MOEA/D algorithm (i.e., MOEA/D-RS) can outperform both the tested multi-objective genetic algorithms with the given problem scenario.

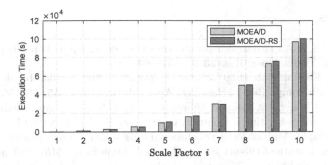

Fig. 6. Execution Times of original MOEA/D and MOEA/D-RS by scaling an example scenario

This observation is further confirmed by an experiment that scales the problem size. From this experiment, we also observe that MOEA/D-RS requires similar execution time in comparison with MOEA/D, but can provide better optimisation results in terms of less time required for production and less amount of overproduced commodities (resulting in a reduced storage cost).

7   Conclusion

In this paper, a real-world factory scheduling problem has been described whose goal is not only to minimise the manufacturing makespan but also to minimise the production surplus via selecting recipes multisubset to be executed. As this problem was difficult to be solved by typical multi-objective genetic algorithms, a modification of MOEA/D has been proposed, which applies customised mutation and elitism operators developed specially for the studied problem. The experiments have demonstrated the superiority of the proposed algorithm (MOEA/D-RS) in comparison with state-of-the-art NSGA-II and original MOEA/D, where for the analysed cases, MOEA/D-RS produces solution that strictly dominates any solution obtained by both NSGA-II and MOEA/D (i.e., DCI values $(0, 0, 1)$ for NSGA-II, MOEA/D and MOEA/D-RS respectively). In particular, the proposed mutation operator improved the results with DCI values $(0, 1)$ of the original MOEA/D and MOEA/D with the customised mutation for all experiments. The proposed algorithm demonstrates similar scalability as the original MOEA/D does in general when being applied to relatively large factories and high quantity of production.

Acknowledgement. The authors acknowledge the support of the EU H2020 SAFIRE project (Ref. 723634).

References

1. Ali, C.I., Ali, K.A.: A research survey: review of flexible job shop scheduling techniques. Int. Trans. Oper. Res. **23**(3), 551–591 (2015)
2. Ham, A.: Flexible job shop scheduling problem for parallel batch processing machine with compatible job families. Appl. Math. Model. **45**, 551–562 (2017)
3. Wang, K., Choi, S.: A holonic approach to flexible flow shop scheduling under stochastic processing times. Comput. Oper. Res. **43**, 157–168 (2014)
4. Gen, M., Lin, L.: Multiobjective evolutionary algorithm for manufacturing scheduling problems: state-of-the-art survey. J. Intell. Manufact. **25**(5), 849–866 (2014)
5. Ozguven, C., Ozbakir, L., Yavuz, Y.: Mathematical models for job-shop scheduling problems with routing and process plan flexibility. Appl. Math. Model. **34**(6), 1539–1548 (2010)
6. Thomalla, C.: Job shop scheduling with alternative process plans. Int. J. Prod. Econ. **74**(1–3), 125–134 (2001)
7. Ishibuchi, H., Murata, T.: A multi-objective genetic local search algorithm and its application to flowshop scheduling. Trans. Sys. Man Cyber Part C **28**(3), 392–403 (1998)
8. Zhang, Q., Li, H.: MOEA/D: a multiobjective evolutionary algorithm based on decomposition. IEEE Trans. Evol. Comput. **11**(6), 712–731 (2007)
9. Li, L., Huo, J.Z.: Multi-objective flexible job-shop scheduling problem in steel tubes production. Syst. Eng. Theor. Pract. **29**(8), 117–126 (2009)
10. Méndez, C.A., et al.: State-of-the-art review of optimization methods for short-term scheduling of batch processes. Comput. Chem. Eng. **30**(6–7), 913–946 (2006)
11. Huynh, N.T., Chien, C.F.: A hybrid multi-subpopulation genetic algorithm for textile batch dyeing scheduling and an empirical study. Comput. Ind. Eng. **125**, 615–627 (2018)
12. Amjad, K.M., et al.: Recent research trends in genetic algorithm based flexible job shop scheduling problems. Math. Probl. Eng. 1–32 (2018)
13. Michalewicz, Z.: Genetic Algorithms + Data Structures = Evolution Programs, 3rd edn. Springer, Heidelberg (1996). https://doi.org/10.1007/978-3-662-03315-9
14. Indrusiak, L.S., Dziurzanski, P.: An interval algebra for multiprocessor resource allocation. In: International Conference on Embedded Computer Systems: Architectures, Modeling, and Simulation (SAMOS), pp. 165–172, July 2015
15. Miettinen, K.: Nonlinear Multiobjective Optimization, vol. 12. Springer, New York (2012). https://doi.org/10.1007/978-1-4615-5563-6
16. Deb, K., et al.: A fast and elitist multiobjective genetic algorithm: NSGA-II. IEEE Trans. Evol. Comput. **6**(2), 182–197 (2002)
17. Li, M., Yang, S., Liu, X.: Diversity comparison of pareto front approximations in many-objective optimization. IEEE Trans. Cybern. **44**(12), 2568–2584 (2014)

Quantifying the Effects of Increasing User Choice in MAP-Elites Applied to a Workforce Scheduling and Routing Problem

Neil Urquhart[✉], Emma Hart, and William Hutcheson

School of Computing, Edinburgh Napier University, 10 Colinton Road,
Edinburgh EH10 5DT, UK
n.urquhart@napier.ac.uk

Abstract. Quality-diversity algorithms such as MAP-Elites provide a means of supporting the users when finding and choosing solutions to a problem by returning a set of solutions which are diverse according to set of user-defined features. The number of solutions that can potentially be returned by MAP-Elites is controlled by a parameter that discretises the user-defined features into 'bins'. For a fixed evaluation budget, increasing the number of bins increases user-choice, but at the same time, can lead to a reduction in overall quality of solutions. *Vice-versa*, decreasing the number of bins can lead to higher-quality solutions but at the expense of reducing choice. The goal of this paper it to explicitly quantify this trade-off, through a study of the application of Map-Elites to a Workforce Scheduling and Routing problem, using a large set of realistic instances based in London. We note that for the problems under consideration 30 bins or above maximises coverage (and therefore choice to the end user), whilst reducing the bins to the minimal size of 5 can lead to improvements in fitness between 23 and 38% in comparison to the maximum setting of 50.

Keywords: MAP-Elites · Transportation · Illumination · WSRP

1 Introduction and Motivation

The Workforce Scheduling and Routing Problem (WSRP) is a commonly encountered real-world problem in which there is a requirement to allocate tasks to individuals within a workforce, and to provide an optimised routing-plan between allocated tasks. Typical domains in which these problems arise include healthcare (e.g scheduling home-visits from care-workers) and maintenance scheduling (e.g. scheduling service-engineers to jobs). As with many real-world problems, the ability to provide an end-user with a set of potential solutions is of great importance, enabling them to select between alternatives with respect to specific company priorities or objectives.

© Springer Nature Switzerland AG 2019
P. Kaufmann and P. A. Castillo (Eds.): EvoApplications 2019, LNCS 11454, pp. 49–63, 2019.
https://doi.org/10.1007/978-3-030-16692-2_4

One approach to generating a set of potential solutions is to use a quality-diversity algorithm [1] which returns a set of diverse but high fitness solutions. In previous work, we applied one such algorithm, MAP-Elites[1], to the WSRP in order to generate multiple solutions that were diverse with respect to a set of four user-defined features [2]. This enabled a planner to see how a WSRP solution could be tailored to specific requirements, for example examining the effects of implementing a policy that increased public transport usage, and noting the effects on other objectives such as financial cost or emissions.

The MAP-Elites algorithm requires the user to define n features (in this case CO_2 produced, car use, travel cost and staff cost) of interest which are mapped to an n-dimensional grid. The grid is discretised according to a user-defined parameter d which controls the number of *bins*, i.e. discrete cells, on each axis of the grid. The maximum number of cells and therefore solutions that can be obtained is therefore d^n. Given a fixed evaluation budget, then increasing d is very likely to lead to a reduction in solution quality given that the algorithm is forced to maintain solutions across a larger space, with a consequent reduction in selection pressure. For the end-user, this introduces a trade-off between selecting a grid at one end of the spectrum that provides a *large* number of *lower-quality* solutions and at the other, a grid providing a *small* number of *high-quality* solutions. The goal of this paper is to quantify this trade-off, by conducting a thorough empirical investigation of the influence of the number of bins (d) on the quality of solutions produced and the coverage obtained when solving instances of the WSRP given a fixed optimisation budget.

2   Background and Previous Work

The WSRP was defined in [3] as a scenario that involves the scheduling and routing of personnel in order to perform activities at different locations. Although similar to vehicle routing problems, the focus of the WSRP is on individuals rather than vehicles. For a comprehensive introduction to the WSRP and an overview of the latest developments, the reader is directed towards [4–6]. A number of previous researchers have examined the scheduling and routing of workforces, including home-care scheduling [7], security personnel scheduling [8] and technician scheduling [9]. An attempt to model the WSRP as a bi/multi-objective problem can be found in [5], the authors use cost and patient convenience as the twin objectives. The solution cost is the travel cost and staff overtime costs, patient convenience is defined as to whether the member of staff allocated is preferred, moderately preferred or not preferred, with penalties allocated as appropriate. The results presented show a strong relationship between convenience and cost; the more convenient a solution the higher the cost is likely to be.

Quality-Diversity algorithms (QD) [1] produce an array of high-quality solutions with respect to a set of user-defined features. Although a number of QD algorithms now exist, including Novelty-Search with Local Competition [10], and MAP-Elites [11], here we focus only on the latter given its prevalence in the

---

[1] Multi-dimensional Archive of Phenotypic Elites.

literature. The Multi-dimensional Archive of Phenotypic Elites (MAP-Elites) was first introduced by Mouret *et al.* [11] and provides a mechanism for illuminating search spaces by creating an archive of high-performing solutions mapped onto solution characteristics defined by the user. The majority of applications of illumination algorithms have been to *design* problems [11,12]. The basic approach has been extended by surrogate-assistance to reduce the computation time associated with real-world evaluations [13] and more recently by including a user in an evolutionary loop to focus search on particular regions of the feature-space of interest [14]. There are very few applications of QD algorithms apparent in the combinatorial optimisation domain; to the best of our knowledge, we were the first to show that MAP-Elites could be successfully applied to combinatorial optimisation using the WSRP problem as an example [2]. The algorithm was applied to produce multiple solutions to WSRP instances modelled using real data based upon the City of London, with four dimensions of user interest. The discretisation parameter d was arbitrarily chosen to give 160,000 cells in this work however, motivating the need for further exploration.

3 Methodology

3.1 WSRP Problem Description

The WSRP considered is defined as follows. An organisation has to service a set of clients, who each require a single visit. Each visit v must be allocated to an employee, such that all client visits are made by an employee. Each visit v is located at g_v, where g represents an actual UK post-code, and has a visit length d_v and a time-window in which it must commence $\{e_v, l_v\}$. Visits are grouped into *journeys*, where each journey is allocated to an employee and contains a subset V_j of the V visits, starting and ending at a central office. In this formulation an unlimited number of employees are available.

Two modes of travel are available to employees, private transport (car) or public-transport, encouraging more sustainable travel, each journey is carried out using one of these modes for the entire journey. The overall goal of our WSRP is to minimise the total distance travelled across all journeys completed. Discussions with end-users [15] highlights four characteristics of solutions that are of interest:

- **Emissions** incurred by all employees on their journeys
- **Employee Cost** the cost (based on £/hour) of paying the workforce for the duration of the journeys and visits
- **Travel Cost** the cost of all of the travel activities undertaken by the workforce
- The % of Employees using **car travel** for their journeys

We use a problem-representation described in [15]. The genotype defines a permutation of all v required visits, divided into individual feasible journeys using a *decoder*. For each visit, the genotype also includes an additional gene that denotes the preferred mode of transport to be used for the visit (public or private).

The decoder converts the genome into a set of employee journeys by examining each visit in order. Initially, the first visit in the genotype is allocated to the first journey. The travel mode (car or public transport) associated with this visit in the genome is then allocated to the journey. The travel mode associated with the first visit is adopted for the entire journey (regardless of the information associated with subsequent visits in the genome). The decoder then examines the next visit in the permutation—this is added to the current journey if it is *feasible*. Feasibility requires that the employee arrives from the previous visit using the mode of transport allocated to the journey within the time window associated with the visit. Subsequent visits are added using the journey mode until a hard constraint is violated, at which point the current journey is completed and a new journey initiated.

3.2 Problem Instances

We use a set of problem instances based upon the city of London, divided into two problem sets, termed *Lon* (60 visits) and *BLon* (110 visits). These instances were first introduced in [15] and also used later in [2]. For each of the problem sets, 5 instances are produced in which the duration of each visit is fixed to 30 min. Visits are randomly allocated to one of t time-windows, where $t \in \{1, 2, 4, 8\}$. For $t = 1$, the time-window has a duration of 8 h, for $t = 2$, the time-windows are "9 am–1 pm" and "1 pm–5 pm" etc. These instances are labelled using the scheme *<set>-numTimeWindows*, i.e. *Lon-1* refers to an instance in the London with one time-window and *Blon-2* refers to an instance of the BigLondon problem with 2 time windows. The 'rnd' instance (e.g. BLon-rnd) represents a randomly chosen mixture of time windows based on 1, 2, 4 and 8 h.

When a journey is undertaken by car, the distance and time is calculated according to the real road-network using the GraphHopper library[2]. This relies on Open StreetMap data[3]. Car emissions are calculated as 140 g/km based upon values presented in [16]. For journeys by public-transport, data is read from the Transport for London (TfL) API[4] which provides information including times, modes and routes of travel by bus and train. Public transport emissions factors are based upon those published by TfL [16].

3.3 MAP-Elites

The implementation of MAP-Elites used in this paper was used previously by [2] and was taken directly from [11]. The algorithm commences with an empty, N-dimensional map in which solutions $\mathcal{X}$ and their performances $\mathcal{P}$ are subsequently placed. A new solution is generated either by random initialisation or through reproduction operators, depending on the phase of the algorithm. If the number of solutions is $>= G$, then a solution (or solutions) is randomly

---

[2] https://graphhopper.com/.
[3] https://openstreetmap.org/.
[4] https://api.tfl.gov.uk/.

selected from the map. The *Random Variation()* method applies either crossover followed by mutation, or just mutation, depending on the experiment. All operators utilised are borrowed from the authors' previous work on these problems [17]. The *mutation* operator moves a randomly selected entry in the chromosome to a randomly selected point in the tour. The *crossover* operator selects a random section of the tour from parent-1 and copies it to the new solution. The missing elements in the child are copied from parent-2 in the order that they appear in parent-2. For each child solution x', a *feature-descriptor b* is obtained by discretising the four features of interest associated with the solution. The upper and lower bounds required for discretisation are taken as the maximum and minimum values observed within these data sets in [15]. A new solution is placed in the cell in the archive corresponding to b if its fitness p (calculated as total distance travelled) is better than the current solution stored, or the cell is currently empty.

Algorithm 1. MAP-Elites Algorithm, taken directly from [11]

---

procedure MAP-ELITES ALGORITHM
   $(\mathcal{P} \leftarrow \emptyset, \mathcal{X} \leftarrow \emptyset)$
   for iter $= 1 \rightarrow$ I **do**
      if iter $<$ G **then**
         $x' \leftarrow$ randomSolution()
      else
         $x' \leftarrow$ randomSelection($\mathcal{X}$)
         $x' \leftarrow$ randomVariation($\mathcal{X}$)
      end if
      $b' \leftarrow$ featureDescriptor(x')
      $p' \leftarrow$ performance(x')
      if $\mathcal{P}(b') = \emptyset$ **or** $\mathcal{P}(b') < p'$ **then**
         $\mathcal{P}(b') \leftarrow p'$
         $\mathcal{X}(b') \leftarrow x'$
      end if
   end for
   return feature-performance map($\mathcal{P}$ and $\mathcal{X}$)
end procedure

3.4   Experimental Set up

We vary the number of bins (d) from 5 to 50 in steps of 5 (see Table 1). The number of cells is therefore $cells = d^n$ where n is the number of dimensions within the problem. Note that the *range* for each feature axis remains constant regardless of the number of bins that it is discretised into.

MAP-Elites was executed on each problem instance 10 times for each bin configuration. The function evaluation budget is fixed at 5,000,000 evaluations all experiments. We use the coverage metric to measure the area of the feature-space covered by a single run of the algorithm, i.e. the number of cells filled. For a single run x of algorithm y, $coverage = noOfCellsFilled/C_{Max}$ where C_{Max} is the total number of cells filled by combining all of the solutions produced to the problem under consideration.

Table 1. Numbers of cells in each map. The number of cells is calculated as $cells = d^n$

Bins (d)	Cells	Bins (d)	Cells
5	625	30	810000
10	10000	35	1500625
15	50625	40	2560000
20	160000	45	4100625
25	390625	50	6250000

4   Results

4.1   The Effects of Bin Quantity on Fitness

The average, maximum and minimum fitness values found for each problem instance are given in Table 3: entries in the table are of the format <avg>(min/max) over the 10 runs undertaken on each problem instance. It is clear that as the number of bins increases the average and best fitness increases in value, i.e. represents inferior performance. The drop off in performance is relatively consistent as evidenced by Table 4: the increase in average fitness is in the range of 23–34% and increases as the problems become more constrained (e.g. with smaller time windows). The relationship between performance (as evidenced by the lowest fitness found over 10 runs) and the number of bins is shown in Fig. 5. Correlation coefficients of $R = 0.8943$ for the Lon data and $R = 0.9619$ for the BLon data indicate a strong positive relationship between the best solution found and the number of bins.

4.2   The Effects of Bin Quantity on Coverage

Figures 3 and 4 show the average levels of coverage achieved. The reader is reminded that coverage measures the proportion of the cells in the archive that contain a solution. We notice in both figures that the highest coverage is obtained with 25–30 bins, with the average coverage dropping before that.

When looking for a relationship between coverage and the number of bins (d) we find that a correlation coefficient of $R = 0.5314$ and $R = 0.291$ for BLon and Lon. This suggests that the relationship between Bins and Coverage is not strong. We do note that for BLon lower coverage is achieved with smaller values of d. The differing problem instances show different levels of coverage, but follow the same overall trends in the case of BLon. It is worth noting that when a small number of bins are used the coverage can drop as low as 50% (Blon): this translates to user-choice being limited to very few solutions. At a discretisation level of 5 bins, this results in   312 solutions; this is in stark contrast to using e.g. 25 bins which gives a potential of 390625 solutions of which 80% are covered (albeit at lower quality). The best/worst coverage scores (averaged over 10 runs) for each value of d may be seen in Figs. 1 and 2. The larger datasets (Blon) show a more consistent performance in relation to d with the values of 30 or over giving the best coverage. The smaller dataset (Lon) gives a less consistent performance, generally smaller values of d produce less coverage, but there is not a value of d which differentiates consistently between best and worst performances. In order to confirm that there is a significant difference between the best and worst performance we apply a t-test to the coverage values obtained from the individual runs (Table 2). We note that a significant difference is obtained in every case, and in 6 cases the difference is classed as extremely significant.

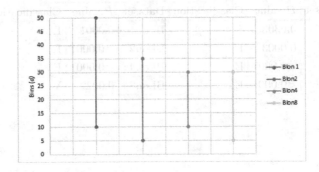

Fig. 1. The best/worst average coverage obtained for the BLon datasets. The worst is consistently obtained with 5–10 bins, the best coverage is obtained with ≥ 30 bins.

4.3   Effects of Bin Quantity on Other Solution Characteristics

A visual indication of the solutions found is shown in Figs. 6 and 7 which examine results achieved from the BLon-rnd dataset. The figures chart the results obtained with differing numbers of bins: note that each row represents the final result of a specific run. We note that although the number of cells increases,

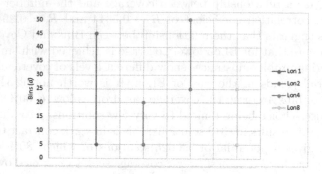

Fig. 2. The best/worst average coverage obtained for the Lon datasets. The least coverage is obtained with 5 bins, except for Lon4 the best coverage is obtained with bins ranging from 20 to 50.

Table 2. The t-test results for the best/worst coverage (see Figs. 1 and 2) based on the coverage obtained on each individual run. The t-tests were carried out using (https://www.graphpad.com/), the classifications based on https://www.graphpad.com are 'S' - significant, 'E' - extremely significant and 'V' very significant.

Dataset	P value	Classification	Dataset	P value	Classification
Lon1	0.0393	S	BLon1	0.0001	E
Lon2	0.0003	E	BLon2	<0.0001	E
Lon4	<0.0001	E	BLon4	<0.0001	E
Lon8	<0.0001	E	BLon8	0.0022	V

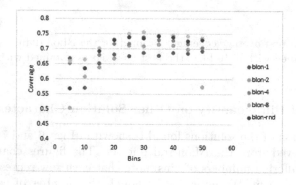

Fig. 3. The coverage achieved using with BLon datasets. Each data point represents the average coverage over 10 runs.

Table 3. The fitness (distance) values achieved on each of the problem instances, the results are in the format of \<avg\>/\<lowest\>/\<highest\> based on 10 runs. The results in **bold** show the lowest overall and lowest average fitness values found.

Problem	Number of bins				
	5	10	15	20	25
Lon-1	**343.17 (169.72**/582.58)	368.32 (177.72/604.06)	392.55 (181.45/640.38)	404.09 (194.59/651.94)	419.19 (221.44/670.35)
Lon-2	316.14 (189.68/500.40)	335.26 (**185.45**/617.62)	361.00 (191.99/639.24)	382.90 (219.27/662.77)	398.83 (243.32/668.99)
Lon-4	289.52 (201.27/489.72)	302.10 (**187.22**/551.87)	328.39 (214.04/585.18)	350.54 (231.61/615.31)	369.23 (247.72/631.39)
Lon-8	267.03 (209.85/415.40)	277.14 (**202.67**/504.63)	301.58 (218.08/552.50)	323.88 (233.18/573.96)	342.29 (250.64/598.11)
Lon-rnd	302.87 (202.48/516.74)	323.81 (**196.94**/607.91)	347.86 (219.05/647.06)	371.41 (245.39/654.72)	391.31 (263.34/662.31)
	30	35	40	45	50
Lon-1	426.57 (225.94/678.49)	438.23 (231.51/705.43)	440.74 (240.23/693.97)	449.25 (245.00/690.69)	450.32 (264.40/689.91)
Lon-2	412.29 (253.64/664.07)	424.12 (264.65/672.79)	433.92 (277.33/692.45)	442.50 (271.10/685.20)	450.24 (292.77/677.21)
Lon-4	386.14 (256.57/626.87)	401.86 (281.49/647.94)	414.90 (284.74/642.31)	425.24 (291.31/644.13)	433.86 (307.19/642.20)
Lon-8	359.06 (266.18/602.82)	374.84 (275.39/601.51)	388.46 (287.01/618.53)	399.58 (286.76/609.64)	409.79 (307.17/607.84)
Lon-rnd	409.99 (278.47/677.60)	424.29 (289.09/679.12)	437.99 (284.55/658.81)	448.22 (308.62/665.29)	456.55 (313.57/669.29)
	5	10	15	20	25
BLon-1	908.88 (576.88/1353.63)	984.16 (**575.24**/1485.79)	1049.33 (678.25/1544.02)	1077.76 (722.07/1629.96)	1100.84 (777.11/1660.29)
BLon-2	843.40 (**595.33**/1258.17)	921.98 (644.37/1426.47)	979.64 (698.01/1536.59)	1027.11 (707.28/1577.54)	1061.74 (797.55/1603.90)
BLon-4	812.18 (**593.46**/1218.16)	850.28 (616.68/1384.28)	914.15 (688.46/1447.85)	962.87 (719.31/1556.15)	1005.62 (781.23/1547.89)
BLon-8	739.43 (575.02/1123.34)	762.15 (**558.48**/1264.02)	814.39 (621.05/1388.22)	876.25 (688.50/1425.02)	921.19 (745.42/1441.38)
BLon-rnd	772.14 (**560.96**/1318.77)	841.77 (614.57/1355.84)	906.13 (546.29/1465.39)	963.29 (721.45/1599.49)	1010.42 (770.88/1580.51)
	30	35	40	45	50
BLon-1	1127.61 (794.66/1639.93)	1144.45 (787.04/1707.63)	1155.95 (828.50/1712.46)	1171.13 (830.16/1714.22)	1179.35 (874.86/1709.89)
BLon-2	1089.34 (814.67/1604.73)	1115.92 (856.11/1634.92)	1135.05 (862.98/1664.27)	1148.45 (889.40/1660.21)	1168.40 (919.96/1653.63)
BLon-4	1039.31 (802.53/1528.20)	1071.97 (847.85/1562.03)	1094.58 (861.59/1581.44)	1114.82 (867.15/1585.64)	1132.02 (903.20/1578.84)
BLon-8	961.76 (745.99/1505.71)	996.44 (808.32/1529.49)	1024.60 (832.17/1503.73)	1050.35 (858.94/1515.39)	1069.19 (893.62/1495.86)
BLon-rnd	1044.26 (828.00/1565.58)	1077.62 (859.40/1593.16)	1099.46 (847.55/1598.61)	1121.43 (900.96/1639.58)	1141.09 (906.73/1604.65)

Table 4. The increase (i.e. loss) in average and best fitness between the lowest and highest number of bins (d) used. Recall this is a minimisation problem so lower objective values are preferred

	Avg			Best		
	5	50	% Increase	5	50	% Increase
Lon-1	343.17	450.32	23.79%	169.72	264.4	35.81%
Lon-2	316.14	450.24	29.78%	189.68	292.77	35.21%
Lon-4	289.52	433.86	33.27%	201.27	307.19	34.48%
Lon-8	267.03	409.79	34.84%	209.85	307.17	31.68%
Lon-rnd	302.87	456.55	33.66%	202.48	313.57	35.43%
BLon-1	908.88	1179.35	22.93%	576.88	874.86	34.06%
BLon-2	843.4	1168.4	27.82%	595.33	919.96	35.29%
BLon-4	812.18	1132.02	28.25%	593.46	903.2	34.29%
BLon-8	739.43	1069.19	30.84%	575.02	893.62	35.65%
BLon-rnd	772.14	1141.09	32.33%	560.96	906.73	38.13%

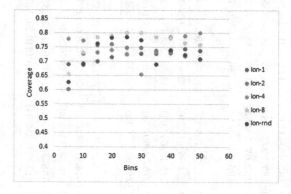

Fig. 4. The coverage achieved using with Lon datasets. Each data point represents the average coverage over 10 runs.

the shape of the filled in area, largely remains the same. The resolution of the heat map increases, but the image remains largely unaltered. Those areas which are not filled in on the 5 bin heat maps are largely the same as those not filled in on the 50 bin heat maps. The lightest cells (green) represent those solutions with the least distance cost and we see largely the same distribution of colours (Figs. 8, 9 and Table 5).

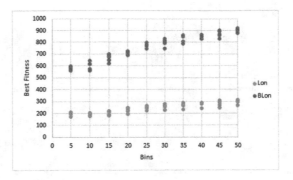

Fig. 5. Best fitness achieved (over 10 runs) plotted against the number of bins. The correlation coefficients obtained were $R = 0.8943$ and $R = 0.9619$ for the Lon and BLon data respectively.

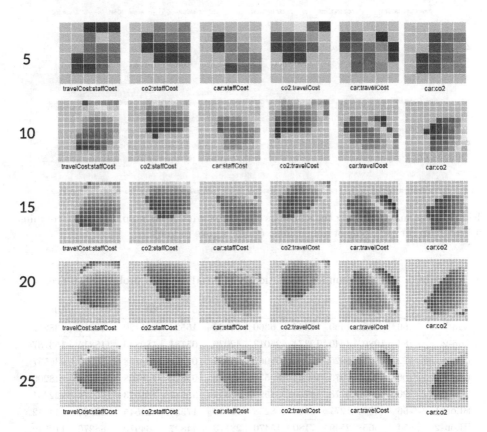

Fig. 6. Heat maps showing the solutions contained within the map. Each row contains a heat map for each pair of dimensions. (Color figure online)

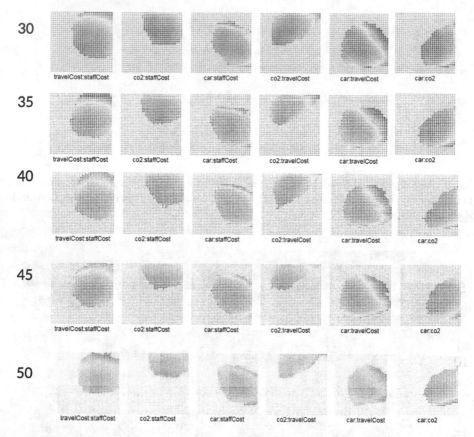

Fig. 7. Heat maps showing the solutions contained within the map. Each row contains a heat map for each pair of dimensions. (Color figure online)

Table 5. Average number of solutions produced for each problem instance.

	5	10	15	20	25	30	35	40	45	50
Lon-1	69	432	1450	3344	6360	10328	15524	21129	28759	38252
Lon-2	84	1080	4640	12374	25013	42039	65781	89835	118815	153787
Lon-4	122	1501	6926	19325	38961	65684	101561	140768	185533	236537
Lon-8	106	1490	6979	19144	39533	67495	103840	145798	194424	248200
Lon-rnd	120	1552	6878	18749	37590	62441	95737	131973	175174	222401
BLon-1	56	362	1271	3263	6258	11038	17539	24545	33898	44642
BLon-2	54	565	2549	7180	15470	28173	44827	64970	88276	112549
BLon-4	68	752	3305	9376	20473	35916	56615	81887	109628	141184
BLon-8	72	859	3821	10993	22940	41354	63907	92374	123681	157905
BLon-rnd	84	888	4113	11478	24330	42021	65049	92556	122091	156211

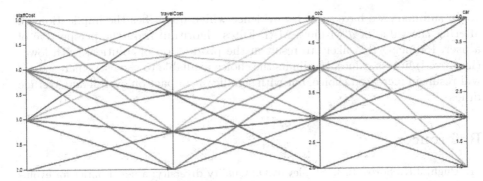

Fig. 8. A parallel coordinates plot showing the solutions found for a run with the BLon-rnd data set with $d = 5$.

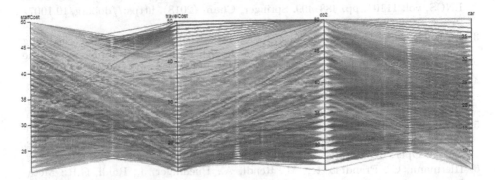

Fig. 9. A parallel coordinates plot showing the solutions found for a run with the BLon-rnd data set with $d = 50$.

5   Conclusions

In this paper we set out to quantify the trade-off between user choice (coverage) and solution quality, by varying the number of bins (d).

The potential for user-choice is reflected by the size of the map, which is determined by the number of bins. Table 4 shows a decrease in the performance of the algorithm (measured as best and average fitness) as the number of bins (d) increases. For these instances, a user interested in maximising objective performance can expect an average loss in performance of between 32% and 38% if the value of d is increased in order to maximise choice. On the other hand, the coverage results show that choice is maximised by setting d to higher values: an approximate gain in coverage of between 10 and 20% can be found by judicious choice of d depending on the problem when compared to setting it to 5—the value which maximises the objective function.

In summary we have provided evidence which quantifies the implications of altering the value of d within a MAP-Elites algorithm from the perspective of a user. Reducing d is likely to result in the production of solutions with lower (improved) fitness values, but limits the choice available to the user. Increasing d will result in increased choice, but with significant impact on overall and average fitness.

References

1. Pugh, J.K., Soros, L.B., Stanley, K.O.: Quality diversity: a new frontier for evolutionary computation. Front. Robot. AI **3**, 40 (2016)
2. Urquhart, N., Hart, E.: Optimisation and illumination of a real-world workforce scheduling and routing application (WSRP) via Map-Elites. In: Auger, A., Fonseca, C.M., Lourenço, N., Machado, P., Paquete, L., Whitley, D. (eds.) PPSN 2018. LNCS, vol. 11101, pp. 488–499. Springer, Cham (2018). https://doi.org/10.1007/978-3-319-99253-2_39
3. Castillo-Salazar, J.A., Landa-Silva, D., Qu, R.: A survey on workforce scheduling and routing problems. In: Proceedings of the 9th International Conference on the Practice and Theory of Automated Timetabling, pp. 283–302 (2012)
4. Castillo-Salazar, J.A., Landa-Silva, D., Qu, R.: Workforce scheduling and routing problems: literature survey and computational study. Ann. Oper. Res. **239**(1), 39–67 (2016). https://doi.org/10.1007/s10479-014-1687-2
5. Braekers, K., Hartl, R.F., Parragh, S.N., Tricoire, F.: A bi-objective home care scheduling problem: analyzing the trade-off between costs and client inconvenience. Eur. J. Oper. Res. **248**(2), 428–443 (2016)
6. Hiermann, G., Prandtstetter, M., Rendl, A., Puchinger, J., Raidl, G.R.: Metaheuristics for solving a multimodal home-healthcare scheduling problem. Cent. Eur. J. Oper. Res. **23**(1), 89–113 (2015). https://doi.org/10.1007/s10100-013-0305-8
7. Rasmussen, M., Justesen, T., Dohn, A., Larsen, J.: The home care crew scheduling problem: preference-based visit clustering and temporal dependencies. DTU Management (2010)
8. Misir, M., Smet, P., Verbeeck, K., Berghe, G.V.: Security personnel routing and rostering: a hyper-heuristic approach. In: Proceedings of the 3rd International Conference on Applied Operational Research, ICAOR11 (2011)
9. Günther, M., Nissen, V.: Application of particle swarm optimization to the British telecom workforce scheduling problem. In: Proceedings of the 9th International Conference on the Practice and Theory of Automated Timetabling (PATAT 2012), Son, Norway (2012)
10. Lehman, J., Stanley, K.O.: Evolving a diversity of virtual creatures through novelty search and local competition. In: Proceedings of the 13th Annual Conference on Genetic and Evolutionary Computation, pp. 211–218. ACM (2011)
11. Mouret, J., Clune, J.: Illuminating search spaces by mapping elites. CoRR (2015)
12. Vassiliades, V., Chatzilygeroudis, K., Mouret, J.B.: Using centroidal voronoi tessellations to scale up the multi-dimensional archive of phenotypic elites algorithm, pp. 1–1 (2017)
13. Gaier, A., Asteroth, A., Mouret, J.B.: Data-efficient design exploration through surrogate-assisted illumination. Evol. Comput. **26**(3), 381–410 (2018)

14. Hagg, A., Asteroth, A., Bäck, T.: Prototype discovery using quality-diversity. In: Auger, A., Fonseca, C.M., Lourenço, N., Machado, P., Paquete, L., Whitley, D. (eds.) PPSN 2018. LNCS, vol. 11101, pp. 500–511. Springer, Cham (2018). https://doi.org/10.1007/978-3-319-99253-2_40

15. Urquhart, N.B., Hart, E., Judson, A.: Multi-modal employee routing with time windows in an urban environment. In: Proceedings of the Companion Publication of the 2015 Annual Conference on Genetic and Evolutionary Computation, pp. 1503–1504. ACM (2015)

16. TFL: Travel in London: key trends and developments. Techical report, Transport for London (2009)

17. Urquhart, N.B., Hart, E., Judson, A.: Multi-modal employee routing with time windows in an urban environment. In: Proceedings of the Companion Publication of the 2015 Annual Conference on Genetic and Evolutionary Computation, pp. 1503–1504. GECCO Companion 2015. ACM, New York (2015). https://doi.org/10.1145/2739482.2764649

A Hybrid Multiobjective Differential Evolution Approach to Stator Winding Optimization

André M. Silva[1]([envelope]) [iD], Fernando J. T. E. Ferreira[1,3] [iD],
and Carlos Henggeler Antunes[2,3] [iD]

[1] ISR Coimbra, DEEC, University of Coimbra,
Polo 2, 3030-290 Coimbra, Portugal
andre.msilva@isr.uc.pt
[2] INESC Coimbra, DEEC, University of Coimbra,
Polo 2, 3030-290 Coimbra, Portugal
[3] Department of Electrical and Computer Engineering, University of Coimbra,
Polo 2, 3030-290 Coimbra, Portugal

Abstract. This paper describes a multiobjective differential evolution approach to the optimization of the design of alternating current distributed stator windings of electric motors. The objective functions are minimizing both the machine airgap magnetomotive force distortion and the winding wire length. Constraints are related to the physical feasibility of solutions. Four distinct winding types are considered. Three mutation variations of the multiobjective differential evolution algorithm are developed and assessed using different performance metrics. These algorithmic approaches are able to generate well-distributed, uniformly spread solutions on the nondominated front. The characterization of the nondominated fronts conveys helpful information for aiding design engineers to choose the most suitable compromise solution for a specific machine, embodying a balanced trade-off between machine efficiency and manufacturing cost.

Keywords: Multiobjective differential evolution ·
Optimization of distributed windings · Mutation operators ·
Machine efficiency

1 Introduction

Alternating Current (AC) distributed windings are widely used in rotating motors and generators. The main advantage of this type of windings, when compared to concentrated windings, is the lower airgap magnetomotive force (MMF) harmonic

A. M. Silva acknowledges the support by the Portuguese Science and Technology Foundation (FCT).
C. H. Antunes acknowledges the support of projects UID/Multi/308/2019, ESGRIDS (POCI-01-0145-FEDER-016434) and MAnAGER (POCI-01-0145-FEDER-028040).

© Springer Nature Switzerland AG 2019
P. Kaufmann and P. A. Castillo (Eds.): EvoApplications 2019, LNCS 11454, pp. 64–71, 2019.
https://doi.org/10.1007/978-3-030-16692-2_5

content. The main drawbacks are the higher amount of copper necessary, the larger impedance and the higher manufacturing complexity. To overcome these first two issues, and further improve the airgap MMF distribution, it is common practice to shorten the coil pitch. Another possibility is to have coils with different number of turns, although finding the optimal coil turns distribution is not a simple task. Few solutions have been proposed in literature to solve the stator winding optimization based on general, systematic, and fully automated approaches [1–3].

The problem is intrinsically of multiobjective nature, since different technical aspects should be considered. The characterization of the nondominated fronts conveys helpful information for aiding design engineers to choose the most suitable compromise solution for a specific machine, embodying a balanced trade-off between those evaluation aspects. In this paper, the stator winding optimization is defined as a two-objective optimization problem: minimizing both the machine airgap magnetomotive force distortion and the winding wire length. To solve this optimization problem a new hybrid multiobjective algorithm combining differential evolution with NSGA-II is developed. The performance of three mutation variations is assessed using different metrics. Computational results with real-world problems show that the algorithmic approach is able to generate well-distributed, uniformly spread solutions on the nondominated front.

This paper is organized as follows. Section 2 describes the mathematical model of the optimization problem. Section 3 presents the algorithmic approach. Section 4 presents a case study of four different windings and the corresponding computational results.

2   Mathematical Model of the Optimization Problem

The decision variables $x = \{x_1, ..., x_n\}$, $x_j \in [0, T_{\max}]$, $j = 1, ..., n$, are the number of turns of the jth coil, whose variation, up to a maximum number of turns ($T_{\max}$), does not affect the winding symmetry.

The two objective functions (minimization of airgap MMF total harmonic distortion (THD) and winding wire length) herein considered for the machine winding optimization problem are conflicting and incommensurate. The minimization of the airgap MMF THD (Eq. 1) may improve the machine efficiency, since lower MMF spatial harmonics are contributing to the iron and rotor losses [1].

$$\text{THD}_{\text{MMF}}(x) = \frac{\sqrt{\sum_{\substack{n=1 \\ n \neq fund}}^{\infty} \left[\frac{C_n(x)}{n}\right]^2}}{C_{fund}(x)/fund}, \tag{1}$$

with

$$C_n(x) = \sqrt{\left[\sum_{k=1}^{Z} t_k(x)\cos(n\theta_k)\right]^2 + \left[\sum_{k=1}^{Z} t_k(x)\sin(n\theta_k)\right]^2}, \tag{2}$$

where $fund$ is the airgap MMF fundamental term index, Z is the number of stator slots, $\theta_k = 2\pi k/Z$ is the angular position of the k-slot (from an arbitrary reference), and the function $t_k(x)$ returns the number of conductors, connected in series, of a phase [4].

The winding wire length per phase (Eq. 3) is directly proportional to the winding material cost, therefore it can be used as a manufacturing economic measure of the stator winding. In addition, reducing the wire length reduces the stator winding resistance, thus reducing the stator winding Joule losses, and improving the machine efficiency.

$$l_{\text{wire}}(x) = 2 \sum_{j=1}^{C_{\text{wp}}} t_{\text{w},j}(x) \left(K_h \cdot \tau_j + l_{core}\right), \tag{3}$$

where C_{wp} is the number of coils per phase, the function $t_{\text{wp},j}(x)$ returns the number of turns of the jth phase coil, K_h is the coefficient of the winding heads shape, τ_j is the jth coil side-to-side arc length, and l_{core} is the stator core length.

Two sets of technical constraints should be considered. The first one restrains the magnetizing flux per pole $\phi_p(x)$ produced by the stator winding to a value within a lower and an upper bound (Eq. 4), so the machine is able to develop the desired torque and avoid undesirable magnetic saturation levels, respectively.

$$\phi_p^{\text{lower}} \leq \phi_p(x) = K_{\text{flux}}/C_{fund}(x) \leq \phi_p^{\text{upper}}, \tag{4}$$

where K_{flux} is the machine magnetizing flux constant.

The second set of constraints restrains the slot fill factor $K_{\text{sff}}(x)$ to guarantee that the winding can be assembled to the stator (Eq. 5). It calculates the maximum slot occupancy as the fraction of the conduction area and the slot useful cross section and compares it to the maximum admissible value ($K_{\text{sff}}^{\text{max}}$).

$$K_{\text{sff}}(x) = \max\left(t_{\text{w},k}(x) \cdot s_{\text{cond}}\right)/s_{\text{slot}} \leq K_{\text{sff}}^{\text{max}}, k \in \{1, 2, ..., Z\}, \tag{5}$$

where the function $t_{\text{w},k}(x)$ returns the total number of conductors inside the kth slot, s_{cond} is the conductors cross section, and s_{slot} is the slots useful cross section.

Then, the optimization problem is formulated as follows:

$$\begin{cases} \min f_1(x) = \text{THD}_{\text{MMF}}(x) \\ \min f_2(x) = l_{\text{wire}}(x) \\ \\ \text{s.t.}: \phi_p^{\text{lower}} \leq \phi_p(x) \leq \phi_p^{\text{upper}} \\ \quad K_{\text{sff}}(x) \leq K_{\text{sff}}^{\text{max}} \end{cases} \tag{6}$$

3   A Hybrid Multiobjective Differential Evolution Approach

The computational simplicity and effectiveness of Differential Evolution (DE) makes it an adequate algorithmic approach to deal with difficult nonlinear continuous problems [5,6]. When dealing with multiobjective optimization problems, the choice between two solutions becomes more difficult. The dominance criterion may be used, although it fails to assess two nondominated solutions. In [7], a hybrid algorithm combining DE with NSGA-II [8] displayed improved convergence while maintaining diversity of the solution population, when compared to other three multiobjective evolutionary algorithms. The approach herein developed is also based on the hybridization of DE with NSGA-II, making the most of the physical characteristics of the problem.

There are several variations for the creation of the mutant population. The mutation operator creates a mutant population by applying a mutation variation to the parent population, with scaling factor F and crossover probability C_r. In this paper, three mutation operators are used: the standard DE/rand/1/bin, the DE/mean/1/bin, and the DE/rand/2/bin. These variations may have distinct base vectors or multiple differential-vectors, but the process of creation of new candidates is similar.

The variation DE/mean/1/bin (Eq. 7) employs as base vector the arithmetic mean of the population individuals.

$$u_i = \frac{1}{N} \sum_{k=1}^{N} y_k + F \left(y_{r_2} - y_{r_3} \right), \tag{7}$$

where u_i and y_i are the ith mutant and parent current individuals, respectively; and r_* is a randomly selected index of the parent population.

The variation DE/rand/2/bin (Eq. 8) employs two differential-vectors, which is expected to improve the diversity of the solutions generated.

$$u_i = y_{r_1} + F_1 \left(y_{r_2} - y_{r_4} \right) + F_2 \left(y_{r_3} - y_{r_5} \right) \tag{8}$$

An offspring population of size $2N$ results from the combination of the parent and mutant populations. With the aim to induce a uniformly distributed, well-spread, nondominated front in the offspring population, the selection operator developed is based on the elitist NSGA-II algorithm. Hence, two metrics are used to rank the population individuals: *Border Index* (I_f) assigns the index 0 to all nondominated solutions. These solutions are then removed and the new nondominated solutions are identified within the remaining population, thus a new front is identified with the index 1. The next front is assigned the index 2, and so forth until there is no remaining dominated solutions; *K-neighbourhood* (δ_k) for each individual y_i, the Euclidean distance to its nearest neighbour, with equal I_f, is calculated, and the farthest individuals are indexed first. In the case two individuals are equally distanced, the two are subjected to a draw to select which one is indexed first.

Finally, individuals of the offspring population are sorted by increasing I_f indexes first, and then by increasing δ_k, thus preferring diversified nondominated solutions. The next generation consists of the individuals that occupy the first N positions in the rank.

The first issue to handle regarding constraint satisfaction concerns the DE mutation process. Some differential-vectors generated may fall out of bounds and, consequently, they have to be regenerated into the feasible region. In this case, this is done by using the bounce-back reinitialization [9]. This method selects randomly a parameter value that lies between the base parameter (the parent y_i) value and the bound being violated. Hence, the progress toward best values is taken into account.

A self-adaptive technique developed in [10] is used to handle the magnetizing flux per pole and slot fill factor constraints (Eqs. 4 and 5). The method applies modified objective functions values, through a two penalty function, based on the balance between the severity of the constraints' violations and the number of feasible individuals in a population for dominance assessment in the population.

4   Case Studies

In this section, the metrics used for comparing the performance of the results of the experiments are described. Then, the case studies consisting of four stator windings to be optimized considering the three mutation variations are presented.

In order to evaluate and compare the performances of each mutation variation (DE/rand/1/bin, DE/mean/1/bin, and DE/rand/2/bin), the following metrics are used: *Efficient Set Space* (ESS) [11] and *Hypervolume* [12]. The ESS indicator is used to evaluate the spread of the solutions in the nondominated front. Smaller values mean that the solutions are distanced uniformly to their closest neighbors, and therefore are better distributed. The hypervolume measures the space of the dominated solutions (an area for two-dimensional problems), where higher values mean better performance. This indicator is the most used performance metric [13]. *Statistical Interpolation* (SI) [14] is used to compare statistically different algorithms instead of comparing only scalar values. It is calculated by the sum of linear segments of a dominant attainment surface of a given algorithm, then divided by the length of the total dominant attainment surface. Non-parametric statistical procedures can be applied to assess which algorithm outperforms the other. *Purity* [15, 16] calculates the fraction between the number of nondominated solutions and the total number of solutions that result from the unified set of solutions found at the end of each run for a given algorithm. A value near to 1 means better performance, since more solutions contributed to the unified nondominated front.

Table 1 shows the main characteristics of the four case studies before optimization and the algorithm parameters. The algorithm scaling factors were 0.8 for the variations DE/rand/1/bin and DE/mean/1/bin, and 0.4 for both scaling factors of the variation DE/rand/2/bin. The crossover probability was set to 0.9 for all variations. These parameters have been tuned through extensive experimentation for all mutation variations. The algorithm ran 20 times for each mutation variation; the results of each variation were then combined, and the dominated solutions were removed (Fig. 1).

Compared with the initial characteristics of the four winding designs (Table 1), the final nondominated frontiers show quite good solutions for all cases. In fact, even the extreme solutions of the nondominated fronts, which are the individual optimal objective function values, are quite better than the initial values for all four cases. From an engineering design point of view, these results are practically indistinguishable for the different mutation variations.

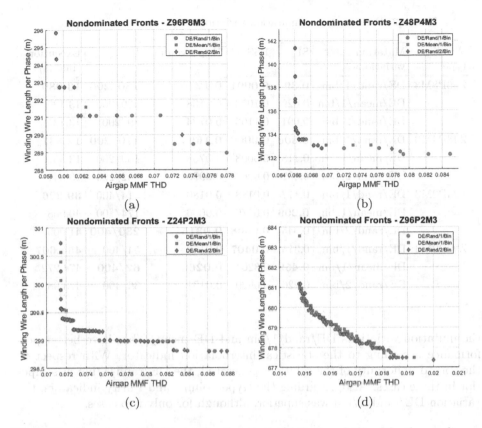

Fig. 1. Final nondominated fronts of each mutation variation obtained for the windings: (a) Z96P8M3, (b) Z48P4M3, (c) Z24P2M3, and (d) Z96P2M3.

Table 1. Main characteristics of the winding designs before optimization.

Winding type	Decision variables	Population size	Generations	Combinations	THD$_{MMF}(x)$ (%)	$l_{wire}(x)$ (m)
Z96P8M3	4	10	100	3418801	9.057	330
Z48P4M3	4	10	200	6765201	8.584	153
Z24P2M3	6	20	400	6.32752E12	9.057	311
Z96P2M3	16	20	1800	2.03805E35	4.921	725

A comparative analysis of the performances of the different mutation variations has been done (Table 2). Regarding the execution time, the fastest mutation variation was the DE/rand/1/bin, its advantage becoming more relevant for higher dimension problems, the other two variations having practically identical computational times. Regarding the quality of the solutions generated,

Table 2. Performance of mutation variations for the 4 winding types.

Winding type	Mutation variations	SI	ESS	Hypervolume	Purity	Execution time (s)
Z96P8M3	DE/rand/1/bin	**0.394**	**0.0001**	**0.1240**	**130/200**	**13.089**
	DE/mean/1/bin	0.029	0.1073	0.0108	109/200	13.129
	DE/rand/2/bin	0.091	0.1193	0.0724	89/200	13.131
Z48P4M3	DE/rand/1/bin	**0.306**	**0.0000**	**0.1694**	**170/200**	**12.837**
	DE/mean/1/bin	0.118	0.0003	0.0758	139/200	13.352
	DE/rand/2/bin	0.002	0.0068	0.0998	162/200	13.341
Z24P2M3	DE/rand/1/bin	0.177	0.0188	0.0180	314/400	**39.226**
	DE/mean/1/bin	**0.206**	0.0179	0.0006	174/400	40.968
	DE/rand/2/bin	0.045	**0.0008**	**0.0315**	**230/400**	41.006
Z96P2M3	DE/rand/1/bin	0.315	**0.0407**	0.0150	54/400	**455.607**
	DE/mean/1/bin	**0.459**	0.3204	**0.0266**	**62/400**	474.272
	DE/rand/2/bin	0.226	0.0929	0.0191	36/400	473.983

the mutation variations DE/rand/1/bin and DE/mean/1/bin have better performance according to the statistical interpolation indicator. With respect to the spread uniformity of the solutions generated, the DE/rand/1/bin was superior in three cases. Also, regarding the Hypervolume and Purity indicators, the variation DE/rand/1/bin was superior, although for only two cases.

5   Conclusion

In this paper, the optimization of AC distributed stator winding has been investigated through the development of a hybrid multiobjective algorithm combining differential evolution with NSGA-II. In addition, the performances of three mutation variations were evaluated. The results showed that, for four representative case studies, the variations DE/rand/1/bin and DE/mean/1/bin have better performances, the former outperformed the later for the smaller dimension problems and the contrary happened for larger problems. The DE/rand/2/bin variation had the poorest performance. From an engineering design point of view, the quality of the solutions found by the three variations is very good, being practically indistinguishable from a practical implementation point of view. Thus, for high dimension problems the DE/rand/1/bin variation, which was the fastest, seems adequate, since computation speed may be relevant. Future work involves further algorithmic refinement to make the most of the physical characteristics of the problem and developing a decision support system based on the algorithms to support design engineers.

References

1. Tessarolo, A.: A quadratic-programming approach to the design optimization of fractional-slot concentrated windings for surface permanent-magnet machines. IEEE Trans. Energy Convers. **33**(1), 442–452 (2018)
2. Smith, A.C., Delgado, D.: Automated AC winding design. In: 5th IET International Conference on Power Electronics, Machines and Drives (PEMD 2010), pp. 1–6, April 2010
3. Bekka, N., Zaïm, M.E.H., Bernard, N., Trichet, D.: A novel methodology for optimal design of fractional slot with concentrated windings. IEEE Trans. Energy Convers. **31**(3), 1153–1160 (2016)
4. Silva, A.M., Ferreira, F.J.T.E., Falcáo, G.F., Rodrigues, M.: Novel method to minimize the air-gap MMF spatial harmonic content in three-phase windings. In: 2018 XIII International Conference on Electrical Machines (ICEM), pp. 2504–2510, September 2018
5. Vesterstrom, J., Thomsen, R.: A comparative study of differential evolution, particle swarm optimization, and evolutionary algorithms on numerical benchmark problems. In: Proceedings of the 2004 Congress on Evolutionary Computation (IEEE Cat. No. 04TH8753), vol. 2, pp. 1980–1987, June 2004
6. Salvatore, N., Caponio, A., Neri, F., Stasi, S., Cascella, G.L.: Optimization of delayed-state kalman-filter-based algorithm via differential evolution for sensorless control of induction motors. IEEE Trans. Industr. Electron. **57**(1), 385–394 (2010)
7. Pan, X., Zhu, J., Chen, H., Chen, X., Hu, K.: A differential evolution-based hybrid NSGA-II for multi-objective optimization. In: 2015 IEEE 7th International Conference on Cybernetics and Intelligent Systems (CIS) and IEEE Conference on Robotics, Automation and Mechatronics (RAM), pp. 81–86 (2015)
8. Deb, K., Pratap, A., Agarwal, S., Meyarivan, T.: A fast and elitist multiobjective genetic algorithm: NSGA-II. IEEE Trans. Evol. Comput. **6**(2), 182–197 (2002)
9. Price, K.V., Storn, R.M., Lampinen, J.A.: Differential Evolution: A Practical Approach. NCS. Springer, Heidelberg (2005). https://doi.org/10.1007/3-540-31306-0
10. Woldesenbet, Y.G., Yen, G.G., Tessema, B.G.: Constraint handling in multiobjective evolutionary optimization. IEEE Trans. Evol. Comput. **13**(3), 514–525 (2009)
11. Sarker, R., Coello Coello, C.A.: Assessment Methodologies for Multiobjective Evolutionary Algorithms, vol. 48, pp. 177–195. Springer, Boston (2002). https://doi.org/10.1007/0-306-48041-7_7
12. Zitzler, E., Thiele, L.: Multiobjective evolutionary algorithms: a comparative case study and the strength pareto approach. IEEE Trans. Evol. Comput. **3**(4), 257–271 (1999)
13. Riquelme, N., Lücken, C.V., Baran, B.: Performance metrics in multi-objective optimization. In: 2015 Latin American Computing Conference (CLEI), pp. 1–11, October 2015
14. Fonseca, C.M., Fleming, P.J.: On the performance assessment and comparison of stochastic multiobjective optimizers. In: Voigt, H.M., Ebeling, W., Rechenberg, I., Schwefel, H.P. (eds.) Parallel Problem Solving from Nature – PPSN IV, pp. 584–593. Springer, Heidelberg (1996). https://doi.org/10.1007/3-540-61723-X_1022
15. Ishibuchi, H., Murata, T.: A multi-objective genetic local search algorithm and its application to flowshop scheduling. IEEE Trans. Syst. Man Cybern. Part C (Appl. Rev.) **28**(3), 392–403 (1998)
16. Bandyopadhyay, S., Saha, S., Maulik, U., Deb, K.: A simulated annealing-based multiobjective optimization algorithm: amosa. IEEE Trans. Evol. Comput. **12**(3), 269–283 (2008)

GA-Novo: *De Novo* Peptide Sequencing via Tandem Mass Spectrometry Using Genetic Algorithm

Samaneh Azari[1]($\boxtimes$), Bing Xue[1], Mengjie Zhang[1], and Lifeng Peng[2]

[1] School of Engineering and Computer Science, Victoria University of Wellington,
P.O. Box 600, Wellington 6140, New Zealand
{samaneh.azari,bing.xue,mengjie.zhang}@ecs.vuw.ac.nz
[2] Centre for Biodiscovery and School of Biological Sciences,
Victoria University of Wellington, P.O. Box 600, Wellington 6140, New Zealand
lifeng.peng@vuw.ac.nz

Abstract. Proteomics is the large-scale analysis of the proteins. The common method for identifying proteins and characterising their amino acid sequences is to digest the proteins into peptides, analyse the peptides using mass spectrometry and assign the resulting tandem mass spectra (MS/MS) to peptides using database search tools. However, database search algorithms are highly dependent on a reference protein database and they cannot identify peptides and proteins not included in the database. Therefore, *de novo* sequencing algorithms are developed to overcome the problem by directly reconstructing the peptide sequence of an MS/MS spectrum without using any protein database. Current *de novo* sequencing algorithms often fail to construct the completely matched sequences, and produce partial matches. In this study, we propose a genetic algorithm based method, GA-Novo, to solve the complex optimisation task of *de novo* peptide sequencing, aiming at constructing full length sequences. Given an MS/MS spectrum, GA-Novo optimises the amino acid sequences to best fit the input spectrum. On the testing dataset, GA-Novo outperforms PEAKS, the most commonly used software for this task, by constructing 8% higher number of fully matched peptide sequences, and 4% higher recall at partially matched sequences.

Keywords: Genetic algorithm · Tandem mass spectrometry ·
De novo sequencing · Proteomics

1 Introduction

In mass spectrometry, *de novo* peptide sequencing is the process of determining the amino acid sequence of peptides directly from MS/MS spectra. There are 20 amino acids represented by the letters A, C, D, E, F, G, H, I, K, L, M, N, P, Q, R, S, T, V, W, and Y. Peptide sequences are generally considered to be short chains of amino acids (from 2 to 50 amino acids). A peptide P with length

© Springer Nature Switzerland AG 2019
P. Kaufmann and P. A. Castillo (Eds.): EvoApplications 2019, LNCS 11454, pp. 72–89, 2019.
https://doi.org/10.1007/978-3-030-16692-2_6

Table 1. An example of a mass fragmentation ladder.

Mass	Ion	b-ions	y-ions	Ion	Mass
114	b1	L	GVTLYK	y6	680
171	b2	LG	VTLYK	y5	623
270	b3	LGV	TLYK	y4	524
371	b4	LGVT	LYK	y3	423
484	b5	LGVTL	YK	y2	310
647	b6	LGVTLY	K	y1	147

l contains a sequence of amino acids, $P = a_1, a_2, a_3...a_l$, where each amino acid has a mass. The mass of the peptide, which is called parent mass, equals to the total mass of its amino acids plus mass of water (see Eq. 1).

$$PM(P) = \sum_{i-1}^{l} mass(a_i) + mass(H_2O) \tag{1}$$

An MS/MS spectrum S consists of a list of peaks each having a mass-to-charge ratio (m/z) value and an intensity value (peak height). The m/z values are results of ionizing the biological samples and their intensities indicate the abundance of ions. Assume the spectrum is represented by two vectors of m/z values and intensities $S = (M, I)$, where $M = (m_1, m_2, m_3, ..., m_n)$ and $I = (I_1, I_2, I_3, ..., I_n)$. The experimental parent mass or precursor mass is calculated based on Eq. 2.

$$Prec._{mass} = \text{pepmass} \times \text{charge} - \text{charge} \times mass(\text{Proton}) \tag{2}$$

where pepmass is mass of the fragmented ion, charge is the precursor charge state and mass of Proton equals to 1.00727647 atomic mass units (amu).

Collision-induced dissociation (CID) is known to be highly suitable technique for the identification of peptide sequences [1]. In this technique, fragmentation happens at the peptide bonds, producing b-/y-ions. The fragment containing only the first amino acid from left side (N-terminus) of the peptide is termed b_1, while the one that contains the first two amino acids is called the b_2 ion, and so forth. Y-ions extend from the right side or C-terminus of the peptide. In the CID fragmentation technique the amino acid sequence of an MS/MS spectrum can be determined by the mass differences between b- and y-ions.

The complete CID peptide fragmentation gives a contiguous series of ion types called "ladder". The sum of their masses equal to the mass of the pre-fragmented peptide. Having the complete ion ladder, the *de novo* sequencing algorithm selects pairs of peaks and labels them if their mass differences are within the tolerance ranges of the amino acids masses.

However, it is often that peptide fragmentations are neither sequential nor complete. Moreover, peptides may not fragment at some positions and resulting in missing data. Also, a real MS/MS spectra with hundreds of peaks normally contain background noise. Therefore, while exactly 1 of 20^l amino acid sequences can be considered as the potential correct prediction (l is the peptide length), *de novo* sequencing with internal fragment ions is recognized as a combinatorial problem and known to be NP-hard [2].

There have been attempts to solve the *de novo* sequencing problem using different approaches. PAA3 [3], as the first *de novo* sequencing algorithm, generated exhaustively all possible peptide sequences and compared each candidate with the spectrum. However, the method is only feasible for very short peptides, because the time complexity grows exponentially in terms of peptide length.

Dynamic programming has been widely used for *de novo* sequencing [4]. The major approach is generating a graph from an MS/MS spectrum where peaks are the vertices and edges are defined as the corresponding amino acids to the mass differences between two vertices. A probability based fitness function is used to score the paths and dynamic programming is used to traverse through the spectrum graph [5,6]. However, this approach results in having a huge graph due to the noise peaks caused by internal cleavages or post-translational modifications (PTM)s. Another problem is the lack of full path due to the missing ion types caused by incomplete fragmentation and low instrument accuracy. Therefore, *de novo* sequencing of full-length peptides remains a challenge.

De novo sequencing can be formulated as an optimisation problem where the objective is to discover the most likely amino acid sequence that can be generated by the input spectrum [7]. *De novo* sequencing has been performed via stochastic optimisation using a genetic algorithm (GA) [8,9], where a GA tries to optimise the amino acid sequence in respect to a scoring function. However, the existing *de novo* sequencing methods using GA often fail to discriminate the mismatches because the fitness functions could not capture various aspects of peak matching [10]. Moreover, the basic genetic operators used in these works are not capable enough to guide GAs during the evolutionary process to construct the fully matched sequence.

1.1   Goal

The goal of this work is to develop an effective *de novo* sequencing algorithm using GAs to construct the full length amino acid sequences of MS/MS spectra. Unlike exhaustive approaches, GA does not need to generate all possible amino acid sequences for a give spectrum. A set of initial amino acid sequences using an effective initialisation method is generated and during the evolutionary process these sequences are manipulated by appropriate domain dependant genetic operators until finding the one that best fits to the spectrum in respect to the fitness function. Unlike spectrum graph based algorithms, it is expected that the performance of GA does not deterred by discontinuities in the search space (lack of full path in the graph) due to missing ions. In addition, among other evolutionary computation (EC) techniques, GA is the most suitable method to solve the problem of *de novo* sequencing. GA represents an individual as variable-length bit string called it a chromosome which is appropriate for keeping a peptide sequence containing a series of amino acids. Therefore, the following objectives are investigated in this work:

1. Developing a new fitness function that captures important spectral features and enables GA to discriminate the mismatches.
2. Developing an effective set of mutation and crossover operators that help GA to construct the full length amino acid sequence.
3. Designing an effective GA algorithm that can perform the *de novo* sequencing task, and achieving a high number of fully matched sequences out of the input spectra.

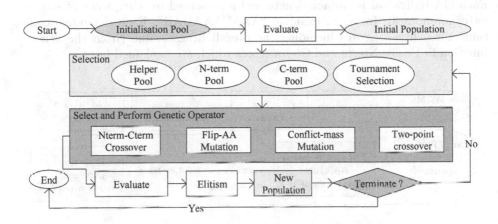

Fig. 1. The workflow of GA-Novo.

2   The Proposed Method

Figure 1 presents the workflow of GA-Novo. Given the raw MS/MS experimental spectrum S, first a tag-based initialisation method is applied in order to create a set of candidate initial individuals for the GA algorithm. The candidate individuals are kept in a big initialisation pool. The individuals are evaluated and based on three criteria including the fitness value, Nterm score and Cterm score are selected to generate the initial population for GA. Then the evolutionary process starts with applying selection in order to create four pools for different purposes and the size of each pool is a third of the total population size. Starting from left, the helper pool contains top best individuals in terms of fitness values. The individuals in N-term and C-term pools are the top best individuals in terms of Nterm and Cterm scores, respectively. The individuals in last pool are selected using tournament selection based on their fitness values. There are four genetic operators, two crossovers and two mutations. The individuals for Nterm-Cterm crossover are selected from the first three pools. Other genetic operators get their individuals directly from the tournament pool. Nterm-Cterm crossover is designed to construct individuals with correct matches from both sides and possibly from middle, whereas two-point crossover aims to repair the individuals from middle. The mutation operators randomly flip flop the each bit/amino acid in the sequence. In each generation, elitism keeps the best three individuals in terms of overall fitness value, Nterm and Cterm score. The evolutionary process repeats until the termination criterion which is the number of generations is met. The method returns the best individual in terms of overall fitness value. More details about the components in this flowchart are as follow.

2.1   Representation

Each GA individual is variable-length and represented by a sequence of single-letter amino acids from, for example "AAALAAADAR". Each individual contains three fitness scores including the overall fitness value (from the fitness function in Eq. 3), Nterm and Cterm scores which are explained later.

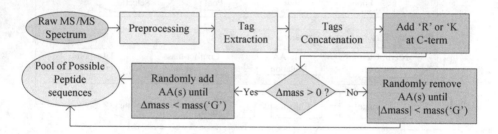

Fig. 2. The workflow of the tag-based initialisation method.

2.2   Tag-Based Initialisation Method

A domain dependant initialisation method is used to generate initial individuals for GA. The workflow of this method is illustrated in Fig. 2. The overall goal of this method is to construct full length peptide sequences which are preferably partially matched with the spectrum and having as small as possible mass difference ($\Delta mass$).

The input of the workflow is an MS/MS spectrum (experimental spectrum) and the output is a set of peptide sequences corresponding to the spectrum. The workflow starts with preprocessing the input spectrum. Then all 3-letter tags are extracted from the preprocessed spectrum in tag extraction step. In tags concatenation step, each time 2, 3 or 4 tags are randomly selected and concatenated to construct a sequence with length 6, 9 or 12. These numbers are in the range of the peptides' length that fall in the precursor mass range of spectra used in this study. Since all tryptic peptides have either amino acids 'R' or 'K' at the end, these two amino acids are randomly added to the end of the sequences from tags concatenation step. Since mass difference is a constraint, it is important to construct the sequences with $|\Delta masses| \leq 0$. So the rest of the workflow checks whether or not the mass difference between each constructed sequence and the spectrum is less than the mass of amino acid 'G' which has the smallest mass. Therefore, based on the $\Delta mass$ value, appropriate amino acids are randomly added to/removed from the sequence and the resulting peptide sequence is sent to the pool of possible peptide sequences corresponding to the input experimental spectrum. The preprocessing step and the tag extraction are explained in the following.

Spectrum Preprocessing. The MS/MS noise reduction step has been done based on the noise reduction method proposed in SEQUEST [11], which is a dominant database search tool in proteomics. Given a spectrum, at first the whole m/z range is divided into 10 windows (regions). In each window, if the number of existing peaks exceeds 9, there should be some possible noise, which needs to be eliminated from that window. The peak intensity with the highest frequency is considered to be the noise threshold. Therefore, all peaks whose intensities are smaller than the noise threshold will be removed from that window. After removing these noisy peaks, the next step is normalising peak intensities. In each window, each peak's intensity is replaced with its square root and then all intensities are normalised by dividing into the highest intensity. Then each peak in the spectrum is checked for the existence of its complementary peak which will be added if required. The sum of the two complementary ions' masses should be equal to the precursor mass of the spectrum. Now the next step is extracting all 3-letter tags from the spectrum.

Tag Extraction. In tag extraction, all 3-letter tags from the N-terminus to the C-terminus are extracted from the spectrum [12]. As previously mentioned, here a spectrum is represented by two vectors of mz values and intensities $S = (M, I)$. Considering the M vector $M = (m_1, m_2, m_3, ..., m_n)$, two peaks construct a peak pair if their m/z values satisfy $|m_i - m_j - mass(a)| \leq \tau$ where $1 \leq i \leq j \leq n$, $mass(a)$ is the mass of one of the 20 popular amino acids and τ is the MS/MS mass tolerance. A tag with length one is represented by $t(i, j)$ and a label of a corresponding to its amino acid. Two tags $t(i, j)$ and $t(i\prime, j\prime)$ are considered sequential if $j = i\prime$. So all 3-letter tags from the spectrum will be extracted and are used in the initialisation method.

2.3 Fitness

Fitness Function. The fitness function evaluates the quality of matching between an input experimental spectrum and a peptide sequence constructed by GA-Novo. For being able to match the peptide sequence against the experimental spectrum, a theoretical spectrum T based on the known CID fragmentation rules of doubly charged peptides [13] is constructed from the peptide sequence.

The theoretical spectrum only contains m/z values with no intensities. Both b-/y-ion ladders in Table 1 along with internal fragments are constructed in the theoretical spectrum. Then peaks in the theoretical spectrum is matched against the peaks in the experimental spectrum within the MS/MS mass tolerance of τ.

Equation 3 presents the new fitness function for measuring the goodness of the peptide spectrum match (PSM).

$$fitness(PSM) = \frac{\sum I_{matched}}{\sum\limits_{i=1}^{n} I_i} - \frac{|\Delta mass|}{Prec._{mass}} + \frac{Nterm + Cterm - \sum N_{unmatched}}{length(P)} \qquad (3)$$

where $I_{matched}$ is the sum of intensities of those peaks in the experimental spectrum S which are matched with theoretical spectrum T corresponding to the peptide P. Then total intensities of matched peaks is normalised by dividing into the total intensities of the whole spectrum S. $\Delta mass$ is the mass difference between parent mass of peptide P and the spectrum precursor mass ($Prec._{mass}$). Since the total mass of the predicted peptide by GA is expected to be equal to the precursor mass of the spectrum, the absolute value of $\Delta mass$ is considered as a penalty to avoid getting undesirable short or long peptides. $Nterm$ is the number of sequential b-ion matches from N-terminus (left to right) and $Cterm$ is the number of sequential y-ion matches from C-terminus (right to left) of the theoretical spectrum T. These terms check the quality of match from both sides of the theoretical spectrum and reward the match. As normally those b-/y-ions in the middle part of the spectrum tend to have higher intensities, whereas those on the other two sides particularly N-terminus have lower intensities, without having these two terms in the fitness function there is a chance of ending up to a peptide sequence which is partially matched with the spectrum only from middle. Therefore, with having these two terms, a peptide which has a few b-/y-ions matched from two sides but not from middle, still has the chance to survive. In this case, the peptide gets a reasonable fitness value and has a chance to remain in the population, going through the evolutionary process for further improvement. $N_{unmatched}$ indicates the number of b-/y-ions in the theoretical spectrum T which are not a match against the spectrum S. The three terms are divided into the length of peptide.

Apart from the fitness value produced by the fitness function above, the two terms $Nterm$ and $Cterm$ (without being divided into the peptide length), are also kept as additional fitness scores for each individual. These values later are used to apply a new crossover operator and are explained in the following section.

Nterm and Cterm Scores. The idea of calculating these two terms comes from the ion ladder of sequences and the CID fragmentation rules. The mass of any theoretical b-ion can be calculated based on Eq. 4, where $1 \leq j \leq l - 1$, l is the length of the peptide P, b_j is the j-*th* b-ion of P, and a_i is the i-*th* amino acid in P. Similarly theoretical y-ions can be calculated based on Eq. 5. Also as mentioned in Table 1, the complementary theoretical b- and y- ions in each row of the table have the mathematical relation presented in Eq. 6.

$$b_j = \sum_{i=1}^{j} mass(a_i) + 1 \tag{4}$$

$$y_j = \sum_{i=l-j}^{l} mass(a_i) + 19 \tag{5}$$

$$b_j + y_{l-j} = PM(P) + 2 \tag{6}$$

To calculate the b-ions in the theoretical spectrum Eq. 4 is used. Having the total mass of the peptide (parent mass in Eq. 1), the y-ions can be calculated either by Eq. 6 or Eq. 5. Therefore, for calculating the Nterm score, fist all b-ions are calculated. Then, in Eq. 6 instead of $PM(P)$ which the mass of the peptide, $Prec._{mass}$ which is the precursor mass (from Eq. 2) is replaced, and y-ions are calculated. Let's call these y-ions as experimental y-ions (becasue we used mass of the spectrum). The experimental y-ions are compared with theoretical y-ions from Eq. 5. Starting from y_1, if any two sequential experimental y-ions are equal to the corresponding theoretical y-ions, the Nterm score increases by one.

The Nterm score is able to check whether a matched b-ion is a random match or not. Similarly, Cterm is calculated by using Eq. 5 and applying the similar process. The values of Nterm and Cterm scores do not necessarily indicate the exact amino acid matches in the sequence. For example a sequence with Nterm = 2, does not indicate that the first 2 amino acids from N-terminus are exact matches compared to the ground truth peptide. The reason is that we are not aware of the ground truth during the matching process. However, these two scores are able to check the quality of match from each side of the spectrum, and check whether it is a random match from one side or a potential correct match from two sides of the spectrum.

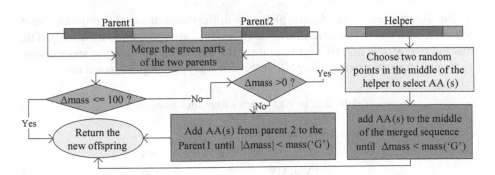

Fig. 3. The workflow of Nterm-Cterm crossover operator.

2.4 Nterm-Cterm Crossover

A new domain specific crossover is designed for this problem. The crossover mates two parents each having at least one exact match one b-/y-ion from N-terminus and C-terminus. The goal is to mate these two parents in the way that the new offspring would have exact b-/y-ion matches from both sides and possibly from the middle as well. Figure 3 illustrates the Nterm-Cterm crossover workflow. The input of the crossover is three GA individuals, two individuals as parents and one as a helper, and the output is a new offspring. At first, the exact match parts (the green parts) from both parents are concatenated. Here the $\Delta mass$ condition is more relaxed, allowing up to 100 Da mass difference. If the $\Delta mass$ is more than absolute value of 100, then the new concatenated sequence is checked whether it needs to remove/add amino acids from/to the sequence. A negative $\Delta mass$ indicates that the sequence is long and needs removing a few amino acids from it and vice versa for a positive value. The reason is that based on Eq. 1 a long sequence has more amino acids and possibly it could have a bigger parent mass compared to a shorter sequence with less number of amino acids. Since there might be some overlap between the green parts of the two parents, these $\Delta mass$ conditions help the operator to avoid constructing a bad offspring having a big $\Delta mass$ penalty in its fitness value. Therefore, when $\Delta mass$ is negative, for removing the overlap the green part of the parent 1 is considered as the N-terminus of the new sequence and each time one amino acid from C-terminus (the most right) of the parent 2 is added to the new sequence until the $\Delta mass$ criterion is met.

If $\Delta mass$ is positive, it is required to add a few amino acids in the middle of the green parts of the two parents. Here instead of adding random amino acids, another individual as helper is used. The helper parent has a high fitness value which possibly could indicate having more matched peaks in the middle. So two crossover points are picked randomly from the middle of helper parent and the amino acids in between those two points are added to the middle of the new sequence one by one until the mass difference criterion is met.

2.5 Flip-AA Mutation

The flip-AA mutation randomly pick one amino acid from the sequence and replaced it with one of the 19 amino acids ('I' and 'L' are considered identical). The mutation operator is not allowed to mutate the last amino acid in the sequence as it is always supposed to be either 'R' or 'K' for a tryptic peptide sequence.

Table 2. The dictionary of conflict masses.

Single AA	di-peptide	Mass
W	DA, AD, EG, GE, VS, SV	186
R	VG, GV	156
Q	AG, GA	128
N	GG	114

Table 3. The set of peptide spectrum matches used in this study.

No.of PSMs	Peptide length range	Avg. length of peptides	Charge No.	Precursor mass range	Fragmentation (Da)
120	7–12	9.5	2	<1150	0.5

2.6 Conflict-Mass Mutation

There are situations where the mass of a single amino acid conflicts with the mass of two amino acids (di-peptides). For example, the mass('W') = mass ("DA") = 186. A dictionary of such conflict masses is provided and shown in Table 2. The conflict-mass mutation operator checks whether the sequence contains any amino acid in the conflict mass dictionary and randomly replaces the amino acid with any of the corresponding di-peptides.

3 Experiment Design

3.1 Dataset

The comprehensive full factorial LC-MS/MS benchmark dataset, which is particularly designed for evaluating MS/MS analysis tools, containing 50 protein samples extracted from Escherichia coli K12, is used in this study [14]. The MS/MS spectra in this dataset have been already searched against a curated *E.coli* database by using Mascot v2.2 [15]. From the peptide identification results provided by this dataset, a set of 120 doubly charged peptide-spectrum matches (PSMs) with a minimum Mascot peptide identification score of 45, minimum peptide length of 7 amino acids and maximum length of 12 is selected. Based on Table 3, the average length of the peptides is 9.5. The spectra have a precursor of less than 1150 Da and the fragment ion of 0.5 is used as the value of tolerance τ. The so-called "ground truth" is used to test the performance of *de novo* sequencing algorithms.

3.2 Parameters, Evaluation and Benchmark Algorithm

The parameters in Table 4 are used to setup the GA algorithm. A-Novo is implemented in Python 3.6 and uses DEAP (Distributed Evolutionary Algorithms in Python) package [16]. To evaluate the accuracy of *de novo* sequencing results, the *de novo* peptide sequences constructed by the algorithm are compared with the real peptide sequences from the ground truth dataset. The total recall and precision metrics are calculated based on the following equations:

$$\text{precision} = \frac{\text{total number of matched amino acids}}{\text{total length of predicted peptide sequences}} \tag{7}$$

$$\text{recall} = \frac{\text{total number of matched amino acids}}{\text{total length of ground truth peptide sequences}} \tag{8}$$

Table 4. GA-Novo parameters

Parameter	Value	Parameter	Value
Initialisation pool size	1000	Population size	300
Size of sub-pools	100	Generations, runs	50, 30
Flip-AA mutation rate	0.1	Conflict-mass mutation rate	0.15
Crossover rate (2point)	0.35	Elitism rate	0.01
Nterm-Cterm crossover rate	0.40	Selection	Tournament, 7

The performance of GA-Novo is compared with PEAKS, which is a popular benchmark *de novo* sequencing algorithm [5] and with PepyGen [9] which is a freely available *de novo* sequencing tool using GA. The metrics in both Eqs. 7 and 8 measure the accuracy of the results in amino acid level. The following metric is also used to evaluate the results of both algorithms in peptide level.

$$\text{recall}_{peptide\ level} = \frac{\text{total number of fully correctly predicted peptide sequences}}{\text{total number of ground truth peptides}}$$

(9)

4   Results and Discussions

This section presents three different experiments. The first experiment uses GA-Novo for *de novo* sequencing of 120 MS/MS spectra in the dataset and the results are compared with those of PEAKS and PepyGen. The rest of this section analyses the effectiveness of two main components used in GA-Novo namely tag-based initialisation method and the domain dependant Nterm-Cterm crossover.

4.1   Performance Comparison Between GA-Novo, PEAKS and PepyGen

This section compares the overall performance of GA-Novo with PEAKS and PepyGen. All spectra in the dataset (Table 3) are used to assess the performance of both algorithms. Among these spectra some of them are noisy and some might have incomplete ion ladders.

Given an MS/MS spectrum to PEAKS, the output is a set of peptide sequences each having a confidence score level between 0 and 100 [5]. The score indicates how likely the complete sequence is correct. For each spectrum, the top scored sequence is taken as the output of *de novo* sequencing by PEAKS. PEAKS was run with an error tolerance of 0.5 Da and tryptic digestion.

For GA-Novo and PepyGen, the experiments are repeated for 30 independent runs with 30 different random seeds. For each spectrum in each run, the best fit sequence constructed by GAs are taken as the outputs of both GA methods.

Table 5. The results of sequencing 120 MS/MS spectra by GA-Novo and PEAKS.

Algorithm	Precision	Recall	recall$_{pep.\ level}$	Avg. len. of partial matches	Avg. len. of predicted sequences
GA-Novo	0.89 ± 0.03 (+)	0.88 ± 0.03 (+)	0.64 ± 0.06 (+)	8.4 ± 0.27 (+)	9.4 ± 0.1 (=)
PEAKS	0.85	0.84	0.56	8.06	9.43
PepyGen	0.42 ± 0.05	0.41 ± 0.05	0.14 ± 0.04	3.9 ± 0.2	9.1 ± 0.2

To compare the results of GA-Novo in 30 runs with PEAKS, one sample statistical t-test with 95% confidence interval and two-tailed P value less than 0.0001 is used to compare the performance of two methods. Table 5 presents the results of *de novo* sequencing by these two methods. (+) in the table indicates the difference between the results of GA-Novo and PEAKS is considered to be statistically significant and (=) indicates not statistically significant. It can be seen that the results of GA-Novo in most cases are statistically significant. GA-Novo outperforms PEAKS by 4% increase in precision and 4% increase in recall. Moreover, the accuracy of fully matched peptide sequences predicted by GA-Novo, in peptide level, is 8% higher than PEAKS. The reason of having lower recall compared to the precision in the results of both algorithms is that, they mainly construct either equal or slightly shorter peptide compared to the real peptide in terms of length. Also, the results show that in overall GA-Novo is able to find more partially matched sequences compared to PEAKS, as the average length of partially matched sequences for GA-Novo is 8.4 and statistically significant than the result of PEAKS.

As shown in Table 3 that the average length of peptides in this dataset is 9.5, sequences predicted by GA-Novo and PEAKS have the average length of about 9.4. No doubt that this value is close to the average length of the peptides in ground truth as the goal of both algorithms is constructing full length individuals. The sequences "AMVEVFLER" and "DAGTLLWLGK" are two examples of when PEAKS failed to predict the whole sequences, whereas GA-Novo could successfully construct the fully matched peptides. The sequences were predicted by PEAKS as "TT**VEVFLER**" and "W**GTLLWLGK**" while the first two amino acids in both sequences were predicted wrongly. More analysis on the results of PEAKS shows that it sometimes fails to predict the conflict masses from Table 2, whereas GA-Novo gets benefit of its domain dependant mutation operator, conflict-mass mutation to avoid these types of mismatches.

Also from the results shown in Table 5, it can be seen that GA-Novo outperforms PepyGen by 45% increase in precision and 47% increase in recall. Moreover, the accuracy of fully matched peptide sequences predicted by GA-Novo, in peptide level, is 50% higher than PepyGen. The reason of low performance of PepyGen is its GA design which is not able to construct the correct peptide sequences. PepyGen does not apply the tag-based initialisation.

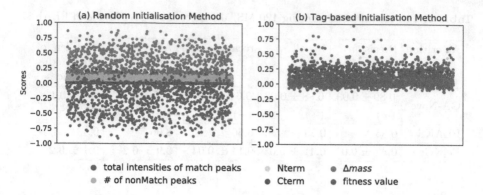

Fig. 4. Plots of 1,000 individuals generated by random and tag-based initialisation.

Also, in the design of its fitness function the two terms Nterm and Cterm do not exist and this makes the algorithm to fail constructing the fully matched sequences. PepyGen, uses only simple mutation and crossover operators and this makes the algorithm not to be able to create good individuals for the next generations.

Although the results show that GA-Novo is able to construct the full length of sequences (9.4 relatively close to 9.5), GA-Novo also sometimes fails to construct the fully matched sequences (8.4). However, comparing the difference between the average length of ground truth peptides, 9.5, and the results of average length of partial matches for GA-Novo, 8.4, the result shows that in overall GA only fails to fully match either one or two amino acids. The reason of this mismatch is the conflict mass of di-peptides. As mentioned previously in Table 2 where the mass of di-peptides conflicts with the mass of one single amino acid, there are other situations where the mass of two di-peptides conflict with each other.

4.2 Tag-Based Initialisation Vs. Random Initialisation

Figure 4 illustrates two plots presenting the overall fitness value and the values of its 5 terms included in the fitness function (see Eq. 3). As the random initialisation method does not use any domain knowledge and randomly generates sequences between length 7 and 12, it can be seen from Fig. 4(a) that the fitness values of majority of population is below zero. The reason of such low fitnesses is that the random initialisation does not pay attention to $\Delta mass$, mass difference, which is a penalty in fitness function. Generating short or long peptide sequences results in a big $\Delta mass$ penalty. However, the tag-based initialisation plot in this figure, shows how the $\Delta mass$ values are small in this method and the overall fitness values are bigger than random method.

The results in Table 6 show that the best individual out of 1000 individuals in a single run of random method is "YVMNEAR" with a fitness value of 0.25. In this table, each sequence is shown by its overall fitness value and fine different terms from fitness function, including I, D and N which indicate the total

Table 6. The best individual in a single run tag-based and random initialisation methods using the spectrum of "AAALAAADAR" peptide.

	Sequence	Fitness scores					
		Fitness	I	D	N	Nterm	Cterm
Ground-truth	AAALAAADAR	2.1950	0.595	0.000003	0.0	0.8	0.8
Random initial	YVMNEAR	0.25	0.057	0.020099	0.071	0	0.28
Tag-based initial	RVAAAAWR	1.14	0.528	0.000027	0.0	0	0.625

Table 7. The statistics on three fitness scores in 30 different runs of tag-based and random initialisation methods using the spectrum of "AAALAAADAR" peptide.

	Fitness value				Nterm				Cterm			
	Min	Max	Avg.	Std.	Min	Max	Avg.	Std.	Min	Max	Avg.	Std.
Random initial	−0.97	0.32	−0.29	0.24	0	1.57	0.002	0.06	0	2.57	0.01	0.15
Tag-based initial	−0.15	1.07	0.1	0.17	0	4.83	0.04	0.35	0	5.93	0.45	0.98
Significance	(+)				(=)				(+)			

intensities of matched peaks, $\Delta mass$ and the number of unmatched peaks, respectively. Nterm and Cterm are normalised here. Based on the tag-based method, the best individual is "RVAAAAWR" with fitness value of 1.14. Therefore, the fitness value of the best individual produced by tag-based initialisation method is 4.7 times bigger than the one in random initialisation. As mentioned above the fitness value of the ground truth is 2.19, therefore the tag-based initialisation method could be a better start point for GA.

The statistics results in Table 7 show the significance of comparison between the results of two methods. An unpaired statistical t-test with 95% confidence interval is used to compare the performance of two methods. This table presents the statistics on overall fitness value, Nterm and Cterm scores. Please notice that Nterm and Cterm scores are not normalised here. From this table it can be seen that the average fitness values and Cterm scores of tag-based initialisation method are statistically significant than random based method. However, Nterm scores are not statistically significant than random method, thanks to the tag extraction step which sometimes is not able to extract partially matched 3-letter tags from N-terminus of the spectrum due to the missing b-ions in this area. During peptide fragmentation, peptides may not fragment at some positions and leave no information, resulting in missing data. That is why the first two fragments b_1 and b_2 ions are seldom observed in the spectrum.

In overall based on the results in both Tables 6 and 7, tag-based method constructs better/fitter individuals compared to random initialisation, as tag-based method focuses on concatenating randomly 2, 3 or 4 tags. Then the method reduces the absolute mass differences between the constructed sequences and the spectrum by randomly inserting/removing random amino acids into the sequences. As a known domain knowledge, each tryptic peptide ends in either

Table 8. An example of applying Nterm-Cterm crossover on two long partially matched parents that have matched amino acids overlap.

	Sequence	Fitness	I	D	N	Nterm score	Cterm score
$Nterm_{parent}$	**AAALA**GGWR	0.79	0.21	0.031	0.05	4	2
$Cterm_{parent}$	NVL**AAADAR**	1.34	0.58	0.000002	0.02	0	7
$Helper_{parent}$	RG**LAAAD**VK	0.58	0.59	0.00003	0.01	0	0
$Offspring$	**AAALAAADAR**	2.19	0.59	0.000003	0.000	8	8

Table 9. An example of applying Nterm-Cterm crossover on two long partially matched parents that have matched amino acids overlap.

	Sequence	Fitness	I	D	N	Nterm score	Cterm score
$Best\ Nterm_{parent}$	**AAA**PEPSEQK	0.1173	0.118	0.14	0.060	2	0
$Best\ Cterm_{parent}$	PEPSEQ**AR**	0.4477	0.237	0.014	0.025	0	2
$Best\ helper_{parent}$	RG**LAAAD**TK	0.2952	0.309	0.002	0.011	0	0
$Offspring$	**AAALAAADAR**	2.1950	0.595	0.000003	0.000	8	8

'K' or 'R', so this heuristic has been applied randomly on the sequences constructed by this method as well. As the result this method decreases the mass differences and increases the number of match ions, resulting in an increase in the total intensities of the match ions. Back to the best sequence produced by the tag-based initialisation, "RVAAAAWR" in Table 6, it is expected this sequence goes through the GA evolutionary process and after a few generations converts to the exact match.

4.3 Analysis the Effectiveness of Nterm-Cterm Crossover

This section presents two examples when Nterm-Cterm Crossover is applied on different individuals and also shows the performance of this operator across 30 different runs. Tables 8 and 9 show how new Nterm-Cterm Crossover can result in whole sequence exact match. By looking at the Nterm and Cterm scores of Nterm and Cterm parents in Table 8, it can be seen that these parents have quite big values that could indicate potential exact amino acid matches from each side. Considering the sequence of amino acids of these parents and knowing the ground truth, it can be seen that the two parents have a few number of exact amino acid matches. However, concatenating the exact match amino acids (shown in bold), results in a false sequence "AAALALAAADAR" which is not desired. As the technique was explained previously, the two parents are concatenating with consideration of removing the overlap and this results in a whole sequence exact match as the offspring. As both parents have enough Nterm and Cterm match amino acids, the third parent, helper, is not used here.

The second example in Table 9 shows two parents with only a few number of exact match amino acids. As the concatenated sequence still does not meet the mass difference criterion, the third parent is used to fill the gap. It can be seen from the fitness values of the helper parent that it is not necessary to have a high Nterm or Cterm score, as the helper parent is chosen based on its overall fitness value. Here also an exact match is obtained, but it is worth mentioning that applying this operator does not always results in whole sequence exact match, but mainly there is an improvement in the fitness value of the new offspring.

Table 10 presents the overall performance of Nterm-Cterm Crossover on a number of individuals produced by tag-based initialisation method. In the first row, the tag-based initialisation method is used in 30 independent runs, each run producing 1000 individuals. In each run, out of 1000 individuals three individuals with having the best Nterm score, Cterm score and fitness value are chosen to be Nterm parent, Cterm parent and helper parent, respectively. Then the Nterm-Cterm Crossover is applied on the parents of each run and the average delta fitness values are calculated for all the runs. It can be seen that in overall the fitness values of the offsprings improved by 62% compared to the Nterm parent, 37% to Cterm parent, and 28% compared to the helper parent.

Similarly, the second row of Table 10 presents the results of improvement in the fitness values of new offsprings produced by Nterm-Cterm Crossover in a single run, but randomly choosing 30 individuals as parents which are not necessarily the best scored parents. The results show that in this case also in average there is 4% improvement in the fitness score of the new offspring compared to its Nterm parent, 11% compared to Cterm parents and 1.4% compared to the helper parent. One reason of not having a significant improvement in this results is that the parents are not filtered. That is why in design of the GA algorithm, presented in Fig. 1, the individuals in two Nterm and Cterm pools must have at least an Nterm/Cterm scores of one.

Table 10. Performance evaluation of Nterm-Cterm Crossover operator using the spectrum of "AAALAAADAR" peptide in different scenarios.

	$\Delta f_{cx,N_{parent}}$	$\Delta f_{cx,C_{parent}}$	$\Delta f_{cx,H_{parent}}$
30 runs "Best" individuals	0.62	0.37	0.28
Single run "Random" individuals	0.4	0.11	0.014

5   Conclusions and Future Work

The goal of this paper was developing an effective *de novo* sequencing algorithm that constructs full length sequences. The goal has been successfully achieved by developing an effective GA algorithm that gradually and rapidly construct the peptide sequences that match the input MS/MS spectra.

Other developments presented in this work are a new domain dependant fitness function, new initialisation method and two new genetic operators that were particularly designed for the GA algorithm. The GA fitness function was able to capture main spectral features and guided GA to produce the fully matched peptides. The initialisation method was an excellent start point to accelerate the evolutionary process. The tag-based initialisation method helped GA to start with better/fitter initial population, accelerating its convergence speed, and providing high quality individuals for the GA components. The genetic operators helped GA to maintain the diversity in the population and gradually convert partial matches to fully matched sequences. The results showed that GA-Novo achieved higher number of fully matched sequences compared to PEAKS, the most commonly used *de novo* sequencing software. GA-Novo achieved both higher recall and precision than PEAKS. GA-Novo outperformed PEAKS by 4% higher precision, 4% higher recall in amino acid level and 8% higher recall in peptide level. Also GA-Novo got twice performance of PepyGen in terms of precision and recall in amino acid level and 4.5 higher in peptide level.

As future work, we will investigate the performance of GA-Novo using more spectra from different types of mass spectrometers. We will also design a ranking based algorithm to refine the results of GA-Novo aiming at finding the true match among the top five best candidates from the final generation of GA.

References

1. Papayannopoulos, I.A.: The interpretation of collision-induced dissociation tandem mass spectra of peptides. Mass Spectrom. Rev. **14**(1), 49–73 (1995)
2. Xu, C., Ma, B.: Complexity and scoring function of MS/MS peptide de novo sequencing. Comput. Syst. Bioinform. Conf. **5**, 361–369 (2006)
3. Sakurai, T., Matsuo, T., Matsuda, H., Katakuse, I.: PAAS 3: a computer program to determine probable sequence of peptides from mass spectrometric data. Biol. Mass Spectrom. **11**(8), 396–399 (1984)
4. Ma, B.: Novor: real-time peptide de novo sequencing software. J. Am. Soc. Mass Spectrom. **26**(11), 1885–1894 (2015)
5. Ma, B., Zhang, K., Hendrie, C., Liang, C., Li, M., Doherty-Kirby, A., Lajoie, G.: PEAKS: powerful software for peptide de novo sequencing by tandem mass spectrometry. Rapid Commun. Mass Spectrom. **17**(20), 2337–2342 (2003)
6. Nielsen, M.L.: Characterization of polypeptides by tandem mass spectrometry using complementary fragmentation techniques. Ph.D. thesis, Acta Universitatis Upsaliensis (2006)
7. Webb-Robertson, B.J.M., Cannon, W.R.: Current trends in computational inference from mass spectrometry-based proteomics. Brief. Bioinform. **8**(5), 304–317 (2007)
8. Heredia-Langner, A., Cannon, W.R., Jarman, K.D., Jarman, K.H.: Sequence optimization as an alternative to de novo analysis of tandem mass spectrometry data. Bioinformatics **20**(14), 2296–2304 (2004)
9. Kistowski, M., Gambin, A.: Optimization algorithm for de novo analysis of tandem mass spectrometry data. BioTechnologia J. Biotechnol. Comput. Biol. Bionanotechnol. **92**(3), 296–300 (2011)

10. Allmer, J.: Algorithms for the de novo sequencing of peptides from tandem mass spectra. Expert Rev. Proteomics **8**(5), 645–657 (2011)
11. Eng, J.K., McCormack, A.L., Yates, J.R.: An approach to correlate tandem mass spectral data of peptides with amino acid sequences in a protein database. J. Am. Soc. Mass Spectrom. **5**(11), 976–989 (1994)
12. Yu, F., Li, N., Yu, W.: PIPI: PTM-invariant peptide identification using coding method. bioRxiv, p. 055806 (2016)
13. Herrmann, R.L.B., Hilderbrand, A.: Peptide fragmentation overview. In: Principles of Mass Spectrometry Applied to Biomolecules, vol. 10, p. 279 (2006)
14. Wessels, H.J., et al.: A comprehensive full factorial lc-ms/ms proteomics benchmark data set. Proteomics **12**(14), 2276–2281 (2012)
15. Cottrell, J.S., London, U.: Probability-based protein identification by searching sequence databases using mass spectrometry data. Electrophoresis **20**(18), 3551–3567 (1999)
16. Fortin, F.A., De Rainville, F.M., Gardner, M.A., Parizeau, M., Gagné, C.: DEAP: evolutionary algorithms made easy. J. Mach. Learn. Res. **13**, 2171–2175 (2012)

Ant Colony Optimization for Optimized Operation Scheduling of Combined Heat and Power Plants

Johannes Mast[1]([✉]) [iD], Stefan Rädle[1] [iD], Joachim Gerlach[1] [iD],
and Oliver Bringmann[2] [iD]

[1] Albstadt-Sigmaringen University, 72458 Albstadt, Germany
{mast,raedle,gerlach}@hs-albsig.de
[2] University of Tübingen, 72076 Tübingen, Germany
oliver.bringmann@uni-tuebingen.de

Abstract. In the worldwide expansion of renewable energies, there is not only a need for weather-dependent plants, but also for plants with flexible power generation that have the potential to reduce storage requirements by working against fluctuations. A highly promising technology is provided by Combined heat and power (CHP) plants, which achieve high efficiencies through the simultaneous generation of electricity and heat. This is why they are also being promoted by the European Union. Also, the construction of biogas plants is usually linked to the construction of CHP plants in order to generate energy from the emission-free produced biogas. However, until now CHP plants have mostly been operated by heat demand (just like boilers), causing the generated electricity often to put additional stress on the power grid. The planning of a CHP plant, whose generated heat always finds a consumer and the generated electricity is simultaneously optimized with regard to an optimization objective, requires nonlinear optimization approaches due to the physical effects in the heat storage. This paper presents a methodology for optimized planning of CHP plants using Ant Colony Optimization. The selected optimization objectives are the power exchange, the tenant electricity and CO_2. It could be shown that all optimizations are at least 10% better than the heat-led operation. The best results were achieved with the electricity exchange optimization that can be up to 24% more profitable than a CHP in a heat-led mode.

Keywords: Ant Colony Optimization · Combined heat and power · Operation scheduling · Electricity exchange · Tenant electricity

1 Introduction

In the European Union, CHP plants are promoted by the EU Directive 2004/08/EC and are being further promoted by many countries through separate grants [1]. The reason for the promotion is that CHP plants produce electricity and

© Springer Nature Switzerland AG 2019
P. Kaufmann and P. A. Castillo (Eds.): EvoApplications 2019, LNCS 11454, pp. 90–105, 2019.
https://doi.org/10.1007/978-3-030-16692-2_7

heat at the same time, resulting in a high efficiency (between 75 and 85%) which makes it possible to reduce primary energy consumption and greenhouse gas emissions. In addition, they are usually operated as decentralized energy sources, which reduces energy loss in the grid, reduces bottlenecks in the transmission grid and increases the quality and reliability of the power supply.

One challenge with technologies that generate heat and electricity simultaneously is that the demand for heat and electricity varies over time. Since it is easier to distribute electricity via the power grid than is the case with heat, decentralized CHP plants are today mostly operated according to heat demand. From the point of view of the power grid, this is contradictory, since electricity is a product whose value is highly dependent on current electricity demand (see fluctuating prices on the electricity exchange). This is why CHP plants operating in this way are not in a profitable position to sell their electricity on the electricity exchange. When electricity is sold on the electricity exchange, optimal planning of the operation of the CHP plant is very important and it is necessary to have an optimized operating plan for a short period of time on electricity price, local heat and electricity demand forecasts.

Chen et al. have recognized that the heat-led mode of operation is one of the main causes for the wind power curtailment in winter. Therefore, they have designed an optimized operation scheduling for CHP plants with heat storages, which lowers the curtailed wind power and production costs [2].

Mongibello et al. have developed an optimized operation scheduling for CHP plants with different engines (including Stirling engine and combustion engine), which allows maximizing the revenue in terms of separate generation. For this purpose, they created a model that maps the heat loss in the storage unit. They use the pattern search algorithm for optimization [3].

Majic et al. have carried out optimized operational scheduling for trading on the electricity market using linear programming. As pointed out by the authors, they do not consider non-linear factors such as efficiency as a function of heat and electricity output [4].

In contrast to the known approaches, we use a highly accurate simulation model of a heat storage within the optimization, which also considers factors that are nonlinear [5]. In addition to optimized operational planning for the electricity exchange, the methodology should also allow optimization for tenant electricity and CO_2 emissions in the electricity mix. Besides, no evolutionary algorithms have been used in the presented papers for the planning of an optimized operation scheduling. The Ant Colony Optimization (ACO) is considered as one of the best solutions for many graph problems like the traveling salesman problem and the vehicle routing problem and is the subject of numerous recent publications [6]. Since the optimized planning of a CHP plant can also be represented as a graph, whose throughput costs should be minimized, the attempt to use ACO for optimized operational planning is reasonable. For this reason, this paper examines in which way the optimization problem can be converted into a representation suitable for the application of ACO.

The paper is structured as follows: Sect. 2 describes the ACO and the basic characteristics and operating modes of CHP plants. Section 3 explains the application of optimized operational scheduling to ACO and how it can be used to create optimized operational plans for various optimization objectives. Section 4 examines the methodology on the basis of a daily and annual review of all optimization goals.

2   Background

This section first describes the ACO algorithm followed by an explanation of the characteristics and types of CHP plants and then outlines the experimental settings of the tests.

2.1   Ant Colony Optimization

ACO is an ant behavior inspired algorithm used to solve difficult combinatorial optimization problems by trying to find the shortest paths in a graph.

ACO is an iterative algorithm. At each iteration the graph is traversed by N artificial ants. When ants walk a path in the graph, they leave a pheromone track that can affect the probability that other ants will use this path in the coming iterations. The edges in the graph that are walked by an ant are determined with a so called Pseudo-Random-Proportional Action Choice Rule whose depends on the current pheromone value $\tau_{i,j}$ of the edge and the inverse of the distance between two nodes $\eta_{i,j} = \frac{1}{d(i,j)}$:

$$P_{i,j} = \frac{[\tau_{i,j}]^\alpha [\eta_{i,j}]^\beta}{\sum_{j \in allowed}[\tau_{i,j}]^\alpha [\eta_{i,j}]^\beta} \qquad (1)$$

Where α and β are fixed settings that can be used to control the importance of distance and pheromone [7].

There are many variants of the ACO for strengthening pheromone values. The first variant ever proposed is Ant System (AS). This variant adjusts the amount of pheromone of the pheromone trail after the completion of a complete tour:

$$\tau_{i,j}(t_{iter}) = (1 - \rho)\tau_{i,j}(t_{iter} - 1) + \sum_{i=1} \frac{Q}{L_i} \qquad (2)$$

Where ρ is the evaporation constant to reduce the attractiveness of an edge, Q is a constant and L_i is the length of the tour constructed by ant i [7].

Just like *Ant System (AS)* the *Elitist AS* variant strengthen the pheromone value of all run values, additionally *Elitist AS* strengthens the global best trail with an extra factor. While the *MAX-MIN AS* variant only strengthens the global best trail through pheromones. In addition, it sets maximum and minimum pheromone limits for a higher exploration of solutions. Another variant is the *Rank-Based AS*. After each ant has passed through the graph, the ants are sorted according to the length of their completed tours. The strengthening of the trail is done according to the rank of each ant [8].

2.2   Characteristics of CHP Plants

In order to achieve the high efficiencies between 75 and 85%, a CHP plant consists of a prime mover, a generator and a heat recovery system. The prime mover supplies the driving force for the power generator and generates the heat. Usually it is a gas turbine, a steam turbine or combustion engine. Due to the different types of engines, the CHP can be operated with a variety of fuels such as natural gas, wood pellets, heating oil and biogas. Especially in biogas there is an opportunity to further reduce emissions. The heat recovery plant recovers the heat produced by the generator and the prime mover. Usually, a heat storage and a boiler are linked to a CHP plant. The recovered heat can be stored either in form of hot water or steam. If more heat is required than the CHP plant can supply, the boiler is additionally powered [9].

Typically, the combustion engine of a CHP plant is only operated in two modes: off or on (at nominal power). As a result, the CHP plant runs in a low-wear operation with the highest possible efficiency. CHP plants reach their nominal output within a short time (seconds to a maximum of a few minutes). The operation scheduling of CHP plants is usually adapted to the heat demand, but can also be done according to the electricity demand or a mixed operation. A CHP plant only brings economic and ecological savings during ongoing operation, so that it is only profitable with a high and constant heat demand - usually at least 4.500 h/year in a heat-led mode.

2.3   Experimental Setting

Figure 1 shows the experimental setting in which the optimization of the operation scheduling takes place. The CHP plant is a "SenerTec Dachs" with a single-cylinder four-stroke engine as prime mover. It has a displacement of approx. $580 \, cm^3$ and is operated with natural gas as standard. Thus it can be assumed that the CHP has a typical CO_2 emission of 425 g/kWh [10]. It has an electrical output of 5.0 kW and a thermal output of 12.5 kW. In the heating circuit, a heat storage is used which is filled with water. Its storage capacity is 900 L [11].

In addition, a supporting boiler is placed in the heating circuit, whose thermal output is able to meet the load at any time. The heat generation of a boiler is usually greater than that of a CHP plant with the same use of fuel. Therefore, the start-up of a CHP plant is only profitable compared to a boiler as long as the profit from the generated electricity exceeds the additional fuel costs. If an additional boiler is available, an optimized operation scheduling can save fuel costs by switching off the CHP at low exchange prices. At these times, the boiler alone covers the heat demand. It is assumed that when the electricity from the CHP is sold for 0.018 €/kWh or more, the operation of the CHP is more profitable than the operation of the boiler (hereafter referred to as marginal costs).

The maintenance contract of a CHP usually specifies time conditions for low-wear operation. This usually includes a minimum running time and minimum cooling time to prevent the CHP from clocking. In our case, both are defined as one hour.

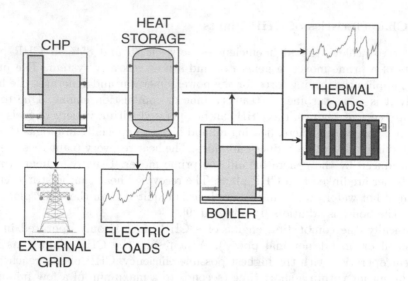

Fig. 1. Experimental setting for the optimization

The optimization for the experimental setting takes place using a highly accurate simulation model with a storage that interacts with the ACO algorithm [5].

3 Methodology

This section describes the adaption of the ACO algorithm to our optimization problem and shows the implementation of the path length calculation for the different optimization objectives is done.

3.1 Applying the Optimization Problem to Ant Colony Optimization

Figure 2 shows the graph by which the ants have to find an optimized way with regard to the optimization objective. In the following, the optimization of a day with a resolution of five minutes is considered as time period, i.e. the number of points is $n = 288$ and the time range between two edges is $r_t = 5$ min. Since there are two possible CHP operation modes (off or on) for each time point, $2^{288} \approx 4.973e^{86}$ possible scheduling states exist for one day at the selected resolution. Due to the keeping of minimum running times and minimum cooling times, these are in reality somewhat less, but still such a large number that cannot be handled without systematic traversing through the state space.

Within each new iteration, the ants start in the starting state, which is immediately before the start time. At the beginning, it must be decided whether the CHP plant is to be switched on or off. This decision does not only affect a time range r_t, but since according to the maintenance contract a minimum

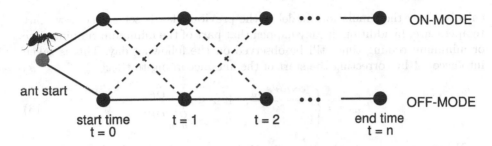

Fig. 2. Graph of the ants for the operation schedule

running time or minimum cooling time of r_{min}^{ON} and r_{min}^{OFF} (here: one hour) must be observed, $\frac{r_{min}^{ON/OFF}}{r_t}$ of such ranges are defined at once (here: $\frac{60}{5} = 12$).

Because from each node V_i two edges $E_{i,j}$ run to the next time point t_{i+1}, there are always two possibilities to choose from for which the probability $P_{i,j}$ is calculated. The calculation of the probability for selecting one of the two possible edges is only influenced by the current pheromone value $\tau_{i,j}$. The inverse of the distance $\eta_{i,j}$ in Eq. (1) is set to one.

After the first decision the ant has been made, a distinction is made when calculating the probabilities for the next edges. If the last operation mode is maintained, the minimum running time or minimum cooling time is already respected, so that the decision only affects another time range r_t. When changing the operating mode, the minimum running time or minimum cooling time must be maintained again, so that the decision affects again several time points $\frac{r_{min}^{ON}}{r_t}$ or $\frac{r_{min}^{OFF}}{r_t}$. This process continues until the ant arrives at node V_n at the end of the day. When an ant has passed through the graph, the score or length L_i is calculated for the ant i by executing a simulation model from a CHP. The simulation model receives, in addition to the time series for the heat load and the ambient temperature, the operating plan run by the ant. To determine the score, the outputs of the simulation model are evaluated according to the optimization objective. This is described in detail in Sect. 3.2.

Afterwards, the next ants run through the graph until all ants of the iteration have arrived at node V_n and their length L_i has been determined with the outputs of the simulation model.

For adjustment of the pheromone values of the global trail the variants *Ant System (AS)*, *Elitist AS*, *MAX-MIN AS* and *Rank-Based AS* were tested. The best results are achieved when the pheromone values have limits, as is the case with *MAX-MIN AS*.

At the start it is assumed that the storage tank is half full, i.e. has a temperature between the minimum and maximum value. If the optimization is done for consecutive days, it must be ensured that the last storage tank temperature calculated by the simulation model of the previous day is set as the new start temperature. In addition, it may happen that part of the minimum running time or minimum cooling time still be observed on the following day. This is taken into account by correcting the start of the ants according to this:

$$t_{start} = \begin{cases} \frac{r_{min}^{ON/OFF} - u}{r_t}, & \text{if } u \leq r_{min}^{ON/OFF} \\ 0, & \text{if } u > r_{min}^{ON/OFF} \end{cases} \qquad (3)$$

Where u represents the time duration performed with the last operation mode on the previous day.

Enhancements for Faster Problem Solving. For the acceleration of the optimization, it may be advisable to initially strengthen the pheromones of the edges that fit on the heat-led CHP plant. On many days, the optimized operation schedules are only slight adjustments of the heat-led mode of operation, therefore the ants get faster to good operation schedules. In addition, the heat-led mode of operation ensures that not too much heat is generated.

Another possible acceleration point may be to increase the probability for the edge of the current operation mode. It is beneficial if a CHP plant stays in an operation mode for as long as possible. If the decision is checked in five minute intervals as described, this usually leads to many operation mode changes. This can be improved by increasing the likelihood of the operation mode being maintained.

3.2   Evaluation of Operation Scheduling with Simulation Model

In Sect. 3.1 the application of ACO to the optimization problem was described. The calculation of the length of a travelled path was not deepened. This subsection describes the methodology developed for this purpose.

The length of a path L_i is calculated by a score function that evaluates the operation scheduling. It is divided into two parts: In the first part, the temperatures in the heat storage are analyzed to ensure that the CHP does not generate too much heat energy and in the second part the fulfillment of the optimization objective is evaluated. If only one optimization goal is considered, a stronger penalty for the storage has proven to be effective:

$$L_i = 0.8 * score_{heat\_storage} + 0.2 * score_{objective} \qquad (4)$$

In order to correlate the parts together, a normalized value between 0 and 1 is produced from both parts.

Boundary Condition: Storage Temperatures. The time series provided by the simulation model contains storage temperatures $\boldsymbol{s} = (s_1, \ldots, s_n)$ with $s_i \in \mathbb{R}^n$ for each time point t in the simulated time period. If a day is considered with a five-minute resolution, the final evaluation of an operation schedulings returns $n = 288$ temperature values. For operational scheduling, there is a value s_{max} that is not allowed to be exceeded. In the case of the considered CHP (see 2.3) the emergency shutdown takes place at a temperature of 95 °C. An operation scheduling, which puts the CHP in danger of coming close to an emergency shutdown, should therefore result in a lower score (which results in a higher length). In order to have a buffer of 5 °C, it makes sense for such a CHP to consider all temperatures above 90 °C as critical and to include them negatively in the score [12].

The evaluation and normalization was done via a hyperbolic function which returns a value between 0 and 1. A hyperbolic function has the advantage over a jump function that the ACO algorithm sets the pheromones better since even small scheduling changes in the direction of the non-critical temperatures lead to a better score and thus to more pheromones of the chosen edges. The following hyperbolic function (see also Fig. 3) has proven to be suitable:

$$n(x) = \begin{cases} -\frac{1.0}{(x-89)} + 1, & \text{if } x \geq 90 \\ 0, & \text{if } x < 90 \end{cases} \qquad (5)$$

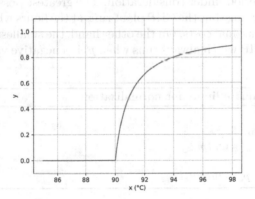

Fig. 3. Hyperbolic function for normalization of storage temperatures

The normalized values are added to a vector $\boldsymbol{f} = (n(s_1), \ldots, n(s_N))$. After all temperatures in the heat storage have been evaluated, the average value in the vector is formed:

$$Score_{Storage} = \frac{\sum_{i=0}^{n} f_i}{n} \qquad (6)$$

As a result, not only the strength of the critical range but also the number of temperatures in the critical range are weighted. In the best case all temperature values are below the critical range, then this part of the score function is evaluated with the best possible value of 0 [12].

Optimization Objective: Electricity Exchange. In order to optimize the economic efficiency at the electricity exchange, it is necessary to obtain prices for the time period to be optimized. The time series for past dates up to 2005 of the european electricity exchange (EPEX SPOT) can be obtained from the ENTSO-E database. If an operation schedule for the CHP plant is to be prepared for a future point in time, forecast data from the electricity exchange must be used, which are also provided by ENTSO-E [13].

If one day is considered, the vector for the hourly fluctuating electricity prices is $\boldsymbol{p} = (p_1, \ldots, p_{m=24})$. The marginal costs $g \in \mathbb{R}$ are defined, which indicate at which price on the electricity exchange, the switching on of the CHP is profitable compared to the boiler (see 2.3). With the help of these data a vector $\boldsymbol{p}' \in \mathbb{R}^m$ is created, which contains the difference between the prices of the electricity exchange and the marginal costs:

$$\boldsymbol{p}' = (p_1 - g, \ldots, p_m - g) \tag{7}$$

These values are negative if the price on the electricity exchange is below the marginal costs, otherwise, they are positive [12].

For the normalization, the largest and smallest possible economic profit must be known for the period under consideration. The greatest possible value can be determined by assuming that the CHP plant runs at the hours when the electricity price exceeds the marginal costs. On the other hand, the smallest possible value is determined by a CHP which always runs when p'_i is a negative value:

Algorithm 1. Evaluate limits for normalization

1: **For** all p'_i in $\boldsymbol{p}'$:
2:     **if** $p'_i \leq 0$ **then**
3:         l_{lower} += $p'_i * Output_{el}$
4:     **else**
5:         l_{upper} += $p'_i * Output_{el}$

By the two limits, l_{lower} and l_{upper} a normalization of the state can be done, which always lies between 0 and 1. Due to the storage conditions, the best possible operational scheduling usually cannot be used. The optimization problem is therefore consists of the identification of an operation schedule that achieves the highest possible profit in the time range while respecting the specified storage temperatures [12].

When the simulation model is called with an operational schedule from the ACO algorithm it also generates the electrical power of the CHP. It is a time series, where $\boldsymbol{y} \in \mathbb{R}^n$ represents the values, with n describing the number of discrete time points. To calculate the economic efficiency, the hourly generated energy amounts $\boldsymbol{Y}$ are needed, which can be calculated from the vector $\boldsymbol{y}$ by integrating and accumulating it hourly. It applies:

For all Y_i in $\boldsymbol{Y}$:

$$Y_i = \begin{cases} 0, & \text{if } i = 1 \\ Y_{i-1} + \frac{1}{2}(Y_i + Y_{i-1})(t_n - t_{n-1}), & \text{if } i > 1 \end{cases} \tag{8}$$

Now the economic efficiency $e \in \mathbb{R}$ can be determined from $\boldsymbol{p'}$ and $\boldsymbol{Y}$ for the period under consideration by adding up the profits or losses for each hour.

$$e = \sum_{i=1}^{n} p_i' a_i \tag{9}$$

This value indicates how much Euro profit or loss is made compared to the pure boiler operation with the operation schedule. This is normalized with initially calculated limit values to a value between 0 and 1:

$$score_{objective} = \frac{(l_{upper} - l_{lower}) - (e - l_{lower})}{l_{upper} - l_{lower}} \tag{10}$$

where a lower value is considered better [12].

Optimization Objective: Tenant Electricity. In Germany, there is a remuneration model under which local neighbors can purchase their electricity directly from plant operators (hereafter referred to as tenant electricity model) [14].

By this model, two prices are assumed for the optimization of the tenant electricity. If the tenant purchases the generated electricity, the operator receives P_{grant} for the generated amount of electricity, if not it is sold via a fixed price P_{no_grant} to the distribution system operator. Normally P_{grant} is much larger than P_{no_grant}. The tenant is additionally connected to the grid so that if less electricity is generated, he can obtain the required difference from the grid. The optimization tries to generate as much electricity as possible during times when the tenant demand is high. This means that the intersection area between the generation curve of the CHP and the consumption curve of the tenant should be as large as possible.

First, the limits of the maximum possible profit are determined. Since both P_{grant} and P_{no_grant} should always be positive, the limit l_{upper} can be set by assuming that the CHP is switched on for the entire period and vice versa. Now it is tried to find an operation mode which, under respect of the storage conditions 3.2, achieves the highest overlaps.

For the CHP output $\boldsymbol{y} \in \mathbb{R}^n$ and the tenant demand $\boldsymbol{h} \in \mathbb{R}^n$ all intersection points $\boldsymbol{z} \in \mathbb{R}^k$ are calculated. Where $k \in \mathbb{Z}$ is the number of intersection points and z_i the time when intersection i takes place.

Due to the discrete time points of the other vectors, a function for interpolation $int(t)$ must be used, which returns the corresponding interpolation value $y \in \mathbb{R}$ at a time $t \in \mathbb{R}$ (here the time can lie between two time points, so $t \notin \mathbb{Z}$).

For the calculation of the profit, we use Eq. (8) to calculate the two integrated and accumulated time series $\boldsymbol{Y}$ and $\boldsymbol{H}$ from the CHP output $\boldsymbol{y}$ and the tenant requirement $\boldsymbol{h}$. Now the following logic can be used to calculate the economic profit e:

Algorithm 2. Calculating the profit for the tenant electricity scheduling

1: **For** all k_i in $\boldsymbol{k}$:
2:    if $int_Y(k_i) \geq int_H(k_i)$ **then**
3:        $e \mathrel{+}= (int_Y(k_i) - int_H(k_i)) * P_{no\_grant}$
4:        $e \mathrel{+}= int_H(k_i) * P_{grant}$
5:    **else**
6:        $e \mathrel{+}= int_Y(k_i) * P_{grant}$

The score $score_{objective}$ (the normalized value) can now be calculated with the economic profit e and the limits l_{lower} and l_{upper} according to Eq. (10).

Optimization Objective: CO_2 of Electricity Mix. With this optimization objective, the CHP plant tries to minimize the specific CO_2 emissions of the electricity mix. The score indicates how much CO_2 was saved compared to the electricity mix in g.

For many countries, the ENTSO-E database contains the hourly energy amounts of the various technologies that feed into the power grid [13]. From these values and a table of specific CO_2 emissions in g/kWh [10], it is possible to calculate for which CO_2 emissions the average plant feeds into the grid per hour. The hourly CO_2 emissions are stored in the vector $\boldsymbol{c} \in \mathbb{R}^m$.

If the CO_2 emission in the electricity mix are higher than the ones of the CHP (425 g/kWh for natural gas [10]), it is most economical to feed electricity into the grid. Conventional plants with typically high CO_2 emissions are nowadays often only put into operation if the renewable energies do not generate enough electricity due to weather conditions. By optimizing the CO_2 emissions in the power grid, a contribution is made to a clean electricity mix, because these conventional plants can be switched off again or remain switched off.

The limits l_{lower} and l_{upper} are calculated by always running the CHP at l_{upper} if the CO_2 emission in the electricity mix c_i is greater than that of the CHP c_{CHP}:

Algorithm 3. Evaluate CO_2 limits for normalization

1: **For** all c_i in $\boldsymbol{c}$:
2:    if $c_i \geq c_{CHP}$ **then**
3:        $l_{lower} \mathrel{+}= c_i$
4:        $l_{upper} \mathrel{+}= c_{CHP}$
5:    **else**
6:        $l_{lower} \mathrel{+}= c_{CHP}$
7:        $l_{upper} \mathrel{+}= c_i$

The score function calculates from the output of the CHP simulation model $\boldsymbol{y} = (y_1, \ldots, y_N)$ how long the CHP was switched on in the corresponding hours. When the CHP was not running the entire hour, the CO_2 value of the electricity mix c_i is added to the CO_2 value for the off times of the CHP:

Algorithm 4. Calculate the CO_2 savings

1: **For** all y_i in $\boldsymbol{y}$:
2:     **if** $y_i > 0$ (if CHP is turned ON) **then**
3:         $e \mathrel{+}= (t_{i+1} - t_i) * z_{CHP}$
4:     **else**
5:         $e \mathrel{+}= (t_{i+1} - t_i) * z_i$

This means that e is smaller if the CHP ran at times when the CO_2 level in the electricity mix was large and vice versa. With the calculated limits, the total value $score_{objective}$ can be calculated according to Eq. (10).

4   Validation

For the validation of the methodology, the advantages of the optimized operational scheduling for one day (13.08.2016), as well as for one year (2016) will be demonstrated. While the daily view illustrates the principle of optimized operation scheduling, the annual view allows to observe the economic benefits and CO_2 savings of the optimized operation for a longer period of time.

The location of the setting is in Baden-Württemberg, Germany. For the heat load, measured data of a multi-family house are used with a measurement resolution of 15 min. More detailed information about the structure of the setting can be found in Sect. 2.3.

4.1   Observation of a Day

The observation of a day is done for the 13 August 2016. It was chosen because it provides good properties to show the optimization of all optimization objectives: First of all, it is a summer day with a lot of scheduling flexibility, because the heat load is comparably low. Furthermore, the prices on the electricity exchange were partly above and partly below the assumed marginal costs of 0.018 €/kWh and also differences can be seen in the CO_2 in the German electricity mix. The results of all operation schedules of the specific day can be seen in Fig. 4. Figure 4a shows the heat-led mode of operation which has to be optimized. It can be seen that it always tries to keep the heat storage half full by switching on the CHP as soon as the storage temperature is lower than 65 °C by cooling and load coverage and turns the CHP out again as soon as the temperature has been heated up to 76 °C. Thus the CHP runs on this day over the day on an evenly distributed level.

Figure 4b shows the electricity exchange optimized operation scheduling. It can be seen that the CHP tries to maximize its running time at times when the price is above the marginal costs. Between 10 and 15 o'clock the water in the heat storage cools down so far that the supporting boiler covers the heat load. Due to the marginal costs, the operation of the CHP would not be profitable here. One sees that the heat-led mode of operation would have run in a loss during these hours.

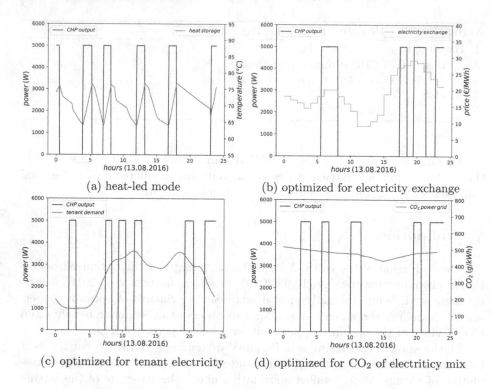

(a) heat-led mode

(b) optimized for electricity exchange

(c) optimized for tenant electricity

(d) optimized for CO_2 of electriticy mix

Fig. 4. Comparison for one day with same heat demand

Figure 4c shows the optimized operation schedule for the tenant electricity model. It can be seen that the scheduling tries to maximize the area between the tenant's demand and the generation of the CHP. So the CHP runs more in the morning and at noon when the electricity demand is typically higher.

Figure 4d shows the optimized operation schedule for reducing the CO_2 emission of the electricity mix. In principle, the CHP plant can always reduce the CO_2 level in the electricity mix on this day, as it is continuously above the 425 g/kWh of the CHP plant. Nevertheless, it is visible that the CHP remains switched off in the afternoon hours. This is due to the fact that the CO_2 emission are there at a low point, which is presumably due to an increased feed-in of solar and wind energy. As the CHP plant remains switched off during these hours, it can achieve longer running times in the evening, when CO_2 emissions are higher again.

4.2   Observation of a Year

The results for the year are presented in Table 1. In the rows, objective criteria are listed and the columns represent the different operation schedules.

Table 1. Results for the year 2016

	Heat-led	Optimized for electricity exchange	Optimized for tenant electricity	Optimized for CO_2 of electricity mix
Operating hours (hours/year)	4952.7	4716.5	5114.3	3498.4
Profit electricity exchange (€)	293.04	**364.63 (124.43%)**	320.63	194.94
Profit tenant electricity (€)	3811.23	3731.99	**4195.74 (110.08%)**	2699.728
CO_2 savings (gram)	128366	139043	128239	**156182 (121.67%)**

In the first line, the annual operating hours of the CHP are considered. The aim is to achieve the highest possible value in order to justify the purchase of the CHP plant. As a rule, a CHP plant is dimensioned and operated in such a way that it reaches at least 4500 h/year. With the heat-led mode of operation, the CHP plant is above this value. This shows that the CHP plant is well dimensioned for the heat load it is intended to cover. The running time of the CHP plant, which is optimized for the electricity exchange, is about 5% below the heat-led mode, which is an acceptable value taking to account that it avoids times when the CHP plant is not profitable due to low electricity prices. With the tenant flow optimized mode of operation, even longer annual operating hours can be achieved, as the heat storage is better utilized. The optimized mode of operation on the CO_2 emission of the electricity mix with about 3500 h/year achieves very short annual operating hours. This is due to the fact that there are more often phases in which the CO_2 emission in the electricity mix is below that of a natural gas-fired CHP.

The second line indicates how much profit can be made compared to a pure boiler operation if all the electricity is sold on the electricity exchange. The comparison between heat-led and electricity exchange-optimized operation is of particular interest here. The optimized mode of operation achieves over 24% more profit compared to boiler operation than is possible with the heat-led mode of operation.

The third line indicates how much profit can be made if a tenant buys the electricity when he needs it for assumed 0.21 €/kWh and otherwise the electricity is sold to the distribution network operator at a fixed price (assumed 0.08 €/kWh). The comparison between heat-led operation and optimized schedule for tenant electricity is of particular interest here. The optimized mode of operation achieves over 10% more profit compared to boiler operation than is possible with the heat-led mode of operation. These values cannot be directly compared with those of the optimized operation scheduling for the electricity exchange, as no marginal costs for the boiler were deducted here.

The fourth line indicates how much CO_2 was saved by operating the CHP compared to obtaining the power from the grid with its specific electricity mix. All operating modes were able to achieve a certain degree of CO_2 savings, as the average CO_2 emission in the electricity mix was with our calculations with the ENTSO-E time series at 432.83 g/kWh for 2016 in the German grid. Nevertheless, the optimized mode of operation is almost 22% better than the heat-led mode of operation. The mode of operation optimized for the electricity exchange is also significantly better than the heat-led mode of operation. This is because the price on the electricity exchange is highly dependent on the supply of sun and wind. If the electricity price is low, it can therefore be assumed that renewable energies are currently supplying a lot of electricity and vice versa. This means that the electricity exchange's optimized mode of operation also indirectly optimizes the CO_2 in the electricity mix.

In summary, it can be pointed out that optimization to the electricity exchange and tenant flow are highly recommended if the electricity is to be distributed in one of these two ways. When optimizing the CO_2 emission of the electricity mix, this form of optimization is less attractive for private users due to the short running times for a natural gas-powered CHP, but could still be interesting for public utilities and transmission system operators who benefit from saving CO_2 emissions. Since past measurement data was used for all optimizations, no forecasts had to be made. If in practice, the data for heat load, ambient temperature, electricity price, tenant requirement etc. have to be predicted, the results must be expected to be somewhat worse due to forecast inaccuracies.

5   Conclusion

This paper works out that in many situations an optimized operation scheduling of a CHP plant is preferable compared to a heat-led operation mode. With all optimizations, the optimization objective was achieved at least 10% better than with the heat-led mode of operation. The optimized operation scheduling for the electricity exchange can even achieve up to 24% more profit over the year and even reduce stress on the grid indirectly.

ACO provides suitable mechanisms for finding an optimized operation schedule as quickly as possible and it allows to integrate non-linear factors in combination with a heat storage simulation model. In this way, it was shown that the competitiveness of ecological, energy-related components can be further increased compared to conventional power plants by developing optimized operational scheduling using advanced optimization algorithms.

References

1. Directive 2004/08/EC of the European parliament and of the council. Off. J. Eur. Union **47**(52), 50–60 (2004)
2. Chen, H., Yu, Y., Jiang, X.: Optimal scheduling of combined heat and power units with heat storage for the improvement of wind power integration. In: 8th IEEE PES Asia-Pacific Power and Energy Engineering Conference (APPEEC), pp. 1508–1512. IEEE (2016). https://doi.org/10.1109/APPEEC.2016.7779742
3. Mongibello, L., Graditi, G., Bianco, N., Musto, M., Caliano, M.: Optimal operation of residential micro-CHP systems with thermal storage losses modelling. In: 2014 International Symposium on Power Electronics, Electrical Drives, pp. 1027–1033. IEEE (2014). https://doi.org/10.1109/SPEEDAM.2014.6872090

4. Majic, L., Krzelj, I., Delimar, M.: Optimal scheduling of a CHP system with energy storage. In: 36th International Convention on Information and Communication Technology, Electronics and Microelectronics (MIPRO), pp. 1253–1257. IEEE (2013)

5. Sauter, A., Gerlach, J., Bringmann, O.: Simulationsbasierte analyse energietechnischer systemszenarien. In: 19th Workshop Methoden und Beschreibungssprachen zur Modellierung und Verifikation von Schaltungen und Systemen (MBMV), pp. 139–150. Freiburg (2016)

6. Blum, C.: Ant Colony Optimization: introduction and recent trends. Phys. Life Rev. **2**(4), 353–373 (2005). https://doi.org/10.1016/j.plrev.2005.10.001

7. Katiyar, S., Nasiruddin, I., Ansari, A.Q.: Ant Colony Optimization: a tutorial review. In: National Conference on Advances in Power and Control, pp. 99–110. Manav Rachna International University, Faridabad (2016)

8. Stephen, A.A., Misra, S.: A comprehensive study on the Ant Colony Optimization algorithms. In: 11th International Conference on Electronics, Computer and Computation (ICECCO), pp. 1–4. IEEE (2014). https://doi.org/10.1109/ICECCO.2014.6997567

9. Nowak, W., Arthkamp, J.: BHKW-Grundlagen, 1st edn. ASUE Arbeitsgemeinschaft, Berlin (2010)

10. Günther, M.: Energieeffizienz durch Erneuerbare Energien. Springer, Wiesbaden (2015). https://doi.org/10.1007/978-3-658-06753-3

11. Senertec: Dachs G/F und HR - Technical data sheet. https://www.senertec-info.de/downloads/. Accessed 4 Nov 2018

12. Mast, J., Rädle, S., Gerlach, J.: Multikriterien-Optimierung energietechnischer Komponenten unter Anwendung von Methoden der Künstlichen Intelligenz. In: 59th MPC-Workshop, pp. 59–66. IEEE German Section Solid-State Circuit Society (2018)

13. Andrychowicz, M., Przybylski, J.: Evaluation of the flexibility approach in construction of Scenario Outlook & Adequacy Forecast 2015 by ENTSO-E. In: 13th International Conference on the European Energy Market (EEM), pp. 1253–1257. IEEE (2016). https://doi.org/10.1109/EEM.2016.7521183

14. Landlord-to-tenant electricity supply. https://www.bmwi.de/Redaktion/EN/Artikel/Energy/landlord-to-tenant-electricity-supply.html. Accessed 5 Nov 2018

A Flexible Dissimilarity Measure for Active and Passive 3D Structures and Its Application in the Fitness–Distance Analysis

Maciej Komosinski$^{(\boxtimes)}$ and Agnieszka Mensfelt

Poznan University of Technology, Piotrowo 2, 60-965 Poznan, Poland
maciej.komosinski@cs.put.poznan.pl

Abstract. Evolutionary design of 3D structures – either static structures, or equipped with some sort of a control system – is one of the hardest optimization tasks. One of the reasons are rugged fitness landscapes resulting from complex and non-obvious genetic representations of such structures and their genetic operators. This paper investigates global convexity of fitness landscapes in optimization tasks of maximizing velocity and height of both active and passive structures. For this purpose, a new dissimilarity measure for 3D active and passive structures represented as undirected graphs is introduced. The proposed measure is general and flexible – any vertex properties can be easily incorporated as dissimilarity components. The new measure was compared against the previously introduced measure in terms of triangle inequality satisfiability, changes in raw measure values and the computational cost. The comparison revealed improvements for triangle inequality and raw values at the expense of increased computational complexity. The investigation of global convexity of the fitness landscape, involving the fitness–distance correlation analysis, revealed negative correlation between the dissimilarity of the structures and their fitness for most of the investigated cases.

Keywords: Evolutionary design · 3D structure ·
Dissimilarity measure · Optimization · Global convexity

1 Introduction

Three-dimensional structures can be divided into two main groups based on whether they have a system that actively controls their physical body, or they don't have such a system. Passive designs do not perform any action (they are static) and therefore only their physical structure exists and can be subject to evolutionary optimization – examples of such structures are an antenna [1],

The second author was supported by the Faculty of Computing, Poznan University of Technology, through the funds provided by the Ministry of Science and Higher Education.

ⓒ Springer Nature Switzerland AG 2019
P. Kaufmann and P. A. Castillo (Eds.): EvoApplications 2019, LNCS 11454, pp. 106–121, 2019.
https://doi.org/10.1007/978-3-030-16692-2_8

a bridge [2,3] or a truss [4]. Conversely, active designs do have a control system, and the system is evolved along with the body – such structures are often encountered in robotics [5–7] and in artificial life [8,9].

Designing 3D structures can be considered one of the hardest optimization problems for a number of reasons: the search space is usually infinitely big, the optimization process must combine both continuous and discrete aspects, and evaluation of solutions is costly, non-deterministic, and involves multiple criteria [10]. Additionally, a complex, non-obvious representation of solutions along with its genetic operators usually leads to highly rugged fitness landscapes [11,12]. There are many metaheuristic approaches that can be used for automated design, but evolutionary algorithms turned out to be the most successful [11]. Despite the progress in the area of evolutionary design, many aspects have to be still improved before it will be possible to routinely apply this technique for successful automated design of 3D structures.

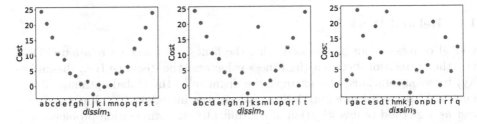

Fig. 1. Three sample fitness landscapes with the set of the same 20 solutions denoted by letters 'a' to 't' (in this work, these letters correspond to different 3D structures). Each of these fitness landscapes is induced by a different dissimilarity measure, from $dissim_1$ – the most convex, to $dissim_3$ – the most chaotic. For best performance of the optimization algorithm and the exploitation of the global convexity property during search, the neighborhood and reconfiguration operators should preserve the topology induced by $dissim_1$.

The difficulty of any optimization task relies heavily on the shape of its fitness landscape, with rugged landscapes generally corresponding to substantially harder optimization tasks than smooth landscapes [13,14]. For combinatorial optimization problems (e.g., for the traveling salesman problem), it was demonstrated that their fitness landscape may possess the property of global convexity [15]. If the fitness landscape exhibits this property, better solutions are more similar to each other and to the global optimum than they are similar to worse solutions. This property can be exploited to facilitate the process of evolutionary search. Note that the fitness landscape may possess the property of global convexity or not – this depends on how the landscape is constructed, i.e., on genetic operators that are used to traverse the landscape [16] (see Fig. 1). In order to construct a smooth, globally convex landscape, efficient genetic operators need to be designed. This is where the dissimilarity measure is helpful: it allows to evaluate the correlation between the similarity of solutions and the similarity of their fitness values. If this correlation is high (i.e., the dissimilarity measure captured some important characteristics of solutions), then designing genetic operators such that neighboring solutions in the search space are similar will result in a smooth fitness landscape.

Despite the successes in the combinatorial optimization domain, global convexity was not extensively researched so far in the area of optimization of active and passive 3D structures. One of the reasons is the difficulty of dissimilarity calculation for such structures. Therefore, the first aim of this paper is the introduction of a new measure for active and passive 3D structures. The second aim is the application of this measure in the analysis of global convexity of fitness landscape for different optimization tasks, such as maximizing the velocity of active 3D structures, and maximizing the height of passive 3D structures. The results of such analyses may help design genetic representations and operators that will increase the efficiency of solving the demanding tasks of evolutionary design.

1.1 Related Work

Global convexity can be assessed using the fitness–distance correlation (FDC), i.e., the correlation between the fitness value and the distance (i.e., dissimilarity) between solutions – for example the distance to the global optimum, if it is known. Fitness–distance correlation analysis was introduced by Jones and Forrest as a method of investigation of the difficulty of optimization problems [17]. As previously mentioned, global convexity depends on the specific distance (dissimilarity) measure used. Therefore, the FDC analysis using different dissimilarity measures can be employed to identify properties of solutions that correlate with their fitness, and then to devise genetic operators that preserve such properties. One example of such an operator is the distance-preserving crossover (DPX). Such an operator attempts to guarantee that the distance between each parent solution and a child solution is not higher than the distance between parents [18,19]. So far, global convexity tests were successfully used for the development of distance-preserving crossover operators in combinatorial optimization problems [20–22].

Apart from the FDC analysis, dissimilarity measures for 3D structures have plenty of other applications [10], including automated classification, discovering clusters in solutions, population analysis and inferring dendrograms. Dissimilarity measures for 3D structures already exist in many domains such as computer vision [23], bioinformatics [24] or chemical informatics [25], but they are usually too domain-specific to be applied to comparing arbitrary 3D designs. Additionally, the measure should take into account not only the "body" of the solution, but in the case of active structures, also their control systems. Moreover, the algorithm for calculating the measure needs to be efficient to handle complex 3D structures. The dissimilarity measure introduced by Komosinski and Kubiak [26] was designed to fulfill these demands, however it has a number of disadvantages described in Sect. 2.3.

2 Dissimilarity Measure for Active and Passive 3D Structures

The dissimilarity measure introduced in this paper was designed to overcome the problems of the simpler measure devised earlier [26]. The calculation algorithms of both measures, as well as a model of a 3D structure required by the measures, are described in the following sections. The proposed measure has been implemented in the Framsticks simulation environment [9,27] and the C++ source code is available as a part of the Framsticks SDK [28]. The value calculated by the algorithm for two structures is interpreted as follows:

- 0 means that both structures are identical (i.e., no dissimilarity),
- A positive value reflects the dissimilarity (the "distance") between both structures.

2.1 3D Active Structure Model

The 3D structures considered in this work are modeled as undirected graphs (Fig. 2), and such structures are simulated in the Framsticks environment [9,27]. The structures can be either active or passive. Active structures are equipped with working control units – artificial neurons – including sensors and effectors. The neurons may be optionally attached to vertices. Therefore, each vertex in a structure can be described with the following properties:

- Its degree (i.e., the number of edges incident to the vertex),
- The number of neurons attached to the vertex,
- Its position in the three-dimensional coordinate system.

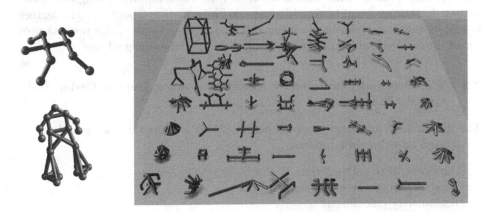

Fig. 2. Sample 3D structures compatible with the model considered in this work. Left: a close-up of two structures with visible vertices and edges. Right: an 8 × 8 sample of structures from the $w+s+o$ test set described in Sect. 2.5.

2.2 Dissimilarity Measure Assumptions

The measure should allow for the comparison of different properties of any two 3D structures. Therefore, the value of the measure consists of four components:

- d_V – the absolute difference in the number of vertices in both structures,
- d_D – the absolute difference in the degree of matched vertices,
- d_N – the absolute difference in the number of neurons attached to the matched vertices,
- d_G – the Euclidean distance between matched vertices.

Since in the simplest case these components are aggregated into a single value using the weighted sum, the user can adjust the importance of each component by setting the weight (w_V, w_D, w_N and w_G) of this component to a value higher than or equal to zero.

If the w_G weight is higher than zero, the two structures should be aligned in 3D space before calculating Euclidean distances between matched vertices. For this purpose, a multidimensional scaling [29] procedure (MDS) has been used for each structure separately. After the application of this procedure, centers of both structures are located in the origin of the coordinate system. The axis with the highest variance of coordinates is chosen as the first axis of the structure, and the axis with the second highest variance of coordinates is chosen as the second axis of the structure.

Since a standard MDS procedure takes as an input the distance matrix based on original vertex coordinates, vertex degrees are not considered during spatial alignment of structures. As a result, vertices with a similar vertex degree in both structures may not be aligned properly. To overcome this problem, instead of the standard MDS procedure, we use the weighted MDS (wMDS). In wMDS, during the alignment, the distance matrix between vertex coordinates is weighted using vertex degrees as weights. In this way, the information about vertex degrees is incorporated and exploited during the alignment process – this process uses richer information which ultimately leads to better alignment of the structures.

2.3 The Original Dissimilarity Measure: Vertex Degree Order and Greedy Matching

The dissimilarity measure proposed earlier [26] is a heuristic. Its algorithm can be divided into three main parts:

- Alignment of the structures,
- Construction of the matching function,
- Dissimilarity calculation.

The alignment procedure has been described in Sect. 2.2. The main part of the algorithm is the construction of the matching function. In order to build the matching, the vertices of each structure are sorted according to the vertex degree, and then according to the number of neurons within the groups of the same vertex degree. The matching procedure starts with vertices with the highest vertex degree in both structures, and tries to find the pairs of matching vertices. The within-group matching ends when there are no unmatched vertices with a given degree in one or both of the structures. Then the algorithm proceeds to handle the group of vertices with the second highest vertex degree, and then continues and handles groups with lower and lower vertex degree. Within groups with the same vertex degree, the matching is built according to the minimum distance calculated as

$$w_D \cdot d_D(v_{i1}, v_{j2}) + w_N \cdot d_N(v_{i1}, v_{j2}) + w_G \cdot d_G(v_{i1}, v_{i2})$$

where weight w_x corresponds to the x component of the measure, v_{i1} denotes the i-th vertex of the first structure, and v_{j2} denotes the j-th vertex of the second structure. For details of the procedure, see [26].

One problem of this measure is the fixed order in which the vertices are matched. Even if the weight of the d_D component is equal to zero, the matching procedure still starts from the vertices with the highest vertex degree and follows the logic described above. Another issue with the matching procedure is that it is a greedy algorithm. It always chooses the matching which provides the minimal distance between currently considered vertices, however this choice does not have to result in the minimal overall distance between complete structures. This algorithm will be referred to in the following sections as $dissim_{DegGreedy}$ (vertex degree order and greedy matching).

2.4   The Improved Dissimilarity Measure: Flexible Criteria Order and Optimal Matching

The dissimilarity measure proposed in this paper overcomes the disadvantages of $dissim_{DegGreedy}$ described above. The calculation of the improved dissimilarity measure is also preceded by the alignment procedure described in Sect. 2.2. In order to avoid the greediness of the matching procedure, the Kuhn-Munkres algorithm [30, 31] (also known as the Hungarian algorithm) is applied. The goal of the matching procedure is to find the matching of vertices that will minimize the overall distance between two structures. The overall distance consists of two components: the sum of the distances between matched vertices, and the penalty for the unmatched vertices (only present in the case of the structures differing in the number of vertices). The distance between each pair of matched vertices is calculated as follows:

$$dist_{v_{i1}v_{j2}} = w_D \cdot d_D(v_{i1}, v_{j2}) + w_N \cdot d_N(v_{i1}, v_{j2}) + w_G \cdot d_G(v_{i1}, v_{i2})$$

The penalty for each unmatched vertex v_i is the sum of the following components:

- $penalty_D(v_i) = w_D \cdot \text{vertex\_degree}(v_i)$,
- $penalty_N(v_i) = w_N \cdot \text{number\_of\_neurons}(v_i)$,
- $penalty_G(v_i) = w_G \cdot \text{distance\_to\_the\_origin}(v_i)$.

In order to take the penalty into account during the Hungarian matching procedure, additional rows or columns are created in the distance matrix for the smaller structure. Such additional rows or columns are filled with the penalty for the inability to match the parts of the structure with more vertices. This dissimilarity measure will be referred to as $dissim_{FlexOpt}$ (flexible criteria order and optimal matching). The matching procedure along with the distance calculation procedure is outlined in Listing 1.1.

```
GS = structure with more vertices ("greater")
SS = structure with less vertices ("smaller")
nGreat = vertices_number(GS)
nSmall = vertices_number(SS)
dist_matrix = matrix(size=(nGreat, nGreat))
for i in range(0, nGreat-1):
  for j in range(0, nGreat-1):
    if i >= nSmall:
          dist_matrix[i][j] = penalty(v_GSj)
      else:
          dist_matrix[i][j] = dist(v_SSi,v_GSj)
matched = HungarianAlgorithm(dist_matrix)
distance = ∑_i ∑_j dist(matched(v_SSi,v_GSj))
distance = distance + difference_in_unattached_neurons(GS, SS)
distance = distance + w_V · d_V(GS,SS)
```

Listing 1.1. The outline of the $dissim_{FlexOpt}$ vertex matching and dissimilarity measure calculation algorithm.

The measure described above is very general and can be used for calculation of dissimilarity of any 3D objects that can be represented as undirected graphs. Furthermore, $dissim_{FlexOpt}$ is more flexible than $dissim_{DegGreedy}$. In $dissim_{DegGreedy}$, some of the components were more important than others independently of the weight values – vertex degree and neuron count were always used to sort the vertices before matching. In $dissim_{FlexOpt}$, all of the components are processed in a uniform way. This allows any properties of the vertices in the model to be easily incorporated into dissimilarity calculation as subsequent measure components. The proposed measure will be referred to in the following sections as $dissim_{FlexOpt}$ (flexible criteria order and optimal matching). The matchings of the same two 3D structures obtained using $dissim_{DegGreedy}$ and $dissim_{FlexOpt}$ algorithms are compared in Fig. 3.

2.5   Comparison of Dissimilarity Measures

Both measures, the original one ($dissim_{DegGreedy}$) and the one introduced in this paper ($dissim_{FlexOpt}$), were compared on four different sets of three-dimensional models using three evaluation criteria. For each of the test sets, the dissimilarity was calculated for all pairs of the structures in the set – for example, for a set with 400 structures, the dissimilarity was computed $400 \times 400 = 160\,000$ times.

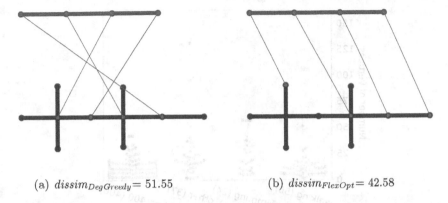

(a) $dissim_{DegGreedy} = 51.55$ (b) $dissim_{FlexOpt} = 42.58$

Fig. 3. The matching and dissimilarity between the same two structures obtained using $dissim_{DegGreedy}$ and $dissim_{FlexOpt}$ measures with all the weights equal, $w_V = w_D = w_N = w_G = 1$. The $dissim_{DegGreedy}$ matching procedure (a) starts from vertices with the highest vertex degree in both structures. Therefore, vertices with the degree of 4 from the bottom structure have to be matched with vertices with the degree of 2 from the top structure, and then vertices with the degree of 2 from the bottom structure have to be matched with vertices with the degree of 1 from the top structure. In the case of $dissim_{FlexOpt}$ (b), the matching procedure tries to find the matching that minimizes the total dissimilarity value.

Test Sets. Four test sets of 3D structures were used. Three of these sets (*walking*, *swimming* and *other*) are sample sets provided in the Framsticks distribution [27] and consist mostly of structures designed by humans, or pre-designed manually and then evolved to meet some specific goal. The fourth set (*best400*) is the result of an evolutionary experiment, with four fitness criteria and 100 structures optimized for each criterion using evolutionary algorithms [10] (the visualization of this experiment is available at https://www.youtube.com/watch?v=lo4vL7gOuYk). The test sets differ in the number of structures they contain and in the size (i.e., the number of vertices) of the structures. The distribution of the structure size in each test set is shown in Fig. 4.

Comparison Criteria and Results. Both measures – $dissim_{DegGreedy}$ and $dissim_{FlexOpt}$ – were compared using three criteria; these criteria are described below.

Triangle inequality violation. It is desirable for a dissimilarity measure to be a metric – with this property, it is possible to construct a metric space for the set of analyzed structures. In order to be a metric, the measure must satisfy the conditions of non-negativity, identity of indiscernibles, symmetry, and triangle inequality. It can be easily shown that both of the analyzed measures satisfy the first three conditions.

The satisfiability of the triangle inequality condition has been tested computationally. For this purpose, the three sets: *walking*, *swimming*, and *other* were

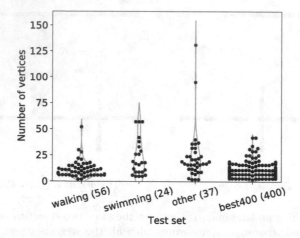

Fig. 4. The distribution of the number of vertices in 3D structures for each test set. The number in parentheses after the name of the test set indicates the number of structures in that set.

merged into one set denoted as $w+s+o$. In order to test the influence of different components and their combinations on the triangle inequality violations, all possible combinations of binary weights: $w_V, w_D, w_N, w_G \in \{0, 1\}$ were tested. The results of the investigation on triangle inequality are presented in Table 1. It can be seen that the number of weight sets for which triangle inequality was violated is lower for $dissim_{FlexOpt}$. Also, the percentage of the cases for which violation occurred is significantly lower for $dissim_{FlexOpt}$ than for $dissim_{DegGreedy}$. For the cases in which only one component of the dissimilarity measure was taken into account, violations occurred only for the geometrical distance component using $dissim_{FlexOpt}$.

It is worth noting that for $dissim_{FlexOpt}$, the number of non-zero triangle inequality violations is the lowest when all four components of the dissimilarity measure are taken into account. These results suggest that the number of violations can be decreased by using all four components of the measure, i.e., by exploiting all the information about compared 3D structures the measure can access.

Computational cost. The results of time measurement for calculation of the full square distance matrix for different test sets are shown in Fig. 5. In all cases, the calculation time for $dissim_{FlexOpt}$ is higher than for $dissim_{DegGreedy}$. The ratio of both calculation times is the highest for test sets containing structures with higher number of vertices – the calculation time for $dissim_{FlexOpt}$ is almost 4 times longer for *other*, and almost 3 times longer for *swimming* test set.

Values of dissimilarity measures. Both measures were also compared in terms of the raw distance (dissimilarity) value – lower dissimilarity values suggest that the measure is able to find a better match between the structures. For this purpose,

Table 1. The percentage of triangle inequality violations for the original and the new dissimilarity measure and for different sets of weights. Combinations of weights for which there were no triangle inequality violations in both measures are not shown.

Test set	w_V	w_D	w_N	w_G	$dissim_{DegGreedy}$ [%]	$dissim_{FlexOpt}$ [%]
$w+s+o$	0	0	1	0	0.479	0.0
	0	0	0	1	0.158	0.004
	1	0	1	0	0.271	0.0
	1	0	0	1	0.154	0.0
	0	1	1	0	0.131	0.0
	0	1	0	1	0.123	0.010
	0	0	1	1	0.106	0.018
	1	1	0	1	0.115	0.021
	0	1	1	1	0.090	0.011
	1	1	1	0	0.130	0.0
	1	0	1	1	0.102	0.013
	1	1	1	1	0.089	0.003
$best400$	0	0	1	0	1.503	0.0
	0	0	0	1	2.604	0.810
	1	0	1	0	1.112	0.0
	1	0	0	1	2.549	0.807
	0	1	1	0	0.928	0.0
	0	1	0	1	2.352	0.794
	0	0	1	1	1.438	0.267
	1	1	0	1	2.566	0.788
	0	1	1	1	1.390	0.248
	1	1	1	0	0.927	0.0
	1	0	1	1	1.510	0.257
	1	1	1	1	1.277	0.246

again, the three test sets: *walking, swimming* and *other* were merged into one set denoted as $w+s+o$. The number of negative, zero, and positive differences was calculated for distance matrices obtained using $dissim_{DegGreedy}$ and $dissim_{FlexOpt}$ with different weight combinations.

The results of this analysis are shown in Fig. 6. It can be seen that there were no positive differences, which means that distances obtained using $dissim_{FlexOpt}$ were always equal or lower than the distances obtained using $dissim_{DegGreedy}$. Distances obtained using both measures are the same only when the d_G and the d_N components are not taken into account. The number of negative differences is the highest when the d_G is included as a component of the measure. These results demonstrate that $dissim_{FlexOpt}$ can yield lower distance values

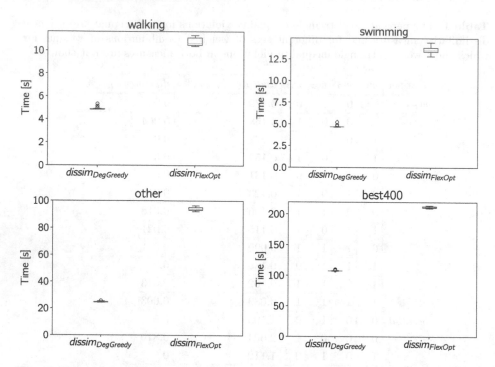

Fig. 5. Time of the full distance matrix calculation for each of the four test sets and both dissimilarity measures. All tests were performed on a computer with the Ubuntu Linux OS, the Intel Core i5-4200U processor and 4 GB of RAM.

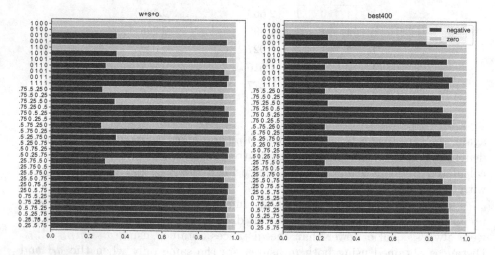

Fig. 6. The fraction of negative (black) and zero (gray) differences between the new and the previous measure for different test sets and different weight values w_V, w_D, w_N, and w_G. The differences were calculated as $dissim_{FlexOpt} - dissim_{DegGreedy}$.

then $dissim_{DegGreedy}$ in terms of the d_G and the d_N components, and such lower dissimilarities are the consequence of a better matching.

3   Fitness–Distance Correlation Analysis

As mentioned in Sect. 1.1, the value of the fitness–distance correlation can be used to assess the global convexity of the fitness landscape. The results of the global convexity analysis can later be used to guide the design of efficient genetic representations and operators.

Both dissimilarity measures, $dissim_{DegGreedy}$ and $dissim_{FlexOpt}$, were used for fitness–distance correlation analysis, for different fitness functions, as shown in Fig. 7. Since the global optimum is not known and there are many local optima, mean distance to structures with the same or better fitness was computed, instead of the distance to the optimal structure. The analysis was conducted on four subsets of $best400$, each consisting of 100 structures evolved using one of the following maximized fitness functions:

- Velocity on land (active structures: neural network activated),
- Velocity in water (active structures: neural network activated),
- Height (passive structures: no neural network),
- Height (active structures: neural network activated).

For velocity on land, all the correlations were negative, meaning that the structures most similar to better or equally good structures (in terms of fitness) had in general higher fitness. This was the case for all three components, with the lowest correlation strength for d_N and the highest for d_D. This result suggests that velocity on land is correlated the most with the dissimilarity of the vertex degree.

For velocity in water, negative correlations were even stronger for d_G and d_N, showing correlation of dissimilarity in terms of these components with the fitness value. Surprisingly, for the d_N component, weak positive correlations were obtained.

Even more surprisingly, for maximizing the height of passive structures, high positive correlations were revealed for the d_G and the d_D components (since passive structures were considered, the d_N component was not taken into account). In this seemingly simple task, no global convexity was discovered. On the contrary, the more dissimilar the structure was to the better or equally fit structures, the higher was its fitness. This suggests the need to develop another dissimilarity measure (or another component of the measure) to help improve the efficiency of optimizing the height of passive structures. Such improvement would be possible if properties considered in the measure whose dissimilarity correlates with fitness were preserved by genetic operators that modify solutions during optimization.

For maximizing the height of active structures, moderate to high negative correlations were obtained. One difference is the correlation for the d_G component using $dissim_{DegGreedy}$, for which the correlation strength is near zero. Interestingly, correlation is significantly stronger for the d_G component using $dissim_{FlexOpt}$.

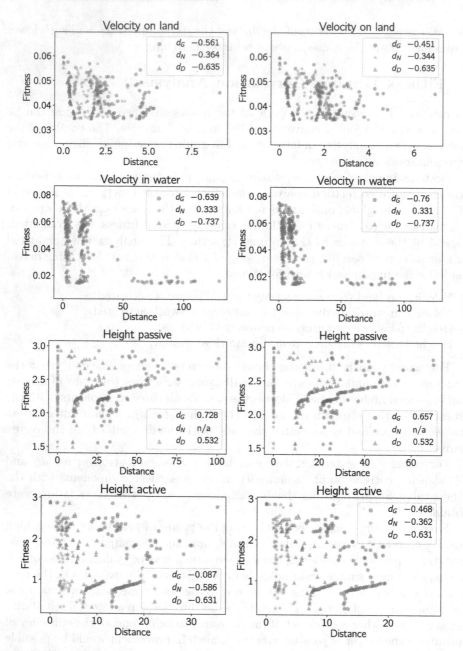

Fig. 7. The relationship between fitness value of the structure and a mean distance (dissimilarity) to structures with the same or better (i.e., higher) fitness value. Results for each component of the dissimilarity measure are presented separately on each plot, except for the d_V component, which is not taken into account. Spearman's r_s rank correlation coefficient values are shown in the legend for each of the components. Results obtained using $dissim_{DegGreedy}$ are shown in the left column, and results obtained using $dissim_{FlexOpt}$ are shown in the right column.

4   Conclusions and Further Work

The $dissim_{FlexOpt}$ measure introduced in this paper is an improvement over $dissim_{DegGreedy}$ in terms of the satisfiability of triangle inequality. It is not perfect, however – violations of this important property still occur, albeit very rarely. The measure can be further improved by eliminating the triangle inequality violation. Results of the analysis suggests that the geometrical distance component may be the main source of the violations. Therefore, the solution to this problem could be the change of the geometrical distance penalty for unmatched vertices. However, the penalty for the geometrical component is not so obvious as for the remaining components of dissimilarity measures. Currently, the penalty for the geometrical component is calculated as the distance of the unmatched vertex to the origin of the coordinate system; nonetheless, there are other possible approaches, some of which may fulfill the satisfiability of the triangle inequality condition.

In terms of the measure value, $dissim_{FlexOpt}$ introduced in this work is also an improvement over $dissim_{DegGreedy}$. The change of a greedy method of matching to the optimal matching lowered primarily the value of the geometrical distance component. Lower measure value indicates that $dissim_{FlexOpt}$ is able to find a better matching of both compared structures than $dissim_{DegGreedy}$. However, the price for those improvements is the increase in computational cost. While the execution time is still reasonable for sets containing structures with a moderate mean number of vertices, it may be prohibitive for data sets comprised of very complex structures.

Another advantage of the $dissim_{FlexOpt}$ measure is its flexibility. The earlier matching procedure based on the sorting of the vertices according to their vertex degree and neuron count was replaced by the Hungarian algorithm. Because of this change, all of the components of the dissimilarity measure are now processed uniformly. In consequence, the measure can take into account any property of the vertices and the influence of this property on the dissimilarity value will be proportional to the corresponding weight.

The fitness–distance correlation analysis revealed global convexity for most of the considered optimization tasks, as long as the vertex degree and the geometrical distance components were employed in the dissimilarity measure. While these results are promising, some unexpected relationships were discovered – like positive correlation between dissimilarity of the number of neurons and the velocity in water, or strong positive correlation between geometrical dissimilarity and the height of passive structures. These discoveries indicate that the development of other dissimilarity measures for 3D structures would be beneficial, as such measures would likely help identify properties of the optimized structures that yield high FDC. Incorporating such properties in the development of genetic representations and operators will in turn increase the efficiency of the optimization process in the area of evolutionary robotics and evolutionary design.

References

1. Hornby, G.S., Globus, A., Linden, D.S., Lohn, J.D.: Automated antenna design with evolutionary algorithms. In: AIAA Space, pp. 19–21 (2006)
2. Funes, P., Pollack, J.: Evolutionary body building: adaptive physical designs for robots. Artif. Life **4**(4), 337–357 (1998)
3. Byrne, J., et al.: Combining Structural Analysis and Multi-Objective Criteria for Evolutionary Architectural Design. In: Chio, C., et al. (eds.) EvoApplications 2011. LNCS, vol. 6625, pp. 204–213. Springer, Heidelberg (2011). https://doi.org/10.1007/978-3-642-20520-0_21
4. Kicinger, R., Arciszewski, T., DeJong, K.: Evolutionary design of steel structures in tall buildings. J. Comput. Civil Eng. **19**(3), 223–238 (2005)
5. Lipson, H., Pollack, J.B.: Automatic design and manufacture of robotic lifeforms. Nature **406**(6799), 974–978 (2000)
6. Nolfi, S., Floreano, D.: Evolutionary Robotics: The Biology, Intelligence, and Technology of self-Organizing Machines. MIT Press, Cambridge (2000)
7. Corucci, F.: Evolutionary developmental soft robotics: towards adaptive and intelligent soft machines following nature's approach to design. In: Laschi C., Rossiter J., Iida F., Cianchetti M., Margheri L. (eds) Soft Robotics: Trends, Applications and Challenges. Biosystems & Biorobotics, vol 17, pp. 111–116. Springer, Cham (2017). https://doi.org/10.1007/978-3-319-46460-2_14
8. Sims, K.: Evolving 3D morphology and behavior by competition. Artif. Life **1**(4), 353–372 (1994)
9. Komosinski, M., Ulatowski, S.: Framsticks: Creating and Understanding Complexity of Life, 2nd edn. pp. 107–148. Springer, New York (2009). http://www.springer.com/978-1-84882-284-9. Chapter 5
10. Komosinski, M.: Applications of a similarity measure in the analysis of populations of 3D agents. J. Comput. Sci. **21**, 407–418 (2017)
11. Bentley, P.J.: An introduction to evolutionary design by computers. In: Bentley, P.J. (ed.) Evolutionary Design by Computers, 1st edn. Morgan Kaufmann Publishers Inc., San Francisco (1999)
12. Hornby, G.S., Pollack, J.B.: Generative representations for evolutionary design automation. Brandeis University (2003)
13. Smith, T., Husbands, P., O'Shea, M.: Fitness landscapes and evolvability. Evol. Comput. **10**(1), 1–34 (2002)
14. He, J., Reeves, C., Witt, C., Yao, X.: A note on problem difficulty measures in black-box optimization: classification, realizations and predictability. Evol. Comput. **15**(4), 435–443 (2007)
15. Boese, K.D.: Cost versus distance in the traveling salesman problem. UCLA Computer Science Department (1995)
16. Pitzer, E., Affenzeller, M.: A comprehensive survey on fitness landscape analysis. In: Fodor J., Klempous R., Suárez Araujo C.P. (eds) Recent Advances in Intelligent Engineering Systems, vol. 378, pp. 161–191. Springer, Heidelberg (2012). https://doi.org/10.1007/978-3-642-23229-9_8
17. Jones, T., Forrest, S.: Fitness distance correlation as a measure of problem difficulty for genetic algorithms. In: Proceedings of the 6th International Conference on Genetic Algorithms, pp. 184–192. Morgan Kaufmann Publishers Inc., San Francisco (1995)
18. Freisleben, B., Merz, P.: New genetic local search operators for the traveling salesman problem. Parallel Problem Solving from Nature - PPSN IV, pp. 890–899 (1996)

19. Freisleben, B., Merz, P.: A genetic local search algorithm for solving symmetric and asymmetric traveling salesman problems. In: Proceedings of IEEE International Conference on Evolutionary Computation, pp. 616–621. IEEE (1996)
20. Jaszkiewicz, A., Kominek, P.: Genetic local search with distance preserving recombination operator for a vehicle routing problem. Eur. J. Oper. Res. **151**(2), 352–364 (2003)
21. Jaszkiewicz, A., Kominek, P., Kubiak, M.: Adaptation of the genetic local search algorithm to a car sequencing problem. In: 7th National Conference on Evolutionary Algorithms and Global Optimization, Kazimierz Dolny, Poland, pp. 67–74 (2004)
22. Kubiak, M.: Systematic construction of recombination operators for the vehicle routing problem. Found. Comput. Decis. Sci. **29**(3) (2004)
23. Cyr, C.M., Kimia, B.B.: 3D object recognition using shape similiarity-based aspect graph. In: Proceedings Eighth IEEE International Conference on Computer Vision, ICCV 2001, vol. 1, pp. 254–261 (2001)
24. Barthel, D., Hirst, J.D., Błażewicz, J., Burke, E.K., Krasnogor, N.: ProCKSI: a decision support system for protein (structure) comparison, knowledge, similarity and information. BMC Bioinform. **8**(1), 416 (2007)
25. Bero, S.A., Muda, A.K., Choo, Y.H., Muda, N.A., Pratama, S.F.: Similarity measure for molecular structure: a brief review. J. Phys. Conf. Ser. **892**(1), 012015 (2017). http://stacks.iop.org/1742-6596/892/i=1/a=012015
26. Komosinski, M., Kubiak, M.: Quantitative measure of structural and geometric similarity of 3D morphologies. Complexity **16**(6), 40–52 (2011)
27. Komosinski, M., Ulatowski, S.: Framsticks web site (2019). http://www.framsticks.com
28. Komosinski, M., Ulatowski, S.: Framsticks SDK (Software Development Kit). http://www.framsticks.com/sdk
29. Cox, T.F., Cox, M.A.A.: Multidimensional Scaling. Chapman and Hall/CRC, New York (2000)
30. Kuhn, H.W.: The hungarian method for the assignment problem. Nav. Res. Logist. Q. **2**(1–2), 83–97 (1955)
31. Munkres, J.: Algorithms for the assignment and transportation problems. J. Soc. Ind. Appl. Math. **5**(1), 32–38 (1957)

Games

Free Form Evolution for Angry Birds Level Generation

Laura Calle[1], Juan J. Merelo[1(✉)], Antonio Mora-García[1],
and José-Mario García-Valdez[2]

[1] CITIC, Universidad de Granada, Granada, Spain
laucalle09@gmail.com, {jmerelo,amorag}@ugr.es
[2] Instituto Tecnológico de Tijuana, Calzada Tecnológico, s/n, Tijuana, Mexico
mario@tectijuana.edu.mx

Abstract. This paper presents an original approach for building structures that are stable under gravity for the physics-based puzzle game Angry Birds, with the ultimate objective of creating levels with the minimum number of constraints. This approach consists of a search-based procedural level generation method that uses evolutionary algorithms. In order to evaluate the stability of the levels, they are executed in an adaptation of an open source version of the game called *Science Birds*. In the same way, an open source evolutionary computation framework has been implemented to fit the requirements of the problem. The main challenge has been to design a fitness function that, first, avoids if possible the actual execution of the simulator, which is time consuming, and, then, to take into account the different ways in which a structure is not structurally sound and consider them in different ways to provide a smooth landscape that eventually achieves that soundness. Different representations and operators have been considered and studied. In order to test the method four experiments have been carried out, obtaining a variety of stable structures, which is the first path for the generation of levels that are aesthetically pleasing as well as playable.

Keywords: Search-based Procedural Content Generator ·
Evolutionary algorithm · Game development · Angry Birds ·
Level generation

1 Introduction

Angry Birds is a mobile game created by the Finnish company Rovio Entertainment Corporation [1], first launched in 2009. In the game, there is a variety of defensive structures made out of blocks which house *pigs* (which we can call *pigsties* for that reason) and the player has to fire birds from a slingshot so the structure is destabilized or, in some cases, destroyed. What is interesting to us is how heavily the game relies in gravity to create interesting puzzles, making the structures very close to reality in its dynamic behavior. The main challenge from the Procedural Content Generation (PCG) perspective is to build

© Springer Nature Switzerland AG 2019
P. Kaufmann and P. A. Castillo (Eds.): EvoApplications 2019, LNCS 11454, pp. 125–140, 2019.
https://doi.org/10.1007/978-3-030-16692-2_9

stable structures that are robust enough to take more than a single shot before crumbling to the ground. Prior to that they have to obviously stand on their own and not crash immediately, even before a single shot has been fired.

There is a number of competitions centered around playing AI agents for specific games or genres, but it is less common to find PCG competitions like the *Angry Birds Level Generation Competition*, whose third edition was celebrated in 2018 during the conference on Computational Intelligence and Games (CIG) [2]. Participants must build computer programs that are able to generate levels for the *Angry Birds* game; these levels must first be valid (i.e. not crash), and then, they are evaluated and judged for playability and aesthetics.

The ultimate objective of the work presented in this paper is to build a program capable of creating levels for *Angry Birds* and which would be able to eventually compete in the above mentioned competitions. We can break it down to the following, more concise, objectives:

- Explore the expressiveness and variability of Search-Based PCG (SBPCG) using evolutionary algorithms (EA).
- Adapt the game code to extract data from execution to evaluate the levels.
- Produce stable structures under gravity.

In this paper we will focus on these objectives, with sight on producing free-form structures that do not collapse; in a nutshell, we will be describing a PCG method that efficiently generates free form and stable Angry Birds *pigsty* structures.

For the purpose of this paper, we take the definition of Procedural Game Generation (PCG) as *the algorithmic creation of game content with limited or indirect user input* [3]. Aesthetic elements, game rules, levels, items, stories and characters among others are considered content in this definition [3]. But the fast pace at which the game industry grows poses a challenge to developers. How do we create a vast amount of content that suits players' expectations with low investment? PCG can tackle this by increasing replayability, offering adaptive content or reducing designers' workload [4].

In this paper we target replayability by insisting in the generation of free-form structures. If all content generated follow a simple pattern, this is easily spotted and the player disengages. By creating structures whose form is only constrained by gravity, levels generated will be much more engaging.

There are many PCG methods and it is necessary to look at some traits that characterize and differentiate them from each other [4]. The method proposed in this paper is described as *offline*, *necessary*, *generic*, *stochastic generate-and-test* and potentially be either *automatic* or *mixed authorship*. From all these features, the actual challenge in the game we are occupied with in this paper, Angry Birds, is to generate potentially correct solutions.

A special kind of *generate-and-test* approach to PCG is Search-based Procedural Content Generation (SBPCG), which is usually tackled with Evolutionary Algorithms [5] like the one proposed in [6]. The problems faced by SBPCG are not very far from those encountered in Evolutionary Algorithms; since evolutionary algorithms are search methods, they can be a good match to perform this kind of procedures.

Since in Angry Birds birds are launched against structures that include pigs, we will be searching for these structures, whose playability and appearance has to be optimized. We try to address playability by eliminating the constraint that they must follow a pattern, and do not really address aesthetics in this paper, not even constraining the structures to be symmetric. However, the free form implies that structural integrity is not guaranteed, and the structures have a realistic gravity, so we must avoid its collapse when they are erected at the beginning of a level. We will have to include this factor into the *fitness function*, which grades how good a solution is, or, as it might be the case, how far away it is from actually being good. In SBPCG, how to evaluate the quality of a solution has no straightforward answer. It requires formalizing as an evaluation function how much fun, exciting or engaging certain content is, which are usually based in subjective assumptions. However, before being fun, the generated content must be valid; this is the main issue we are tackling in this paper.

There are three main classes of fitness functions in SBPCG [5], but we will use a *direct fitness function*. In this kind of function, we extract measurable features from the generated content and map them to a fitness value.

Since we are interested in structural integrity of the generated structures, neither actual gameplay nor players are taken into account. However, this is a time-consuming procedure since it actually involves the graphic representation of the structure and application of falling motion to its different parts. If we have to subject every single individual in the population to this, the size of the search space (which is huge since it is free form) explored is going to be very small. So we will have to change it to minimize the actual number of structures that are simulated by applying heuristics to the data structure and assigning it a fitness even before simulation.

The fact that we are focusing on structures with a free form also implies that the data structure we will evolve has to take this into account. However, there is no single way to create this representation from first principles, so several options will have to be evaluated.

The rest of this paper is organized as follows: next section presents the state of the art in this type of level generation, as well as its relationship to the actual problem of generation of structurally sound structures. The problem of generating Angry Birds levels is described next in Sect. 3. Next experiments are presented in Sect. 4. We present our conclusions in Sect. 5.

2 Background and State of the Art

PCG is becoming more frequently used in new games as one of the main tools for supporting designers and developers. It has a growing relevance which has been also translated into the creation of new international competitions aimed to design algorithms to generate interesting/funny/enjoyable contents for different games, such as Super Mario (a clone of it indeed) [7], not specified General Video Games [8,9], or recently, for Angry Birds [10].

Obviously, all the entries presented in previous editions of this last competition are strongly related with this work. Unfortunately just a few authors have published their proposals in conferences or journals. The latest edition [10] was focused on finding the best entry generating entertaining levels. Fun was the main factor in the evaluation of the participants; creativity and difficulty were then also taken into account as secondary factors. Six entries competed on it, being the most representative the one by J. Yuxuan et al., able to generate random quotations and formulae with the different components of a level; J. Xu et al. proposal which generates levels that look like pixel images; or a third approach (by C. Kocaogullar) that generates levels by translating music patterns to structures.

The winner was a submission called Iratus Aves, an new iteration of the works by Stephenson and Renz [11,12], which is a *constructive method* where the structures displayed on the level are built from top to down, in several phases, recursively building a structure, each row composed of a single type of block (with a fixed rotation). The likelihood of selecting a certain block is given by a probability table, which was tuned using an optimization method. Then, the blocks are placed using a tree structure, where the first selected block is the peak and blocks underneath it are split into subsets that support the previous row with one, two or three blocks. This ensures local stability, but not global stability, which is tested once the whole structure is completed. After that, other objects (pigs and TNT) are placed, following an scheme of potential positions for them according to a rank based on some terms such as structural protection or dispersion.

This approach was improved by adding a 'layer' devoted to the selection of the material (stone, ice or wood) based on different strategies. Materials are important given that certain birds are more efficient against particular ones, so they will have a high impact on the game dynamics. Thus, a strategy sets stone to detected weak points. Trajectory analysis-based strategy sets to the same material all blocks in the trajectory of a shot aimed at a particular pig. Clustering strategy takes a random block, sets its material and propagates to the surrounding blocks that have not been assigned yet. Row grouping and structure blocking apply the same material to a whole row or structure respectively.

The main problem with this and other constructive approaches is that the range of different structures created is going to be relatively small; monotony leads to boredom, decreasing playability. On the other hand, generated structures are guaranteed to be structurally sound, and constructive approaches are generally faster than search-based procedures.

Search-based approaches propose a good alternative to deal with these limitations. Thus, Ferreira and Toledo [13] presented a solution – ranked 4th on the last competition – based on SBPCG, which uses a Genetic Algorithm (GA) and a game clone named *Science Birds* developed to evaluate the levels, the same one we use in this paper.

In a GA individuals correspond to levels, each represented by an array of columns. Each column is a sequence of blocks, pigs and predefined compound blocks, using an identification integer. This representation also includes the distances between different columns. The population is initialized randomly following a probability table which defines the likelihood of a certain element being placed in a certain position inside a column. This implies that a shape is chosen beforehand, once again to guarantee stability, but decreases playability by generating structures whose only differences are which blocks are placed on top of which. Levels are evaluated executing them in the simulator and checking their average stability, considering the speed of every block when erected – which must tend to be zero for having a stable structure. The authors designed specialized crossover and mutation operators, aiming to maintain some consistency in the new solutions generated.

However, this paper proposes a different approach: the *free form evolution*. To our knowledge there are no other PCGs that use this approach for the generation of Angry Birds levels. Thus, we will mainly have to look outside computational intelligence in games to focus on optimization in architecture, where there are several approaches to structural optimization using search-based algorithms. Gandomi et al. [14] proposed a bioinspired algorithm called Cuckoo Search and tested its performance with several structural optimization problems. The algorithm is inspired by brood parasitic behaviour of some cuckoo species using Levy flight behaviour, a random walk that follows a certain probability distribution that can be modelled mathematically. One of the test cases for the algorithm is a structural design of a pin-jointed plane frame. In this case the forces and the length of the base is predefined and the frame is parametrized with two angles. The lengths of the members need to be minimized but they depend on those two angles. The cuckoo search locates two global optima for this problem specification. However this kind of structural optimization is not flexible enough to fit our optimization problem, since we are looking for evolution of structures that do not follow any pattern.

A different take on the design of structures is using Generative Grammatical Encodings as Hornby and Pollack present in [15]. In this case, structures are formed by voxels and build following instructions given as commands to a LOGO-style turtle. They implemented an EA which evolved tables as structures. The commands are generated by a context-free L-system. For the EA, an L-system and its production rules are considered individuals. In order to allow these systems to evolve they are parametrized and constrained.

All individuals have a predetermined number of production rules with a fixed number of arguments. Each of them is then initialized with random build commands grouped in blocks. The mutation of the system is done by making a small change in it such as replacing one command, changing the parameter by adding a constant or changing the condition equation, among others. The crossover takes two individuals, copies one of them and inserts some small parts of the other one in it, like a subsequence of commands for example. Still, this increases the number of patterns that can be generated, but still precludes the formation of disjoint structures, for instance, a defensive tower before a simple pigsty in our scope.

In general, we will try to follow a realistic structure generation approach, without constraining it to a fixed form, thus advancing the state of the art by allowing the creation of Angry Birds level without any structure. The next section will describe how we characterize this problem and our approach to it.

3 Problem Description

Science Birds is nowadays the main open source Angry Birds simulator. It was developed by Ferreira and Toledo [13], and is available at GitHub [16]. We used it as a starting point; however we had to patch it for producing usable output and automate its work. The modified version is available on GitHub [17]. It produces an output containing the position and average magnitude of the velocity of each block that was not broken after the simulation. It can be run from beginning to end without user interaction and minimizes the amount of time spent on simulating each level.

Once the simulator that is going to be used to measure fitness is ready, we have to design the fitness function. As obvious as it might seem, the main feature of a stable level is that it is not in motion, so it seems reasonable to evaluate its whole stillness as opposed to its speed – considering every single block.

$$fitness_{ind} = \frac{1}{|V|} \sum_{i=0}^{b} V_i + P_{broken} \cdot (b - |V|)$$

The modified game simulator provides the average magnitude of velocity for each block. This set is noted as V, with $|V|$ being the cardinality of the set. The number of blocks in an individual is b and it can differ from the cardinality of V since broken blocks are not tracked. The number of broken blocks is $b - |V|$ and it is multiplied by a penalization factor, since a level whose blocks break without user interaction would not be considered valid. This happens when a block free-falls from a certain height or collides with a falling object. P_{broken} is set to 100 since objects in a level do not usually reach that velocity, therefore it will separate non-valid levels from potentially good ones.

However, simulating a level is quite time consuming, on the order of seconds, which makes it almost unfeasible for our purpose, so we decided to take additional factors on the fitness so that not all levels are actually simulated. For that reason, before testing a level, there are some indicators, like distance to the ground and number of overlapping blocks, that a level would not be a valid solution and thus simulation can be skipped.

If the lowest object is not close to the ground it is very likely that it, along with all the others blocks above it, will break from the impact. Levels that have all their blocks higher than a certain threshold will not be simulated in the game. The threshold used is 0.1 in game units and the penalty applied to the distance in order to compute the fitness value is 10:

$$f_{distance} = \begin{cases} P_{distance} \cdot D_{lowest}, & \text{if } D_{lowest} > threshold \\ 0, & \text{otherwise} \end{cases}$$

The other measure is the number of overlapping blocks. The separating axis theorem [18] determines if two convex shapes intersect. It is commonly used in game development for detecting collisions. A level with blocks that occupy the same space is not likely to be stable, as the Unity Engine underlying the simulator will solve the issue moving the blocks until there is no collision, but this is done by Unity proprietary code and it is not possible to know what it does. So, a penalty is applied and the level is not simulated either. In this case it is $f_{overlapping} = P_{overlapping} \cdot N_{overlapping}$ where the first factor is a penalty set to 10 and the second is the number of blocks that overlap with each other.

If both $f_{distance}$ and $f_{overlapping}$ are 0 then the level is suitable for simulation and fitness is calculated as $fitness_{ind}$. This would be considered *overpenalization* but exploring infeasible regions entails a serious overhead that we need to minimize [19]. On the other hand, levels with multiple blocks broken during the simulation are not feasible either but running the simulation is necessary. In this case, the penalization does not prevent the region to be explored.

Since one of the objectives of this work is to explore the expressiveness and variability of SBPCG, it seems reasonable to use a flexible representation. We will try these solutions to allow a less directed search than previous solutions while keeping a simple representation.

Individuals are composed by a list of blocks; platforms, TNT boxes or pigs are not considered in this paper, since we are focused on the generation of structures. These building blocks have the following attributes:

- Type: there are eight regular blocks that can be placed in the level with distinct shapes or sizes; they are represented as an integer between 0 and 7.
- Position: coordinates x and y of the centre of the block in game units.
- Rotation: rotation of the block in degrees. Here four different rotations are considered, 0, 45, 90 or 135° represented as integers between 0 and 3.
- Material: three types, which determine the durability of the block. However, this does not affect their stability, so it will remain constant for now as *wood* material.

Using this representation a gene representing a single block will be formed by two integers and two floating point numbers.

Individuals are a collection of genes, in the same way a level is a collection of building blocks. The number of blocks is variable and the order in which they are listed is not important.

The fitness of the worst individual that has been tested in game is stored, so that the value of not tested levels is always above—it is a minimization problem—the in-game tested levels; the starting point for fitness of such individuals is the worst in-game score.

This penalization is calculated using the distance of the lowest block to the ground, which can be easily obtained, and the number of blocks that collide. This requires a bit more of computation, so it will be stored and set in the initialization of the individual. When a gene is modified, the number of overlapping blocks is recalculated for that specific change.

Considering all of the above, the chromosome object is composed by:

- A list of genes.
- A fitness value.
- A penalty (set to `False` for in-game evaluated levels).
- Number of overlapping blocks (calculated).

Initialization is done randomly, with each individual having a random number of genes, which are initialized by several methods:

- Random: selects a random number for each attribute of the gene.
- No Overlapping: also selects a random number but the gene is only added to the chromosome if it does not overlap with an already existing gene.
- Discrete: selects a random number for type and rotation, but the position must be multiple of the dimensions of the smallest block (blocks will be aligned).
- Discrete without overlapping: it combines the second and third initialization method.

Candidates for reproduction are selected using tournaments. Two individuals are chosen from the population and the best will be a parent in this generation. This is repeated until a certain percentage of pairs have been reached. It is important to note that individuals chosen are not removed from the population and therefore they can appear several times in the list of parents.

Once the parents have been selected, we implement two different methods of combination:

- Sample Crossover: gives a single individual per parent pair. It takes all genes from both parents—excluding genes that are repeated—and randomly takes a number of them to create the new individual. The number of blocks is the minimum between the maximum number of blocks allowed, the mean of the two parent individuals and the number of distinct genes.
- Common Blocks: produces two individuals. The common genes to both parents are passed on to both children. The remaining genes are randomly distributed to each child, half to one and half to the other.

There are four different mutations:

- Rotation: rotation is represented as an integer (it is discretized), so it adds or subtracts one to the current value.
- Type: similarly to rotation mutation.
- Position X: a real value between 0 and 1—excluding 0—is added or subtracted from the value of the position X.
- Position Y: same as position X mutation, for position Y.

The new generation is selected using an elitist strategy. Best individuals in both the old population and their offspring pass on the next generation, maintaining the size of the population.

The information that describes a level can be too complex to have a binary representation as pure genetic algorithms suggest, so the framework should be flexible enough to support complex data structures. This prevented us from using frameworks such as DEAP [20], PyEvolve [21] or Evolopy [22] and therefore a new framework was created. The source code can be found on GitHub [23].

In order to evaluate the different options and check if they meet our objective, we performed a series of experiments presented in next section.

4   Experimentation and Results

We set out to evolve free-form structures, but we need to test, one by one, the different parts of our algorithm: fitness function, some evolutionary algorithm parameters and the genetic operators. In order to do so, we performed a set of experiments to test our hypotheses, whose result is shown in Table 1 shows an overview of the results.

Table 1. Summary of the results of the last generation in 15 runs for each experiment. G: number of generations, E: experiment number

	E1	E2	E3	E4
Time (h)	0.89	1.002	1.76	5.03
G	100.0	155.087	76.625	365.929
Best	61.334	110.66	0.0015	0.0018
Avg	383.701	327.547	0.54	0.203
Worst	510.515	367.895	0.828	0.2997

4.1   Baseline Experiment

The premise of the first experiment is that our basic EA should be able to minimize the movement of the blocks placed on the level and the flexibility of the representation should allow variety in the structures. This will be used as a baseline.

The EA in this experiments uses:

- Initialization with the discrete method described earlier.
- Basic sample crossover.
- All four mutations.
- Elitist replacement.
- Tournament selection using.

The parameters are:

The results suggest that the hypothesis was not correct. The average best solution has a fitness value of 61.334 (as shown in Table 1) which indicates that probably most levels have blocks falling (and even breaking) when loaded. The standard deviation of this measure is 133.0209 which implies that while some executions

Table 2. Parameters used in the first experiment

Population size	100
Number of generations	100
Percentage of parents	0.5
Percentage of type mutations	0.5
Percentage of rotation mutations	0.5
Percentage of axis x mutations	0.5
Percentage of axis y mutations	1

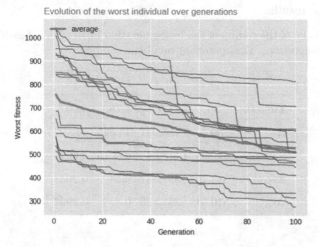

Fig. 1. In grey, different executions of the first experiment (E1)

performed poorly, some others may be good. Even the best levels have blocks that break after loading so they would not be valid. However, we can tell that there is variety in the structures, since they clearly differ from each other.

In the experiment the only termination condition was reaching the maximum number of generations. However, looking at the results, it seems that the execution ended before the population stabilized or converged. Since mutation percentages are high it is normal that convergence where every single individual is the same one, is not reached. However, it could be possible that the population is stable, where every child has a greater fitness value than its parents, therefore no new members are allowed. If there are no new individuals and the population is completely *stuck*, the fitness value of the worst individual should be the same over several generations. In Fig. 1 we can see that this is not the case in average. Although some populations do remain the same for several generations close to the maximum, most of them do not.

Most executions from the previous experiment reached the maximum number of generations without stabilizing or converging. This means the EA may need a larger number of generations to fully evolve a solution. This is the hypothesis for the second experiment, which we describe next.

4.2 Changing Termination Conditions

The main change in this implementation is the addition of two new stop conditions: being 10 generations without changes (stable population) or best fitness value below 0.01. The set of operators is the same as the previous one, and the parameters remain unchanged except for the *Number of generations* which is equal to 1000. The increase in the number of generations allows each run to fully develop their individuals. This would not be feasible if we had to simulate all the individuals, but thanks to the fitness function and the new stop conditions we can afford this number of generations.

Although the best levels obtained with this experiment are better than those evolved in the first experiment, the bad solutions have a really high fitness value. In Table 1, we can see that average fitness of levels produced with this version of the EA is worse than the ones generated in the first experiment. It suggest that populations can be stuck for many generations before making any type of improvement. Any of the generated levels have a fitness below 0.01 or reached maximum number of generations, which means the termination criteria that stopped the evolution was that the population was stable, without any new individuals added for 10 generations.

Figure 2 represents the evolution of the best individual of each execution. Most of them have no more than 50 generations, therefore the hypothesis for

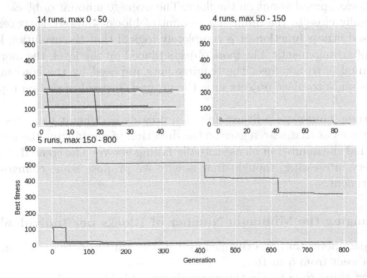

Fig. 2. Best individual evolution for all executions, grouped by number of generations

this experiment is not correct. Short evolutions show that the best individual at initialization is very similar to the last one. Slight improvements may be achieved by small mutations but it seems difficult for new generations to outperform previous ones. We can appreciate that significant improvements are most common in those executions with a poor initial population. Even the ones with several hundreds of generations struggle to improve the initial population.

Our hypothesis that the number of generations could be the main factor is then not correct; it is more likely that there this EA is biased towards exploration rather than exploitation. The genetic operators are failing to create new individuals that inherit good traits from their parents. A new crossover operator could shift the focus to exploitation, and we will examine this in the next experiment.

4.3 Improving Exploitation by a Better Crossover Operator

For the third experiment, the change introduced is in the crossover operator used which was previously described as *Common blocks* crossover. The rest of the operators as well as the termination criteria are kept the same.

Table 1 shows that the results have radically improved as the average fitness of the best solutions drops to 0.0015, a decrease of almost 100%. Additionally, it took less generations in general to reach those results. However, executions took longer on average, which makes sense given that a greater number of individuals would have been simulated. The average fitness of the population and the worst individual have similar values now, which suggest that in most executions the population did converge.

The levels are stable and the blocks do not fall when loading the level, but it is arguable that they would be considered structures, since most of them consist in a few blocks spread about on the floor. The average amount of blocks is 6.26, which is really close to the minimum amount of blocks allowed. However, given the proposed fitness function, it is completely logical that the evolution leads to this kind of arrangements. The more objects placed on the level, the more likely the individual is to not meet the requirements imposed by the constraints. It also makes sense to place objects near the ground, instead of one on top of the other.

The fitness landscape a fitness function creates is difficult to assess, but in this case it is clear that search goes on the direction of minimizing the number of blocks so that the number of blocks actually falling is zero. The created structures are not very interesting, though, that is why we propose some changes in the next experiment.

4.4 Changing the Minimum Number of Blocks per Individual

The only parameter that was changed for this experiment is the minimum number, which went from 5 to 10.

The first thing to notice in these experiments (Table 1) is the increase of the average execution time: it went from 1.76 h in the third experiment to 5.03 h in this one. The time spent running the simulation for each population in this experiments and in the previous one should be similar. However, the number of actual executions of the simulator, which only kick in if there is actually some block on the floor and there is no overlap, drastically increased too. There is no doubt that placing at least ten objects in a structure that does not collapse is more difficult than placing just five. The average best fitness value increased slightly, while the average and worst values are lower. This suggest that the latest generations of this EA are less diverse than those from the third experiment.

However, the results are more interesting visually. We can find some blocks being stack together, not only lied on the ground like in the previous experiment.

5   Conclusions and Future Work

This paper was developed with the main objective of implementing a competitive entry for the Angry Birds level generation contest, and a set of sub-objectives in mind: exploring the expressiveness and variability of SBPCG with evolutionary techniques, adapting the game to extract data from execution and producing stable structures under gravity.

For this aim we have implemented an Evolutionary Algorithm able to optimize level structures to meet some criteria, being the main one the stability of the constructions. Considering this, the method studied was sufficiently general and flexible to draw some conclusions about the topic. SBPCG methods are a potential good solution to offline content generation but it requires a great amount of problem-specific knowledge. Like any other form of creative work, the biggest issue may be how to measure how good, creative or enjoyable is the piece. The more rules the author adds, expressiveness starts to get lost as the results are variations of the same idea. However, it is crystal clear from the experiments run in this paper that a lack of knowledge will lead to unexpected outcomes. In our case, the fitness function used the stability of the structure and only considered other features—overlapping blocks and distance to the ground—to ensure the levels would be valid.

The second objective, adapting the game to extract data from execution, was certainly achieved. It was also a basic requirement to proceed with the rest of them. The game does provide the data, as long as the input is correctly structured. In order to conduct our experiments the Angry Birds simulator (Science Birds) has been adapted to our necessities, yielding now some other information required to evaluate the individuals of the implemented EA. It would be interesting to obtain other kind of data from the simulator, such as the height of the structure and, eventually, its resistance to bombardment by angry birds. However, this is left as future work.

Producing stable structures under gravity was the third objective and the closest to the ultimate one. That objective was effectively achieved, but the consequence of evolving in a path of minimum movement or maximum stability results in structures that are close to the ground and do not have more than a few floors, which could not be very exciting for the players. The main issue is, then, how we evaluate the levels. In fact this is a matter of how we define what an Angry Birds' level *is* and if that definition matches the fitness function. How this definition of level plays with the paths of evolution is also a problem. Since we define as level as a structure that is stable, evolution will maximize stability finding, as in the beginnings of architectural practice, squat structures that are neither aesthetically pleasing nor playable, although undeniably sturdy and stable. We have been successful in, evolving free form structures, to find these type. But once we get there, we need to go into a different evolution mode that takes into account several features as a multiobjective optimization problem: stability is the first, but we can sacrifice a bit of stability for height or some other aesthetic quality.

To conclude on an optimistic tone, this work provides an interesting insight into the SBPCG, through the completion—and failure—of the goals we set out to achieve at the beginning.

In order to improve the results of the method, different constraints could be expressed as multiple objectives. Overlapping blocks and velocity could be treated as minimization objectives and height as a maximization one.

If we pay attention at the stages of evolution in this work, there is also room for improvement in the genetic operators. For example, the initialization produces a small amount of valid individuals which suggested that an elitist strategy for selection would work best. However, new experiments will help to better balance exploration and exploitation. An interesting addition would be to add *building* operators that pile blocks on structures that are already stable.

Another important issue that needs to be addressed is the time performance. Right now the simulation is the main bottleneck in the execution. One way to speed up the process can be *cleaning* up the current simulation, getting rid of any unnecessary assets while maintaining the bare minimum to perform evaluation.

The communication between the simulation and the content generator is through read and write operations on disk, instead of memory. This could be avoided if both tools were integrated in one. If we aim for an online automatic generator, the generator should be integrated in the game. However, if our goal is to generate levels for mixed authorship, as an assistance to developers, it may be a better idea to integrate the simulation in the generator. This could be done by approximating in-game physics with real physics, as described in [24].

Acknowledgements. This paper has been supported in part by projects TIN2014-56494-C4-3-P s (Spanish Ministry of Economy and Competitiveness) and DeepBio (TIN2017-85727-C4-2-P).

References

1. Rovio Entertainment Corporation: Angry Birds official site. http://www.rovio.com/games/angry-birds. Accessed 22 May 2018
2. aibirds.org: AIBIRDS CIG 2018 level generation competition (2018). https://aibirds.org/other-events/level-generation-competition.html

3. Togelius, J., Kastbjerg, E., Schedl, D., Yannakakis, G.N.: What is procedural content generation? Mario on the borderline. In: Proceedings of the 2nd International Workshop on Procedural Content Generation in Games, p. 3. ACM (2011)
4. Togelius, J., Shaker, N., Nelson, M.J.: Introduction. In: Shaker, N., et al. (eds.) Procedural Content Generation in Games. CSCS, pp. 1–15. Springer, Cham (2016). https://doi.org/10.1007/978-3-319-42716-4_1
5. Togelius, J., Yannakakis, G.N., Stanley, K.O., Browne, C.: Search-based procedural content generation. In: Di Chio, C., et al. (eds.) EvoApplications 2010. LNCS, vol. 6024, pp. 141–150. Springer, Heidelberg (2010). https://doi.org/10.1007/978-3-642-12239-2_15
6. Hastings, E.J., Guha, R.K., Stanley, K.O.: Evolving content in the galactic arms race video game. In: IEEE Symposium on Computational Intelligence and Games, CIG 2009, pp. 241–248. IEEE (2009)
7. Shaker, N., Togelius, J., Yannakakis, G.: Mario AI championship - level generation track (2012). http://www.marioai.org/LevelGeneration. Accessed 05 Nov 2018
8. Khalifa, A.: The general videogame AI competition - level generation track (2018). https://github.com/GAIGResearch/gvgai/wiki/Tracks-Description#level-generation-track. Accessed 05 Nov 2018
9. Khalifa, A., Perez-Liebana, D., Lucas, S.M., Togelius, J.: General video game level generation. In: Proceedings of the Genetic and Evolutionary Computation Conference, GECCO 2016, pp. 253–259. ACM, New York (2016)
10. Stephenson, M., Renz, J., Ferreira, L., Togelius, J.: 3rd Angry Birds level generation competition (2018). https://project.dke.maastrichtuniversity.nl/cig2018/competitions/#angrybirds. Accessed 05 Nov 2018
11. Stephenson, M., Renz, J.: Generating varied, stable and solvable levels for Angry Birds style physics games. In: IEEE Conference on Computational Intelligence and Games (CIG), pp. 288–295. IEEE (2017)
12. Stephenson, M., Renz, J.: Procedural generation of complex stable structures for Angry Birds levels. In: IEEE Conference on Computational Intelligence and Games (CIG), pp. 1–8. IEEE (2016)
13. Ferreira, L., Toledo, C.: A search-based approach for generating Angry Birds levels. In: IEEE Conference on Computational Intelligence and Games (CIG), pp. 1–8. IEEE (2014)
14. Gandomi, A.H., Yang, X.S., Alavi, A.H.: Cuckoo search algorithm: a metaheuristic approach to solve structural optimization problems. Eng. Comput. 29(1), 17–35 (2013)
15. Hornby, G.S., Pollack, J.B.: The advantages of generative grammatical encodings for physical design. In: Proceedings of the 2001 Congress on Evolutionary Computation, vol. 1, pp. 600–607. IEEE (2001)
16. Ferreira, L.: Science birds. https://github.com/lucasnfe/Science-Birds. Accessed 17 June 2018
17. Thor, A.U.: Science birds adaptation. https://github.com/a-u-thor/repo-name
18. Ericson, C.: Real-Time Collision Detection. CRC Press, Boca Raton (2004)
19. Runarsson, T.P., Yao, X.: Evolutionary search and constraint violations. In: The 2003 Congress on Evolutionary Computation, CEC 2003, vol. 2, pp. 1414–1419. IEEE (2003)
20. Fortin, F.A., Rainville, F.M.D., Gardner, M.A., Parizeau, M., Gagné, C.: DEAP: evolutionary algorithms made easy. J. Mach. Learn. Res. 13, 2171–2175 (2012)
21. Perone, C.S.: PyEvolve: a Python open-source framework for genetic algorithms. ACM SIGEVOlution 4(1), 12–20 (2009)

22. Faris, H., Aljarah, I., Mirjalili, S., Castillo, P.A., Merelo, J.J.: EvoloPy: an open-source nature-inspired optimization framework in Python. In: Guervós, J.J.M., et al. (eds.) Proceedings of the 8th International Joint Conference on Computational Intelligence, IJCCI 2016, Volume 1: ECTA, Porto, Portugal, 9–11 November 2016, pp. 171–177. SciTePress (2016). http://dx.doi.org/10.5220/0006048201710177

23. Calle, L.: Angry Birds level generator GitHub repository. https://github.com/Laucalle/AngryBirdsLevelGenerator. Accessed 30 Oct 2018

24. Blum, M., Griffith, A., Neumann, B.: A stability test for configurations of blocks. Technical report, MIT (1970). http://hdl.handle.net/1721.1/5861

Efficient Online Hyperparameter Adaptation for Deep Reinforcement Learning

Yinda Zhou, Weiming Liu, and Bin Li[✉]

CAS Key Laboratory of Technology in Geo-spatial Information Processing
and Application Systems, University of Science and Technology of China,
Hefei, China
{zhouyd,weiming}@mail.ustc.edu.cn, binli@ustc.edu.cn

Abstract. Deep Reinforcement Learning (DRL) has shown its extraordinary performance on a variety of challenging learning tasks, especially those in games. It has been recognized that DRL process is a high-dynamic and non-stationary optimization process even in the static environments, their performance is notoriously sensitive to the hyperparameter configuration which includes learning rate, discount coefficient, and step size, etc. The situation will be more serious when DRL is conducting in a changing environment. The most ideal state of hyperparameter configuration in DRL is that the hyperparameter can self-adapt to the best values promptly for their current learning state, rather than using a fixed set of hyperparameters for the whole course of training like most previous works did. In this paper, an efficient online hyperparameter adaptation method is presented, which is an improved version of Population-based Training (PBT) method on the promptness of adaptation. A recombination operation inspired by GA is introduced into the population adaptation to accelerating the convergence of the population towards the better hyperparameter configurations. Experiment results have shown that in four test environments, the presented method has achieved 92%, 70%, 2% and 15% performance improvement over PBT.

Keywords: Reinforcement learning · Hyperparameter adaptation · Game

1 Introduction

Deep reinforcement learning (DRL) [1], combining reinforcement learning [2] with the deep neural network [3], shows extraordinary performance in solving some complex and challenging problems ranging from Games [4–8], Robotics [9,10] to Computer Vision [11–13], etc. However, DRL's performance is notoriously sensitive to the choice of their hyperparameters configuration [14,15]. In the most practice of conventional supervised learning, researchers usually spend

© Springer Nature Switzerland AG 2019
P. Kaufmann and P. A. Castillo (Eds.): EvoApplications 2019, LNCS 11454, pp. 141–155, 2019.
https://doi.org/10.1007/978-3-030-16692-2_10

a lot of time and computing resources on hyperparameter selection and adjustment in order to obtain satisfactory model performance. But due to the long learning process and the dynamic nature of reinforcement learning, it is infeasible to conduct a comprehensive search for a single optimal configuration of hyperparameters [16].

In DRL, the training data are obtained by the agent interacting with the environment under the guidance of its own strategy. The agent's strategy is updated/changed continuously during the reinforcement learning process, the conduct sampled by the agent is continuous. Meanwhile, in most games, the environment/scenario will change from time to time. All of these changes will make the distribution of data samples constantly changing, resulting in unstable end-to-end learning [17]. There is no single fixed optimal hyperparamater configuration that could be suitable for the whole course of the training process of DRL. The ideal state is that the hyperparameters can be self-adapted promptly to fit the current learning process.

Some previous works implemented hyperparameter adaptation in DRL by simple rules or manually defined schedules [18–21]. But such kind of adaptation often requires a lot of empirical knowledge. Jaderberg et al. [22] proposed Population-based Training (PBT), a simple population-based hyperparameter adaptation neural network training method. Different from all previous methods, PBT has realized the adaptation of hyperparameter in the whole reinforcement learning process without human intervention. However, PBT uses simple stochastic perturbations to achieve hyperparameter adaptation, which we believe is inefficient to track the change of potential temporary optimal hyperparameter configuration. The motivation of this work is to improve the efficiency of hyperparameter adaptation by PBT and then improve the performance of DRL.

In the genetic algorithm, the recombination operation can make the population converge to the local optimal quickly. Inspired by this, in this paper, we propose a new population based hyperparameter adaptation training method, the key point is to introduce the recombination operation into the original PBT framework to accelerate the convergence of candidate hyperparameter configuration towards the local temporary optimal. The method is tested on four classic video games. Experiment results have shown that the efficiency of hyperparameter adaptation was improved by our method, and the DRL model trained with our method obtained higher final performance than that with the PBT. We also found that the frequency of the adaptation operation had a greater impact on the performance of our method, we call this *timeliness of hyperparameter adaptation* in DRL, we also gave reasonable suggestions obtained from the empirical analysis. In addition, we have designed and implemented a distributed queue computing system, which is very friendly to laboratories and researchers who do not have large-scale computing clusters. The contributions of this paper can be summarized as follows:

1. A more efficient population-based hyperparameter adaptation method is proposed to improve the hyperparameter adaptation efficiency, then improve the performance of DRL;

2. The phenomenon of *timeliness of hyperparameter adaptation* in Deep Reinforcement Learning was found and reasonable suggestions were given;
3. Design and implement a distributed queue computing system, it provides a viable solution for small laboratories who want to combine research on population algorithms and reinforcement learning, but do not have enough computing resources.

The rest of the paper is organized as follows. In Sect. 2, we describe the background and related work about our paper. In Sect. 3, we proposed our method and made a detailed description. Section 4 shows the experiment details. We analyzed the experimental results in Sect. 5. Conclusion and future work are given in Sect. 6.

2   Background and Related Work

2.1   Reinforcement Learning

Standard reinforcement learning (RL) usually considers such scenes where an agent interacts with an environment $\mathcal{E}$ over a series of discrete time steps. At each time step t, the agent receives a state s_t from $\mathcal{E}$ then the agent generates an action a_t from some set of possible actions $\mathcal{A}$ to $\mathcal{E}$ according to its own current policy π, where π is a mapping from states s_t to actions a_t [23]. Meanwhile, the agent receives the next state s_{t+1} and a reward r_t. This interaction process continues until the agent receives a terminal signal from $\mathcal{E}$, then we can get an episode sequence $(s_0, a_0, r_0, ..., s_{end}, a_{end}, r_{end})$ of state, action, and reward at each timestep. $R_t = \sum_{t=0}^{end} \gamma^t r_t$ is the total cumulative reward from each time step t with discount coefficient $\gamma \in (0,1]$. The agent's goal is to maximize the expected cumulative reward $\mathbb{E}_\pi[R_t]$ from each state s_t under the policy π. So the optimal π^* is

$$\pi^* = \arg\max_\pi \mathbb{E}_\pi[R_t]. \tag{1}$$

In high-dimensional reinforcement learning settings, the policy π is often modeled by using a deep neural network π_θ with parameters θ called deep reinforcement learning (DRL). The goal is to solve for θ^* that attains the highest expected episodic return.

$$\theta^* = \arg\max_{\theta \in \Theta} \mathbb{E}_{\pi_\theta}[R_t]. \tag{2}$$

There are two classic methods in reinforcement learning. The first one is the value-based method [6,24–29], which uses a manner similar to dynamic programming to estimate a value function $V(s;\theta)$ or action-value function $Q(s,a;\theta)$ and then to improve the policy based on the function. The value-based method is mainly applied to the task of discrete action space. When the tasks are within continues action space, policy-based method [23,30,31,31–33], which directly optimizing the parameterized policy $\pi(a|s;\theta)$ can do better than the value-based method.

In addition to model parameters θ, DRL algorithm has its own parameters. For instance, discount coefficient γ, exploration coefficient ϵ, learning rate r, etc, we call them hyperparameter h. These hyperparameters are critical because they directly control the behavior of RL and have a significant effect on the performance of the model being trained. In early DRL works, some traditional hyperparameter tuning methods (e.g. Grid search, Random search and Automatic hyperparameter Tuning [34–38]) were adopted to search for an approximate optimal fixed hyperparameter configuration h^* for DRL (see Eq. 3, $eval$ is the performance function that can give a performance evaluation (cumulative rewards in RL scenario) of the current model), then apply it to the whole course of optimization process.

$$h^* = \arg\max_{h \in \mathcal{H}} eval(\pi_\theta | h). \tag{3}$$

Note that reinforcement learning is a high-dynamic and non-stationary optimization process, these hyperparameters should self-adapted promptly to fit the current learning process (see Eq. 4, t indicates the current stages of reinforcement learning), rather than using fixed values. For example, the discount coefficient should be varied as the learning progress, in order to force the learning to gradually focus on long-term returns. Exploration coefficient should also be varied as it can affect the dynamic of RL heavily and do great effect on the learning progress. It is well-known that the learning rate is important in neural network optimization, and it should decrease as the learning progress. In the RL scenario, as the learning is highly dynamic, we speculate that the varying learning rate should help RL to learn more efficiently and stable, so the learning rate should not be decremented or fixed at all times.

$$h_t^* = \arg\max_{h_t \in \mathcal{H}} eval(\pi_\theta | h_t). \tag{4}$$

2.2 Hyperparameter Adaptation for Deep Reinforcement Learning

Some researchers [16, 18–21] have proposed hyperparameter adaptation in DRL according to simple rules or schedules that are manually defined or based on Bayesian method. But such methods often require a lot of experimentation and prior knowledge. In addition, such methods can only achieve adaptation of single hyperparameter, and cannot achieve joint adaptation of multiple hyperparameters.

Stefan et al. [16] proposed the OMPAC method, which is based on the evolutionary approach, to perform the online adaptation of hyperparameter in reinforcement learning. OMPAC is the first method to implement DRL's multiple hyperparameters adaptation by the population-based approach. Soon, Jaderberg et al. [22] proposed the population-based neural network training (PBT) method, which effectively utilizes a fixed computational budget to jointly optimize a population of models and their hyperparameters configuration to maximize performance, and it has achieved good results in DRL, Machine Translation and GANs.

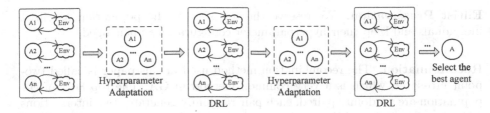

Fig. 1. Schematic diagram of the online hyperparameter adaptation method. A_n is the agent in DRL, Env is the environment.

3 Method

In this section, we will introduce our online hyperparameter adaptation method in detail. It should be noted that we are presenting an online hyperparameter adaptation method, not an automatic hyperparameter tuning method (e.g. Bayesian Optimization for Hyperparameter Tuning [36]). The goal of our method is to self-adapt the DRL's hyperparameters efficiently throughout the whole reinforcement learning process in a changing environment and to enhance the performance of DRL.

3.1 Online Hyperparameter Adaptation Method

For Deep Reinforcement Learning task, there is a model/agent π_θ and a set of hyperparameter configurations h, where θ is model parameters, h is the hyperparameter configuration $[h_1, h_2, ..., h_n]$ of the DRL algorithm.

In our method, the learning task is accomplished via 2 alternately adaptation process (see Fig. 1 and Eq. 5): hyperparameter adaptation to set hyperparameters h for more efficient reinforcement learning, and reinforcement learning to update model parameters θ to enrich the knowledge of agent. We use gradient-based approach to updating the model parameters θ at the end of each iteration of the sampling process, and present an improved efficient population-based approach to adapt the hyperparameters h online.

$$h_t^* = \arg\max_{h_t \in \mathcal{H}} eval(\pi_\theta | h_t), \theta^* = \arg\max_{\theta \in \Theta} eval(\pi_\theta | h_t). \tag{5}$$

Firstly, we are given N agents $\{\theta_i\}_{i=1}^N$ and their respective hyperparameter configurations $\{h_i\}_{i=1}^N$. The N agents interact with the environment respectively and obtain training data in parallel. Then, each agent performs gradient optimization to update its model parameters $\{\theta_i\}_{i=1}^N$ with its current hyperparameter configuration $\{h_i\}_{i=1}^N$ (corresponding to the solid line box in Fig. 1). Once the number of interactions reaches N_{step}, we evaluate the performance $\{s_i\}_{i=1}^N$ of the N agents.

Secondly, we perform adaptation operation (corresponding to the dotted box in Fig. 1). The adaptation operation includes Elitist Preservation, Recombination, and Mutation.

Elitist Preservation. We reserve the top 20% of the population as elitist individuals and keep their hyperparameter configurations unchanged.

Recombination. The recombination method used in this paper is called two-point crossover, which is a very common method in GA. The top 80% of the population are randomly paired, each pair randomly generates two intersections $index1$ and $index2$, and then exchanges some of the hyperparameters (see Eq. 6) according to a certain probability p_c. We also tried to use roulette to generate a new population and then recombination, but the results were not satisfactory.

$$
\begin{bmatrix}
[h_i^1, \ldots, h_i^{index1} \ldots h_i^{index2}, \ldots, h_i^n] \\
[h_j^1, \ldots, h_j^{index1} \ldots h_j^{index2}, \ldots, h_j^n]
\end{bmatrix}
\Rightarrow
\begin{bmatrix}
[h_i^1, \ldots, h_j^{index1} \ldots h_j^{index2}, \ldots, h_i^n] \\
[h_j^1, \ldots, h_i^{index1} \ldots h_i^{index2}, \ldots, h_j^n]
\end{bmatrix}
\tag{6}
$$

Mutation. Individuals in the population (except elitist individuals) perform mutation operations with a certain probability p_m (see Eq. 7). The magnitude (0.8 or 1.2) of the mutation is the same as in PBT. It should be noted that if the hyperparameter is an integer value, the mutated value is rounded.

$$
h^i =
\begin{cases}
h^i \times (0.8 \ or \ 1.2) & \text{if } p < p_m \\
h^i & \text{if } p \geq p_m
\end{cases}
, \ i = 1, \ldots n
\tag{7}
$$

Then, a new $\{h_i^{new}\}_{i=1}^N$ is generated, and each agent performs environmental interaction and gradient optimization under their respective new hyperparameter configurations. When the number of adaptive operations reaches $N_{frequency}$, the entire optimization process ends and we can get the best model θ^* from $\{\theta_i\}_{i=1}^N$. Algorithm 1[1] describes these 2 alternately adaptation learning system in more detail and Algorithm 2 describes adaptation operation.

3.2 Distributed Queue Computing System

In the related work of combining reinforcement learning with population-based approaches, there is a huge computational need inevitably. In addition, current deep reinforcement learning algorithms are increasingly using multi-process models to shorten the training time, which leads to further increases in computational requirements (especially CPU resources). This is a huge obstacle to small laboratories or researchers with limited computing resources. In order to alleviate this issue, in this paper, we have designed and implemented a distributed queue computing system[2] (see Fig. 2).

For no large-scale GPU or CPU clustering, but several small servers (with inconsistent performance) and PCs in the same LAN, we designed a distributed

---

[1] https://github.com/zhoudoudou/online-hyperparameter-adaptation-method.
[2] https://github.com/zhoudoudou/distributed-queue-computing-system.

Algorithm 1. Online Hyperparameter Adaptation Method

Input: initial $\{\theta_i\}_{i=1}^N$, $\{h_i\}_{i=1}^N$, $count_1 = 0$, $count_2 = 0$
Output: θ_i with highest s_i
1: **while** $count_1 <= N_{frequency}$ **do**
2:    **for** i=1 to N (synchronously in parallel) **do**
3:       **while** $count_2 < N_{step}$ **do**
4:          $\theta_i \leftarrow \text{DRL}(\theta_i, h_i)$
5:          $count_2 \leftarrow count_2 + 1$
6:       **end while**
7:    **end for**
8:    $count_2 = 0$
9:    $\{s_i\}_{i=1}^N \leftarrow eval(\{\theta_i\}_{i=1}^N)$
10:   $\{h_i^{new}\}_{i=1}^N \leftarrow$ Adaptation Operation$(\{h_i\}_{i=1}^N)$
11:   $count_1 \leftarrow count_1 + 1$
12: **end while**
13: **return** θ_i with highest s

Algorithm 2. Adaptation Operation

Input: $\{h_i\}_{i=1}^N$
Output: $\{h_i^{new}\}_{i=1}^N$
1: $\{h_i^{new}\}_{i=1}^{N*0.2} \leftarrow$ Elitist Preservation Operation$(\{h_i\}_{i=1}^N)$
2: $\{h_i^{new}\}_{N*0.2+1}^N \leftarrow$ Recombination and Mutation Operation$(\{h_i\}_{i=1}^N)$
3: **return** $\{h_i^{new}\}_{i=1}^N$

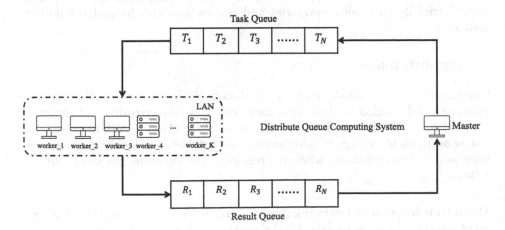

Fig. 2. Distribute Queue Computing System

queue computing system to utilize the parallel synchronization characteristics of our method. Assuming that the number of agents is N and there are K computing devices $(K<N)$. N agents' weight matrix θ, hyperparameter configuration h, accumulative reward s are packaged as N tasks, and inserted into the queue and exposed on the LAN. Each of the K computing devices takes tasks from the task queue sequentially and independently performs environment interaction and gradient optimization. Once a device has completed the current task, put the optimization result into the result queue immediately, and check whether the task queue still has tasks not being executed. If there are still unexecuted tasks in the queue, the idle device will continue to get the task to execute.

Once all tasks have completed one iteration, we perform adaptation operations then generates new N tasks and push them into the task queue for the next iteration until the termination condition is met. In that way, the system maximizes the efficient use of limited resources.

3.3   Difference Between PBT and Our Method

The main difference between our method and PBT is the adaptation operation. In PBT, if the individual's current performance ranking is at the bottom 20%, it is uniformly sampled from the top 20% of the individuals for replacement and performs a simple perturbation of its hyperparameters. In our approach, we retain elitist in the top 20% of the population, then recombine and mutate the top 80% of the population to create a new population. In addition, PBT adopts asynchronous implementation, and we find that if this method is implemented in a distributed way, it is necessary to ensure the consistency of the performance of the computing device as much as possible, otherwise, the difference in individual optimization speed will be caused. In this regard, we adopt synchronous implementation and skillfully design a distributed queue computing system, which is very friendly to small laboratories and researchers with limited computing resources.

4   Experiments

The experiments were designed to verify the effectiveness of our Online Hyperparameter Adaptation Method. In order to ensure the universality and repeatability of our method as much as possible, we used the classical A2C algorithm as our benchmark algorithm. In addition, we used stable and efficient open source code and test environments whenever possible. The experimental setups are as follows:

Deep Reinforcement Learning Algorithm. In the experiment, the reinforcement learning algorithm uses the A2C[3] algorithm. A2C is a synchronous, deterministic variant of Asynchronous Advantage Actor Critic (A3C) by OpenAI, which is

---

[3] https://blog.openai.com/baselines-acktr-a2c.

more cost-effective than A3C, but can give equal performance. According to the actual situation of our computing equipment, we set the number of A2C processes to 16. In the experiment, we used the open source A2C code[4] provided by OpenAI directly.

Network Architecture. A2C's deep network model [6] includes three convolutional layers and one fully-connected layer and an output layer. The first convolution layer has 32 filters of size 8×8 with stride 4. The second convolution layer has 64 filters of size 4×4 with stride 2. This is followed by a third convolutional layer 64 filters of size 3×3 with stride 1. The final layer is fully-connected with 512 rectifier units. The output layer output policy and value. In addition, the gradient optimizer is RMSProp [39].

Hyperparameters. we used our method for online adaptation of the following hyperparameters: [ent_coef, vf_coef, lr, n_step, alpha, gamma], where ent_coef is the coefficient in front of the policy entropy in the total loss function and vf_coef is the coefficient in front of the value function loss in the total loss function, lr and alpha is the learning rate and decay coefficient in RMSProp optimizer, nstep is the step size of the agent, gamma is the reward discount coefficient. The expert-level hyperparameter configuration is given by [23] (see Table 1), we configure it as an initialization hyperparameter for all agents. In addition, the values of alpha and gamma are limited to no more than 0.99. It should be emphasized that we do not do adaptation operation on network structure hyperparameters.

Table 1. Expert-level hyperparameter configuration.

Hyperparameter	ent_coef	vf_coef	lr	n_step	alpha	gamma
Value	0.01	0.5	7e-4	5	0.99	0.99

Adaptation Operation. In our experiments, the number of agents N is set to 10, the total computational budget C_{total} for single agent is 40M training step, the adaptation operating frequency $N_{frequency}$ is set to 50, N_{step} calculated by Eq. 8, the probability of recombination p_c is set to 0.7 and the probability of mutation p_m is set to 0.2. The PBT's experimental setups as in [22] and the number of agents is set to 10 too.

$$N_{step} = \frac{C_{total}}{N_{frequency}} \tag{8}$$

---

[4] https://www.github.com/openai/baselines.

Test Environment. We chose four relatively complex and challenging video games as our test environment, they are MsPacman, SpaceInvaders, Seaquest, BeamRider. In the field of video games, any of them can be used as a research object alone. In the experiment, the environment test platform is available from Gym[5], you only need to provide actions a, the environment will return state s and rewards r.

Eval. We evaluate the performance of each current agent with the last 100 episodic scores during training, each episodic scores is a cumulative reward.

Distributed Queue Computing System. We implement our Online Hyperparameter Adaptation Method with our distribution queue computing system. We have four small servers (each server has an NVIDIA GTX 1080Ti GPU and an Inter(R) Core(TM) i9-7900X CPU) and a PC. We use the PC as a master controller to perform hyperparameters adaptation operation and queue management. Four small servers interact with the environment and perform gradient optimization of the agent.

5   Results and Analysis

In this section, we present our experimental results and conduct corresponding analysis and discussion. The experiment compares the results of A2C with fixed hyperparameter configuration (expert-level hyperparameter configuration recommended in the [23], see Table 1), A2C with PBT and A2C with the hyperparameters adaptation method proposed in this paper. As shown in Table 2 and Fig. 3, our adaptation method can achieve higher model performance. Our method achieved 92% performance improvement over the PBT in MsPacmam, 70% in SpaceInvaders, 2% in the Seaquest, 15% in BeamRider. We have reason to believe that hyperparameter adaptation can be more effectively achieved through hyperparameter recombination operation between population. We can also see that the performance of using a fixed set of hyperparameters throughout the optimization process is much lower than using the hyperparameter adaptation method. A more intuitive comparison of the PBT and our methods is presented in Fig. 4. Figure 5 shows the adaptation process of six hyperparameters in MsPacman.

To further emphasize, our approach is to allow hyperparameters to be adaptation throughout the DRL optimization process, not to obtain a fixed set of hyperparameter configurations. We also conducted a comparative test, using the hyperparameters configuration that our method found by the end of training (see Fig. 6). Its performance is far lower than the results obtained through our adaptation method. The result future illustrates the necessity and efficiency of hyperparameter adaptation in DRL.

---

[5] Specific details can be found: https://gym.openai.com.

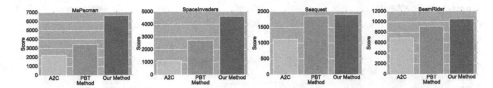

Fig. 3. Per game result with A2C(red), PBT(blue) and our method(green). (Color figure online)

Table 2. Per game result of A2C with fixed hyperparameters, A2C with PBT and A2C with our Method. Each experimental condition was repeated three times, and the numbers in the table show the mean and median score of the single best performing population member at the end of the training run.

Atari game	A2C		PBT		Our method	
	Mean	Median	Mean	Median	Mean	Median
MsPacman	2245(101)	2205	3444(496)	3771	**6613(1062)**	**7115**
SpaceInvader	1123(26)	1139	2720(230)	2634	**4631(846)**	**4088**
Seaquest	1740(6)	1740	1859(32)	1881	**1900(1)**	**1900**
BeamRidert	6929(128)	6975	9114(700)	9223	**10497(643)**	**10397**

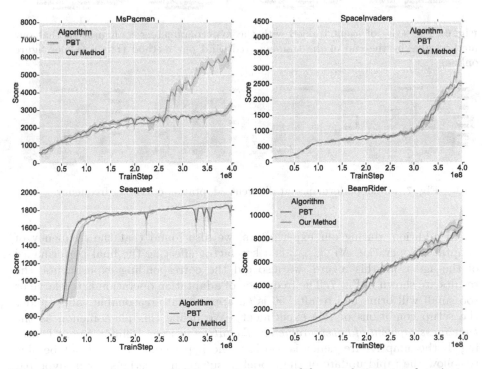

Fig. 4. Average training curves of the top five agents of the population at each trainstep with our method (orange) and with PBT (blue). (Color figure online)

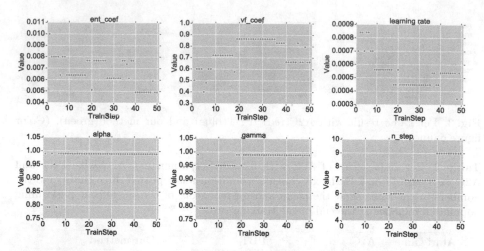

Fig. 5. The adaptation process of six hyperparameters on MsPacman.

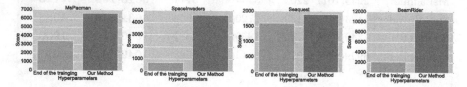

Fig. 6. The score of agent trained with the hyperparameters configuration that our method give by the end of the training (red) and our method (blue). (Color figure online)

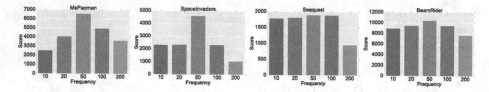

Fig. 7. Model performance at different adaptation operating frequency.

In addition, during our experiments, we also found that the frequency of adaptation operation $N_{frequency}$ is a key factor affecting the final performance of the agent. In this regard, we also did the corresponding experiments, the results are shown in Fig. 7. The frequency of adaptation operation is too large or too small will bring bad results, 50 is an approximate reasonable value (under the setup conditions of this experiment). We define this phenomenon as the *timeliness problem of hyperparameter adaptation* in Deep Reinforcement Learning. If the adaptive frequency is too low, the hyperparameter will not be able to follow the rapid update of the model, resulting in a serious lag in hyperparameter adaptation. If the frequency of the adaptive operation is too high, the same calculation cost will cause the time of the single hyperparameter to act on the model to be sharply shortened, so the subsequent evaluation of the hyperparameters will be biased (noise), ultimately affecting the effect of adaptation operation.

6   Conclusions and Future Works

In this paper, we presented an online hyperparameter adaptation method for deep reinforcement learning. Using the population-based approach, the hyperparameter adaptation in the DRL optimization process was achieved through Elitist Preservation, Recombination, and Mutation. And through experiments, we proved the efficiency of our method and better performance than PBT, which is the state-of-art method similar to our work. In addition, we found the *timeliness problem of hyperparameter adaptation* in Deep Reinforcement Learning, which was not discovered by PBT and OMPAC. Meanwhile, we designed and implemented a distributed queue computing system.

Recently, the related work of the combination of reinforcement learning and evolutionary algorithms has attracted widespread attention. How to better take advantages of evolutionary algorithms to assist reinforcement learning is the issue we will explore in the future work.

Acknowledgment. The work is supported by the National Natural Science Foundation of China under grand No. 61836011 and No. 61473271.

References

1. Li, Y.: Deep reinforcement learning: An overview. arXiv preprint arXiv:1701.07274 (2017)
2. Sutton, R.S., Barto, A.G., Bach, F., et al.: Reinforcement Learning: An Introduction. MIT Press, Cambridge (1998)
3. LeCun, Y., Bengio, Y., Hinton, G.: Deep learning. Nature **521**(7553), 436 (2015)
4. Silver, D., et al.: Mastering the game of go with deep neural networks and tree search. Nature **529**(7587), 484 (2016)
5. Silver, D., et al.: Mastering the game of go without human knowledge. Nature **550**(7676), 354 (2017)
6. Mnih, V., et al.: Human-level control through deep reinforcement learning. Nature **518**(7540), 529 (2015)
7. Mnih, V., et al.: Playing atari with deep reinforcement learning. arXiv preprint arXiv:1312.5602 (2013)
8. Justesen, N., Bontrager, P., Togelius, J., Risi, S.: Deep learning for video game playing. arXiv preprint arXiv:1708.07902 (2017)
9. Levine, S., Finn, C., Darrell, T., Abbeel, P.: End-to-end training of deep visuomotor policies. J. Mach. Learn. Res. **17**(1), 1334–1373 (2016)
10. Mirowski, P., et al.: Learning to navigate in complex environments. arXiv preprint arXiv:1611.03673 (2016)
11. Yoo, S., Yun, K., Choi, J.Y.: Action-decision networks for visual tracking with deep reinforcement learning. In: CVPR, pp. 2711–2720 (2017)

12. Ren, Z., Wang, X., Zhang, N., Lv, X., Li, L.J.: Deep reinforcement learning-based image captioning with embedding reward. arXiv preprint arXiv:1704.03899 (2017)
13. Zhang, J., Wang, N., Zhang, L.: Multi-shot pedestrian re-identification via sequential decision making. arXiv preprint arXiv:1712.07257 (2017)
14. Henderson, P., Islam, R., Bachman, P., Pineau, J., Precup, D., Meger, D.: Deep reinforcement learning that matters. arXiv preprint arXiv:1709.06560 (2017)
15. Islam, R., Henderson, P., Gomrokchi, M., Precup, D.: Reproducibility of benchmarked deep reinforcement learning tasks for continuous control. arXiv preprint arXiv:1708.04133 (2017)
16. Elfwing, S., Uchibe, E., Doya, K.: Online meta-learning by parallel algorithm competition. In: Proceedings of the Genetic and Evolutionary Computation Conference, pp. 426–433. ACM (2018)
17. Melo, F.S., Meyn, S.P., Ribeiro, M.I.: An analysis of reinforcement learning with function approximation. In: Proceedings of the 25th international conference on Machine learning, pp. 664–671. ACM (2008)
18. François-Lavet, V., Fonteneau, R., Ernst, D.: How to discount deep reinforcement learning: Towards new dynamic strategies. arXiv preprint arXiv:1512.02011 (2015)
19. Downey, C., Sanner, S., et al.: Temporal difference bayesian model averaging: A bayesian perspective on adapting lambda. In: ICML, pp. 311–318. Citeseer (2010)
20. Ishii, S., Yoshida, W., Yoshimoto, J.: Control of exploitation-exploration metaparameter in reinforcement learning. Neural Netw. **15**(4–6), 665–687 (2002)
21. Mann, T.A., Penedones, H., Mannor, S., Hester, T.: Adaptive lambda least-squares temporal difference learning. arXiv preprint arXiv:1612.09465 (2016)
22. Jaderberg, M., et al.: Population based training of neural networks. arXiv preprint arXiv:1711.09846 (2017)
23. Mnih, V., et al.: Asynchronous methods for deep reinforcement learning. In: International Conference on Machine Learning, pp. 1928–1937 (2016)
24. Van Hasselt, H., Guez, A., Silver, D.: Deep reinforcement learning with double q-learning. AAAI **16**, 2094–2100 (2016)
25. Schaul, T., Quan, J., Antonoglou, I., Silver, D.: Prioritized experience replay. arXiv preprint arXiv:1511.05952 (2015)
26. Wang, Z., Schaul, T., Hessel, M., Van Hasselt, H., Lanctot, M., De Freitas, N.: Dueling network architectures for deep reinforcement learning. arXiv preprint arXiv:1511.06581 (2015)
27. Anschel, O., Baram, N., Shimkin, N.: Averaged-dqn: Variance reduction and stabilization for deep reinforcement learning. arXiv preprint arXiv:1611.01929 (2016)
28. Fortunato, M., et al.: Noisy networks for exploration. arXiv preprint arXiv:1706.10295 (2017)
29. Hessel, M., et al.: Rainbow: Combining improvements in deep reinforcement learning. arXiv preprint arXiv:1710.02298 (2017)
30. Lillicrap, T.P., et al.: Continuous control with deep reinforcement learning. arXiv preprint arXiv:1509.02971 (2015)
31. Schulman, J., Levine, S., Abbeel, P., Jordan, M., Moritz, P.: Trust region policy optimization. In: International Conference on Machine Learning, pp. 1889–1897 (2015)
32. Wu, Y., Mansimov, E., Grosse, R.B., Liao, S., Ba, J.: Scalable trust-region method for deep reinforcement learning using kronecker-factored approximation. In: Advances in Neural Information Processing Systems, pp. 5285–5294 (2017)
33. Schulman, J., Wolski, F., Dhariwal, P., Radford, A., Klimov, O.: Proximal policy optimization algorithms. arXiv preprint arXiv:1707.06347 (2017)

34. Bergstra, J., Bengio, Y.: Random search for hyper-parameter optimization. J. Mach. Learn. Res. **13**, 281–305 (2012)
35. Snoek, J., Larochelle, H., Adams, R.P.: Practical bayesian optimization of machine learning algorithms. In: Advances in neural information processing systems, pp. 2951–2959 (2012)
36. Hutter, F., Hoos, H.H., Leyton-Brown, K.: Sequential model-based optimization for general algorithm configuration. In: Coello, C.A.C. (ed.) LION 2011. LNCS, vol. 6683, pp. 507–523. Springer, Heidelberg (2011). https://doi.org/10.1007/978-3-642-25566-3_40
37. Bergstra, J.S., Bardenet, R., Bengio, Y., Kégl, B.: Algorithms for hyper-parameter optimization. In: Advances in Neural Information Processing Systems, pp. 2546–2554 (2011)
38. Bergstra, J., Yamins, D., Cox, D.: Making a science of model search: hyperparameter optimization in hundreds of dimensions for vision architectures. In: International Conference on Machine Learning, pp. 115–123 (2013)
39. Tieleman, T., Hinton, G.: Lecture 6.5-rmsprop: divide the gradient by a running average of its recent magnitude. COURSERA Neural Netw. Mach. Learn. **4**(2), 26–31 (2012)

GAMER: A Genetic Algorithm with Motion Encoding Reuse for Action-Adventure Video Games

Tasos Papagiannis[(✉)] [iD], Georgios Alexandridis[iD], and Andreas Stafylopatis

School of Electrical and Computer Engineering,
National Technical University of Athens, Zografou, Greece
{tasos,gealexandri}@islab.ntua.gr, andreas@cs.ntua.gr

Abstract. Genetic Algorithms (GAs) have been predominantly used in video games for finding the best possible sequence of actions that leads to a win condition. This work sets out to investigate an alternative application of GAs on action-adventure type video games. The main intuition is to encode actions depending on the state of the world of the game instead of the sequence of actions, like most of the other GA approaches do. Additionally, a methodology is being introduced which modifies a part of the agent's logic and reuses it in another game. The proposed algorithm has been implemented in the GVG-AI competition's framework and more specifically for the Zelda and Portals games. The obtained results, in terms of average score and win percentage, seem quite satisfactory and highlight the advantages of the suggested technique, especially when compared to a rolling horizon GA implementation of the aforementioned framework; firstly, the agent is efficient at various levels (different world topologies) after being trained in only one of them and secondly, the agent may be generalized to play more games of the same category.

Keywords: Intelligent agent · Video games · Genetic algorithms · Action encoding · GVG-AI

1 Introduction

In the last decades, video games have become a very popular application area of modern methods that stem from the broader *Artificial Intelligence* (AI) research field. Starting from *Deep Blue* [1], the first chess-playing computer system, to the most recently released *AlphaGo* [2], there have been many different approaches to *Game AI*. *Heuristics, Neural Networks, Search Trees* and *Evolutionary Algorithms* (EAs) [3,4] have been the most prominent techniques used so far in an attempt to develop intelligence competitive to the human level. In this direction, a very challenging goal is the creation of agents capable of playing many different video games, which is also the main objective of this work.

© Springer Nature Switzerland AG 2019
P. Kaufmann and P. A. Castillo (Eds.): EvoApplications 2019, LNCS 11454, pp. 156–171, 2019.
https://doi.org/10.1007/978-3-030-16692-2_11

More specifically, *Genetic Algorithms* (GAs), a subclass of EAs inspired by the biological model of natural selection, are used for the creation of an efficient agent for action-adventure type of video games. It is not uncommon for games in this category to share some similar characteristics and mechanics (e.g. in the way agents move) and therefore intelligent approaches utilizing the context of transfer learning could be developed, where part of an agent's training in one game may be reused in other.

At present, the area of game AI is dominated by a set of *Reinforcement Learning* (RL) algorithms' variations. However, *Evolutionary Strategies* (ES) have been proven to be a quite competitive alternative in terms of performance, while being much simpler to implement [5]. The main advantage of the latter family of techniques is the high parallelism and the low memory requirements, which result in faster training times. Nevertheless, RL algorithms are more data efficient, as they require a smaller number of "seen" states in order to converge.

In general, GAs have already been applied in intelligent game agents, but mostly as a part of broader methods. For instance, they are commonly used to evolve neural networks during the training stage, providing the desirable agent's actions for a certain case of the game. This is because the sequence of actions generated offline by the GA only works for a specific level of the game. In this approach, on the contrary, GAs are used as a stand-alone AI methodology.

This work introduces a technique in which the actions taken according to the state of the world of the game are encoded instead. In this way, the agent can be applied to more than one levels of the corresponding game, as the same basic states are encountered in different game levels. Additionally, relevant game states are observed in games of the same type, which means that after the agent is trained in one of them, it can be adapted to be used in similar games.

The basic intuition is to use discrete encodings for the actions related to the movement of the agent and the ones related to facing the enemies. The motion encoding can then be applied to other games' encodings (after it has been determined once) and reduce the states that have to be explored by the GA. This course of action may be critical in complex video games which can not be effectively solved by the GA due to the large state space, as well as time saving. Lastly, a representation of the world's state by processing groups of blocks is also presented.

The overall approach is subsequently implemented on *Zelda* and *Portals*, two action-adventure games of the framework of the *General Video Game AI* (GVG-AI) annual competition [6] and it is compared to a *rolling horizon GA* (RHGA) implementation of the said framework. The rest of this paper is structured as follows; Sect. 2 outlines relevant approaches to the proposed method. Section 3 details the proposed agent's functionality and Sect. 4 describes the platform used in the experiments as well as the configuration of the tested games. Section 5 presents the obtained results and finally the paper concludes in Sect. 6, where possible extensions and future work are discussed.

2   Related Work

The concept of employing autonomous GAs on adventure-type video games has originally been introduced, to the best of our knowledge, in the OLEMAS system [7]. Action-state pairs are formed by an evolutionary process and, after being enhanced with online learning, each state is matched to the best possible action. This procedure has been applied to a multi-player pursuit game, in which four agents cooperate trying to surround an enemy (prey) by moving left, right, up and down, and the results reported by the authors have been very promising.

Similar approaches have already been proposed for other categories, including fighting and board games. For example, in a fighting game of the *M.U.G.E.N framework* [8], genetic programming has been used in order to determine the best strategy (e.g. sequence of actions), with respect to the current state [9]. After the process was completed, the created agents were tested against humans, achieving a win rate of nearly 60% in the best case. A relevant methodology has also been implemented for the real-time strategy game of *Starcraft* [10]. A three-scenario case of the game, where the player's options field consisted of three possible actions, has been studied. The algorithm prioritized these actions given an input of several parameters relevant to the player's and enemies' units' attributes and eventually the avatar learned to select the appropriate actions for each unit.

Another attempt to create agents with GAs has been made for the very popular, classic *Tetris* game [11]. An evaluation function is responsible for scoring each possible move for a certain piece and the action with the best score is then selected. This function is the result of the combination of some feature functions, with the GA producing a set of values that define the respective weights. This method accomplishes quite fast convergence and approximates the best known results to an acceptable point.

In [12], an agent based on macro-actions logic for the *Physical Traveling Salesman Problem* (PTSP) game is developed. PTSP is a 2D game in which the world consists of a maze and the goal is to visit a set of points as soon as possible. A macro action refers to a sequence of actions of specific length and in this technique each individual of the GA corresponds to such an action. Every single action is repeated N times, meaning that a macro action defines $N\times$(macro's length) single actions. The algorithm runs online, producing new individuals in every time window and evaluating them based on a forward model, so as to select the best fitting one. The obtained results indicate a dependency between the length of the macro actions and the effectiveness of the agent, as well as the better performance of the controller over the 2012 PTSP competition winner *Monte Carlo Tree Search* (MCTS) algorithm, in terms of time, when the optimum length is selected.

Other than the previous methods, evolutionary learning has also been used in an indirect way for the development of game playing bots. Fine tuning of AI is a quite common approach, as shown in [13], where an evolutionary system is used in order to determine the parameters and architecture of a neural-network controlled opponent. Furthermore, [14] depicts the use of GAs for the optimization of a set of values, as well as the creation of rule-based decision trees with Genetic Programming, achieving in both cases to improve the default AI.

3   The GAMER Algorithm

The goal of the present algorithm is to find, through a trial and error evolutionary process, the best possible action for an agent at every time step, depending on the game world's state i.e., the positions and orientations of the game's sprites. A GA runs offline in order to determine the most suitable action for each state and during the game the "suggested" actions are checked through an online simulation to decide the final best action. Ideally, this method should include every game object on the map in order to take advantage of all the information provided in every frame. Unfortunately, the search space grows exponentially on the examined world's blocks with the number of different types of sprites being the growth factor, as shown in Eq. 1 below:

$$S(b, t) = t^b \tag{1}$$

where S is the set of different possible game states, b is the number of blocks taken into consideration and t is the number of different sprite types. Obviously, it is practically impossible to encode every game state as, for a common 2D adventure game with about 100 world blocks and 5 different sprite types, there are 5^{100} possible game states. Therefore, a particular state is hardly going to be observed twice in a game exactly as it is and consequently it is necessary to devise a way of treating similar states likewise.

3.1   Motion Encoding

In order to achieve a trade-off between the available information and the computational cost, a window of M blocks around the agent is considered at each time step and the world's blocks are processed in N-groups. In the example of Fig. 1, M is set to 28 and N to 3 (triplets). There are seven triplets on each direction (*up*, *down*, *left* and *right*) but only the ones in the direction of the current target (e.g., the exit door), contribute to the corresponding action's decision. As shown in Fig. 1, triplets consist mainly of blocks in the current orientation but also of blocks in other directions around the avatar; thus significant information is provided for the representation of the game state, while fewer chromosomes are used for the encoding. If the target is positioned diagonally relative to the agent, then all 14 triplets are used. Each triplet is encoded in a chromosome consisting of genes equal to the number of possible different states for the triplet as a whole. Gene values indicate the best action that should be taken according to the state of the triplet (which means only one gene is selected from each chromosome every time) and as a result, each gene contains an integer in the [0, 5] range, since there are six available actions. Finally the action that was indicated most times is chosen and in case of an equilibrium, one of the equals is selected uniformly at random.

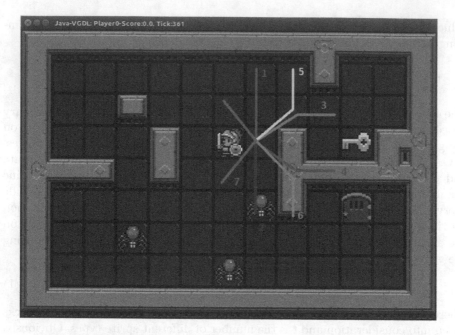

Fig. 1. Selected block triplets in the right direction for the Zelda game (Color figure online)

Even for a N-group consisting of three blocks, a large amount of states may occur. Further reduction of the state space can be achieved by considering less sprite types (Eq. 1). For this reason, apart from the walls and the empty blocks, all other *non-player characters* (NPCs) are encoded in separate chromosomes. In this way, and for the example at hand, each triplet's chromosome may consist of up to 8 genes. Since the triplets are symmetric in the 4 directions, 7 chromosomes (encoding triplets of one direction) are enough to encode all 28 available triplets. These chromosomes are used for the selection of an action when there are no NPCs close to the avatar and they are the same for all games.

3.2 NPCs Encoding

As outlined above, the encoding of a state is different when there are NPCs around the agent. The reason for this design choice is twofold; to reduce the states of the N-groups and therefore make it more feasible to train the motion chromosomes on the one hand, and to be able to process separately the more critical in-danger situations on the other. Since the NPCs' behavior is, obviously, not the same in different games (even of the same category), the encoding has to be separate for each game. Thus, in contrast to the motion chromosomes (which are trained on one of the games and are then reused), the NPCs' chromosomes are trained individually for every game. In the experimental part of this work

(Sect. 4), a more detailed description of the NPCs encoding for the examined games is given.

Figure 2 depicts the structure of the genotype. Each row corresponds to a chromosome and the cells (genes) represent the possible states of the triplets or the NPCs the particular chromosome encodes. Blue, green and red colored chromosomes correspond to motion, and the different NPCs encoding, respectively.

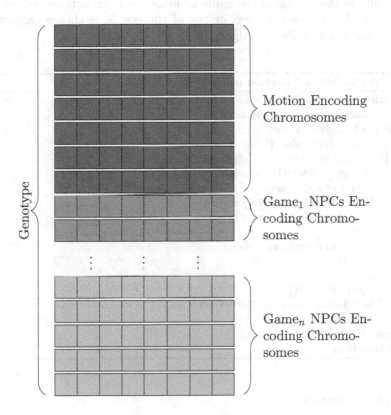

Fig. 2. Genotype structure

3.3   Action Selection

The proposed technique consists of an offline execution of the genetic algorithm (training phase) in which the chromosomes are defined and an online use of the given results combined with an additional check during the gameplay.

After the training phase is over, the agent is ready to play, using the genotype generated by the GA. In the testing phase, the forward model is also included in the decision-making process for each action. The forward model enables the agent to access a future state by simulating an action and is used in this approach as an additional check. More explicitly, in cases where the avatar is in danger, the best action given by the corresponding chromosome is simulated first to ensure that it will not lead to fatal condition (since this check is a one step look-ahead, the forward model is polled at most once every game tick and it complies with the framework's computing time limitations). This is because the NPCs are encoded individually (due to the size of the state space) and consequently the suggested action might not always result to the expected state. If the future state results in the ending of the game or more than one actions get an equal score, a completely random move, or one of the equals, is chosen respectively. This procedure is described in Algorithm 1.

Algorithm 1. Action Selection using Chromosomes

1: **function** ACT(StateObservation)
2:     initialize *actionsValues* to zeros
3:     **if** npcInRange() **then**
4:         *NpcUpdateValues()*
5:         *bestAction* ← *actionsValues.getMax()*
6:         *stCopy* ← *StateObservation.copy()*
7:         *advance(bestAction)*
8:         **if** stCopy.isAgentAlive() **then**
9:             **return** *bestAction*
10:        **else**
11:            **return** *randomAction()*
12:        **end if**
13:    **else**
14:        *updateValues()*
15:        *bestAction* ← *actionsValues.getMax()*
16:        **return** *bestAction*
17:    **end if**
18: **end function**

3.4   GA's Parameters

The general form of the fitness function, which evaluates all individuals of each generation, is presented in Eq. 2 below. The resulting rank is used for the selection phase and obviously for finding a satisfactory solution.

$$fitness = a\left[\mathbf{h}\left(D(subgoal), D(subgoal)_{\max}\right)\right] \tag{2}$$

where a is a reward factor (if the agent is still alive) in the range $[0, 1]$ (determined though experimentation), $D(x)$ is the distance from sprite x and $\mathbf{h}$ is a function that calculates the score based on the avatar's distance from the current goal (or subgoal), depending on each game's mechanics.

Regarding the rest of the GA's hyper-parameters, various selectors and alterers were tested during the experimentation phase, concluding to a set of some widely-accepted defaults. More specifically, the roulette-wheel selector, single-point crossover and uniform mutation operators are chosen for the reproduction process. Uniformly distributed random integers (genes) in the given interval of possible numbers (please recall that each action is encoded by an integer) are sampled in the population initialization phase, so that the algorithm does not converge too fast [13].

4 Experiments

4.1 The GVG-AI Framework

The GVG-AI Competition [6], firstly held in 2014, takes place annually, aiming at the development of general agents (*controllers*), capable of playing any kind of game. A JAVA framework that loads games described in *Video Game Description Language* (VGDL) [15] is provided by the competition organizers, consisting of more than 100 arcade 2D games for single and multi-player modes [16]. Along with the interface, the game engine provides a default set of functions that retrieve and provide information about the state of the world. The most important of them deal with the positions of the sprites, the available actions, the score and the end of the game. In addition, there exists the possibility of simulating the next states, as if certain actions were taken, and "predict" their outcome.

According to the rules of the competition, the agents have a time budget of 40 ms for the selection of each action and 1 s for initialization at the start of the game. The games are divided into three subsets [15]. The first one is the training subset and is openly provided to the contestants. The second subset is hidden and the contestants can only access the results of their agents on these games to check their performance. Finally the third subset is the one that is being used for the testing and ranking of the agents.

The controller presented in this work has been developed and tested on the *Zelda* and *Portals* games[1]. They are both typical adventure-type video games, where the goal is to reach an exit door while surviving from the enemies. More specifically, in Zelda (Figs. 1 and 3a) the agent (*avatar*) has to, at first, find a key and subsequently reach the door. Six actions are available in total, with the action *use* serving as a sword attack, making it possible to also kill the enemies. In Portals (Fig. 3b), there are several blue and purple portals scattered around different rooms of the game world. The blue ones transport the avatar to purple ones randomly and the goal is to reach the correct room and finally exit through the door. This game is more complex and difficult, as there are many enemies of different types and the action *use* is not supported, which means that the avatar is not able to attack and can only avoid enemies. After an initial experimentation procedure, the optimal values of M and N for both games were determined to be 28 and 3, respectively.

---

[1] Available at https://git.islab.ntua.gr/tasos/gamer/.

Initially, the proposed approach is applied to Zelda in order to determine both the motion and the NPCs' chromosomes. Then the motion chromosomes are reused in Portals and the GA runs from scratch, so that the enemy chromosomes are encoded in this game as well. This strategy has been chosen for two reasons; firstly, there are more NPC states to encode in Portals, making it more difficult and time consuming to calculate the motion parameters and secondly it is more likely to use motion chromosomes in Zelda (since there are much less NPCs) and therefore this game is more suitable for this type of training.

4.2 Zelda Configuration

In Zelda, there are 3 or 4 NPCs with similar behavior (as depicted in Fig. 1) at each level. Consequently, it is possible to encode all of them in a single chromosome, since the agent should deal with them in the same way. A notable aspect of this particular game is that when a motion action is chosen, the agent moves in that way only in case the move' s direction matches its orientation; otherwise the action causes a change to the orientation, but the avatar remains still. This means two consecutive executions of a motion action are necessary for the avatar to move in a different direction. The avatar's orientation, therefore, needs to be considered every time an NPC is observed in a range around it.

The described agent's particularity makes it also mandatory to encode the NPCs according to their distance from the avatar. Figure 3a illustrates the window of blocks that are checked for this purpose. In the presented technique, two chromosomes are used; one that is related to the NPCs next to the avatar (blocks $2, 5, 6$ and 9 in Fig. 3a) and one related to the more distant ones. Only NPCs at a maximum distance of two blocks are considered, as the others can not directly threaten the avatar. These chromosomes' genes indicate the best action, depending on the avatar's orientation and the sprite placed to the oriented block.

The main goal of this game is to reach the exit door, but there is also a subgoal which lies on finding the key. Considering this, the fitness function (Eq. 3) is split into two equally weighted parts (0.5), respective to the two goals. The fitness initially depends solely on the distance from the key, thus being able to reach a maximum value of 0.5 and after the key is found, it is incremented according to the distance from the door, leading to a maximum total value of 1 when the level is completed.

$$f_{\text{zelda}} = a \left[0.5 \left(2 - \frac{D(e)}{D(e)_{\max}} \right) c + 0.5 \left(1 - \frac{D(k)}{D(k)_{\max}} \right) (\neg c) \right] \qquad (3)$$

where e is the exit door, k is the key and c is either 1 or 0, depending on whether the avatar has found the key or not. The reward parameter a (survival of the agent itself) has been determined, after experimentation, to be 1.0 if the agent is still alive and 0.7 otherwise.

4.3 Portals Configuration

Portals is a much more complex game than Zelda because of the large number of NPCs, as well as the way they move. Depending on their motion, they can be classified in the following categories:

- Vertical lasers: laser beams with up-down movement that only change direction after collision with a wall
- Horizontal lasers: laser beams with left-right movement that only change direction after collision with a wall
- Golden spheres: small spheres that move disorderly around in space

Apart from the various existing types, another characteristic of the game is the possible blocks for the NPCs to move to. Although the avatar can only move from one single block to a neighboring one at a time (just like in Zelda), NPCs can also move in-between. Laser-type ones need two actions in order to move to the next block, meaning they are placed between the starting and destination blocks (Fig. 3b) after the first action is taken, while the sphere-type ones can be placed on any coordinates of the world, as long as this is permitted by the topology.

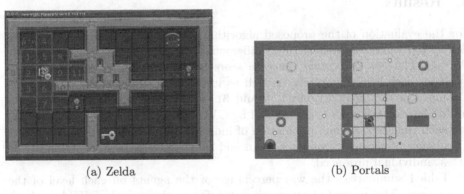

(a) Zelda   (b) Portals

Fig. 3. Blocks checked for NPCs (Color figure online)

Considering the encoding in this case, it would not be effective to encode all NPCs on a single chromosome, as their behavior is significantly different. As a result, a number of chromosomes equal to the various types of NPCs is necessary for the encoding of each type individually. The laser categories presented above can be further separated, depending on the direction of the sprite, which means that two subclasses arise from each of the first two categories, resulting in a total of five different types. These are:

- Vertical up-oriented lasers
- Vertical down-oriented lasers
- Horizontal left-oriented lasers
- Horizontal right-oriented lasers
- Golden spheres

As in Zelda, the symmetry permits the encoding of each type in one direction and then the modification of the selected action, so that it responds to the actual enemy position. Therefore, five chromosomes are to be defined by the GA, each for one of the possible types, indicating the best action depending on the NPC's direction and position.

Concerning the fitness function (Eq. 4) in this game, there are several different rooms the avatar has to go through until it reaches the exit. Thus, every room reached increases the total fitness score by a certain amount, depending on the number of rooms in the level. When moving around a room, the fitness is also increased proportionally to the distance from the portal that leads to the next room. All the values sum up to 1 if the avatar wins.

$$f_{\text{portals}} = a \left[r_{\max}^i - \left(r_{\max}^i - r_{\max}^{i-1} \right) \frac{D(g^i)}{D(g^i)_{\max}} \right] \tag{4}$$

where r_{max}^i is the maximum fitness that can be gained in the ith room and g^i is the ith room's goal. The reward parameter a is set to the same values as in Sect. 4.2.

5   Results

For the evaluation of the proposed algorithm, the agent played 500 games for Zelda, 100 times for each of the 5 different levels, and another 500 games for Portals. The *win percentage, average score* and *timesteps played* were calculated for each level separately, as well as in total, to indicate the overall performance of the agent (Tables 1, 2 and 3). The results are compared with a rolling horizon GA approach provided by GVG-AI. This controller evolves, during each timestep, a small population of individuals (each individual represents a sequence of actions), evaluates them and then chooses the first action of the fittest individual produced.

Table 1 summarizes the win percentage of the agents on each level of the two games. The presented approach outperforms the sample RHGA in both games in the total of five levels, even though the latter agent performs better in some levels of Portals. More specifically, with the exception of the last level in Portals, GAMER is capable of winning in all of the other tested levels. Portals' L4 difficulty lies in the map configuration which is full of NPCs, pushing the agent to continuously prioritize survival and fail to efficiently plan a path to the final goal. The particular level's peculiarity is also exhibited in RHGA's low win percentage. The total win ratio difference between Zelda and Portals confirms the complexity of the latter due to the wide variety of sprites, which makes the representation of the world harder. Furthermore, apart from the various worlds in a single game, the agent also seems to adapt well to both games using the same motion encoding.

In Portals, the final score, is binary (1 for a win, 0 for a loss) and therefore the average score is actually equal to the win percentage. In Zelda, more aspects need to be considered. In particular and according to the GVG-AI framework,

Table 1. Win Percentage at each level (100 runs) for the Zelda and Portals games for both approaches. Values in bold indicate the highest performance in the corresponding level.

Win percentage (100 games per level)						
Zelda						
Levels	L0	L1	L2	L3	L4	Overall
RHGA	30.0%	9.0%	7.0%	15.0%	67.0%	25.6%
GAMER	**38.0%**	**11.0%**	**98.0%**	**64.0%**	**77.0%**	**57.6%**
Portals						
Levels	L0	L1	L2	L3	L4	Overall
RHGA	19.0%	**69.0%**	68.0%	**34.0%**	**3.0%**	38.6%
GAMER	**29.0%**	60.0%	**94.0%**	14.0%	0.0%	**39.4%**

Table 2. Normalized average scores at each level (100 runs) for the Zelda and Portals games for both approaches. Values in bold indicate the highest performance in the corresponding level.

Normalized average scores (100 games per level)												
Zelda							Portals					
Levels	L0	L1	L2	L3	L4	Overall	L0	L1	L2	L3	L4	Overall
RHGA	**8.46**	**6.41**	**6.55**	**7.48**	**8.63**	**7.51**	1.9	**6.9**	6.8	**3.4**	**0.3**	3.86
GAMER	5.44	4.47	6.13	7.43	6.48	5.99	**2.9**	6	**9.4**	1.4	0.0	**3.94**

2 points are assigned for each killed enemy, 1 for finding the key, 1 for winning the game and 1 point is subtracted for dying. Since the computed scores are based on different criteria, they are first normalized in the [0, 10] scale in order to be comparable.

A strange, at a first glance, conclusion drawn from the matching of wins and scores on each level of the Zelda game (Tables 1 and 2), is that they do not seem to be proportional. On the contrary, even though GAMER achieves a higher win rate, it has a lower average score. A logical expectation would be for the score to increase along with the amount of wins, however this depends on the fitness function. Equation 3 shows that the individuals' fitness scores are determined only by the key finding subgoal along with the distance from the exit door. That explains why the agent does not seek to kill NPCs and thus the final score gets lower as the agent gets more successful. This fact is also confirmed by the average timesteps taken by the two agents, as shown in Table 3. GAMER places the emphasis on completing the levels as fast as possible, while the rolling horizon GA approach spends more time playing and as a result it scores more points. Similar assumptions can not be made for Portals, since the score depends exclusively on the avatar's victory or defeat.

Considering the evolution process of the GA, the best individual's fitness and the win percentage per generation for both games are presented in Figs. 4 and 5. As expected, the starting fitness scores are higher in the Portals' execution. That is because the motion chromosomes that were calculated in Zelda are used directly in Portals, meaning a part of the needed information is already known and the individuals perform better. However, the first winning agent in Portals appears at generation 38 while in Zelda it needs only 20 generations, indicating the excess in difficulty of the first. As highlighted in Fig. 6, in which the two approaches (the one reusing the motion part and the one which does not) for Portals are compared, the reuse of the motion chromosomes speeds up the process (the first winning agent in the non-reusing case is a member of the 68th generation) and improves the performance of the GA, by reducing the number of chromosomes that have to be determined.

Table 3. Average timesteps at each level (100 runs) for the Zelda and Portals games for both approaches. Values in bold indicate the highest performance in the corresponding level.

Average timesteps (100 games per level)						
	Zelda					
Levels	L0	L1	L2	L3	L4	Overall
RHGA	855.2	951.3	918.9	797.8	411.2	786.9
GAMER	**765.7**	**882.5**	**204.3**	631.0	**395.2**	**575.7**
	Portals					
	L0	L1	L2	L3	L4	Overall
RHGA	453.8	252.5	597.2	471.5	49.1	364.8
GAMER	**129.0**	**80.9**	**89.7**	**49.9**	**20.9**	**74.1**

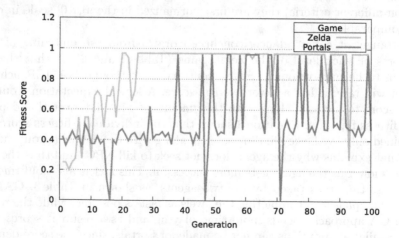

Fig. 4. Best fitness score per generation

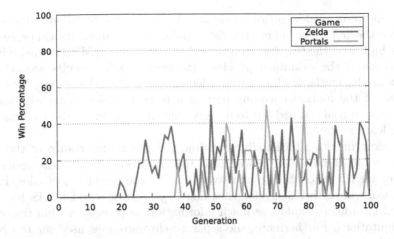

Fig. 5. Win percentage per generation

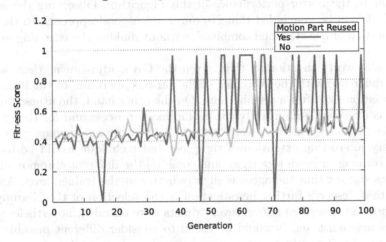

Fig. 6. Best fitness score per generation (Portals)

6 Conclusion

In this paper, a procedure for encoding actions depending on the world's states in action-adventure type video games via genetic algorithms has been presented. The agent has been implemented as the combination of two offline executed GAs and an online action evaluation procedure during the game. The GVG-AI framework has been used for the development and testing of the agent and the rolling horizon GA agent (provided by the competition) was taken as the baseline, in order to measure GAMER's performance. GAMER surpassed the aforementioned approach in terms of the overall win percentage, although it was not superior in all levels individually, and proved to be faster in completing the tested levels as well.

The obtained results outline its capability of adapting to different topologies played for the first time. In both the Zelda and Portals games, the win percentage remains higher on the training level (level 2) indicating GAMER's sensitivity on the instance of the examined problem. However, similar results are achieved in most of the levels, leading to an adequate overall performance. Moreover, the reuse of the motion encoding part in a more complex game with similar mechanics and goals turned out to be very efficient in terms of the training time and the learning rate.

The experiments additionally underline the great importance of the state representation. It has been proven that the reduction of the state space was necessary for the algorithm to effectively encode the possible scenarios. In the present work, a methodology for encoding N-groups of world blocks has been proposed in order to exploit as much information as possible within the search space limitations. Furthermore, the separate chromosomes used for the NPCs along with the worlds' symmetry contributed to the groups' states decrease and thereafter to the better performance of this algorithm. Observing the agent's behavior, it can be concluded that N-groups are a good approach to the total representation of a state when combined, without making the encoding unfeasible.

It is also worth mentioning that, given the GA's adjustment, there was no notable difference with the switching of the various operators or the increase of the population size after a certain point. On the other hand, the fitness function affects, as expected, significantly the evolutionary process and for that reason, two different fitness functions were devised for the examined games.

A very interesting extension of the presented technique would be to carry out the training in more topologies and combine the different chromosomes, as the scores suggest that the agent is slightly better at the trained level. Another feature that deserves further investigation is the selection of the N-groups. In the experiment described above, seven triplets were encoded; nevertheless other solutions may exist and it would be useful to consider different possible block groups. Finally, it would be interesting to implement a similar state encoding genetic algorithm on another game category in order to verify that the proposed approach can be generalized to various video game types.

References

1. Campbell, M., Hoane, A., Hsu, F.H.: Deep blue. Artif. Intell. **134**(1), 57–83 (2002). http://www.sciencedirect.com/science/article/pii/S0004370201001291
2. Silver, D., et al.: Mastering the game of Go with deep neural networks and tree search. Nature **529**(7587), 484–489 (2016). http://dx.doi.org/10.1038/nature16961
3. Stanley, K.O., Bryant, B.D., Miikkulainen, R.: Real-time neuroevolution in the nero video game. IEEE Trans. Evol. Comput. **9**(6), 653–668 (2005)
4. Galway, L., Charles, D., Black, M.: Machine learning in digital games: a survey. Artif. Intell. Rev. **29**(2), 123–161 (2008). https://doi.org/10.1007/s10462-009-9112-y

5. Salimans, T., Ho, J., Chen, X., Sutskever, I.: Evolution strategies as a scalable alternative to reinforcement learning. CoRR abs/1703.03864 (2017)
6. Perez-Liebana, D., Samothrakis, S., Togelius, J., Lucas, S., Schaul, T.: General video game AI: competition, challenges, and opportunities, pp. 4335–4337. AAAI Press (2016)
7. Denzinger, J., Kordt, M.: Evolutionary online learning of cooperative behavior with situation-action pairs. In: Proceedings of the Fourth International Conference on MultiAgent Systems, pp. 103–110. IEEE (2000)
8. ElecByte: M.U.G.E.N - make your own 2D fighting game. http://www.elecbyte.com/mugendocs-11b1/mugen.html
9. Martínez-Arellano, G., Cant, R., Woods, D.: Creating AI characters for fighting games using genetic programming. IEEE Trans. Comput. Intell. AI Games 9(4), 423–434 (2017)
10. Hsu, W.L., p. Chen, Y.: Learning to select actions in starcraft with genetic algorithms. In: Conference on Technologies and Applications of Artificial Intelligence (TAAI), pp. 270–277 (2016)
11. da Silva, R.S., Parpinelli, R.S.: Playing the original game boy tetris using a real coded genetic algorithm. In: Brazilian Conference on Intelligent Systems (BRACIS), pp. 282–287, October 2017
12. Perez, D., Samothrakis, S., Lucas, S., Rohlfshagen, P.: Rolling horizon evolution versus tree search for navigation in single-player real-time games. In: Proceedings of the 15th Annual Conference on Genetic and Evolutionary Computation, GECCO 2013, pp. 351–358. ACM, New York (2013). http://doi.acm.org/10.1145/2463372.2463413
13. Spronck, P., Sprinkhuizen-Kuyper, I., Postma, E.: Improving opponent intelligence through offline evolutionary learning. Int. J. Intell. Games Simul. 2(1), 20–27 (2003)
14. Mora, A.M., et al.: Evolving bot AI in unreal^{TM}. In: Di Chio, C., Cagnoni, S., Cotta, C., Ebner, M., Ekárt, A., Esparcia-Alcazar, A.I., Goh, C.-K., Merelo, J.J., Neri, F., Preuß, M., Togelius, J., Yannakakis, G.N. (eds.) EvoApplications 2010. LNCS, vol. 6024, pp. 171–180. Springer, Heidelberg (2010). https://doi.org/10.1007/978-3-642-12239-2_18
15. Perez-Liebana, D., Samothrakis, S., Togelius, J., Schaul, T., Lucas, S.M., Couëtoux, A., Lee, J., Lim, C.U., Thompson, T.: The 2014 general video game playing competition. IEEE Trans. Comput. Intell. AI Games 8(3), 229–243 (2016)
16. Gaina, R.D., Pérez-Liébana, D., Lucas, S.M.: General video game for 2 players: framework and competition. In: 8th Computer Science and Electronic Engineering (CEEC), pp. 186–191, September 2016

Effects of Input Addition in Learning for Adaptive Games: Towards Learning with Structural Changes

Iago Bonnici[✉], Abdelkader Gouaïch, and Fabien Michel

LIRMM, Université de Montpellier, CNRS, Montpellier, France
{iago.bonnici,gouaich,fmichel}@lirmm.fr
http://www.lirmm.fr/

Abstract. Adaptive Games (AG) involve a controller agent that continuously feeds from player actions and game state to tweak a set of game parameters in order to maintain or achieve an objective function such as the flow measure defined by Csíkszentmihályi. This can be considered a Reinforcement Learning (RL) situation, so that classical Machine Learning (ML) approaches can be used. On the other hand, many games naturally exhibit an incremental gameplay where new actions and elements are introduced or removed progressively to enhance player's learning curve or to introduce variety within the game. This makes the RL situation unusual because the controller agent input/output signature can change over the course of learning. In this paper, we get interested in this unusual "protean" learning situation (PL). In particular, we assess how the learner can rely on its past shapes and experience to keep improving among signature changes without needing to restart the learning from scratch on each change. We first develop a rigorous formalization of the PL problem. Then, we address the first elementary signature change: "input addition", with Recurrent Neural Networks (RNNs) in an idealized PL situation. As a first result, we find that it is possible to benefit from prior learning in RNNs even if the past controller agent signature has less inputs. The use of PL in AG thus remains encouraged. Investigating output addition, input/output removal and translating these results to generic PL will be part of future works.

Keywords: Adaptive games · Reinforcement Learning ·
Transfer Learning · Recurrent networks

1 Introduction

Adaptive Games. Video games are part of our digital culture and economy with many applications in ranging from entertainment to training and sport. A video game has a specific feature: It is a software that interacts with players to create a subjective experience that mixes rationality, emotion and aesthetics.

© Springer Nature Switzerland AG 2019
P. Kaufmann and P. A. Castillo (Eds.): EvoApplications 2019, LNCS 11454, pp. 172–184, 2019.
https://doi.org/10.1007/978-3-030-16692-2_12

The degree to which the player values this experience determines how much the video game is accepted and successfully used as an entertainment or training means. The subjective nature of the player's experience reinforces the need for adaptation to better take into account individual characteristics during the play.

This defines an *adaptive game* (AG) as a game that exposes some parameters to be tweaked by an external controller in order to maintain some metrics – often related to player's experience – in an acceptable range of values. AG therefore requires a particular component, called a *control agent*, that continuously feeds from player actions and game state and tweaks a set of game parameters so that player's experience metrics remain optimal. This can be considered a Reinforcement Learning situation (RL) [1,2], so classical machine learning approaches (ML) can be used. With the development of neural networks, in particular Recurrent Neural Networks (RNNs) that are well suited for sequential processing [3,4], new opportunities are offered to develop a game controller using machine learning techniques.

The Protean Situation. Games that aim to create elaborated and original experiences need to provide a rich set of actions and rules. Due to this intrinsic complexity, these games cannot be designed, neither exposed to players, as a monolithic block. In fact, an incremental approach is often required to gradually introduce the gameplay, follow player's learning curve and ease understanding of the game mechanics. This incremental approach is also followed by many training strategies where trainees are trained with only a subset of the whole system and additional concepts are introduced only when this subset is mastered. For instance, Chess can be introduced with sequential subgames involving only pawns and kings, then bishops and rooks, *etc.* Also, the richness of the set of actions and rules is sometimes explored back and forth, with the succession of various game levels where the player's abilities vary from one situation to the other (*e.g.* can fly *vs.* cannot, senses enemies *vs.* does not). In a sense, the game structure changes, leading to changes in the control agent's set of inputs, outputs and feedbacks. We refer to this set as the agent's "signature".

Suppose now that a control agent has been trained using ML techniques and tools, such as RNN, to optimize the player's metrics under the first game signature Δ_0. The question asked is: How can this agent, trained with Δ_0, be used later under other signatures like Δ_1, Δ_2, *etc.*? Is it more interesting to restart the training from scratch on each signature change or is it possible to benefit from previous experience even though the signature is not the same?

In this paper, we trial this question by exploring and evaluating a learning solution which capitalizes on past experiences, even when the signature partially changes. We use the adjective *protean* to characterize this variable nature of the control agent signature, and refer to the situation as a generalization of RL to Protean Learning (PL). After a succinct overview of related works (next section), we shall offer a formalization of PL (Sect. 2.2), and a description of the experimental setting (Sect. 2.3). Section 3 shall expose and discuss our first results before we finally conclude.

Related Work. Situations similar to PL are known in the domain of Transfer Learning (TL). In TL, the agent has already learned tasks called "source" tasks, and the challenge is to benefit from this previous "knowledge" while tackling a new "target" task. In other words, the TL agent is expected to generalize not only *within* tasks, but also *across* tasks [5,6]. This domain is transversal to ML as it applies both to Supervised Learning (SL) and RL. TL may be invoked in various situations: **(1)** The source training process has been successful but costly: one wishes to benefit from transfer to tackle a new target task more efficiently [5,7]. **(2)** The target task is challenging: one wishes to split it up into several easier source tasks, expecting that the overall TL process will be more efficient than direct tackling of the target [5, 8,9]. **(3)** Several tasks must be learned at once: one wishes that TL occurs from the ones to the others, and speeds up the parallel process [10]. **(4)** It is known that the task at hand will undergo future changes: one wishes to design an agent able to adapt these changes and benefit from transfer from one task to the next. AG controllers are in the latter situation. PL is an instance of this situation.

PL is also related to Concept Drift (CD), a situation where the environment is assumed to undergo changes while the agent keeps learning [11,12]. It is also related to Continual Learning (CL) or "lifelong learning", an AI design where the agent keeps learning although its environment is changing and it regularly faces new challenges [13,14]. To our knowledge, even though the environment function is expected to change in TL, CD, CL, the signature of the agent is often assumed to be fixed. Here, we focus on the various changes in signature that may occur during the learning process (*e.g.* new input, change in output domain, input loss), and how transfer can be achieved in each case with the classical RNNs tools.

2   Materials and Methods

2.1   Approach

Studying PL in AGs must first be conducted with small AGs that optimize basic player's metrics. For instance, the metric optimized may be Csíkszentmihályi's flow, a psychological state universally perceived as a positive experience [15,16]. The video game flOw already does this adaptation [17,18], and would easily be adapted so that its signature changes from one level to the other. This will serve as an experimental setup in subsequent works.

To reach this objective, we first need to formalize the system that we call PL. This is done in Sect. 2.2. Before testing PL in AGs, the basic relevance of PL has to be asserted. In Sect. 2.3, we conduct an experiment involving RNNs in a idealized PL situation to assess the first operation among 6 elementary signature changes: input addition, input removal, output or feedback addition or removal. This experiment verifies that adding an input to the agent during the learning process does not dramatically alter learning, and that a protean learner performs better in this situation than a learner resuming from scratch on a signature change. Therefore, it encourages subsequent works towards the PL AG objective.

2.2   Formalization of PL

In this section, we sum up a formalization of PL situation, based on detailed report [19]. It is offered as a generalization of RL to non-fixed signature learning situations.

A RL agent is continuously fed with inputs, so its inputs can be seen as *signals*: values that change over time. It also feeds from feedback signals, and produces output signals. Therefore RL can be viewed as a case of *signal processing*, where the learner continuously transforms the inputs signals into output signals, with the objective that feedback signals are kept high. On the other hand, the environment continuously transforms the output signals into input signals and feedbacks, by strict enforcement of the universe rules.

The agent *signature* is the number and the type of signals it produces and feeds from. In other words, it is the collection of domains the various signals take their value from. The idea with this formalization is to make the signature a signal itself. As such, it may also change in time, which makes the agent *protean* and extends RL to PL.

We formalize signals with plain functions of continuous time that we call *flows*: $h : \mathbb{R}^+ \rightarrow D$. They take their value in arbitrary domains D. Flows can be discretized in time with arbitrary precision $\epsilon \in \mathbb{R}^{+*}$ by sequences $^\varepsilon h : \mathbb{N} \rightarrow D$ such that:

$$\forall t \in \mathbb{N}, \quad {}^\varepsilon h(t) = h(t\epsilon) \tag{1}$$

Flows transform into each other. Viewed another way, a flow can be determined by another flow. We call a *determination function* f a function able to determine an outgoing flow k from an incoming flow h no matter the precision ϵ considered:

$$\forall \epsilon \in \mathbb{R}^{+*}, \forall t \in \mathbb{N}, \quad {}^\varepsilon k(t) = f_\epsilon({}^\varepsilon h(t), {}^\varepsilon h(t-1), \ldots, {}^\varepsilon h(0)) \tag{2}$$

Flow determination has a *memory* in that current value of k may depend on past values of h, so it is not Markovian in general. However, it is always *causal* in that future values of k cannot be determined given only the current and past values of h. We shall use the following graphical alias for the determination relation (2):

$$h - (f) \rightarrow k \tag{3}$$

The symbol in parentheses represents the determination function, the symbol pointed by the arrow head is the consequence flow, and the symbol pointed by the line with no head is the cause flow. For instance, $i - (P) \rightarrow o$ means that the inner agent process P feeds from input signal i to produce the output signal o in a causal, yet potentially non-Markovian way. And $o - (E) \rightarrow i$ means that the environment E works the other way round.

Multiple flows h carry both a flow of domains h^Δ (or "signature") and a flow of actual values h^ν. *Domains* are a tuple of sets $(S_1, S_2, \ldots)$ and *values* are a tuple of elements of these sets $(v_1 \in S_1, v_2 \in S_2, \ldots)$. For instance, at $t = 0.9$, an agent that is sensitive to both "wind direction and player speed" in the game simulation may receive $h^\Delta(0.9) = ([0, 2\pi[, \mathbb{R}^+)$ and $h^\nu(0.9) = (0.2, 50)$.

The particularity of PL is that the flow of domains is not constant. We call *transformation* of the agent any variation of h^Δ resulting in that the agent may later receive values with different domain signatures, *e.g.* "player speed and rain intensity" $h^\Delta(1.1) = (\mathbb{R}^+, [\![1,5]\!])$ and $h^\nu(1.1) = (31, 4)$.

At the highest level, a PL learner agent can be represented by 3 multiple flows (i, o, φ) and 1 flow of determining functions P with the following determination scheme:

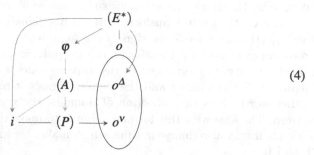

$$(4)$$

The latter scheme can be considered a set of formal equations according to (3) and (2). Each color corresponds to one determination triplet, where:

- E represents the environment in which the agent is immersed. The * denotes that initial values for i, φ, o^Δ are determined by E. For a control agent in AG, E represents both the player and the game engine.
- i represents the agent's *inputs* or *sensors*. In AG, i informs the agent of player's actions and game state. Their nature may change in time as the signature i^Δ evolves (*e.g.* among game levels).
- o represents the agent's *outputs* or *actuators*. In AG, o controls the game parameters that need to be adjusted to maintain good player's metrics. Their nature may also change in time as o^Δ evolves. Note that output values o^ν are determined by the agent, but the output signature o^Δ is determined by the environment.
- φ represents the agent's *feedback*, *rewards* or *objectives*, a continuously fed evaluation of the actions it undertakes. In AG, φ may represent the player's experience metrics like Csíkszentmihályi's flow. They also may change in nature as φ^Δ evolves.

The environment determines i, φ and o^Δ, so the agent cannot directly decide its input or feedback data, nor its output signature.

- P represents the agent current behavior. It is an inner computational procedure that determines current output values based on current input values and all their history.
- A is the abstract strategy of the agent, which is continuously adapting its behavior P based on environmental information.

Only two objects are not depending on time here: the environment E and the inner agent strategy A.

The classical RL problem of maximizing reward [1] can be reformulated in PL as: Given environment E and a corresponding flow of rewards φ with all domains being numerical, find an agent procedure A such that all values taken by signal φ^ν are maximized. A is considered an interesting learner if it can maximize the rewards for a whole family of environments, and no matter the changes in signatures i^Δ, o^Δ or φ^Δ.

In other words, an AG control agent is interesting if it can maintain good experience metrics no matter the player skills, or the game evolution.

2.3 Experiments

As a preliminary to the construction of generic PL agents, we design an experiment to assess viability of the first type of signature change during learning: "input addition". To this end, we restrict ourselves to an ideal situation where we, as experimenters, know the optimal behavior $\hat{P}$. The agent A will be able to access a set of example optimal realizations $T = \{(i_k, \hat{o}_k)\}_{k \in [\![1,1000]\!]}$ with every

$i_k - (\hat{P}) \rightarrow \hat{o}_k$ (see Sect. 2.2). This will constitute a *training set* so the agent

search for an approximation of $\hat{P}$ can be solved by a supervised learning approach (SL). To simulate the signature change, we stop the SL procedure, change the dimension of each i_k along with the signature of $\hat{P}$, and resume SL. Comparison is made with an agent that directly learns with the second signature. The less efficient the second agent compared to the first one, the more encouraging it is in favor the PL approach.

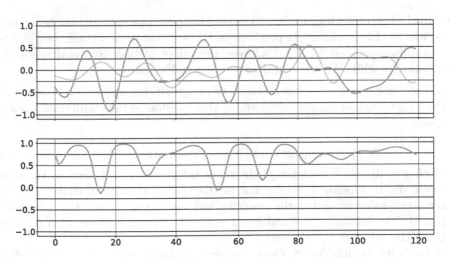

Fig. 1. Example of smooth processed signals. Top: 2D input signal i_k. Bottom: 1D output signal $\hat{o}_k$. In this mem example, $p = 0.8$ (high influence of first dimension of the signal (solid one)) and $c = 1$ (highly skewed output result).

Relative efficiency of these agents is measured in various experimental settings to address results robustness. In the next section, we shall describe protocol for the generation of T, the SL approximation, the signature change simulation and the comparison of the agents, along with the variable experimental settings.

Inputs Generation. Idealized input i_k, are 2-dimensional signals generated in two different ways, depending on the first experimental parameter:

– In the noisy setting, each value is drawn from a uniform distribution so that inputs have no structure: $i_k(t) \hookrightarrow \mathcal{U}([-1, 1]^2)$.
– In the smooth setting, each dimension of the signal is generated independently as a combination of sine waves, so that inputs are autocorrelated. The generation procedure is described as follows (example Fig. 1 top panel):

for each dimension:
  Draw the number of sine waves from a Poisson distribution $\mathcal{P}(500)$
  for each sine wave:
    draw the pulsation ω from uniform $\mathcal{U}([0, 30])$
    draw the phase φ from uniform $\mathcal{U}([0, 2\pi])$
    evaluate $\sin(\omega t + \varphi)$ for 120 values of t evenly spaced between 0 and 2
  end for
  sum all sine waves into one combined signal sequence
  draw the final amplitude a from uniform $\mathcal{U}([0, 1])$
  rescale the sequence so it has desired amplitude:
  sequence $\leftarrow a \times \frac{\text{sequence}}{\max(|\text{sequence}|)}$
end for
join both sequences into one input signal $i_k : [\![1, 120]\!] \rightarrow [-1, 1]^2$

Optimal Outputs Generation. Training outputs examples $\hat{o}_k$ are generated using idealized behavior $\hat{P}$. Depending on the experimental setting, two different formulae are used for $\hat{P}$ to address the effect of task complexity: one is instantaneous while the other exhibits a temporal memory.

– In the nomem setting, $\hat{P}$ has no memory and performs a plain combination of the 2 input signals dimensions i_k^1 and i_k^2 that is affine in the atanh transformed space:

$$\hat{o}_k(t) = \tanh\big(p\,\text{atanh}\big(i_k^1(t)\big) + (1-p)\text{atanh}\big(i_k^2(t)\big) + c\big) \tag{5}$$

With $p \in [0, 1]$. The higher p, the more information contained in the first input signal dimension i_k^1, and not in the second i_k^2. The further c is from zero, the more skewed is the resulting output signal $\hat{o}_k$ (see Fig. 1 bottom panel). The settings $c = 0$ and $c = 1$ are tested, in interaction with p being given the values 0, 0.2, 0.5 and 0.8.
– In the mem setting, the processor is more complex because it exhibits a 1-step memory, so it is non-Markovian. It performs the same combination with the exception that $5 \times (i_k^d(t) - i_k^d(t-1))$ is used instead of $\text{atanh}\big(i_k^d(t)\big)$ in (5). In other words, the derivative of the input signals are used instead of their actual values.

Agent Structure. The agent approximates $\hat{P}$ with its produced actual P. P takes the form of a Recurrent Neural Network (RNN) [3,4]. We use a standard Gated Recurrent Unit (GRU) [20]. P produces the agent actual outputs according to $i_k - (P) \to o_k$. Two different structures are used for P to address their effect on PL.

- In the flat setting, 1 GRU cell is used with 1 internal state, used as network output. So P is Markovian.
- In the deep setting, 2 GRU cells are used as different network layers with 6 internal states, the last one being used as the network output. So P may be non-Markovian.

Loss Function. The loss function to optimize is a classical Mean-Squared-Error (MSE) between the agent predictions o_k and the expected results $\hat{o}_k$.

The learning procedure A processes training examples by batches of 10, and updates the weights parameters of P with a stochastic gradient descent respecting Adam update rule [21] (learning rate $= 0.01$).

Convergence is achieved using pytorch [22] for 1000 iterations.

Realization of Signature Change. Three convergences are achieved on each run (see Fig. 2), according to the following protocol:

- first, one protean "first-form" agent A_{f_1} is constructed with a 1D-1D signature. Its parameters are randomly initialized, then it is trained against the training set T but it only feeds from the first dimension i_k^1 of input signals, being blind to i_k^2.
- second, one protean "second-form" agent A_{f_2} is constructed with a 2D-1D signature. Its parameters are initialized by copying the final value of all homologous parameters from A_{f_1}, and setting additional parameters to zero. Then, it is trained against the whole training set T, not ignoring the second dimension i_k^2 anymore.
- third, one classical "direct" agent A_d is constructed with a 2D-1D signature. Its parameters are randomly initialized, then it is directly trained against the whole training set T.

The (A_{f_1}, A_{f_2}) agent is the experimental model of a PL agent in the idealized learning situation. It experiences the signature change "input addition".

1000 such replicates are run for each combination of experimental settings.

Measure of the Advantage. The advantage of PL compared to direct learning is estimated on learning curves l by a score (see Fig. 2). Score is calculated on each run as a *gain* according to:

$$\text{score} = \text{mean}\left(\log_2\left(\frac{l_{A_d}}{l_{A_{f_2}}}\right)\right) \tag{6}$$

A score $= 1$ means that the second-form agent error is twice as low as the direct agent on average. A score $= 0$ means that the second-form agent and the direct agent perform similarly.

3   Results

The scores obtained on each run are summarized and illustrated in Fig. 3.

To address relevance of the observed variations, we fitted a linear model on the data, testing all interactions between experimental settings considered as factors and one nested slope for parameter p considered numerical (degrees of freedom: 63698, residual stde: 0.7689). Convergence and analysis of the model have been achieved with R-Cran software [23]. All effects were different from zero with high significance $p\text{-}value \leq$ .001 with the only exception of the nested slope flat:mem:smooth:c=0:p ($p\text{-}value = .8118$). We therefore consider that all trends among groups observable on Fig. 3 are relevant to discuss, except the latter slope that must be considered null.

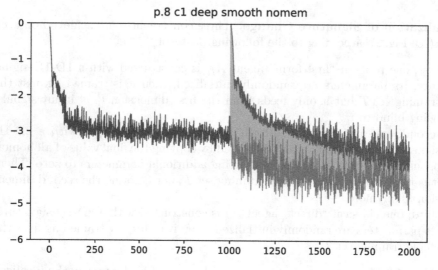

Fig. 2. Example of experimental learning curves l: evolution of the $\log_{10}(\text{MSE})$: the "first form" learner A_{f_1} ($l_{A_{f_1}}$ in blue) is trained to predict a combination of a 2D signal, but it can only access the first dimension as an input so it has incomplete information. The "direct" learner A_d (l_{A_d} in black) is trained on the same task, but it can access the whole information so it performs better. The "second-form" learner A_{f_2} ($l_{A_{f_2}}$ in green) can also access the whole information so it has the same signature as A_d. But A_{f_2} parameters are grown from previous A_{f_1} instead of being randomly initialized. The gain score (in red ≈ 5.242) measures the benefit of PL (based on previous experience) compared to starting the learning from scratch on a signature change. The signature change "input addition" occurs at $t = 1000$. (Color figure online)

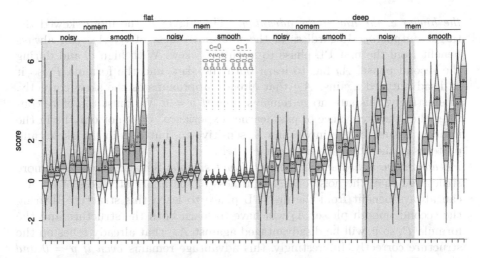

Fig. 3. Violin plot: Comparison of scores in various PL situations differing by the RNN structure of P (flat or deep), the task complexity $\hat{P}$ (nomem or mem), the input data i_k properties (smooth or noisy) and the relative informativeness of partial information (c, p). The benefit of PL differs depending on the situation, but is overall positive. Within each violin, the dashed line represents mean value for the group, the solid line represents median value, and the two gray areas represent 50% and 90% percentiles. Blue circles represent predictions of the linear model fitted on the data. (Color figure online)

However, convergence of the network and good approximation of P within the setting flat:mem is impossible because P (non-Markovian then) uses a memory information that neither A_{f_1}, A_{f_2} nor A_d (Markovian then) can access. As a consequence, they all poorly approximate P so these score values must be considered carefully. Correctly interpreting these values requires further investigations under conditions where networks fail to converge, which is out of the scope of this paper.

As expected, no matter the experimental setting, the measured score is mostly positive on average. This reflects the advantage of the second-form agent A_{f_2} compared to the direct agent A_d after the signature change. When the event "add input" occurs on 1D-1D A_{f_1}, it is better to transform it into a 2D-1D adapted A_{f_2}—and keep its parameters that have converged so far, than to replace it with a new, naive 2D-1D A_d—and forget everything that has been learned so far. This advantage needs to be qualified depending on the learning situation:

– *the lower p, the lower the advantage*: The lower p, the less informative the first dimension i_k^1 that A_{f_1} is sensitive to, because it has low impact on the desired output. As a consequence, there is less learning gain to harvest for the protean agent during the first phase of PL. This confirms a naive intuition that PL is only interesting if past experience of the learner is somehow relevant.

- *the higher c, the higher the advantage*: The higher c, the more skewed the target output $\hat{o}_k$. Regardless of the informativeness of i_k^1, A_{f_1} can always benefit from the first PL phase to learn this skew. With high c, and during the second phase, A_d has to learn both the skew and the formula $\hat{P}$, so it is disadvantaged against A_{f_2} that already approximates the skew correctly. Interestingly, this advantage remains *even if p = 0*. We must therefore consider that the "relevance of past learner experience" does not only lie in the informativeness of the inputs it was sensitive to, but also in the skews and patterns of the objective behavior $\hat{P}$ itself.
- *PL advantage is higher with **smooth** input signals*: Smooth signals are more structured than the noisy ones. Regardless of the informativeness of i_k^1, A_{f_1} can always benefit from the first PL phase to learn this structure. During the second smooth phase, A_d will have to learn both the structure and the formula $\hat{P}$, so it will be disadvantaged against A_{f_2} that already relies on the structure correctly. Interestingly, this advantage remains *even if p = 0 and c = 0*. Therefore, the "freelance of past learner experience" also lies in the very structure of past processed data.
- *PL advantage is lower with **deep** RNNs* (at least in nomem setting for we do not discuss flat:mem): Deeper RNNs are more flexible in the sense that there exists more combinations of their parameters that approximate the objective function $\hat{P}$ correctly. As a consequence, it is easier for A_d to find a path down the loss function and converge during the second PL phase, which makes A_{f_2} advantage less decisive at this stage.

As a summary, we observe that PL seems mostly beneficial after an "input addition", not only if past inputs were relevant ($p > 0$), but also if the learning data is somehow structured ($c = 1$, smooth). Our interpretation is that prior A_{f_1} learner can learn and transmit partial information about this structure to current A_{f_2} agent *via* RNN parameters. One possibility could be that A_{f_1} sometimes converges towards a "dead-end" direction during the first phase of the experiment, so that it has to "unlearn" during the second phase and be disadvantaged against A_d—a phenomenon known as *negative transfer* [5]—but this is not the average behavior we observe here.

There exists several potential advantageous effects of TL for the second-form agent A_{f_2} that benefits from transfer [5] (see Fig. 2). First, A_{f_2} initial loss may be lower than A_d, because A_{f_1} has already started convergence; this is known as *jumpstart* benefit. Second, A_{f_2} loss may decrease faster than A_d, so A_{f_2} is said to learn *faster*. Lastly, A_{f_2} final loss may be lower than A_d, so A_{f_2} is said to learn *better*. In our experiment, the *gain* measured according to (6) is an aggregated estimation of these 3 potential advantages. Therefore, we cannot distinguish them from each other. However, considering that the only difference between A_d and A_{f_2} is their initial RNN parameter values, we strongly conjecture that the jumpstart effect is the major benefit of PL in this experiment.

This experiment is idealized because the agent can access a whole training set of optimal outputs $\hat{o}_k$, and it is directly guided by the loss gradient relative to its inner P RNN parameters. However, it is an instance of the generic PL procedure formalized in Sect. 2.2. Benefiting from these preliminary results in AG is therefore a matter of relaxing the ideal hypotheses and rely more on RL than SL approaches. In addition, PL can be used in other game-related problems like generation of believable behaviors [24–26].

In subsequent works, consistently with our general approach (Sect. 2.1), PL will be used again to address other types of signature changes like input removal, and output/feedback addition/removal. Then, actual PL agents will be constructed and trained against an abstract PL benchmarking task. When stabilized, they will be tested as control agents in simple adaptive games like experimental extensions of flOw.

4   Conclusion

AG demand that a controller agent continuously adapts to the player. This is a case of RL. However, complex games make this learning special because the agent has to face changes in its signature. We generalize the idea of RL to a broader PL theoretical situation explicitly taking these changes into account. This enables a rigorous assessment of how this kind of learners could be developed. With a controlled idealized PL experiment, we have shown that considering input addition for protean learners can be addressed efficiently at least with RNNs. Moreover, the benefits of PL in this situation are both easy to use (simply pad RNN with zeroes) and robust to changes in the learning context (informativeness of partial input, task complexity, data structure). These first results are encouraging, suggesting that the protean approach can be both simple and robust, and that controlling complex, changing AGs with PL is promising. The assessment of PL is still incomplete, since other basic operations must be tested: input removal and output/feedback addition/removal. Moreover, exporting these results from idealized PL to generic PL tasks still needs to be done.

Like any ML approach, PL is generic and can be applied in any other domain where a signature change in RL is identified. For instance, modular robotics could benefit from PL controllers. Long-term learners that cannot discard their past experience on a signature change also need to capitalize on it. And more generally, any bio-inspired agent that need to transform, split or merge, while keeping on learning, can benefit from a PL approach.

References

1. Sutton, R.S., Barto, A.G.: Reinforcement Learning: An Introduction. Adaptive Computation and Machine Learning Series, 2nd edn. MIT Press, Cambridge (2018)
2. Hanna, C.J., Hickey, R.J., Charles, D.K., Black, M.M.: Modular reinforcement learning architectures for artificially intelligent agents in complex game environments. In: Computational Intelligence and Games. pp. 380–387. IEEE, Copenhagen, August 2010
3. Elman, J.: Finding structure in time. Cogn. Sci. **14**(2), 179–211 (1990)
4. Schmidhuber, J.: Deep learning in neural networks: an overview. Neural Netw. **61**, 85–117 (2015)

5. Taylor, M.E., Stone, P.: Transfer learning for reinforcement learning domains: a survey. J. Mach. Learn. Res. **10**(7), 1633–1685 (2009)
6. Lazaric, A.: Transfer in reinforcement learning: a framework and a survey. In: Wiering, M., van Otterlo, M. (eds.) Reinforcement Learning. Adaptation, Learning, and Optimization, vol. 12, pp. 143–173. Springer, Heidelberg (2012). https://doi.org/10.1007/978-3-642-27645-3_5
7. Tanaka, F., Yamamura, M.: Multitask reinforcement learning on the distribution of MDPs. In: International Symposium on Computational Intelligence in Robotics and Automation. Computational Intelligence in Robotics and Automation for the New Millennium, vol. 3, pp. 1108–1113. IEEE, Kobe (2003)
8. Devin, C., Gupta, A., Darrell, T., Abbeel, P., Levine, S.: Learning modular neural network policies for multi-task and multi-robot transfer. CoRR abs/1609.07088 (2016)
9. Frans, K., Ho, J., Chen, X., Abbeel, P., Schulman, J.: Meta learning shared hierarchies. CoRR abs/1710.09767 (2017)
10. Teh, Y.W., et al.: Distral: robust multitask reinforcement learning. CoRR abs/1707.04175 (2017)
11. Tsymbal, A.: The Problem of Concept Drift: Definitions and Related Work (2004)
12. Wang, H., Abraham, Z.: Concept drift detection for streaming data. In: International Joint Conference on Neural Networks, pp. 1–9. IEEE, Killarney, July 2015
13. Ring, M.B.: Continual learning in reinforcement environments. Ph.D. thesis, University of Texas at Austin, Austin (1994)
14. Xu, J., Zhu, Z.: Reinforced continual learning. CoRR abs/1805.12369 (2018)
15. Sweetser, P., Wyeth, P.: GameFlow: a model for evaluating player enjoyment in games. Comput. Entertainment **3**(3), 3 (2005)
16. Holt, R., Mitterer, J.: Examining video game immersion as a flow state. In: 108th Annual Psychological Association, Washington, DC (2000)
17. Chen, J.: flOw, January 2019. http://jenovachen.info/flow/
18. Chen, J.: Flow in games (and everything else). ACM Commun. **50**(4), 31 (2007)
19. Bonnici, I., Gouaïch, A.: Formalisation of metamorph reinforcement learning. Technical report, LIRMM, Montpellier, November 2018. https://hal-lara.archives-ouvertes.fr/hal-01924642
20. Cho, K., et al.: Learning phrase representations using RNN encoder-decoder for statistical machine translation. CoRR abs/1406.1078 (2014)
21. Kingma, D.P., Ba, J.: Adam: a method for stochastic optimization. CoRR abs/1412.6980 (2014)
22. Paszke, A., et al.: Automatic differentiation in PyTorch (2017)
23. R Core Team: R: A Language and Environment for Statistical Computing. R Foundation for Statistical Computing, Vienna (2018)
24. Le Hy, R., Arrigoni, A., Bessiere, P., Lebeltel, O.: Teaching Bayesian behaviours to video game characters. Robot. Auton. Syst. **47**, 177–185 (2004)
25. Tencé, F., Buche, C.: Automatable evaluation method oriented toward behaviour believability for video games. CoRR abs/1009.0501 (2010)
26. Polceanu, M., Mora, A., Jimenez, J., Buche, C., Fernandez-Leiva, A.: The believability gene in virtual bots. In: 29th International Flairs, p. 4. AAAI Press, Key Largo (2016)

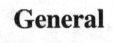

General

Supporting Medical Decisions
for Treating Rare Diseases
Through Genetic Programming

Illya Bakurov[1]($\boxtimes$), Mauro Castelli[1]🆔, Leonardo Vanneschi[1]🆔,
and Maria João Freitas[2]

[1] Nova Information Management School (NOVA IMS),
Universidade Nova de Lisboa, Campus de Campolide, 1070-312 Lisbon, Portugal
{ibakurov,mcastelli,lvanneschi}@novaims.unl.pt
[2] Raríssimas - Associação Nacional de Deficiências Mentais e Raras,
Rua das Açucenas, Lote 1, Loja Dta., 1300-003 Lisbon, Portugal
mjoao.freitas@rarissimas.pt

Abstract. Casa dos Marcos is the largest specialized medical and residential center for rare diseases in the Iberian Peninsula. The large number of patients and the uniqueness of their diseases demand a considerable amount of diverse and highly personalized therapies, that are nowadays largely managed manually. This paper aims at catering for the emergent need of efficient and effective artificial intelligence systems for the support of the everyday activities of centers like Casa dos Marcos. We present six predictive data models developed with a genetic programming based system which, integrated into a web-application, enabled data-driven support for the therapists in Casa dos Marcos. The presented results clearly indicate the usefulness of the system in assisting complex therapeutic procedures for children suffering from rare diseases.

Keywords: Genetic Programming ·
Geometric Semantic Genetic Programming · Medical decisions ·
Rare diseases

1 Introduction

The term *rare disease* is used to identify any disease that affects a tiny percentage of the population. From a regulatory perspective, rare diseases are defined as those diseases where less than 200,000 persons are affected in the USA or no more than one person over 2,000 is affected in the general population of the European Union (EU). Such diseases usually have a genetic basis, often affecting patients early in childhood or even since birth, and are frequently progressive, disabling and life threatening. Nowadays approximately 7,000 different rare diseases have been identified, and the number of people suffering from a rare disease in the USA and EU exceeds 55 million, highlighting the enormous social impact of these diseases. For all these reasons, in the last two decades there has been a substantial

© Springer Nature Switzerland AG 2019
P. Kaufmann and P. A. Castillo (Eds.): EvoApplications 2019, LNCS 11454, pp. 187–203, 2019.
https://doi.org/10.1007/978-3-030-16692-2_13

increase worldwide in the number of specialized medical and therapeutic centers for the care and treatment of patients affected by these diseases [1]. In Portugal there has been a development in the treatment and care of this type of patients, in particular after the foundation of the Portuguese Association of Mental and Rare Diseases in 2002. This association, called *Raríssimas*, is a non-profit organization whose mission is to support people affected by rare diseases and their relatives. In 2010, *Raríssimas*, with substantial contribution of private capitals and donations, gave birth to *Casa dos Marcos*, a medical and residential center. *Casa dos Marcos* is a highly specialized center, with a clinic with the capacity to receive 5,000 patients per year and a Physical Medicine and Rehabilitation unit.

Although extremely active, *Casa dos Marcos* is presently facing several challenges, many of which are shared by the analogous medical centers worldwide: (1) rare diseases are, by their very nature, diverse among each other and thus *unique*, which makes personalized therapies a must; (2) all the actions related to the planning of the therapies, the hospitalization and recovery of the patients and the organization of their everyday life in the medical center are, in large part, performed manually. These issues naturally demand for accurate and efficient computational systems. Many of the activities of the center, in fact, demand for efficient and effective predictive models able to support medical decisions.

The objective of this paper is to contribute towards the achievement of such an ambitious goal, using the development of a Machine Learning (ML) system able to generate six predictive models to forecast the effect of a specialized therapy on one global and five local factors. These models were integrated into a specially designed web-application to support the decision-making processes of *Casa dos Marcos*. The system which generated the models is based on Genetic Programming (GP), and more in particular on one of the newest developments of GP: The Evolutionary Demes Despeciation Algorithm (EDDA). GP holds tremendous potential for this type of application, for at least the following reasons. First, it has the potentiality of generating highly non-linear models of multiple features. Second, it can automatically perform feature selection during the learning phase. Additionally, GP produces models which enable subsequent interpretation and feature importance analysis. To the best of our knowledge, the one presented in this paper is the first GP-based system developed in the context of Rare Diseases.

The document is organized as follows. Section 2 introduces the reader to the context of this study. Section 3 presents the research track regarding practical applications of state-of-the-art ML tools in the field of medicine. Section 4 describes the methodological approach. Section 5 presents the experimental settings and discusses the results obtained. Finally, Sect. 6, summarizes the main findings of this work and provides some suggestions for future work.

2   Problem Description and Data

2.1   The Pediasuit Protocol

Pediasuit is a modern, intensive therapy suit. It is inspired by the *Penguin suit*, developed by the Soviet space program to neutralize the harmful effects of weightlessness and hypokinesis on the body of astronauts during space flights [2]. It is currently used by Casa dos Marcos during some therapeutic procedures. The *Pediasuit Protocol* is a therapeutic approach which uses *Pediasuit*, initially designed for people who happen to have neurological disorders, such as cerebral palsy, developmental delay, autism and other conditions that affect motor development and/or cognitive functions [2]. With the help of tension elastics, than can create an almost anti-gravity effect, patients are able to perform movements that would otherwise seem impossible and will be able to train their muscles and posture in a very specific manner. In this way, the therapy helps minimize pathological reflexes and promote the establishment of new, correct and functional movements.

The Pediasuit Protocol is one of the therapeutic treatments performed by Casa dos Marcos to assist patients, mostly children, with rare, predominantly neurodegenerative, diseases. Concretely, the therapists' goal consists of achieving an improvement, or, at least, a situation of freezing, regarding motor and mental functioning, taking into account the progressive degeneration of nerve cells. The Pediasuit Protocol lasts for four weeks, with daily sessions of two or four hours each. Moreover, since it requires the assistance of highly specialized physiotherapists, the price ranges from 1300 euros to 2500 euros, according to the number of hours per session [3].

Gross Motor Function Measure. The *Gross Motor Function Measure* (GMFM) is a clinical tool designed to evaluate changes in gross motor function in patients (usually children) with cerebral palsy, traditionally using 88 measures (GMFM-88) [4]. The measures of the GMFM-88 standard span a large set of motor activities, each with its own summary-measures (from now on called *factors*): lying and rolling (17 measures), sitting (20 measures), crawling and kneeling (14 measures), standing (13 measures) and walking, running and jumping (24 measures). For each factor, therapists ask patients to perform a set of specific movements and exercises (the measures), which accuracy is then evaluated on an ordinal scale ranging from 0 to 3, where 0 means "does not initiate", 1 means "initiates", 2 means "partially completes", and 3 means "completes". At the end, global assessment measures are calculated: one total and five local scores (one for each factor). The GMFM international standard was adopted by Casa dos Marcos to measure the improvement, in terms of motor functionality, of patients who attended the Pediasuit Protocol. Each patient is evaluated through GMGM-88 twice: one before the start of the therapy and another after its conclusion.

2.2   The Data

Casa dos Marcos provided one dataset, with records of 27 different patients undergoing the therapy. Through data pre-processing procedures, one patient was removed from the analysis because of the high amount of missing values. Each patient underwent the therapy at least once and, for each therapy, two examinations under the GMFM standard were performed. Besides the GMFM measures and six summary indicators, the dataset also contains three socio-demographic attributes: gender, birth date and diagnosis. In total, considering other operational attributes, 108 variables were provided.

Given a dataset with the information related to only 26 patients, developing predictive models with more than 100 attributes is not an easy task. Given this difficulty, to increase the size of the training data, we decided to consider one training instance per therapy. As a result, 41 data instances became available for the analysis. At first glance, it seems inaccurate to consider different therapies of the same patient as independent training data instances. However, given the problem specificity, this decision has two motivational arguments. First, most of the patients who have undergone the therapy present neurodegenerative diseases. Given this fact, a unique and unpredictable deviation is expected, in terms of GMFM, between subsequent therapies taken by the same patient. Second, the same patient is expected to react differently to equivalent therapies performed in different time periods, due to the disease progression status. Furthermore, the GMFM of such a patient, after one year, is expected to vary in a negative fashion.

3   Related Work

This section presents a selection of previous studies where GP was used to tackle challenging problems in the field of health-care and medicine. One of the first studies appeared in [5], where a constrained-syntax GP-based algorithm for discovering classification rules in medical data sets was proposed. To address that problem, the authors defined a GP framework containing several syntactic constraints to be enforced by the system using a disjunctive standard form representation, so that individuals represent valid rule sets that are easy to interpret. GP was compared against a decision-tree-building algorithm over five medical data sets, and it was able to obtain good results with respect to predictive accuracy and rule comprehensibility, by comparison with decision trees.

Another study where GP was used to solve a problem in the field of medicine was proposed in [6], where the authors tackled a problem related to the physio-chemical properties of proteins. More in detail, the problem addressed in their work was the prediction of the physiochemical properties of proteins tertiary structure, with the objective of predicting the size of the residues considering the protein tertiary structure data. The authors employed a semantics-based GP framework to solve the problem successfully, and the system produced a superior performance with respect to other considered techniques like artificial neural networks and support vector machines.

In 2018, Ting et al. [7] used GP to analyze feature importance for metabolomics [8]. In their work the authors analyzed a population-based metabolomics dataset on osteoarthritis and developed a Linear GP (LGP) algorithm to search classification models that can best predict the disease outcome, as well as to identify the most important metabolic markers associated with the disease. The LGP algorithm produced satisfactory performance, also being able to identify a set of key metabolic markers that may be useful for achieving a better understanding of the biochemistry of the disease.

Other successful applications of GP and its variants are in the field of pharmacokinetics, where several contributions appeared in recent years [9–11]. In several of these contributions, one crucial aspect strengthened by the authors is the ability of GP in producing a human-understandable model, a fundamental feature in the field of medicine that also motivates the choice of GP as the elected ML method for solving the problem considered in this study.

The interested reader is referred to [12] for a recent overview of the main contribution of genetic and evolutionary computation in the medical field.

4 Methodology

4.1 Geometric Semantic Genetic Programming

In the current terminology adopted by a considerable part of the Genetic Programming (GP) [13] research community, the term *semantics* indicates the vector of output values of a solution, calculated on the training observations [14,15]. Under this perspective, a GP individual can be seen as a point in a multidimensional space (its semantics). This space, called *semantic space*, has a number of dimensions equal to the number of observations in the training set.

Geometric Semantic Genetic Programming (GSGP) [14] is a recently introduced variant of GP in which standard crossover and mutation are replaced by so-called *Geometric Semantic Operators* (GSOs). The former operators allow the algorithm to exploit semantic awareness and induce precise geometric properties on the semantic space. GSOs, introduced by Moraglio et al. [14], gained popularity in the GP community [15] because of their property of inducing a unimodal error surface (characterized by the absence of locally optimal solutions) for any supervised learning problem. The proof of this property can be found in [14].

Here, we report the definition of the GSOs, as given by Moraglio et al. for real functions domains, since these are the operators that will be used in the experimental phase. For applications that consider other types of data, the reader is referred to [14]. *Geometric Semantic Crossover* (GSC) generates, as the unique offspring of parents $T_1, T_2 : \mathbb{R}^n \to \mathbb{R}$, the expression: $T_{XO} = (T_1 \cdot T_R) + ((1 - T_R) \cdot T_2)$, where T_R is a random real function whose output values range in the interval $[0, 1]$. *Geometric Semantic Mutation* (GSM) returns, as the result of the mutation of an individual $T : \mathbb{R}^n \to \mathbb{R}$, the expression: $T_M = T + ms \cdot (T_{R1} - T_{R2})$, where T_{R1} and T_{R2} are random real functions with codomain in $[0, 1]$ and ms is a parameter called the mutation step.

This work considers the GSOs' implementation presented in [16].

A recognized drawback of GSGP consists of the potential weakness of GSC. Given that GSC generates an offspring whose semantics stands on the segment joining two points representing the parents (in the semantic space), it can only achieve the global optimum solution if the semantics of the individuals in the population "surround" the semantics of the global optimum. Using the terminology of [17,18], GSC only has the possibility of generating a globally optimal solution only if this solution lays within the semantic *convex hull* identified by the population. The need for overcoming this drawback has led to several methods to properly initialize a population of GSGP, like for instance the ones presented in [19–21].

4.2 Evolutionary Demes Despeciation Algorithm

Initialization is known to play a very important role for any population-based algorithm. The same happens in GP, where a wide variety of programs of various sizes and shapes are desirable [13]. With the introduction of GSOs, new techniques taking their particularities into consideration, have been developed [19]. The Evolutionary Demes Despeciation Algorithm (EDDA) is contextualized in this research track.

In Biology, demes are independent populations, or sub-populations, of individuals that actively interbreed and mature, and the term despeciation indicates the combination of demes of previously distinct species into a new population, where distinct biological lineage is blended. The despeciation phenomenon rarely occurs in Nature, but in some cases it is known to fortify populations. In EDDA, the initial population of GSGP is generated using the best individuals obtained from a set of independent sub-populations (demes), that evolved for few generations and under different evolutionary conditions: some demes use standard GP, while others use GSGP and each deme is being evolved under distinct search parameters [20]. EDDA was recently introduced in the GP community [22,23] and owes its success to its simplicity and wide scope of applications. Although EDDA was originally developed to take into consideration the particularities of GSGP, it can also be used to initialize any population-based algorithm.

GSGP using EDDA demonstrated its superiority over GSGP initialized with the traditional Ramped Half-and-Half (RHH) [13] method over six complex symbolic regression applications [20]. More specifically, on all problems, EDDA allowed for generation of solutions with comparable or even better generalization ability and of significantly smaller size than using RHH. The efficacy of EDDA depends on two main parameters: the proportion of GSGP demes in the system (n) and the number of generations to evolve each deme (m). Using an algorithm-specific notation, given two natural numbers n and m, where $n \in [0, 100]$, EDDA-$n\%$ represents a system where demes are left to evolve for m generations such that $n\%$ of the population was initialized using individuals from GSGP demes, while the remaining $(100 - n)\%$ was initialized using standard GP demes. The pseudo-code in Fig. 1 explains the process.

EDDA-n% (evolving demes for m generations):

1. Create an empty population P of size N;
2. Repeat $N * (n/100)$ times:
   (a) Create an empty deme;
   (b) Randomly initialize this deme using a classical initialization algorithm (RHH used here);
   (c) Evolve individuals from 2.b) for m generations using GSGP;
   (d) After finishing 2.c), select the best individual from the deme and store it in P;
3. Repeat $N * (1 - n/100)$ times:
   (a) Create an empty deme;
   (b) Randomly initialize this deme using a classical initialization algorithm (RHH used here);
   (c) Evolve individuals from 3.b) for m generations using standard GP;
   (d) After finishing 3.c), select the best individual from the deme and store it in P;
4. Retrieve P and use it as the initial population of GSGP

Fig. 1. Pseudo-code of the EDDA-n% system, in which demes are left to evolve for m generations.

In the pseudo-code in Fig. 1, points 2(b), 2(c), 3(b) and 3(c) implement the *evolution of demes*, while points 2(d) and 3(d) implement the *despeciation* phase. In the former step, the different demes evolve independently; in the latter phase, individuals coming from different demes, and thus from different evolutionary dynamics and histories, are joined in a new population (P in the pseudo-code). To evolve this new population, GSGP is preferred over standard GP because in several application domains GSGP is known to outperform standard GP [24]. For this reason, in this study, after the *despeciation* phase, we used GSGP to conduct the main evolutionary process (MEP).

5 Experiments

Given that the therapy's impact can be assessed through six possible summary measures (factors), to predict it's impact on a given patient, six different supervised-learning models have to be created for each factor.

5.1 Experimental Settings

The training dataset for each one of the six supervised-learning problems consisted of 41 training data instances and 97 input features. The terminal set consisted of all input features and nine constants defined in $[-1, 1]$ as $\{-1.0, -0.75, -0.5, -0.25, 0.0, 0.25, 0.5, 0.75, 1.0\}$. The function set contained the primitive functions $\{+, -, *, /, sin, cos, \sqrt{}, ln\}$, where $\sqrt{x}$ and $ln(x)$ return x if its value does not fall within their respective domain. Similarly, a/x returns a if x equals to zero.

The fitness was calculated as the Root Mean Squared Error (RMSE) between predicted and expected outputs. Tournament selection with a selection pressure of 10% was used to select the parents. Similarly to [20, 24], the probability of applying a given variation operator was randomly drawn at the beginning of each generation and the mutation step, in the case of geometric semantic mutation, was randomly generated, with uniform probability in [0, 1], at each mutation event. Survival was elitist, i.e., the best individual was copied unchanged into the next population at each generation.

For all the considered algorithm executions, populations of two-hundred individuals were used (both in the initialization and in the main algorithm). Consequently, since we used EDDA to initialize the population, there were two-hundred demes consisting of two-hundred individuals each. Given the restricted number of training data instances allied to the high dimensionality of the problem, each initial population contained a set of 97 individuals, each of which composed by one single terminal (one distinct input feature). This was done in order not to misuse potentially relevant features during the phase of random initialization of the demes. At each deme, tree initialization of the remaining 103 individuals was performed by using RHH, with maximum initial depth equal to 5. Individuals growth was not limited throughout the whole evolutionary process.

For each factor, EDDA was studied in 5 different configurations, as in [20]: EDDA-0, 25, 50, 75, 100%. For each on of these, we studied a version in which each deme was evolved for 5, 10, 20 and 40 generations. As such, 20 benchmarks were considered for each GMFM factor (target), where each benchmark consisted of a pair *(maturation, EDDA-n%)*. This totaled to 120 benchmarks to tune EDDA. For each benchmark, at the end of the evolution of each deme, the best individual (in terms of training error) was selected to seed the initial population of GSGP.

Given that there were only 41 data instances, Leave-One-Out (LOO) cross-validation was applied. More concretely, at the beginning of each run, one different data instance was left out to assess generalization, while the remaining $n - 1$ were used to train the model. For this reason, 41 independent runs of EDDA were performed.

5.2   Experimental Results

Figure 2 exhibits all 120 benchmarks conducted to study the impact of maturation and proportion of the GSGP demes in EDDA for each target (GMFM factor). Each sub-figure summarizes all 41 runs of each benchmark, conducted for a given target, through a median validation error (gray bars with the left vertical axis) and median depth (black lines with the right vertical axis). Both summary statistics were obtained from the analysis of the best individual at the end of each run. Notice that in each sub-figure, the benchmarks were sorted in ascending order by median validation error.

The following paragraph provides a discussion of the results presented in Fig. 2. It is important to notice that our choice of the EDDA parameters was guided not only by validation error but also by the size of the individuals.

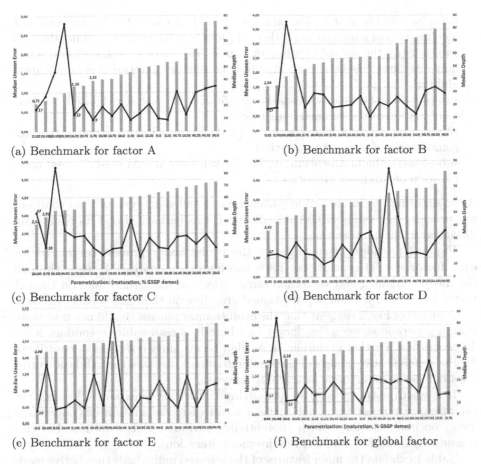

(a) Benchmark for factor A

(b) Benchmark for factor B

(c) Benchmark for factor C

(d) Benchmark for factor D

(e) Benchmark for factor E

(f) Benchmark for global factor

Fig. 2. Evolution of the median best validation error (gray bars measured through left vertical axis) and median depth (black lines measured through right vertical axis) for factors: A (a), B (b), C (c), D (d), E (e) and global factor (f).

Three arguments motivated this. First, Occam's razor, i.e., given models with similar training performance, the simplest (in our case, smallest) model should be preferred. Second, the need for delivering final individuals through a web-application, used by the therapists. Third, models interpretability is important in this application, mainly because the domain experts have to trust the models, and smaller models should be easier to interpret (even though we are aware that this may not always be true). Here are, the chosen parameters with a short motivation for each case reported in Fig. 2:

- sub-plot (a): it was decided to opt for the parameterization provided by bench-mark number 7, i.e. (5, 75), due to its noticeably lower median depth compared to parameterizations with slightly better generalization ability;

- sub-plot (b): the first benchmark, i.e. (5, 25), because it exhibited the best combination of summary statistics, i.e., lowest median error and depth;
- sub-plot (c): the second benchmark, i.e. (5, 75), since it demonstrated a significantly lower median depth at almost no penalty in terms of generalization ability;
- sub-plot (d): the first benchmark, i.e. (5, 25), because it exhibited the best combination of the summary statistics.
- sub-plot (e): the first benchmark, i.e. (5, 0), because it demonstrated the best combination of summary statistics;
- sub-plot (f): the first benchmark, i.e. (5, 50), since it exhibited the best combination of summary statistics.

After selecting the initialization parameters, detailed analysis of the evolutionary process, succeeding EDDA initialization, was performed and its results are discussed in the continuation.

Figure 3 provides visualization of the evolutionary process conducted for each factor, for 50 generations. Each sub-figure summarizes 41 runs through the median training and validation error (gray and black solid lines on the left vertical axis) and median depth (dashed gray line on the right vertical axis).

Analysis of Fig. 3 suggests that the evolutionary process should not take more than 5 generations, excepting factor C, where it is reasonable to conduct it for 20 generation. Further evolution, in median terms, does not seem to contribute for the generalization ability of the final individuals, because the validation error does not decrease anymore or starts to increase.

It is worth noticing that, after applying EDDA initialization technique, the evolutionary process turns out to be a mean of recombination of (potentially local) optimal solutions which (potentially) surround global optima. For this reason, the evolution does not require *many* iterations.

Table 1 exhibits the main features of the selected individuals (predictive models) for each of six factors. The second column of the table provides the generalization ability, measured as the RMSE calculated on validation set and averaged across 41 runs. The third column provides information about efficiency of the models, calculated as standard deviation of RMSE. Finally, the last fourth column provides the depth of each individual.

Table 1. Assessment of generalization ability of evolved individuals for prediction of the six factors.

Factor	$RMSE$	s_{RMSE}	Depth
A	3.83	4.36	13
B	4.54	4.89	21
C	4.71	4.94	20
D	5.93	6.23	13
E	2.95	2.84	17
Global	3.78	3.91	14

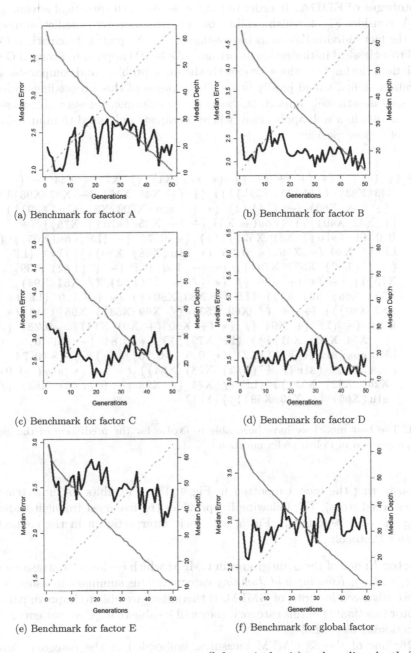

(a) Benchmark for factor A

(b) Benchmark for factor B

(c) Benchmark for factor C

(d) Benchmark for factor D

(e) Benchmark for factor E

(f) Benchmark for global factor

Fig. 3. Evolution of the median best error (left vertical axis) and median depth (right vertical axis), on training and validation data for factors: A (a), B (b), C (c), D (d), E (e) and the global factor (f). The legend for the all sub-plots in the figure is: ——— training error ——— validation error ········ depth.

Advantages of EDDA. In order to fully understand the practical advantage of EDDA, consider Fig. 4, which exhibits one of the six evolved predictive models. Only the fact individuals can fit in one-third of an A4 page is fascinating. Compared to a classical initialization algorithm (like RHH) applied to evolve a GSGP population, the final solutions for exactly the same problem and comparable generalization ability, would hardly fit on all the pages of this Proceedings volume. This issue is extremely important, because, for example, we were able to settle all six models in a web-application, where therapists only need to input GMFM values of a given patient.

```
(+ (+ (+ (+ (* (+ (+ (+ (+ (* sin(X1) (/ X7 (- X44 (- (-
   X43 X83) (^(1/2) X39))))) (+ (+ X45 X13) (+ X37 X96)))
   (+ cos(X66) (- (^(1/2) (* (* sin(X60) X8) sin(X39)))
   (+ X48 X40)))) (cos(+ X33 (+ (- X35 (sin() X75)) (+
   0.0 (+ (sin() X8) X52)))))) (* 0.75 (- (LF cos(X65)) (
   LF (/ X80 (+ X56 (* 0.25 (* sin(X48) X89)))))))))) (LF
   (- (^(1/2) X39) X79))) (* (- 1.0 (LF (- (^(1/2) X39)
   X79))) (+ (* (+ X96 (/ (* (/ (+ (^(1/2) (/ X54 X94))
   X55) X66) X6) X1)) (LF (+ X83 X80))) (* (- 1.0 (LF (+
   X83 X80))) (+ (+ (/ X69 X53) (/ X69 X53)) X96))))) (*
   0.91 (- (LF (+ X94 (/ (* (+ X60 (+ X91 X14)) -0.75) (/
   (- X54 X10) X41)))) (LF X79)))) (* 0.81 (- (LF 0.0) (
   LF (cos(sin(cos(^(1/2) (* 0.5 -0.75))))))))) (* 0.74
   (- (LF (- (sin(/ (^(1/2) X23) X34)) (+ (/ (* sin(-1.0)
   X26) X52) X53))) (LF (/ X74 (/ X68 (- (^(1/2) X62) (/
   sin(X66) (+ X40 X58))))))))))
```

Fig. 4. The best model we have been able to evolve for the prediction of the global factor, expressed in Polish prefix notation

Considering the model reported in Fig. 4, Fig. 5 exhibits its most relevant features (predictors). The following list provides the five most frequent features, among the ones reported in Fig. 5, and their interpretation in the context of Pediasuit Protocol:

- **Factor E:** one of the main groups in GMFM which encloses 24 measures into the *Walking, Running and Jumping* category. This summary measure is the most advanced in terms of GMFM. If therapists are able to improve patients motor functioning in this category, they will be able to improve patients motor functioning as a whole;
- **B26:** one of the 88 GMFM measures, embedded in the category *Sitting*. Following [4], it can be described as the ability of a patient, while sitting on a mat, to touch a toy placed 45° behind his/her right side and return to start;
- **C38:** another GMFM measure, embedded in the category *Sitting*. Following [4], it can be described as the ability of the patient to creep 1.8 m forward;

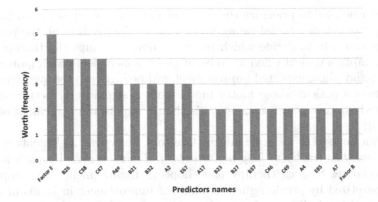

Fig. 5. The features (predictors) that appear more frequently in the model of Fig. 4, together with their worth factor (frequency).

- **C47:** another GMFM measure, embedded in the category *Sitting*. Following [4], it can be described as the ability of the patient to crawl backward down four steps on their hands and knees/feet. The fact B26, C38 and C47 appear to be the most frequent states their importance in therapeutic context. We speculate that a change in therapeutic design, which takes more in consideration improvement in these specific (or similar) motor activities, can produce a better improvement in patients health.
- **Age:** the age of the patient taken before performing the therapy. Our findings demonstrate that patients in different age groups respond differently to the therapy: those whose age is below nine years old show twice as good improvement, measured in terms of the global factor, as the patients whose age is equal or higher than nine years old. This evidence, according to the therapists, has to do with the fact that the body of a younger patient is, naturally, in a more dynamic phase of growth, and this reflects positively on their gross motor functioning improvement after the therapy.

More Than Prediction. In the context of the Pediasuit Protocol, more than merely providing a prediction, the six models that were developed bring other practical benefits:

- they can be a reference for the efficiency of the therapy. Concretely, if, after attempting the therapy, the final GMFM evaluation of a given patient will differ highly from the expectation, then the therapeutic approach at hand (its intensity, aim, coordination, etc.), might be inadequate for that particular patient;

– they can be used to prioritize the queue of patients. Concretely, and hypothesizing the lack of internal resources to address the needs of all the patients, models can help to decide which patients should attempt the therapy first. For example, a patient who may present an extremely debilitated motor functioning and whose expected improvement will be high, can receive higher priority over a patient whose motor functioning is significantly better and the expected improvement is smaller (this should require the agreement of both parties);
– they can be used as a simple informative tool, answering a elementary question: *"will this treatment help my beloved and how much?"* People who take their relatives to such therapy have hope in improvement. This hope can be concretized by providing an estimate of improvement in terms of motor function through different factors;
– they can be used to confirm the diagnosis. In the field of rare diseases, one of the main challenges for the doctors is to identify the diagnosis correctly. This comes from the simple fact that the diseases are rare and, for some of them, there is no concise and extensive framework for their identification;
– they can be used as means of therapeutic design improvement. One of the main features of GP is that individuals can be interpreted. By studying their structure, one can identify the most relevant predictors and their impact. As such, one can say which attributes are determinant for the improvement of the patients at each GMFM factor.

6 Conclusion

Patients affected by rare diseases need very specific and personalized therapies. Casa dos Marcos is the largest specialized medical and residential center for rare diseases in the Iberian Peninsula and has the capacity for hosting several thousands of patients per year. Nowadays, the design and development of personalized therapies and treatments is widely done manually by a team of specialists. Besides conceivably slow, this manual process is also subjective, and thus prone to errors. This motivates the impelling demand for intelligent computational systems, able to support and speedup the decision process. Machine learning is clearly a reasonable option, and predictive models, able for instance to give information about possible reactions of patients to therapies, can be of paramount importance for the development of Casa dos Marcos and the improvement of their everyday work.

This paper summarizes the main outcomes of a project that our research team developed in collaboration with Casa dos Marcos: a Genetic Programming (GP) based system, able to generate predictive models for important motor functioning factors concerning patients who happen to have rare diseases, after some specific therapies.

The presented system uses a very recent development of GP, called Evolutionary Demes Despeciation Algorithm (EDDA). EDDA integrates both the standard version of GP and Geometric Semantic GP (GSGP) in the initialization phase, aiming at capturing the advantages of both these techniques, while at the same time mitigating their respective flaws. One of the major advantages of EDDA is the ability to generate models with comparable accuracy of the ones generated by GSGP, but at the same time with a much smaller size. This is an extremely important characteristic, since the models generated by GSGP are known to be very accurate, but also extremely large. Exploiting this ability, in this paper, we have been able to show and comment on the best model evolved by EDDA, something that would have been unimaginable for GSGP. The reported model is very informative on the effect of the therapy and experts have validated the small subset of features that it uses. Being able to see and, at least partially, interpret the evolved model has been of fundamental importance because it has allowed the personnel of Casa dos Marcos to trust our system. Also, thanks to this increased trust, the models evolved by the presented system are now integrated in a web application, that we have developed, and is nowadays presently in use in Casa dos Marcos.

Acknowledgments. This work was partially supported by national funds through FCT (Fundação para a Ciência e a Tecnologia) under project DSAIPA/DS/0022/2018 (GADgET) and project PTDC/CCI-INF/29168/2017 (BINDER).

References

1. Rare disease resources & FAQs. https://rarediseases.org/for-patients-and-families/information-resources/resources-faqs/
2. Scheeren, E.M., Mascarenhas, L.P.G., Chiarello, C.R., Costin, A.C.M.S., Oliveira, L., Neves, E.B.: Description of the pediasuit protocol^{TM}. Fisioterapia em movimento **25**(3), 473–480 (2012)
3. Centro de desenvolvimento e reabilitação da casa dos marcos. http://rarissimas.pt/centro-de-desenvolvimento-e-reabilitacao-da-casa-dos-marcos/
4. Russell, D.J., Rosenbaum, P.L., Cadman, D.T., Gowland, C., Hardy, S., Jarvis, S.: The gross motor function measure: a means to evaluate the effects of physical therapy. Dev. Med. Child Neurol. **31**(3), 341–352 (1989)
5. Bojarczuk, C.C., Lopes, H.S., Freitas, A.A., Michalkiewicz, E.L.: A constrained-syntax genetic programming system for discovering classification rules: application to medical data sets. Artif. Intell. Med. **30**(1), 27–48 (2004)
6. Castelli, M., Vanneschi, L., Manzoni, L., Popovič, A.: Semantic genetic programming for fast and accurate data knowledge discovery. Swarm Evol. Comput. **26**, 1–7 (2016)
7. Hu, T., Oksanen, K., Zhang, W., Randell, E., Furey, A., Zhai, G.: Analyzing feature importance for metabolomics using genetic programming. In: Castelli, M., Sekanina, L., Zhang, M., Cagnoni, S., García-Sánchez, P. (eds.) EuroGP 2018. LNCS, vol. 10781, pp. 68–83. Springer, Cham (2018). https://doi.org/10.1007/978-3-319-77553-1_5
8. Beger, R.D., et al.: For "Precision Medicine and Pharmacometabolomics Task Group"-metabolomics society initiative: metabolomics enables precision medicine: "a white paper, community perspective". Metabolomics **12**(9), 149 (2016)

9. Castelli, M., Vanneschi, L., Popovič, A.: Parameter evaluation of geometric semantic genetic programming in pharmacokinetics. Int. J. Bio-Inspired Comput. **8**(1), 42–50 (2016)
10. Castelli, M., et al.: An efficient implementation of geometric semantic genetic programming for anticoagulation level prediction in pharmacogenetics. In: Correia, L., Reis, L.P., Cascalho, J. (eds.) EPIA 2013. LNCS (LNAI), vol. 8154, pp. 78–89. Springer, Heidelberg (2013). https://doi.org/10.1007/978-3-642-40669-0_8
11. Vanneschi, L., Castelli, M., Manzoni, L., Silva, S.: A new implementation of geometric semantic GP and its application to problems in pharmacokinetics. In: Krawiec, K., Moraglio, A., Hu, T., Etaner-Uyar, A.Ş., Hu, B. (eds.) EuroGP 2013. LNCS, vol. 7831, pp. 205–216. Springer, Heidelberg (2013). https://doi.org/10.1007/978-3-642-37207-0_18
12. Smith, S.L., Cagnoni, S.: Genetic and Evolutionary Computation: Medical Applications. Wiley, Chichester (2011)
13. Koza, J.: Genetic Programming: On the Programming of Computers by Means of Natural Selection. MIT Press, Cambridge (1992)
14. Moraglio, A., Krawiec, K., Johnson, C.G.: Geometric semantic genetic programming. In: Coello, C.A.C., Cutello, V., Deb, K., Forrest, S., Nicosia, G., Pavone, M. (eds.) PPSN 2012. LNCS, vol. 7491, pp. 21–31. Springer, Heidelberg (2012). https://doi.org/10.1007/978-3-642-32937-1_3
15. Vanneschi, L., Castelli, M., Silva, S.: A survey of semantic methods in genetic programming. Genet. Program Evolvable Mach. **15**(2), 195–214 (2014)
16. Castelli, M., Silva, S., Vanneschi, L.: A c++ framework for geometric semantic genetic programming. Genet. Program Evolvable Mach. **16**(1), 73–81 (2015)
17. Castelli, M., Manzoni, L., Gonçalves, I., Vanneschi, L., Trujillo, L., Silva, S.: An analysis of geometric semantic crossover: a computational geometry approach. In: IJCCI (ECTA), pp. 201–208 (2016)
18. Oliveira, L.O.V., Otero, F.E., Pappa, G.L.: A dispersion operator for geometric semantic genetic programming. In: Proceedings of the Genetic and Evolutionary Computation Conference, pp. 773–780. ACM (2016)
19. Pawlak, T.P., Krawiec, K.: Semantic geometric initialization. In: Heywood, M.I., McDermott, J., Castelli, M., Costa, E., Sim, K. (eds.) EuroGP 2016. LNCS, vol. 9594, pp. 261–277. Springer, Cham (2016). https://doi.org/10.1007/978-3-319-30668-1_17
20. Vanneschi, L., Bakurov, I., Castelli, M.: An initialization technique for geometric semantic GP based on demes evolution and despeciation. In: IEEE Congress on Evolutionary Computation (CEC), pp. 113–120. IEEE (2017)
21. Bakurov, I., Vanneschi, L., Castelli, M., Fontanella, F.: EDDA-V2 – an improvement of the evolutionary demes despeciation algorithm. In: Auger, A., Fonseca, C.M., Lourenço, N., Machado, P., Paquete, L., Whitley, D. (eds.) PPSN 2018. LNCS, vol. 11101, pp. 185–196. Springer, Cham (2018). https://doi.org/10.1007/978-3-319-99253-2_15
22. Bartashevich, P., Bakurov, I., Mostaghim, S., Vanneschi, L.: PSO-based search rules for aerial swarms against unexplored vector fields via genetic programming. In: Auger, A., Fonseca, C.M., Lourenço, N., Machado, P., Paquete, L., Whitley, D. (eds.) PPSN 2018. LNCS, vol. 11101, pp. 41–53. Springer, Cham (2018). https://doi.org/10.1007/978-3-319-99253-2_4

23. Bartashevich, P., Bakurov, I., Mostaghim, S., Vanneschi, L.: Evolving PSO algorithm design in vector fields using geometric semantic GP. In: Proceedings of the Genetic and Evolutionary Computation Conference Companion, GECCO 2018, Kyoto, Japan, 15–19 July 2018, pp. 262–263 (2018)
24. Vanneschi, L., Silva, S., Castelli, M., Manzoni, L.: Geometric semantic genetic programming for real life applications. In: Riolo, R., Moore, J.H., Kotanchek, M. (eds.) Genetic Programming Theory and Practice XI. GEC, pp. 191–209. Springer, New York (2014). https://doi.org/10.1007/978-1-4939-0375-7_11

Evolutionary Successful Strategies in a Transparent iterated Prisoner's Dilemma

Anton M. Unakafov[1,2,3,4,6,7]([✉]), Thomas Schultze[1,3], Igor Kagan[3,4], Sebastian Moeller[1,3,4], Alexander Gail[1,3,4,5], Stefan Treue[1,3,4,5], Stephan Eule[2,3,6,7], and Fred Wolf[2,3,5,6,7]

[1] Georg-Elias-Mueller-Institute of Psychology, University of Goettingen, Gosslerstrasse 14, 37073 Göttingen, Germany
schultze@psych.uni-goettingen.de
[2] Max Planck Institute for Dynamics and Self-Organization, Am Fassberg 17, 37077 Göttingen, Germany
{anton,eule,fred}@nld.ds.mpg.de
[3] Leibniz ScienceCampus Primate Cognition, Kellnerweg 4, 37077 Göttingen, Germany
[4] German Primate Center - Leibniz Institute for Primate Research, Kellnerweg 4, 37077 Göttingen, Germany
{ikagan,smoeller}@dpz.eu, {agail,treue}@gwdg.de
[5] Bernstein Center for Computational Neuroscience, Am Fassberg 17, 37077 Göttingen, Germany
[6] Max Planck Institute for Experimental Medicine, Hermann Rein Strasse 3, 37075 Göttingen, Germany
[7] Campus Institute for Dynamics of Biological Networks, Hermann Rein Strasse 3, 37075 Göttingen, Germany

Abstract. A Transparent game is a game-theoretic setting that takes action visibility into account. In each round, depending on the relative timing of their actions, players have a certain probability to see their partner's choice before making their own decision. This probability is determined by the level of transparency. At the two extremes, a game with zero transparency is equivalent to the classical simultaneous game, and a game with maximal transparency corresponds to a sequential game. Despite the prevalence of intermediate transparency in many everyday interactions such scenarios have not been sufficiently studied. Here we consider a transparent iterated Prisoner's dilemma (iPD) and use evolutionary simulations to investigate how and why the success of various strategies changes with the level of transparency. We demonstrate that non-zero transparency greatly reduces the set of successful memory-one strategies compared to the simultaneous iPD. For low and moderate transparency the classical "Win - Stay, Lose - Shift" (WSLS) strategy is the only evolutionary successful strategy. For high transparency all strategies are evolutionary unstable in the sense that they can be easily counteracted, and, finally, for maximal transparency a novel "Leader-Follower" strategy outperforms WSLS. Our results provide a partial explanation for the fact that the strategies proposed for the simultaneous iPD are rarely observed in nature, where high levels of transparency are common.

© Springer Nature Switzerland AG 2019
P. Kaufmann and P. A. Castillo (Eds.): EvoApplications 2019, LNCS 11454, pp. 204–219, 2019.
https://doi.org/10.1007/978-3-030-16692-2_14

Keywords: Evolutionary game theory · iterated Prisoner's Dilemma · Transparent games

1   Introduction

Game theory is widely used to account for strategic decision-making in rational agents. Classical game theory assumes that players act either sequentially or simultaneously. Yet, for social behavior these assumptions are often not fulfilled since humans and animals rarely act strictly simultaneously or sequentially. Instead, social agents try to observe their partners and adversaries, and use the others' behavior to adjust their own actions accordingly [1,2]. Recently a new game-theoretic setting of "transparent games" has been introduced, taking into account action visibility and providing a more realistic model of interactions under time constraints [3]. In transparent games, each player has a certain probability to observe the partner's choice before deciding on its own action. This probability is determined by the action times of the players. For instance, if Player 1 always acts well before Player 2, the probability to see the partner's choice is zero for Player 1 and one for Player 2 (corresponds to strictly sequential playing). If both players act approximately at the same time, on average, they have equal probabilities p_{see} to see each other's choices. This probability can range from $p_{see} = 0$ (players cannot take choices of each other into account; this case corresponds to the classical simultaneous game) to $p_{see} = 0.5$ (in every round typically one of the two players sees the choice of the partner and can adjust its own decision). The probability that neither player sees the choice of the partner is equal to $1 - 2p_{see}$.

One may expect that action visibility would increase cooperation in non-zero-sum games, such as the iterated Prisoner's dilemma (iPD). However, evolutionary simulations in [3] show that this is not necessarily the case. Evolutionary agents successfully establish cooperation in the iPD for low and moderate transparency by using the classic "Win - stay, lose - shift" (WSLS) strategy [4]. However, for high transparency cooperation drastically decreases (see Fig. 1), and the most frequent strategy is "Leader-Follower", which does not rely on mutual cooperation [3].

This unexpected drop of cooperation in an iPD with high transparency raises new questions. What explains the success of WSLS for the transparent iPD? Why does the Leader-Follower strategy become more frequent for high transparency? Is its success evolutionary stable or is this strategy just transiently successful? To answer these questions, we use evolutionary simulations, since analytic considerations for the transparent games require solving differential equations with many variables. Since most results for simultaneous and sequential versions of the iPD were obtained for strategies taking into account outcomes of only the last interaction ("memory-one strategies"), we also focus on such strategies. We show that the rather complex strategy dynamics associated with $p_{see} = 0$ (simultaneous iPD) [5] is greatly simplified for the transparent settings with $p_{see} > 0$. WSLS is the only non-transient strategy for $0 < p_{see} < 0.5$.

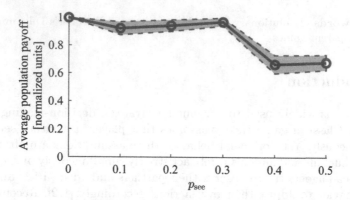

Fig. 1. Average fraction of maximal possible payoff (with 95% confidence interval) achieved by the evolutionary evolved population of players in the transparent iterated Prisoner's Dilemma. Payoff values are computed over 80 runs for probabilities $p_{see} = 0.0, 0.1, \ldots, 0.5$ of a player to see the choice of a partner. Maximal payoff corresponds to mutual cooperation. Stable cooperation was established for low and moderate transparencies, but it was disrupted for $p_{see} > 0.3$ resulting in a significant decrease of average payoff.

For $p_{see} > 0.35$ all memory-one strategies are evolutionary unstable and replace each other in rapid succession, though WSLS and Leader-Follower are more stable than the others.

Our results complement the findings in [3] by providing a detailed analysis of the evolutionarily successful strategies for the transparent iPD. In particular, we emphasize the superiority of WSLS over other memory-one strategies for most transparency levels. Yet, the fact that WSLS and other strategies developed for the simultaneous iPD are unstable for an iPD with high transparency may partially explain why these classic strategies are rarely observed in nature [6,7], where high levels of transparency are prevalent [1].

2   Methods

2.1   Transparent Iterated Prisoner's Dilemma

Prisoner's dilemma [8] is perhaps the most studied game between two players. We refer to [9] for a review. The payoff matrix for this game is presented in Fig. 2. Apparently, the best choice for both players is mutual cooperation, but defection dominates cooperation (i.e., it yields a higher payoff for each individual player regardless of the partner's choice). Thus any self-interested player would defect (D), although mutual defection results in lowest possible joint payoff. This "paradox" of Prisoner's dilemma raises a question: under what conditions is it rational to cooperate (C)?

Player 2 / Player 1	Cooperate (C)	Defect (D)
Cooperate (C)	R=3 R=3	T=5 S=0
Defect (D)	S=0 T=5	P=1 P=1

Fig. 2. Payoff matrix for Prisoner's Dilemma. In this game two players adopt roles of prisoners suspected of committing a crime. Each can either betray the other (Defect), or Cooperate with the partner by remaining silent. The maximal charge is five years in prison, and the payoff matrix represents the number of years deducted from it. For example, if Player 1 defects and Player 2 cooperates, then Player 1 gets payoff $T = 5$ and goes free, while Player 2 gets $S = 0$ and has to spend five years in prison. If both prisoners defect, each gets a slightly reduced sentence of four years ($P = 1$) Finally, with mutual cooperation between the prisoners there is only circumstantial evidence left, which is sufficient to sentence both prisoners to two years in prison ($R = 3$).

One possible answer is provided by playing the game repeatedly: in the iPD cooperation between self-interested players becomes plausible, since they can take into account past outcomes but also have to consider that they will play against each other again in future. When experience shows the partner to be trustworthy, cooperation in repeated game becomes a viable option. Traditionally each round of the iPD is considered to be a simultaneous game [8], where players make choices independently. In the recently suggested transparent iPD [3], both players have a certain chance to learn the current choice of the partner. Three cases are possible:

1. Player 1 sees the choice of Player 2 before making its own choice – with probability p_{see}^1.
2. Player 2 sees the choice of Player 1 before making its own choice – with probability p_{see}^2.
3. Neither of players knows the choice of the partner, probability of this case is $1 - p_{see}^1 - p_{see}^2$.

Note that $p_{see}^1 + p_{see}^2 \leq 1$ meaning that in each round only one of the players can see the partner's choice. In the transparent iPD it is natural to assume that both players act on average at the same time [3]. Indeed, though both players are interested in the partner's choice and would wait for the partner's action, there is always a time constraint preventing players from waiting indefinitely. If this time constraint is explicit and known to both players, in most cases they are motivated to act just before the end of the time allowed for the action, thus - nearly at the same time. Therefore for the rest of the paper we assume that $p_{see}^1 = p_{see}^2 = p_{see} \leq 0.5$ and call p_{see} the transparency level.

As it is usually done for the simultaneous iPD [5], we assume that players take into account outcomes of the previous game round, that is, use "memory-one strategies". Then a strategy of a player in a transparent iPD is represented by a vector $\mathbf{s} = (s_k)_{k=1}^{12}$, where s_k are conditional probabilities to cooperate in 12 different situations. These depend on whether the player and the partner cooperated in the previous round, whether the player can see the current choice of the partner, and what the choice is if it is visible, see Table 1.

Table 1. Representation of strategies in transparent games. Each strategy entry s_i specifies probability of a player to cooperate depending on the outcome of previous round (the first action specifies the choice of the player, and the second the choice of the partner) and whether current partner's choice is visible.

Outcome of previous round	CC	CD	DC	DD
Current partner's choice unseen	s_1	s_2	s_3	s_4
Partner is cooperating (C)	s_5	s_6	s_7	s_8
Partner is defecting (D)	s_9	s_{10}	s_{11}	s_{12}

To find out what strategies are optimal depending on the transparency level p_{see}, we use evolutionary simulations described in the following subsection.

2.2 Evolutionary Dynamics of Memory-One Strategies in Transparent Games

To study evolutionary dynamics of the transparent iPD, we used the techniques developed for the simultaneous iPD in [4,5] with minor adaptation to account for the different strategy representation in transparent games, see also [3].

Consider a population consisting of transparent-iPD players from n "species" defined by their strategies $\mathbf{s}_i = (s_k^i)_{k=1}^{12}$ for $i = 1, 2, \ldots, n$. That is, any species i is a group of players sharing the same strategy $\mathbf{s}_i$, which they use when playing the iPD with a given transparency level $p_{\text{see}} \in [0.0, 0.5]$ against any partner. We assume that the population is infinitely large since a finite number of players results in complex stochastic effects [10], which we do not consider here. The population evolves in generations $t = 1, 2, \ldots$. In each generation every player plays infinitely many rounds of the transparent iPD against a partner assigned randomly according to the current composition of the population. Species getting higher average payoff than others reproduce more effectively, and their relative frequency $x_i(t)$ in the population increases in the next generation. Note that $\sum_{i=1}^{n(t)} x_i(t) = 1$ for any generation t.

Specifically, the evolutionary success of species i is encoded by its fitness $f_i(t)$, computed as the average payoff for a player from this species when playing against the current population:

$$f_i(t) = \sum_{j=1}^{n(t)} x_j(t) \mathrm{E}_{ij},$$

where E_{ij} is the expected payoff of species i playing against species j. If $f_i(t)$ is higher than the average fitness of the population $\overline{f}(t) = \sum_{i=1}^{n(t)} x_i(t)f_i(t)$, then $x_i(t)$ increases with time, otherwise $x_i(t)$ decreases and the species is dying out. This evolutionary process is formalized by the replicator equation [5]:

$$x_i(t+1) = x_i(t) + \frac{f_i(t) - \overline{f}(t)}{\overline{f}(t)} x_i(t) = \frac{f_i(t)}{\overline{f}(t)} x_i(t). \tag{1}$$

It remains to compute the expected payoff E_{ij} for a strategy $\mathbf{s}_i = (s_k^i)_{k=1}^{12}$ against $\mathbf{s}_j = (s_k^j)_{k=1}^{12}$. For this, consider Players 1 and 2 from species i and j, respectively. Since both players use memory-one strategies, their choices in every round of the game depend only on their mutual choices in the previous round. This allows to describe the game dynamics by a Markov chain with four states being the mutual choices of the two players (CC, CD, DC and DD), and transition matrix given by

$$M = (1 - 2p_{\text{see}})M_0 + p_{\text{see}}M_1 + p_{\text{see}}M_2, \tag{2}$$

with matrices M_0, M_1 and M_2 describing the cases when neither player sees the choice of the partner, Player 1 sees the choice of the partner before making own choice, and Player 2 sees the choice of the partner, respectively.

$$M_0 = \begin{pmatrix} s_1^i s_1^j & s_1^i(1 - s_1^j) & (1 - s_1^i)s_1^j & (1 - s_1^i)(1 - s_1^j) \\ s_2^i s_3^j & s_2^i(1 - s_3^j) & (1 - s_2^i)s_3^j & (1 - s_2^i)(1 - s_3^j) \\ s_3^i s_2^j & s_3^i(1 - s_2^j) & (1 - s_3^i)s_2^j & (1 - s_3^i)(1 - s_2^j) \\ s_4^i s_4^j & s_4^i(1 - s_4^j) & (1 - s_4^i)s_4^j & (1 - s_4^i)(1 - s_4^j) \end{pmatrix}, \tag{3}$$

$$M_1 = \begin{pmatrix} s_5^i s_1^j & s_9^i(1 - s_1^j) & (1 - s_5^i)s_1^j & (1 - s_9^i)(1 - s_1^j) \\ s_6^i s_3^j & s_{10}^i(1 - s_3^j) & (1 - s_6^i)s_3^j & (1 - s_{10}^i)(1 - s_3^j) \\ s_7^i s_2^j & s_{11}^i(1 - s_2^j) & (1 - s_7^i)s_2^j & (1 - s_{11}^i)(1 - s_2^j) \\ s_8^i s_4^j & s_{12}^i(1 - s_4^j) & (1 - s_8^i)s_4^j & (1 - s_{12}^i)(1 - s_4^j) \end{pmatrix}, \tag{4}$$

$$M_2 = \begin{pmatrix} s_1^i s_5^j & s_1^i(1 - s_5^j) & (1 - s_1^i)s_9^j & (1 - s_1^i)(1 - s_9^j) \\ s_2^i s_7^j & s_2^i(1 - s_7^j) & (1 - s_2^i)s_{11}^j & (1 - s_2^i)(1 - s_{11}^j) \\ s_3^i s_6^j & s_3^i(1 - s_6^j) & (1 - s_3^i)s_{10}^j & (1 - s_3^i)(1 - s_{10}^j) \\ s_4^i s_8^j & s_4^i(1 - s_8^j) & (1 - s_4^i)s_{12}^j & (1 - s_4^i)(1 - s_{12}^j) \end{pmatrix}. \tag{5}$$

Then we can represent the expected payoff by the following formula

$$E_{ij} = y_R R + y_S S + y_T T + y_P P, \tag{6}$$

where R, S, T, P are the entries of the payoff matrix ($R = 3, S = 0, T = 5, P = 1$ for the standard iPD, see Fig. 2), and y_R, y_S, y_T, y_P represent the probabilities of getting to the states associated with the corresponding payoffs by playing $\mathbf{s}_i$ against $\mathbf{s}_j$. This vector is computed as a unique left-hand eigenvector of matrix M associated with eigenvalue one [5]:

$$(y_R, y_S, y_T, y_P) = (y_R, y_S, y_T, y_P)M. \tag{7}$$

To guarantee the existence of the eigenvector (y_R, y_S, y_T, y_P), strategy entries for any species i should satisfy the following double inequality

$$0 < s_k^i < 1. \tag{8}$$

for all $k = 1, 2, \ldots, 12$. Therefore, it is common to introduce a minimal possible error ε in the strategies such that $\varepsilon \le s_k^i \le 1 - \varepsilon$. This also allows accounting for error-proneness of players (so-called "trembling hand" effect [4,11]).

Equation (6) allows to compute the expected payoff E_{ij} for all strategies i, j in the population. The dynamics of the population is entirely described by the matrix E. This dynamics is relatively simple for two species, when only four cases are possible [10]:

- *Dominance*: one species is unconditionally more fit than the other and replaces it in the population. Strategy i dominates j when $E_{ii} > E_{ji}$ and $E_{ij} > E_{jj}$.
- *Bistability*: either species can take over the whole population depending on the initial relative frequencies $x_i(1)$, $x_j(1)$ and the threshold x^*, given by the (non-stable) equilibrium frequency of species i

$$x^* = \frac{E_{jj} - E_{ij}}{E_{ii} - E_{ij} - E_{ji} + E_{jj}}. \tag{9}$$

If $x_i(1) > x^*$, then species i takes over the population, otherwise species j wins. Bistability occurs when $E_{ii} > E_{ji}$ and $E_{ij} < E_{jj}$.
- *Coexistence*: the population is composed of two species with the asymptotic frequencies $x_i = x^*$, $x_j = 1 - x^*$, where x^* is a stable equilibrium given by (9). This case takes place when $E_{ii} < E_{ji}$ and $E_{ij} > E_{jj}$.
- Finally, *neutrality* takes place when $E_{ii} = E_{ji}$ and $E_{ij} = E_{jj}$, meaning that the frequencies of the species do not change.

We used the analysis of two-strategy dynamics for a pairwise comparison of strategies. However, already for a population consisting of $n = 3$ species, the dynamics can be rather complex, since there are 33 possible types of dynamics [12]. Analytic considerations for $n > 3$ are even more complicated [13] and are usually replaced by evolutionary simulations [5]. Details of our simulations are provided in Sect. 2.3.

2.3   Evolutionary Simulations

For the evolutionary simulations we adopted methods suggested in [4,5]. We studied the populations of transparent iPD players for various transparency levels p_{see}, and for each level we ran multiple evolutionary simulations as described below.

In each run, the population evolved as described in Sect. 2.2, thus the frequencies of species $x_i(t)$ changed according to (1). When $x_i(t) < \chi$, the species i was assumed to die out and was removed from the population; its fraction $x_i(t)$ was distributed among the remaining species proportional to their share in the population. We followed [4,5] in taking $\chi = 0.001$.

In each run a new species could enter the population with probability 0.01, thus new species emerged on average every 100 generations. Following [4], probabilities s_k^i for the new species were randomly drawn from the distribution with U-shaped probability density function (10), favoring probability values around 0 and 1:

$$\rho(y) = \pi\big(y(1-y)\big)^{-1/2} \tag{10}$$

for $y \in (0,1)$. We required $s_k^i \in [\varepsilon, 1-\varepsilon]$ with $\varepsilon = 0.001$ as suggested in [4,5] to satisfy inequality (8). Initial frequencies of the introduced species were set to 1.1χ with $\chi = 0.001$ [4].

The most important strategy entries, especially for low transparency, are s_1, s_2, s_3, s_4 (probability to cooperate when partner's choice is unknown), since players use one of them with probability $1 - p_{\text{see}} > 0.5$. Therefore these entries converge to optimal values quite fast and their values in evolutionary successful strategies are most precise, while convergence of other entries to optimal values may take longer. Since values of s_1, s_2, s_3, s_4 for evolutionary successful strategies were described in [3], here we limited their variability to allow faster convergence of the remaining entries $s_5, \ldots, s_{12}$ to the optimal values. For this we rounded strategy entries s_1, s_2, s_3, s_4 so that they had the values from the set

$$s_1^i, s_2^i, s_3^i, s_4^i \in \{\varepsilon, \frac{1}{6}, \frac{2}{6}, \frac{3}{6}, \frac{4}{6}, \frac{5}{6}, 1-\varepsilon\}. \tag{11}$$

For the iPD with the payoff matrix shown in Fig. 2 commonly discussed strategies are formed by the values from the set $\{\varepsilon, \frac{1}{3}, \frac{2}{3}, 1-\varepsilon\}$ [4,5,14], but in 11 we also consider intermediate values to achieve better discretization.

We carried out evolutionary simulations with random and with pre-defined initial compositions of the population. A random initial population consisted of $n(1) = 5$ species with equal frequencies $x_1(1) = \ldots = x_5(1) = 0.2$ and random strategies. In this case we traced 10^9 generations in each run. We also considered three pre-defined initial populations consisting of a single species with one of the following strategies:

- Win – stay, lose – shift (WSLS): $\mathbf{s} = (1,0,0,1; 1,0,0,1; 0,0,0,0)$;
- Generous tit-for-tat (GTFT): $\mathbf{s} = (1, \frac{1}{3}, 1, \frac{1}{3}; 1, \frac{1}{3}, 1, \frac{1}{3}; 0,0,0,0)$;
- Leader-Follower (L-F):

$$\mathbf{s} = (\frac{1}{3}, 0, 0, 0; \frac{2}{3}, 0, 0, 0; 1, 1, 1, \frac{1}{5}) \tag{12}$$

These strategies were selected as most evolutionary successful in simulations with random initial population for various transparency levels p_{see} (see Sect. 3 for details). For the pre-defined initial populations we traced 10^8 generations in each run since it was not necessary to ensure stabilization of the population dynamics (which was the case for the random initial population). For each of the four initial compositions of the population described above, we performed 80 runs for each value of $p_{\text{see}} = 0.0, 0.1, \ldots, 0.5$. Additional 80 runs of simulations were performed for transparency levels $p_{\text{see}} = 0.26, 0.28, \ldots, 0.50$, for the initial WSLS and L-F populations.

2.4   Measuring the Strategies' Evolutionary Success

To quantify the evolutionary success of a strategy we computed the average of its relative frequency $x_i(t)$ across all generations t and across all runs. Since the strategies in the evolutionary simulations were generated randomly, some of the observed strategies might not entirely converge to the theoretical optimum. Therefore, we used a coarse-grained description of strategies as suggested in [3] with the following notation: symbol 0 for $s_k^i \leq 0.1$, symbol 1 for $s_k^i \geq 0.9$, symbol * is used as a wildcard character to denote an arbitrary probability. We characterized as Always Defect (AllD) all strategies encoded by (0000; **00; **00), meaning that the probability to cooperate when not seeing partner's choice or after defecting is below 0.1, and other behavior is not specified. Similarly, the generalized representations of other strategies were as follows

- Win – stay, lose – shift (WSLS): (100c; 1***; ****) with $c \geq 2/3$;
- Tit-for-tat (TFT): (1010; 1***; ****);
- Generous tit-for-tat (GTFT): (1a1c; 1***; ****), where $0.1 < a, c < 0.9$;
- Generous WSLS (GWSLS): (1abc; 1***; ****), where $c \geq 2/3$, $a, b < 2/3$ and either $a > 0.1$ or $c > 0.1$;
- Firm-but-fair (FbF): (101c; 1***; ****), where $0.1 < c < 0.9$;
- Leader-Follower (L-F): (*00c; ****; *11d), where $c < 1/3$ and $d < 2/3$.

3   Results

Frequencies of strategies for various initial compositions of the population are presented in Fig. 3. In all cases, the fraction of the described strategies drops down for $p_{see} > 0.3$. This happens due to the fact that for higher transparencies population dynamics often enters a "chaotic" mode, where many transient strategies replace each other in a rapid succession [3]. Each of these transient strategies has a low relative frequency, while taken together they constitute a considerable fraction of the population, especially for high p_{see}.

Regardless of the initial population composition, "Win-stay, lose-shift" (WSLS) strategy is a clear winner for $0 < p_{see} < 0.4$. The theoretically optimal form of this strategy is (1001; 1001; 0000), however in simulations it may appear as (1001; 1**1; 000*), where the entries marked by the wildcard character * can take a value from 0 to 1 (see Fig. 5).

For $p_{see} \geq 0.1$ WSLS can be only replaced in the population by a strategy (1001; d***; *00*) with $0 \leq d < 1$ (Fig. 6a–c). In [3] this strategy was called treacherous WSLS since it behaves like WSLS when not seeing the choice of the partner, and defects when seeing that partner cooperates. Treacherous WSLS has low payoff when playing with itself and is easily replaced by other strategies, but it dominates WSLS for any $p_{see} > 0$.

The predominance of WSLS as the single frequent non-transient strategy can be challenged only for $p_{see} \approx 0$ (minimal transparency, nearly-simultaneous iPD) by GTFT and for $p_{see} > 0.4$ (maximal transparency, nearly-sequential iPD) by L-F. Below we discuss both these cases in detail.

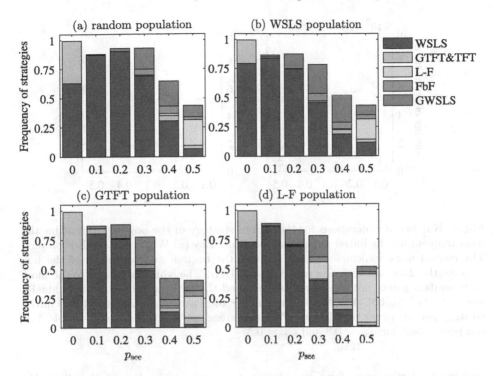

Fig. 3. Average relative frequencies of strategies occurred in the population for its various initial compositions: (a) random; constituted by (b) WSLS, (c) GTFT and (d) L-F strategies. Other strategies are transient (i.e. persist in the population only for a relatively short time). Results for all initial populations are quite similar, yet a few differences are noteworthy. First, the frequencies of WSLS and GTFT depend on the initial composition of the population for $p_{see} = 0$, but not for higher transparency. When one of these strategies was the initial strategy of the population, it is much more abundant than for the random initial composition. Second, when L-F is the initial strategy of the population, it is has considerably higher frequency for $p_{see} = 0.3$ and $p_{see} = 0.5$, but not for $p_{see} = 0.4$. Since for $p_{see} = 0.3$ L-F has a noticeable frequency only when forming the initial population, for this transparency L-F can persist but cannot take over the population when starting with a low frequency (see Fig. 4b). Note that such non-monotonicity in strategy dynamics is not altogether unexpected, since the dynamics depends non-linearly on the transparency level [3].

For minimal transparency the WSLS frequency slightly drops (Fig. 3a) since in this case population of WSLS players can be invaded by AllD. Let $s_1 = (1 - \varepsilon, \varepsilon, \varepsilon, 1 - \varepsilon; 1 - \varepsilon, \varepsilon, \varepsilon, 1 - \varepsilon; \varepsilon, \varepsilon, \varepsilon, \varepsilon)$ (WSLS), $s_2 = (\varepsilon, \varepsilon, \varepsilon, \varepsilon; \varepsilon, \varepsilon, \varepsilon, \varepsilon; \varepsilon, \varepsilon, \varepsilon, \varepsilon)$ (AllD). Estimating matrix of expected payoff by (6), we see that

$$E_{11} \approx 3 - 5\varepsilon, \quad E_{12} \approx \frac{1 + 7\varepsilon}{2} + \frac{1 - 7\varepsilon}{2} p_{see}, \quad E_{21} \approx 3 - \frac{3}{2}\varepsilon - p_{see}, \quad E_{22} \approx 1 + 3\varepsilon.$$

Thus for $p_{see} \leq 3.5\varepsilon$ WSLS is dominated by AllD and for $p_{see} > 3.5\varepsilon$ the two strategies are bistable. AllD invades WSLS in the former case unconditionally

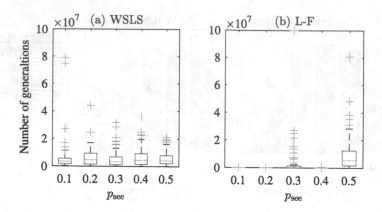

Fig. 4. Number of generations for that initial strategy of the population remains the most frequent for the initial population constituted by (a) WSLS and (b) L-F players. The central mark indicates the median, and the bottom and top edges of the box indicate the 25th and 75th percentiles, respectively. The whiskers extend to the most extreme data points not considered outliers, and the outliers are plotted individually using the '+' symbol. The higher the number of generations, the longer the initial strategy persists in the population. While persistence of WSLS is relatively stable, L-F can persist only for $p_{see} = 0.3$ and $p_{see} = 0.5$.

and in the latter case if its equilibrium frequency given by (9) is sufficiently low. Namely, if AllD with initial fraction 1.1χ is introduced to the population playing WSLS, for $p_{see} < \frac{7\varepsilon + 1.1\chi(1-8\varepsilon)}{2-1.1\chi}$ AllD invades the population. In this case, unconditional defectors can take over the population for a considerable number of generations, and strategies different from WSLS (TFT and GTFT) should be introduced to the population to restore cooperation. These strategies can resist WSLS invasion, therefore once WSLS dies out it may take many generations to re-establish itself.

In our simulations, $\varepsilon = 0.001$, $\chi = 0.001$, so WSLS can resist AllD-invasion for $p_{see} > 0.0041$. In this case WSLS is dominated only by treacherous WSLS. This weak transient strategy is easily replaced by others, which allows WSLS to reappear in the population relatively quickly.

The decrease of WSLS frequency for high transparency (Figs. 3 and 7) is caused by two factors. First, as one can see from Fig. 5, the higher p_{see} the more precisely WSLS should correspond to its theoretically optimal profile. Indeed, for $p_{see} = 0.1$ most entries of WSLS have relatively high standard deviations, meaning that in this case a successful strategy can follow the WSLS principle in a rather general fashion. Meanwhile, for $p_{see} = 0.5$ most entries have low variability, meaning that successful WSLS variants can have only slight deviations from the optimal profile. The closer a strategy must be to the theoretical WSLS profile in order to be successful, the lower the probability that such a strategy is by chance introduced in the population, which results in lower frequency of WSLS for higher p_{see}. Note that the frequency of WSLS would decrease even if we understand WSLS in the most general sense and consider GWSLS as a part

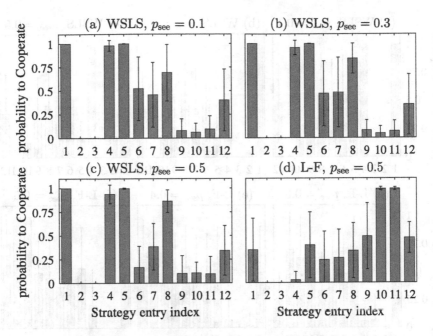

Fig. 5. Average strategy profiles with standard deviations for the winning strategies in a transparent iPD: (a) for WSLS at $p_{see} = 0.1$, (b) $p_{see} = 0.3$, (c) $p_{see} = 0.5$ and for L-F at $p_{see} = 0.5$ (d). Each strategy profile is represented by 12 entries characterizing probability of a player to cooperate. Namely, $s_1, \ldots, s_4$ are probabilities to cooperate when current partner's choice is unknown and the outcome of the previous round was "both cooperated", "self cooperated, partner defected", "self defected, partner cooperated" and "both defected", respectively. Similarly, $s_5, \ldots, s_8$ and $s_9, \ldots, s_{12}$ are probabilities of a player to cooperate seeing that current partner's choice is to cooperate and to defect, respectively. Standard deviations show the difference in variability for various entries. For instance in (a) some strategy entries are constant s_1, s_2, s_3, s_5, some vary only slightly s_4, s_9, s_{10}, s_{11}, and other entries are almost random s_6, s_7, s_8, s_{12}.

of it (Fig. 8). Second, the higher p_{see} is, the faster and easier the treacherous WSLS takes over the population, since it has higher chances to defect WSLS. This also means that a strategy that only roughly resembles the treacherous WSLS profile can be successful against WSLS for high p_{see} (see Fig. 6a–c).

It remains to explain the success of L-F for high transparency. Consider the effect of p_{see} on the frequencies of WSLS and L-F in more detail. Figure 8 shows how these frequencies vary for $p_{see} \in [0.26, 0.50]$ taken with a step of 0.02. for the initial population consisting either of WSLS or L-F players. Note that L-F is only successful for maximal transparency $p_{see} = 0.5$ (Fig. 8a), although for some transparency levels $p_{see} < 0.5$, L-F can remain in the population once introduced (Fig. 8b).

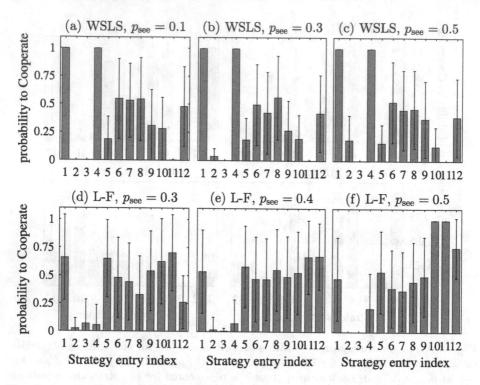

Fig. 6. Average strategy profiles with standard deviations for the strategies replacing initial strategies in simulations with pre-defined initial population: (a)–(c) strategies replacing WSLS for $p_{see} = 0.1$, $p_{see} = 0.3$ and $p_{see} = 0.5$, respectively; (d)–(f) strategies replacing L-F for $p_{see} = 0.3$, $p_{see} = 0.4$ and $p_{see} = 0.5$, respectively.

The success of L-F for maximal transparency can be partially explained by the fact that a variant of this strategy given by Eq. (12) can be replaced in the population only by other variants of the L-F strategy (Fig. 6d). Although these variants are not evolutionary stable and are eventually replaced by other strategies (Fig. 3d), this indicates a good evolutionary potential of L-F for $p_{see} = 0.5$. One particularly interesting modification of L-F is a strategy with a profile (1001; 1001; 1111). As it combines the features of L-F and WSLS we term it "WSLS-like L-F". This strategy dominates normal L-F and easily replaces it in the population. But contrary to the normal L-F, WSLS-like L-F is highly unstable since it is dominated by many strategies including WSLS, treacherous WSLS and AllD. Therefore, WSLS-like L-F stays in population just for a few generations and then is replaced by other strategies (similar to the treacherous WSLS).

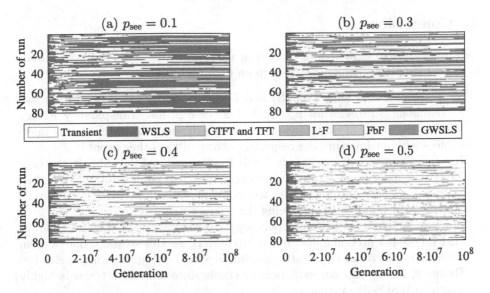

Fig. 7. Dynamics of the most frequent strategy of the population in various runs for initial WSLS-population. (a) For $p_{see} = 0.1$ WSLS is predominant in most runs. (b) For $p_{see} = 0.3$ WSLS frequency wanes, while other TFT, L-F, FbF and GWSLS become more abundant. Fraction of various transient strategies also increases and (c) for $p_{see} = 0.4$ they become most frequent. (d) For $p_{see} = 0.5$ WSLS rarely reestablishes itself and the only frequent strategy is L-F. Population mostly is in a chaotic mode with transient strategies replacing each other.

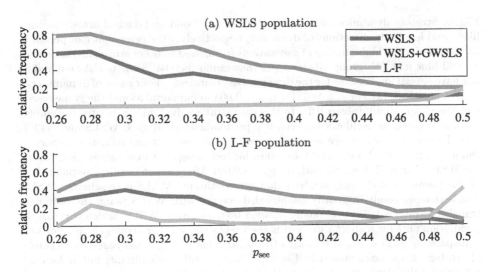

Fig. 8. Average relative frequencies of WSLS, WSLS with GWSLS, and L-F strategies for the initial populations constituted by (a) WSLS player and (b) L-F players. Note that in both cases WSLS frequency decreases considerably for $p_{see} > 0.35$.

4   Conclusion

Here we aimed to explain the success of various strategies in the transparent iterated Prisoner's dilemma [3]. Our main findings are:

- Win – stay, lose – shift (WSLS) and Generous tit-for-tat (GTFT) are two predominant strategies for $p_{see} \approx 0$. This case of low transparency is very close to the classical simultaneous Prisoner's dilemma in which agents make choices without knowing the respective choice of the other agent.
- For most non-zero transparency levels ($0 < p_{see} < 0.5$), WSLS is the only effective memory-one strategy. However, WSLS is not evolutionary stable, since it can be counteracted by its modified version, termed "treacherous WSLS", which defects when seeing the partner's choice.
- For $p_{see} \approx 0.5$ (maximal transparency) a second strategy, Leader-Follower (L-F), becomes predominant. When two players use this strategy, one of them defects and the other cooperates, leading to a kind of "turn-taking" behavior. However, L-F is also not evolutionary stable since it evolves towards highly unstable WSLS-like variants.

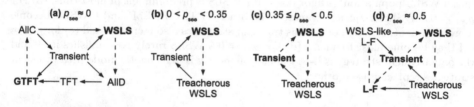

Fig. 9. Strategy dynamics for various values of p_{see}. Solid and dashed arrows indicate likely and less likely directions of dynamics, respectively. Strategies that can persist in the population for high number of generations (compared to other strategies) are shown in bold blue font. We group all strategies that cannot persist in the population and do not have special importance for the dynamics as "transient". Four cases of dynamics can be distinguished. (a) For $p_{see} \approx 0$ (for $p_{see} < 0.004$ in our simulations) the game resembles the classic simultaneous iPD. In this case strategy dynamics is relatively complex [5]. The population oscillates between two predominant strategies, WSLS and GTFT, and the transient states are relatively short. (b) As we increase p_{see}, two important changes take place. First, GTFT becomes ineffective against other strategies, including WSLS, L-F and AllC. Second, for $p_{see} > 0.004$ AllD cannot invade a population of WSLS players, and the only strategy that can "dethrone" WSLS is treacherous WSLS. These changes result in greatly simplified dynamics with WSLS being the only non-transient strategy. It is occasionally invaded by treacherous WSLS, which controls the population only for a short time and is then replaced by other strategies. (c) For high transparency ($0.35 < p_{see} < 0.5$, with threshold value 0.35 selected based on dynamics of strategy frequencies shown in Fig. 8), WSLS is still predominant, but it becomes difficult for this strategy to take over the population. Therefore, the population spends most of the time in a "chaotic" state with various transient strategies quickly replacing each other. (d) Finally, for maximal transparency ($p_{see} \approx 0.5$), L-F becomes the second predominant strategy. Yet, L-F does not have stable control of the population since some modifications of L-F can be invaded by non-L-F strategies. As a result, transient strategies still control the population for most of the time. (Color figure online)

Figure 9 summarizes our results schematically representing the strategy dynamics for various transparency levels p_{see}.

Overall our results provide an important extension to previous studies that have used evolutionary simulations to suggest and test strategies in classic simultaneous and sequential games. By concentrating on transparent games frequently encountered in real life but rarely investigated, we provide a partial explanation for the fact that the strategies developed for the simultaneous iPD are rarely observed in nature [6, 7], where high levels of transparency are common [1].

Acknowledgments. We acknowledge funding from the Ministry for Science and Education of Lower Saxony and the Volkswagen Foundation through the program "Niedersächsisches Vorab". Additional support was provided by the Leibniz Association through funding for the Leibniz ScienceCampus Primate Cognition and the Max Planck Society.

References

1. Dugatkin, L., Mesterton-Gibbonsand, M., Houston, A.: Beyond the Prisoner's dilemma: toward models to discriminate among mechanisms of cooperation in nature. Trends Ecol. Evol. **7**(6), 202–205 (1992)
2. Vaziri-Pashkam, M., Cormiea, S., Nakayama, K.: Predicting actions from subtle preparatory movements. Cognition **168**, 65–75 (2017)
3. Unakafov, A.M., Schultze, T., Kagan, I., Moeller, S., Eule, S., Wolf, F.: Emergence and suppression of cooperation by action visibility in transparent games (2018). https://doi.org/10.1101/314500
4. Nowak, M., Sigmund, K.: A strategy of win-stay, lose-shift that outperforms tit-for-tat in the Prisoner's dilemma game. Nature **364**(6432), 56–58 (1993)
5. Nowak, M.: Evolutionary Dynamics. Harvard University Press, Cambridge (2006)
6. Noë, R.: A veto game played by baboons: a challenge to the use of the Prisoner's dilemma as a paradigm for reciprocity and cooperation. Anim. Behav. **39**(1), 78–90 (1990)
7. Noë, R.: Cooperation experiments: coordination through communication versus acting apart together. Anim. Behav. **71**(1), 1–18 (2006)
8. Axelrod, R., Hamilton, W.: The evolution of cooperation. Science **211**, 13–90 (1981)
9. Doebeli, M., Hauert, C.: Models of cooperation based on the Prisoner's dilemma and the snowdrift game. Ecol. Lett. **8**(7), 748–766 (2005)
10. Traulsen, A., Hauert, C.: Stochastic evolutionary game dynamics. Rev. Nonlinear Dyn. Complexity **2**, 25–61 (2009)
11. Selten, R.: Reexamination of the perfectness concept for equilibrium points in extensive games. Int. J. Game Theor. **4**(1), 25–55 (1975)
12. Hofbauer, J., Sigmund, K.: Evolutionary game dynamics. Bull. Am. Math. Soc. **40**(4), 479–519 (2003)
13. Zeeman, E.C.: Population dynamics from game theory. In: Nitecki, Z., Robinson, C. (eds.) Global Theory of Dynamical Systems. LNM, vol. 819, pp. 471–497. Springer, Heidelberg (1980). https://doi.org/10.1007/BFb0087009
14. Nowak, M., Sigmund, K.: The alternating Prisoner's dilemma. J. Theor. Biol. **168**(2), 219–226 (1994)

Evolutionary Algorithms for the Design of Quantum Protocols

Walter Krawec[1], Stjepan Picek[2], and Domagoj Jakobovic[3(✉)]

[1] Department of Computer Science and Engineering, University of Connecticut,
Storrs, CT 06268, USA
[2] Cyber Security Research Group, Delft University of Technology, Mekelweg 2,
Delft, The Netherlands
[3] Faculty of Electrical Engineering and Computing, University of Zagreb,
Zagreb, Croatia
domagoj.jakobovic@fer.hr

Abstract. In this paper, we use evolutionary algorithm to evolve customized quantum key distribution (QKD) protocols designed to counter attacks against the system in order to optimize the speed of the secure communication. This is in contrast to most work in QKD protocols, where a fixed protocol is designed and then its security is analyzed to determine how strong an attack it can withstand. We show that our system is able to find protocols that can operate securely against attacks where ordinary QKD protocols would fail. Our algorithm evolves protocols as quantum circuits, thus making the end result potentially easier to implement in practice.

Keywords: Quantum cryptography · Evolution strategy ·
Quantum simulator

1 Introduction

Quantum cryptography is a fascinating area of study allowing for the achievement of certain important communication tasks which ordinarily would be impossible through classical communication only. One prominent example of this is *quantum key distribution* (QKD) which permits the establishment of a secret key (a classical bit string which may be used for other cryptographic tasks such as message encryption) between two users A and B, which is secure against an all-powerful adversary (something impossible to achieve through classical communication alone). Beyond this theoretical advantage, *QKD protocols are currently a practical technology which has seen several real-world applications.* Indeed, numerous experimental groups have verified the correctness and applicability of QKD. Additionally, there are currently several companies producing commercial QKD equipment and new QKD networks being established world-wide. For a general survey of QKD, both the theory and practice, the reader is referred to [1].

© Springer Nature Switzerland AG 2019
P. Kaufmann and P. A. Castillo (Eds.): EvoApplications 2019, LNCS 11454, pp. 220–236, 2019.
https://doi.org/10.1007/978-3-030-16692-2_15

One of the unique properties of quantum communication is that there is (assuming the protocol is correct) a direct correlation between the observed noise in a quantum channel and the maximal amount of information any adversary could have gained on the information being sent. QKD protocols are able to operate successfully up until a certain noise threshold is reached (called a protocol's noise tolerance). Before this threshold is reached, a secure key can be distilled - however the process grows less efficient as the noise level increases.

Generally, QKD protocols are constructed and then analyzed mathematically to determine what channels they can operate over (e.g., what are their noise tolerance), and, furthermore, what their efficiency is for particular channels. As an example, the BB84 protocol [2] can work over a symmetric channel so long as the noise is less than 11% [3]. Other protocols exist each with their own noise tolerances (along with other advantages or disadvantages). However, these noise tolerance results generally only hold for symmetric channels - over asymmetric channels, this is not necessarily true! In fact, for certain channels (i.e., attacks against the protocol), none of the current existing protocols may provide optimal noise tolerances and communication efficiency. In this paper, we are interested in the problem of finding QKD protocols optimized to work over given quantum qubit channels so as to maximize the efficiency of the secret key distribution rate beyond what current state-of-the-art QKD protocols may be able to do over this same channel.

In particular, we envision a system whereby users of quantum communication technology may, after running standard quantum tomographic protocols to measure the noise in the quantum channel, insert these measurement results into our algorithm which will then construct a tailor-made QKD protocol specifically designed to counteract the noise in the quantum channel. This quantum tomographic protocol involves users simply sending and receiving quantum bits prepared in a variety of manners so as to produce a *noise signature* of the channel - i.e., a list of several important channel noise statistics which can be used to characterize, at least partially, the adversary's attack. Since adversarial attacks are, in the worst-case, the cause of the noise in the channel, our system is constructing protocols that counteract an all-powerful, quantum capable, adversary. Other applications of our approach may be to counter changes in operating conditions (e.g., changes in environmental conditions which alter the noise in, say, a free-space channel). Such a system may be eventually used to create a more efficient quantum secure communication network. Furthermore, our system will, in fact, give explicit instructions on how to operate quantum devices by providing to users quantum protocols as basic *quantum circuits*. Furthermore, the circuits produced will consist of gates from a user-specified gate set, thus our system can take into account the capabilities, and limitations, of user hardware.

Our approach will utilize evolutionary algorithms to discover optimized QKD protocols. Evolutionary algorithms have been used for some time with success to evolve quantum algorithms [4–6], usually being used to find quantum circuits that are more efficient than human-constructed versions. Some work using (simulated) quantum computers to run classical GAs have been also reported [7].

Evolutionary methods have also been used successfully to study classical cryptography [8,9]. Only recently, they have been applied to the study of quantum cryptography [10,11].

In [11], a genetic algorithm (GA) was proposed to optimize QKD protocols for specific input channels (representing, for example, a particular attack being launched against a system). However, the approach in [11] required the user to provide a fixed template specifying an abstract protocol from which the GA would optimize certain user-specified parameters. Thus, the GA was not free to explore truly innovative approaches - instead, it was forced to search for protocols conforming to this predetermined template. Furthermore, this template needed to be constructed by the user before use.

In this work, we reconsider this problem and use an evolutionary algorithm (more precisely, evolution strategy) to discover optimized QKD protocols, designed to efficiently operate over a given, observed, quantum channel. Unlike prior work, our system will not be forced to use a user-defined template. Instead, our approach will simply be given a quantum communication channel (without explicit rules on how to access it) and must evolve a protocol out of *quantum circuits* allowing our system, in theory, to produce arbitrarily complex quantum communication protocols. While authors have considered using evolutionary algorithms to construct quantum circuits for *algorithms* [12,13], we are the first, to our knowledge, to apply these techniques to the construction of optimized quantum cryptographic protocols using state-of-the-art definitions of security in that field.

There are several advantages to this approach. First, it allows researchers to investigate over what channels QKD is even possible. While theoretical upper-bounds are known, it is not known whether these are tight [14]. Our system may aid researchers in this investigation. Secondly, and more practically, one may eventually envision a future quantum communication infrastructure whereby users have access to adjustable communication equipment. Users A and B may then, on start-up (or intermittently during operation), run a standard quantum tomographic protocol to estimate the channel noise (producing the current "noise signature"), provide these measurement statistics to our algorithm which will then produce a QKD protocol optimized to counter the observed channel noise and maximize efficiency. Users may then configure their quantum communication equipment to run this protocol. This process can be repeated periodically to account for changes in operating conditions (e.g., changes in attack strategy or environmental conditions).

Our system, as we will demonstrate, is able to produce QKD protocols with a higher communication rate than standard state-of-the-art protocols are capable of producing over certain channels. *Indeed, our system can even find protocols where standard protocols would fail.* Our system is also easier to use than prior work in [11] and, since we are evolving quantum circuits, the protocols output by our system may be easier to implement in practice than those obtained in [11]. *Thus, our approach has the potential to greatly increase the efficiency of a future quantum communication network.* As quantum communication is a viable technology now, the methodology we are developing may have the potential to create a more efficient, and robust, secure communication infrastructure.

2   Quantum Communication and Key Distribution

In this section, we introduce some general quantum communication concepts and notation. For more information on this subject, the reader is referred to [15].

A classical bit exists in one of two states 0 or 1; the state of a classical bit can always be determined with certainty and classical bits may be copied arbitrarily. A *quantum bit* or *qubit*, however, can be prepared in infinitely many possible states. More precisely, a qubit is modeled mathematically as a normalized vector in $\mathbb{C}^2$. Thus, any arbitrary (normalized) vector in this space represents a possible qubit state. Furthermore, "reading" a qubit (called *measuring*) is a probabilistic process which potentially destroys the original state; finally, a qubit cannot be copied without potentially destroying it.

An arbitrary qubit is denoted by $|\psi\rangle \in \mathbb{C}^2$. Let $\{|0\rangle, |1\rangle\}$ be an orthonormal basis in which case we may write $|\psi\rangle = \alpha |0\rangle + \beta |1\rangle$. Normalization requires $|\alpha|^2 + |\beta|^2 = 1$. The process of measuring a qubit involves first picking an orthonormal basis and then writing $|\psi\rangle$ as a linear combination of these basis vectors (called a *superposition*). Following this, the actual measurement apparatus will take as input the given quantum state, and output one of the two basis states. The probability of observing a particular basis state is simply the norm squared of the coefficient in front of the basis vector. For example, measuring $|\psi\rangle$ in the $\{|0\rangle, |1\rangle\}$ basis produces an output of $|0\rangle$ with probability $|\alpha|^2$; otherwise the output is $|1\rangle$ with probability $|\beta|^2$. Note that, once a qubit has been measured, it collapses to the observed outcome. Thus, not only are measurements probabilistic, but they also *disturb* the original state, projecting it to the observed basis vector. These measurement operations are irreversible.

Besides the $\{|0\rangle, |1\rangle\}$ basis (called the Z basis), two other important bases are the $X = \{|+\rangle, |-\rangle\}$ and $Y = \{|0_Y\rangle, |1_Y\rangle\}$ basis. These states are defined: $|\pm\rangle = \frac{1}{\sqrt{2}}(|0\rangle \pm |1\rangle)$ and $|j_Y\rangle = \frac{1}{\sqrt{2}}(|0\rangle + i(-1)^j |1\rangle)$.

Qubits are two dimensional systems; more generally, we may model an n-dimensional quantum state as an element in a Hilbert space of dimension n. Since such a space is isomorphic to $\mathbb{C}^n$, an n-dimensional state $|\psi\rangle$ is simply a normalized vector in $\mathbb{C}^n$; i.e., $|\psi\rangle = (\alpha_1, \cdots, \alpha_n)^T$ (transposed as we view these as column vectors). We denote by $\langle\psi|$ to be the conjugate transpose of $|\psi\rangle$. Note that $\langle\phi| \cdot |\psi\rangle$ is simply the inner-product of these two vectors. Since this is such an important operation, the notation is simplified to $\langle\phi|\psi\rangle$.

Given two quantum states $|\psi\rangle \in \mathbb{C}^n$ and $|\phi\rangle \in \mathbb{C}^m$, we model the *joint state* as the tensor product $|\psi\rangle \otimes |\phi\rangle \in \mathbb{C}^n \otimes \mathbb{C}^m \cong \mathbb{C}^{nm}$. As vectors, if $|\psi\rangle = (\alpha_1, \cdots, \alpha_n)^T$, then $|\psi\rangle \otimes |\phi\rangle = (\alpha_1 |\phi\rangle, \cdots, \alpha_n |\phi\rangle)^T$. To simplify notation we often write $|\psi\rangle |\phi\rangle$ or even $|\psi, \phi\rangle$.

While measurements irreversibly cause the quantum state to collapse to the observed basis vector, a second operation allowed by the laws of quantum physics is state evolution via a unitary operator. U is unitary is $UU^* = U^*U = I$ (where we write U^* to mean the conjugate transpose of U). Since we are in the finite dimensional setting, one may view U as an $n \times n$ matrix satisfying this required condition. Given an input state $|\psi\rangle \in \mathbb{C}^n$, the state after evolution via U is modeled simply as the result of the matrix multiplication $U |\psi\rangle$. If U and V are both unitary, then $U \otimes V$ is also a unitary operator acting on the tensor space with its action defined as: $(U \otimes V) |\psi\rangle \otimes |\phi\rangle = U |\psi\rangle \otimes V |\phi\rangle$.

Given a statistical ensemble of states $|\psi_i\rangle$ prepared with probability p_i, we may model this as a *density operator* $\rho = \sum_i p_i |\psi_i\rangle \langle\psi_i|$. Such a state may arise after a measurement is made (since a measurement is a probabilistic process causing the state to collapse to different vectors $|\psi_i\rangle$ with probabilities p_i). More generally, a density operator is a Hermitian positive semi-definite operator of a unit trace.

We may perform various important information theoretic computations on density operators. If ρ_{AE} is a density operator acting on $\mathbb{C}^n \otimes \mathbb{C}^m$, then we write $S(AE)_\rho$ to mean the *von Neumann entropy* of the operator ρ_{AE}. For finite dimensional systems, this is simply: $S(AE)_\rho = - \sum_i \lambda_i \log_2 \lambda_i$, where $\{\lambda_i\}$ are the eigenvalues of ρ_{AE}. The conditional von Neumann entropy is denoted $S(A|E)_\rho = S(AE)_\rho - S(E)_\rho$, where $S(E)_\rho$ is computed using the eigenvalues of ρ_E, where $\rho_E = tr_A \rho_{AE}$ i.e., ρ_E is the result of "tracing out" the A system. If we write $\rho_{AE} = \sum_{i,j} |i\rangle \langle j| \otimes \rho_E^{(i,j)}$, then $\rho_E = \sum_i \rho_E^{(i,i)}$.

2.1 Quantum Key Distribution

A QKD protocol's goal is to establish a shared secret key between two parties A and B, secure against an all-powerful adversary E. To achieve this, A and B are allowed to use a quantum communication channel permitting qubits to be sent between each. Furthermore, a classical authenticated channel is given, which allows A and B to send messages in an authenticated, *but not secret* way. That is, an adversary E may read anything sent on this authenticated channel, but may not write to it. The adversary is allowed to perform any attack on the quantum channel (as allowed by quantum physics).

Such protocols consist of two distinct stages: first a *quantum communication stage* which consists of numerous iterations, each treated independently and identically, whereby A sends qubits to B in a variety of ways according to the rules specified by the protocol; B receives these qubits, performs some measurement on them, and interprets the measurement result. A single iteration can yield at most one raw key-bit (sometimes an iteration is discarded, for example, B's measurement outcome may be "inconclusive" as determined by the protocol). Ultimately, the goal of this stage is to output a *raw key* which is a classical string, N bits long, that is partially correlated, and partially secret. The second stage of a QKD protocol, *information reconciliation* performs an error-correcting protocol (done over the authenticated channel thus leaking more information to E essentially "for free") followed by a privacy amplification protocol. The end result is a $\ell(N)$-bit secret key which may be used for other cryptographic purposes.

We consider *collective* attacks where E treats each iteration of the quantum communication stage independently and identically. Usually this is sufficient to prove security against arbitrary general attacks [16]. Such attacks may be modeled as a unitary operator U acting on the qubit and E's private quantum memory (modeled as a vector space $\mathbb{C}^n$). Without loss of generality, we may assume E's memory is initially cleared to some "zero" state $|0\rangle_E \in \mathcal{H}_E$ and so write U's action as follows:

$$U|0,0\rangle_{TE} = |0,e_0\rangle + |1,e_1\rangle \qquad U|1,0\rangle_{TE} = |0,e_2\rangle + |1,e_3\rangle, \qquad (1)$$

where the states $|e_i\rangle$ are arbitrary elements in $\mathbb{C}^n$ (though unitarity of U imposes important restrictions on them that we will take advantage of later). Due to linearity, this above definition is enough to completely define E's attack on any arbitrary qubit A may send.

Given a protocol and a description of the attack, one may describe a single iteration of the protocol as a density operator ρ_{ABE}. Then, the Devetak-Winter key-rate equation applies [3,17]:

$$r(U) = \lim_{N\to\infty} \frac{\ell(N)}{N} = S(A|E)_\rho - H(A|B), \qquad (2)$$

where $S(A|E)$ is the von Neumann entropy discussed earlier, and $H(A|B)$ is the conditional Shannon entropy. Of course, we must assume the worst case in that E chooses an optimal attack U. However, due to the nature of quantum communication, different types of attacks have, in a way, different "noise signatures" and, so, A and B can determine a set Γ_ν, where ν is a list of certain important measurable noise statistics in the channel. Thus, while it is not known for certain what attack was used, it can be guaranteed that the attack $U \in \Gamma_\nu$. Therefore, the actual key-rate is: $r(\nu) = \inf_{U\in\Gamma_\nu} r(U)$.

It was shown in [11,18] how to construct Γ_ν, to arbitrary levels of precision, in order to compute r. The noise signature ν is constructed from a standard quantum tomography protocol that users may run before using our algorithm. In particular, this signature consists of various probabilities $p_{i,j}$ which denotes the probability that B observes $|j\rangle$ if A sends $|i\rangle$ (and conditioning on A and B choosing the correct basis for such an outcome to occur), where $i \in \{0, 1, +, 0_Y\}$ and $j \in \{0, 1, +, -, 0_Y, 1_Y\}$ (note the asymmetry in the sending set versus the receiving set is intentional).

From this signature, a straight-forward process exists to construct a set $\widetilde{\Gamma}_\nu$ consisting of tuples of the form $(|e_0\rangle, \cdots, |e_3\rangle)$ such that the following properties are satisfied:

1: For every $(|e_0\rangle, \cdots, |e_3\rangle) \in \widetilde{\Gamma}_\nu$, there exists a unitary operator $U \in \Gamma_\nu$ that agrees with Eq. 1. That is, the attack is unitary (so could be implemented in theory) and it agrees with the observed noise signature ν.

2: For every $U \in \Gamma_\nu$, there exists $(|e_0\rangle, \cdots, |e_3\rangle) \in \widetilde{\Gamma}_\nu$ such that the key-rate if E used attack U is equal to the key-rate produced by the attack described by vectors $(|e_0\rangle, \cdots, |e_3\rangle)$ up to an arbitrary, user-defined, level of precision. That is, this construction does not "miss" any important attacks which minimize the key-rate.

For more information on this process of constructing $\widetilde{\Gamma}_\nu$, the reader is referred to [11,18] (in particular Algorithm 1 from [11]).

2.2 Envisioned System

As stated, all attacks by an adversary induce a particular "noise signature" and, ordinarily, QKD protocols are constructed and then analyzed to see which quantum attacks (i.e., what noise signatures) it is secure against. If an attacker is performing an attack with a noise level outside the known acceptable limits of the protocol being implemented (or if, even, just natural noise is inducing this noise), parties must simply abort, or try an alternative protocol and hope it too can at least operate over the channel. Even if a protocol is secure against this attack, however, it may be inefficient, requiring the transmission of thousands of qubits for one single secure key bit.

We are proposing, instead, to produce protocols optimized to counteract a specific noise signature as observed by the users. We envision users A and B having access to standard quantum technology, capable of sending and receiving qubits. Users will begin, after connecting their devices to the quantum channel, by performing a standard quantum tomography protocol, whereby A sends, randomly, qubits prepared in the X, Y, or Z bases and with B measuring, independently of A, in a random basis. Users then use the authenticated classical channel to disclose all measurement results, thus allowing them to determine the noise signature ν.

One of the users (either A or B) will then run our algorithm. The algorithm will take in this noise signature, and, through the use of a genetic algorithm, produce a QKD protocol as a circuit consisting of rudimentary gates. These gates may be specified by A and B - that is, they represent basic, low-level quantum operations which the users' hardware can actually support. With current technology, this gate-set would be limited; however, our system is flexible enough that, should in the future more complicated gates be implemented, the users may simply insert a description of these gates (as unitary matrices) into our algorithm and it will automatically incorporate them in new protocol generations.

After running our system, users are provided with a complete optimized protocol. The user running our algorithm (and thus who holds the description of the protocol) will send the protocol description to the other user, thus allowing both parties to configure their equipment properly. This transmission is done over the authenticated channel so that the adversary cannot tamper with the description to her benefit. The adversary can, however, learn the protocol in its entirety (thus, the protocol itself is never actually secret).

There are two things that E can do to take advantage of this knowledge of the evolved protocol. She can change her old attack (used during the quantum tomography protocol) to a new one such that the noise signature remains the same. Alternatively, she can change her attack to a new attack with a different noise signature. Our algorithm's security analysis will ensure that the protocol evolved is secure against *any* attack with the same noise signature (thus, eliminating the first threat). To ensure security against the second threat (E changing her attack to one with a different noise signature), users must periodically, and randomly (without warning to E), re-run the quantum tomography protocol to ensure that the noise signature did not change.

If the noise levels do change, our algorithm may simply be re-run to produce a new protocol to counter-act this new attack. In practice, one may consider running this system in large blocks of iterations; each block consisting of "real" iterations (i.e., iterations where the constructed protocol was executed) and "test" iterations (those used only for verifying the noise signature did not change). If, after the execution of a large block, the noise signature has changed drastically (small changes can be handled easily due to the continuity of von Neumann entropy) users must discard this block, re-run the algorithm, and try again. The reader may be concerned that this allows E to easily create a denial-of-service attack (where users are constantly discarding and trying again) - however there are easier ways for E to create such denial of service attacks against *any* QKD protocol and so this threat is not unique to our system, but common throughout any QKD protocol. In this work, we do not consider changes to the noise signature (and simply assume E keeps her attack - or rather the noise it induces - constant). There may be very interesting future work directions in discovering a potentially better method of dealing with them.

3   Our Algorithm

In our work, we will be evolving protocols modeled as *quantum circuits*. A quantum circuit operates over m wires, each wire "carrying" a qubit. Thus, the joint state modeling a system running on m wires (i.e., m qubits) is $\mathbb{C}^{2^m}$. On each wire, a *gate* may be placed, which are simply unitary matrices acting on the qubit wire. Common gates include:

$$H = \frac{1}{\sqrt{2}} \begin{pmatrix} 1 & 1 \\ 1 & -1 \end{pmatrix} \qquad\qquad X = \begin{pmatrix} 0 & 1 \\ 1 & 0 \end{pmatrix} \qquad\qquad (3)$$

$$R(p, \theta, \psi) = \begin{pmatrix} \sqrt{p}e^{i\theta} & \sqrt{1-p}e^{-i\theta} \\ \sqrt{1-p}e^{i\psi} & -\sqrt{p}e^{-i\psi} \end{pmatrix}$$

Gates may also be applied in a "control" mode in which case they act on two wires: a *target wire* and a *control wire*. In this mode, the gate will only act on the target wire if the control wire is in a $|1\rangle = (0,1)^T$ state. This operation may be done in a unitary manner. Finally, besides unitary gates, a measurement may be performed on the wire collapsing it, in a probabilistic manner, to a classical "0" or "1" state.

A protocol is, essentially, a probabilistic computation performed by parties. Any classical or quantum computation may be performed on a quantum circuit. Therefore, we will evolve protocols as quantum circuits - one for A and one for

B. Unlike prior work in [11], where protocols are evolved based on solutions to free parameters within a confined template, this new mechanism will allow the EA to discover new solutions not restricted to a given template.

Circuits, as mentioned, operate over several wires. By measuring a wire, it becomes a classical wire (only modeled as a quantum basis state). Of great importance to any QKD protocol are the following:

1. A single wire to carry a qubit from A to B (passing through the adversary E).
2. Each party A and B must have, at the end of the protocol, a classical wire (i.e., a quantum wire that was measured to produce a classical output). This wire will store their key-bit for the iteration.
3. Each party may have access to additional "optional" wires to be used arbitrarily.

A diagram of this scenario is shown in Fig. 1. Note that we do not need to provide additional randomness to our method – indeed, if a protocol requires randomness, it can first apply an $R(p,0,0)$ gate and then a measurement producing a classical output of 0 with probability p and 1 with probability $1-p$. Thus this mechanism is sufficient to model protocols involving quantum and classical computation, along with random choices.

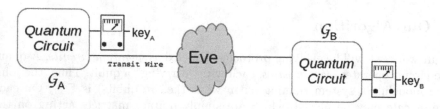

Fig. 1. A QKD protocol as two circuits $\mathcal{G}_A$ and $\mathcal{G}_B$. Circuit $\mathcal{G}_A$ is run first after which, the key$_A$ wire is measured yielding a classical bit. After $\mathcal{G}_A$, Eve is allowed to attack the transit line. Finally, B's circuit is run, acting on the transit wire, and additional wires private to B. Then, B's key$_B$ wire is measured.

3.1   Evolutionary Algorithm Approach and Parameters

A protocol is a specified process for A and B. We restrict ourselves currently to *one-way* QKD protocols whereby A sends qubits to B. In general, the qubit that A sends should depend in some manner (possibly random) on her key-bit choice for the iteration while B's measurement result should lead (again with some potential randomized post-processing) to his key-bit (which should be correlated to A's choice of key-bit). This process then repeats in i.i.d. way over subsequent iterations yielding a raw-key. As discussed earlier, the process for A and B will be described as a quantum circuit.

In particular, a protocol is a pair $\Pi = (\mathcal{G}_A, \mathcal{G}_B)$ where $\mathcal{G}_A = (g_{A,1}, \cdots, g_{A,n_A})$ is a list of gates (i.e., a quantum circuit) which A applies in sequence to her wires (similarly for $\mathcal{G}_B$ which is B's half of the protocol). Gates in $\mathcal{G}_A$ can only be applied to the transit wire and those wires private to A; similarly, gates in $\mathcal{G}_B$ may be applied only to the transit wire and those wires private to B. A gate, abstractly, simply specifies a type (we support H, X, $R(p, \theta, \psi)$, and a measurement operation, however, other gates or operations may be added or removed easily) and what wire it is applied to. Any gate may be added in a "control" mode, thus there is an additional "control" flag which, if true, will cause the gate to only be applied if a specified target wire is in a $|1\rangle$ state; note this target wire is also part of the gate structure.

Translating the above said into a data structure, we encode a potential solution (a potential circuit) as a combination of different data types. More specifically, we use:

1. an integer vector to encode the gate type with values 0 to 3 (see Eq. (3) for a description of gate types with index 3 meaning a measurement operation),
2. an integer vector to encode the gate target (wire),
3. a bit string vector to denote whether a gate is in control mode (value 1) or not (value 0),
4. an integer vector to encode the gate control (wire),
5. a vector of floating-point values to encode the parameters of the R gate: p, θ, ψ. Note, if a different gate type is selected, these values are not used.

The number of elements in each vector is equal to the total number of gates in a circuit, where the first subset of gates is reserved for the A side and the remaining ones for the B side. In the evolutionary computation language, the above data structure represents an individual's chromosome; an evolutionary algorithm will keep a set of these individuals, a population, and perform the search for better individuals using various modification operators and selection methods.

The fitness of a candidate solution $\Pi = (\mathcal{G}_A, \mathcal{G}_B)$ will be its key-rate (Eq. (2)) against any attack with the given noise signature ν. To compute Eq. (2), we must not only simulate the quantum system but we must also simulate all possible $U \in \Gamma_\nu$ (or, rather, all $U \in \tilde{\Gamma}_\nu$ which, as discussed, is sufficient to verify security in the worst-case). To do so, we use a quantum simulator which was specifically designed to work with the combination of quantum cryptography and evolutionary computation, developed originally in [10]. This simulator models arbitrary multi-user quantum states as density operators stored internally as linked-lists of so-called KetBra data structures.

A single KetBra encodes a quantity of the form: $p|i_1, \cdots, i_n\rangle \langle j_1, \cdots, j_n|$, with $p \in \mathbb{C}$ and i_k, j_k integer indices ranging from 0 to the dimension of the k'th subspace. In our case, since we are modeling circuits, each subspace is dimension two (i.e., a quantum wire) and so $i_k, j_k \in \{0, 1\}$ *except for the last subspace* which we assume is held by Eve and so can be higher dimensional. These integer values may represent basis states or arbitrary vectors to be substituted in later. For all but E's wire, these will be basis states (the actual choice of basis is not relevant to entropy computations). For E's wire, these integer indices will actually represent which of the four vectors $|e_i\rangle$ are to be placed there (see Eq. 1). A linked-list of these KetBra structures is taken to mean their sum. Since any density matrix may be written as a sum of terms of the form $p_{i,j} |i\rangle \langle j|$, this mechanism may be used to represent any finite-dimensional quantum system. Given actual vectors for the $|e_i\rangle$ it is a simple method to construct an actual density matrix and then perform the necessary entropy computations to evaluate the key rate $S(A|E) - H(A|B)$ (all of which is already supported by the simulator).

Each wire in the protocol is indexed; for our simulation, we order the wires so that 0 through w_A (inclusive, where w_A is specified by the user) are private to A; furthermore, we enforce the condition that 0 always be considered her key-bit wire (denoted $\text{key}_A = 0$). Wire $T = w_A + 1$ is the "transit" wire which carries a qubit from A to B. This is simulated simply by allowing A to access this wire, followed by the adversary, followed, finally by B. Thus all parties can access wire T, but only in the prescribed order. This ordering is enforced by our fitness calculation function; indeed, if ever a candidate solution were presented for fitness evaluation which allowed A access to B's wires (or B access to A's wires), the fitness is defined to be 0 - i.e., "abort." Wires $T + 1$ through w_B (inclusive) are private to B with wire $\text{key}_B = T + 1$ being his key-bit wire. Finally, subspace $w_B + 1$ is Eve's private quantum memory used during her attack. We do not assume this is a wire, but a higher dimensional subspace.

Our fitness computation for Π begins by resetting the simulator to the "zero" state $1 \cdot |0, \cdots, 0\rangle \langle 0, \cdots, 0|$ (i.e., all wires and E's ancilla begin in a $|0\rangle$ state). Next, all gates in $\mathcal{G}_A$ are applied in sequence (simulating A's protocol). E then attacks (which is abstractly simulated using notation from Eq. (1). Then, the gates in $\mathcal{G}_B$ are applied in order. Finally, we force a measurement of the key_A and key_B wires so that they yield classical outcomes 0 or 1 (with various probabilities). At the conclusion, we have in our simulator a density operator description of our protocol stored as a linked-list of KetBra structures.

At this point, we have a linked-list of KetBra structures representing the density operator of the protocol. Using an algorithm developed in [11], we may enumerate through all potential attack vectors $|e_i\rangle$, substituting them into the density operator, and thus computing its entropy $S(A|E)$ (a simple function of the eigenvalues of the resulting matrix). We take the minimum over all possible $|e_i\rangle$ produced by the algorithm as we must assume the worst case that E chooses an optimal attack. Of course, the goal of our EA is to **maximize** this value over all circuits (protocols).

Evolutionary Algorithm: In our experiments, we use evolution strategy (ES) of the type $(\mu + \lambda)$-ES. In this algorithm, in each generation, parents compete with offspring and from their joint set, μ fittest individuals are kept. In our experiments, offspring population size λ has a value equal to 4, while the parent population size μ equals 1. For further information on ES, we refer interested readers to [19].

In the ES, the offspring may be generated using either a single parent with the mutation operator, which is the most widely used variant. Additionally, the offspring may be created by using two or more parents, which corresponds to the crossover operator. In our experiments with ES we use the mutation operator only, which takes a parent and randomly modifies a part of its genotype. In each mutation operation, first a part of the individual's data structure is selected at random; then, depending on the type of the selected part, the mutation is performed in the following manner:

1. for an integer vector, a single random element in the vector is changed to a new random value (corresponding to either another wire or another gate type being selected);
2. for a bit string vector, a single element is inverted (changing the control nature of the gate);
3. for a vector of floating-point values, a single random element is changed according to the Gaussian distribution with mean 0 and standard deviation of 1 (thus modifying the parameters of the R gate).

Consequently, for each parent individual, four new individuals are created in this way; the best of the five individuals is then selected as the new parent and the process is repeated. In all the experiments, the number of runs for each configuration is 30 and the stopping criterion is either 100 000 evaluations or maximum running time of 10 h per run.

Apart from ES, we also experimented with a genetic algorithm, but we found ES to converge much faster. This is in part due to the simulator: the duration of a single evaluation is not constant and varies greatly depending on the solution quality. The GA tended to generate solutions which take much more time to evaluate, and this would in turn drastically slow down the convergence. The ES, on the other hand, managed to perform many more evaluations in the same amount of time, and consequently reach much better solutions on average.

Summary. To summarize our approach, users begin by using their quantum equipment to run a standard quantum tomographic protocol, resulting in a noise signature ν (in our evaluations this is simulated). From this, one of the parties, either A or B, will run the algorithm, providing as input ν. Our algorithm will produce an optimized protocol, in the form of a quantum circuit consisting of gates which may be *user specified based on the capabilities of their devices*. Whichever party runs the EA will broadcast, through the authenticated public channel, the gate description so that both parties are able to configure their equipment appropriately. Note that this also gives E information on the protocol (i.e., the protocol description is not secret information once it is in operation). This is not an issue for security, so long as the noise in the channel does not alter (as our algorithm builds a protocol based on the optimal attack within Γ_ν). To enforce that E does not change the attack to one outside of Γ_ν, A and B must periodically, and randomly, re-run the tomographic protocol; should the noise signature change, they must simply re-run our EA to produce a new protocol for the new noise signature. Note that we did not consider imperfect parameter estimation, that addition would not be difficult to introduce by increasing the size of Γ_ν based on imperfect ν; we leave this as future work. A schematic diagram of our algorithm and this process is shown in Fig. 2.

4   Experimental Results

We evaluate our algorithm over symmetric channels and arbitrary, asymmetric ones. A symmetric channel is parameterized by a single noise value Q (with $Q = 0$ meaning there is no noise in the channel). In [14], it was shown that the BB84 protocol [2] (which is also generally the protocol used in practice in current-day QKD implementations) cannot be surpassed over such symmetric channels. Thus, this case serves as a useful test to verify the correctness of our algorithm (our system should be able to find a protocol with a key-rate equal to that of BB84). Table 1 shows that our approach does, indeed, find protocols that achieve the optimal BB84 rate.

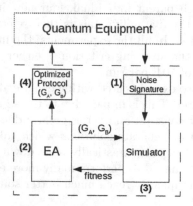

Fig. 2. A diagram of our algorithm and approach. Users begin by providing our system with the channel noise signature ν (1). The EA will evolve candidate solutions, which are pairs of circuits $(\mathcal{G}_A, \mathcal{G}_B)$ (2). Each candidate solution is sent to the simulator (3) for fitness evaluation which requires the noise signature to compute. Finally, an optimized protocol is output (4) from which the users may configure their devices to optimize the secure key distribution rate.

In the experiments presented here, we consider scenarios where there are 5 gates and 5 wires. Two of the wires belong to the A side, two belong to the B side, and 1 wire is a joint one. Finally, out of the 5 gates, we consider a scenario where three gates belong to the A side and two gates belong to the B side. This number of gates is sufficient to construct most standard, state-of-the-art QKD protocols known today (including BB84). One could think that such a limited setting actually gives a small search space size and there is no need for evolutionary algorithms approach. However, even if we do not consider the floating-point encoding in our chromosome, we are still left with a search space size of $4^5 \times 5^5 \times 2^5 \times 3^5$ solutions, which equals to 24 883 200 000 possible configurations.

Additionally, the computational bottleneck is not on the evolutionary side but on the evaluation side due to the quantum simulator complexity, which prohibits any possibility of running an exhaustive search.

All the experiments suggest that the number of 100 000 evaluations is sufficient for the algorithm convergence, after which it is more efficient to restart the run. Convergence for a typical experiment conducted in 30 runs is shown in Fig. 3; at each point up to the maximum number of evaluations, current best fitness values across all the runs in the form of their minimum, maximum, median and mean value are shown.

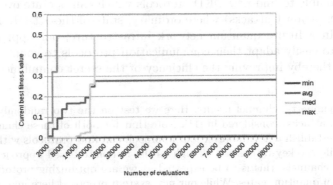

Fig. 3. Convergence rate showing best fitness value over multiple runs.

Table 1. Symmetric channel results for various levels of noise, also comparing with the BB84 protocol (the six-state version) which is known to be optimal on these channels. Our algorithm was able to evolve a protocol matching the optimal BB84 rate.

Noise	Max	Min	Avg	Std dev	BB84 (Opt.)
0.01	**0.864**	0.864	0.864	0	0.864
0.05	**0.497**	$1.45 \cdot 10^{-15}$	0.263	0.2556	0.497
0.1	**0.152**	$1.22 \cdot 10^{-15}$	0.061	0.0766	0.152
0.12	**0.035**	$3.22 \cdot 10^{-15}$	0.007	0.0152	0.035

We also test over arbitrary, asymmetric channels. For such channels, BB84 is not necessarily optimal, but no known theoretical result exists claiming how an optimal protocol should be constructed. Thus, evaluating over asymmetric channels serves as an interesting evaluation test for our system. In particular, we evaluate over the two asymmetric channels that were considered in [11] for comparison purposes. These channels are the result of an attack which could be launched against an actual QKD system; they also cause BB84 to fail (i.e., abort). Since these channels were also the ones considered in [11], we can also compare how our gate-based approach compares with the template-based version. The results of this test are summarized in Table 2. Figure 4 shows a sample protocol output by our algorithm.

5   Closing Remarks

In this paper, we showed how evolution strategy can be used to evolve quantum cryptographic protocols modeled as quantum circuits. Our approach was able to find a protocol matching the optimal BB84 key-rate for symmetric channels. We were also able to find new QKD protocols which can operate over quantum channels (i.e., against attacks) where ordinary, state-of-the-art QKD protocols would fail. In a future quantum network infrastructure, our approach would allow users to easily adapt their communication protocols to counter quantum adversaries, thereby improving the efficiency of the secret communication.

Table 2. Asymmetric channel results. Here we test on the two randomly generated channels (i.e., attacks) considered in [11]. Note that for both of these channels, BB84 would fail to establish a key; however, our approach can find protocols with a positive key-rate. While the key-rate is not as high as the template-based approach from [11] for these two channels, this is to be expected as we are optimizing protocols built of a limited set of quantum gates. While our new system proposed here may not achieve as high a key-rate as the template-based approach, the protocols evolved here are potentially easier to implement in practice as they are based on simple gates.

Description	Max	Min	Avg	Std dev	BB84/Ref. [11]
Channel 1	**0.066**	$2.22 \cdot 10^{-16}$	$5.73 \cdot 10^{-4}$	0.00603	0(abort)/.094
Channel 2	**0.018**	$3.33 \cdot 10^{-16}$	$2.58 \cdot 10^{-4}$	0.00197	0(abort)/.042

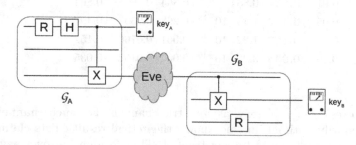

Fig. 4. A protocol output by our algorithm optimized to run on Channel 1.

Many very interesting open problems remain. It would be very interesting to provide our system with increased quantum communication resources (such as two-way channels for example) in order to improve the key-rate. Also providing our system with classical communication resources allowing for the evolution of more complex post-processing strategies, known to be important for improving the key-rate of protocols over noisy channels [3]. We suspect this is one area where our new approach of evolving quantum circuits for QKD protocols would be greatly beneficial (prior work in this area would require users to write out in detail an abstract template which would be challenging for these more powerful quantum and classical resources). It would also be very interesting to take into account practical imperfections in the optical devices used. Other areas of future work could include having the number of gates and wires as part of the solution, however, this would require improvements in the computational speed of the simulator used.

References

1. Scarani, V., Bechmann-Pasquinucci, H., Cerf, N.J., Dušek, M., Lütkenhaus, N., Peev, M.: The security of practical quantum key distribution. Rev. Mod. Phys. **81**, 1301–1350 (2009)
2. Bennett, C.H., Brassard, G.: Quantum cryptography: public key distribution and coin tossing. In: Proceedings of IEEE International Conference on Computers, Systems and Signal Processing, New York, vol. 175 (1984)
3. Renner, R., Gisin, N., Kraus, B.: Information-theoretic security proof for quantum-key-distribution protocols. Phys. Rev. A **72** (2005). https://doi.org/10.1103/PhysRevA.72.012332
4. Spector, L.: Automatic Quantum Computer Programming: A Genetic Programming Approach. Kluwer Academic Publishers, Boston (2004)
5. Wang, X., Jiao, L., Li, Y., Qi, Y., Wu, J.: A variable-length chromosome evolutionary algorithm for reversible circuit synthesis. J. Multiple Valued Log. Soft Comput. **25**(6), 643–671 (2015)
6. Abubakar, M.Y., Jung, L.T., Zakaria, N., Younes, A., Abdel-Aty, A.-H.: Reversible circuit synthesis by genetic programming using dynamic gate libraries. Quantum Inf. Process. **16**(6), 160 (2017)
7. Rylander, B., Soule, T., Foster, J., Alves-Foss, J.: Quantum evolutionary programming. In: Proceedings of the 3rd Annual Conference on Genetic and Evolutionary Computation, pp. 1005–1011. Morgan Kaufmann Publishers Inc. (2001)
8. Picek, S., Golub, M.: On evolutionary computation methods in cryptography. In: Proceedings of the 34th International Convention MIPRO, pp. 1496–1501. IEEE (2011)
9. Picek, S., Mariot, L., Leporati, A., Jakobovic, D.: Evolving s-boxes based on cellular automata with genetic programming. In: Proceedings of the Genetic and Evolutionary Computation Conference Companion, pp. 251–252. ACM (2017)
10. Krawec, W.O.: A genetic algorithm to analyze the security of quantum cryptographic protocols. In: IEEE Congress on Evolutionary Computation (CEC), pp. 2098–2105. IEEE (2016)
11. Krawec, W.O., Nelson, M.G., Geiss, E.P.: Automatic generation of optimal quantum key distribution protocols. In: Proceedings of the Genetic and Evolutionary Computation Conference, pp. 1153–1160. ACM (2017)
12. Rubinstein, B.I.P.: Evolving quantum circuits using genetic programming. In: Proceedings of the Congress on Evolutionary Computation, pp. 144–151. IEEE (2001)

13. Spector, L., Klein, J.: Machine invention of quantum computing circuits by means of genetic programming. AI EDAM **22**(3), 275–283 (2008)
14. Bae, J., Acín, A.: Key distillation from quantum channels using two-way communication protocols. Phys. Rev. A **75**(1), 012334 (2007)
15. Nielsen, M.A., Chuang, I.L.: Quantum Computation and Quantum Information. Cambridge University Press, Cambridge (2000)
16. Christandl, M., Konig, R., Renner, R.: Postselection technique for quantum channels with applications to quantum cryptography. Phys. Rev. Lett. **102** (2009). https://doi.org/10.1103/PhysRevLett.102.020504
17. Devetak, I., Winter, A.: Distillation of secret key and entanglement from quantum states. Proc. R. Soc. A Math. Phys. Eng. Sci. **461**(2053), 207–235 (2005)
18. Krawec, W.O.: Quantum key distribution with mismatched measurements over arbitrary channels. Quantum Inf. Comput. **17**(3), 209–241 (2017)
19. Bäck, T., Fogel, D.B., Michalewicz, Z. (eds.): Evolutionary Computation 1: Basic Algorithms and Operators. Institute of Physics Publishing, Bristol (2000)

Evolutionary Computation Techniques
for Constructing SAT-Based Attacks
in Algebraic Cryptanalysis

Artem Pavlenko[1]([⊠]), Alexander Semenov[2], and Vladimir Ulyantsev[1]

[1] ITMO University, St. Petersburg, Russia
{alpavlenko,ulyantsev}@corp.ifmo.ru
[2] Matrosov Institute for System Dynamics and Control Theory SB RAS,
Irkutsk, Russia
biclop.rambler@yandex.ru

Abstract. In this paper we present the results on applying evolutionary computation techniques to construction of several cryptographic attacks. In particular, SAT-based guess-and-determine attacks studied in the context of algebraic cryptanalysis. Each of these attacks is built upon some set of Boolean variables, thus it can be specified by a Boolean vector. We use two general evolutionary strategies to find an optimal vector: (1+1)-EA and GA. Based on these strategies parallel algorithms (based on modern SAT-solvers) for solving the problem of minimization of a special pseudo-Boolean function are implemented. This function is a fitness function used to evaluate the runtime of a guess-and-determine attack. We compare the efficiency of (1+1)-EA and GA with the algorithm from the Tabu search class, that was earlier used to solve related problems. Our GA-based solution showed the best results on a number of test instances, namely, cryptanalysis problems of several stream ciphers (cryptographic keystream generators).

Keywords: Algebraic cryptanalysis · Guess-and-determine attack ·
SAT · Evolutionary computation

1 Introduction

Algebraic Cryptanalysis (see [1]) is a way of breaking ciphers through solving systems of algebraic equations over finite fields. The corresponding attacks are called algebraic. Systems of algebraic equations constructed for strong ciphers are usually difficult for all known state-of-the-art algorithms. The resulting system of equations can be simplified by guessing the values of some of its variables. Then we can try all possible assignments of such variables, every time obtaining some simplified system. It might happen that the time spent by some algorithm on

The study was funded by a grant from the Russian Science Foundation (project No. 18-71-00150).

P. Kaufmann and P. A. Castillo (Eds.): EvoApplications 2019, LNCS 11454, pp. 237–253, 2019.
https://doi.org/10.1007/978-3-030-16692-2_16

solving all such systems will be significantly smaller than the brute force attack time (e.g., if we test all possible keys of the cipher in question). An algebraic attack that uses some guessed bit set to simplify the system of cryptanalysis equations is called a guess-and-determine attack.

Over the last 20 years a large number of guess-and-determine attacks have been designed. In the vast majority of cases, a guess-and-determine attack is based on the analysis of the cipher features (see Fig. 1 with Trivium-Toy 64 cipher example). Such an analysis usually requires a lot of manual work. The recent papers [2,3] describe an automatic method for constructing guess-and-determine attacks. In the framework of this method, the weakened equations of cryptanalysis are solved using modern Boolean SATisfiability (SAT) solvers. Each guessed bit set is represented as a point in the Boolean hypercube. An arbitrary point is associated with the value of a special function that evaluates the complexity of the corresponding guess-and-determine attack. This function is a black-box pseudo-Boolean function. Thus, the construction of a guess-and-determine attack is reduced to the pseudo-Boolean black-box optimization problem. Various metaheuristic algorithms can be used to solve it and the papers [2,3] employ the simplest local search schemes, such as Simulated Annealing and Tabu Search. The main purpose of this paper is to demonstrate capabilities of evolutionary computation in application to the problem of automatic construction of guess-and-determine attacks in algebraic cryptanalysis. Below is the brief outline of the present work.

In Sect. 2 we introduce basic notations and facts of the presented paper. In particular, we briefly describe construction the known reduction of the problem of cryptographic attacks to the problem of pseudo-Boolean optimization. The corresponding pseudo-Boolean function Φ is not specified analytically and to minimize it metaheuristic algorithms related to local search methods were previously used [2,3]. In the present paper in order to solve this problem we apply two common strategies: (1+1)-Evolutionary Algorithm and Genetic Algorithm. The corresponding algorithms and techniques are described in Sect. 3. Section 4 contains results of computational experiments. In Sect. 5 we summarize the obtained results and outline future research.

2   Preliminaries

In this section, we give some auxiliary information from the Boolean functions theory and cryptanalysis.

2.1   Boolean Functions, Formulas and Boolean Satisfiability Problem (SAT)

Let $\{0,1\}^k$, $k \in \mathbf{N}$ denote a set of all binary words of length k ($\{0,1\}^0$ corresponds to an empty word). The words from $\{0,1\}^k$, $k \geq 1$ are sometimes called Boolean vectors of length k, whereas the set $\{0,1\}^k$, $(k \geq 1)$ is referred to as a Boolean hypercube of dimension k. An arbitrary total function of the form

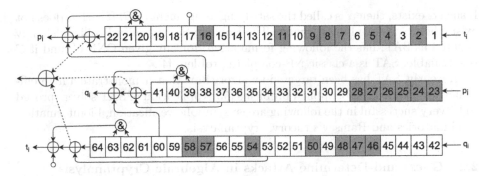

Fig. 1. Visualization of the key stream generator Trivium-Toy 64 with three registers. The guessed bit set, which is considered as an individual of evolutionary algorithms in the present work, is colored grey

$f : \{0,1\}^k \to \{0,1\}$ is called a Boolean function of arity k. An arbitrary function of the form $h : \{0,1\}^* \to \{0,1\}^*$, where

$$\{0,1\}^* = \bigcup_{k=0}^{\infty} \{0,1\}^k$$

is called a discrete function.

Boolean variables are the variables that take values from $\{0,1\}$. A Boolean formula with respect to k variables is an expression built by special rules over the alphabet comprising k Boolean variables $x_1, \ldots, x_k$ and special symbols called Boolean connectives. An arbitrary Boolean formula with respect to k variables defines a Boolean function of the kind $f_k : \{0,1\}^k \to \{0,1\}$. The set of Boolean connectives is called complete if they can be used to create any Boolean function of arbitrary arity. Such a set is called a complete system of connectives or a complete basis. The following set is a complete basis: $\{\land, \lor, \neg\}$, where $\land$ is conjunction, $\lor$ is disjunction and $\neg$ is negation.

Assume that F is an arbitrary Boolean formula, X is a set of variables that can be found in F, B is an arbitrary subset of X ($B \subseteq X$). By $\{0,1\}^{|B|}$ we denote a set of all assignments of variables from B.

Let x be a Boolean variable. The formula that consists of a single variable or a negation is called a literal. Let x be an arbitrary Boolean variable. A pair of literals $(x, \neg x)$ is called a complementary pair. An arbitrary disjunction of different literals, which do not have any complementary pairs among them, is called a clause. An arbitrary conjunction of different clauses is called a Conjunctive Normal Form (CNF). If C is a CNF and $X = \{x_1, \ldots, x_n\}$ is a set of all Boolean variables that can be found in C, then we can say that C is a CNF over the set of variables X.

Let C be a CNF over X, $|X| = k$. Denote $f_C : \{0,1\}^k \to \{0,1\}$ a Boolean function defined by the CNF C. The CNF C is called satisfiable if there exists such $\alpha \in \{0,1\}^k$ (i.e. an assignment of variables from X), that $f_C(\alpha) = 1$ holds.

If such α exists, then α is called the satisfying assignment of C. If such α does not exist, then C is called unsatisfiable. The Boolean Satisfiability Problem, shortly denoted as SAT, has the following formulation: for any given CNF C, find if C is satisfiable. SAT is a classic NP-complete problem [4].

Recently, SAT has been targeted by algorithms that demonstrate a high efficiency for a wide class of applied problems [5]. Application of SAT solvers proved to be very successful in the following areas: symbolic verification, bioinformatics, combinatorics and Ramsey's theory, cryptanalysis.

2.2 Guess-and-Determine Attacks in Algebraic Cryptanalysis

As has been mentioned above, the algebraic cryptanalysis implies solution of systems of algebraic equations (usually over the field $GF(2)$) that describe some cipher.

Any cipher can be considered as a total discrete function of the kind

$$f : \{0,1\}^n \to \{0,1\}^m. \tag{1}$$

Then the cryptanalysis problem can be considered in the context of the problem of finding a preimage for some known value of the function (1): using the known $\gamma \in Range\ f$, $Range\ f \subseteq \{0,1\}^m$, find such $\alpha \in \{0,1\}^n$ that $f(\alpha) = \gamma$. We will call it the inversion problem for the function (1).

It is well-known (see, e.g. [1]) that the algorithm for calculating function (1) can be effectively described by a system of algebraic equations over the field $GF(2)$. Denote this system by $E(f)$. Roughly speaking, it describes the process of finding an output of the function f that corresponds to an arbitrary input. Let X be a set of all variables that can be found in $E(f)$. Denote by X^{in} and X^{out} sets of variables that were assigned to inputs and outputs of the function f, respectively.

Substitution of the values into the system $E(f)$ is determined in a standard way. It can be shown that if we substitute an arbitrary assignment $\alpha \in \{0,1\}^n$ for some variables from X^{in} into $E(f)$, we can derive assignments for all other variables from $E(f)$ by the following simple rules.

Substitute an arbitrary $\gamma \in Range\ f$ into $E(f)$, and let $E(f,\gamma)$ be the resulting system. If we manage to solve $E(f,\gamma)$ then we can extract such $\alpha \in \{0,1\}^n$ from the solution of $E(f,\gamma)$ that $f(\alpha) = \gamma$. However, this is a difficult problem for strong ciphers.

The simplest and most efficient way of solving $E(f,\gamma)$ is a sequential substitution of all possible $\alpha \in \{0,1\}^n$ into $E(f,\gamma)$. We denote the resulting system as $E(f,\alpha,\gamma)$. In line with the above, an arbitrary system $E(f,\alpha,\gamma)$ can be easily solved. If $f(\alpha) \neq \gamma$, then by applying the simple rules mentioned above, we derive a contradiction from $E(f,\alpha,\gamma)$. If $f(\alpha) = \gamma$, the contradiction does not occur and each variable from X will get a certain value. This scheme of trying all possible inputs corresponds to the method of cryptanalysis called the *brute force attack*.

For some ciphers and for functions of the form (1) that describe them, we can select a subset B of the set X with the following properties:

1. $|B| \ll n$;
2. the problem of finding a solution of $E(f, \gamma, \beta)$ or proving its inconsistency can be solved relatively quickly using some algorithm A; here β is an assignment for variables from B;
3. the total time required for finding $\alpha : f(\alpha) = \gamma$ by trying various $\beta \in \{0,1\}^{|B|}$ is considerably smaller than the time of the brute force attack.

If the requirements listed above are fulfilled, then we can talk about a guess-and-determine attack based on the guessed bit set B.

Over the last 15–20 years a substantial number of different guess-and-determine attacks have been proposed. Some of them proved to be fatal for the corresponding ciphers. One of the simplest examples of the guess-and-determine attack is the attack on the A5/1 cipher described by Ross Andersen in 1994 [6]. For a long time, A5/1 served as a standard for encrypting the GSM traffic in cellular telephony. This cipher uses a 64-bit secret key. R. Anderson noted that if we choose in a certain way 53 bits of the internal state of the A5/1 registers and we know a certain number of bits of the key flow, then we can recover 11 unknown bits of the state of the registers by solving a trivial linear system over the field $GF(2)$ (in fact, we just have to solve a triangular system). Therefore, the Anderson attack is based on the guessed bit set $B : |B| = 53$, the role of the A algorithm is played by the algorithm for solving systems of linear equations. The form of these equations enables implementation of this attack on specialized computational architectures, for instance, in [7] the Anderson attack was performed on an FPGA device. The runtime of the corresponding attacks is up to 10 h for one cryptanalysis problem.

A series of works on algebraic cryptanalysis (see e.g., [1,8–10]) implies that algorithm A does not have to be polynomial. The problem of solving systems of the form $E(f, \gamma)$ can be efficiently reduced to combinatorial problems that are difficult in the worst case (NP-hard), but not difficult for most of their particular cases. One of the most computationally attractive problems here is SAT. Accordingly, any SAT-solving algorithm can play the role of A. In all the articles listed above, as well as in a number of other papers, the problem of finding a solution to an arbitrary system of the form $E(f, \gamma)$ was reduced to SAT for CNF $C(f, \gamma)$. To solve the resulting SAT instances, CDCL-based SAT solvers were used [11].

Some of those works describe guess-and-determine attacks where weakened cryptanalysis equations are solved using SAT-solvers. In [8] SAT-solvers were used to build a guess-and-determine attack on the truncated variants of the DES cipher, in [9] similar attacks targeted truncated variants of the GOST 28147-89 cipher. The paper [10] described a SAT-based guess-and-determine attack on the A5/1 cipher that used the guessed bit set $B : |B| = 31$.

2.3 Automatic Methods for Constructing Guess-and-Determine Attacks

In all of the above attacks, the guessed bit set was selected while analyzing the characteristics of the cipher in question. In other words, the corresponding attacks were constructed manually. As mentioned above, papers [2,3] propose approaches to the automatic construction of guess-and-determine attacks. To this end, special functions were introduced: they evaluate the efficiency of the attack and are calculated using a probabilistic experiment. Such functions are black-box pseudo-Boolean functions [12]. In [2,3] the problem of constructing a guess-and-determine attack with the lowest time was reduced to the pseudo-Boolean black-box optimization problem. Next, we briefly describe the corresponding techniques presented in [3].

Consider the problem of finding the preimage for the function $f : \{0,1\}^n \to \{0,1\}^m$, given by some efficient algorithm. Following the ideology of symbolic execution [4,13], we can build the CNF $C(f)$ that possesses some important properties using the algorithm that computes f (for more details see [3]). Let $\gamma \in \{0,1\}^m$ be an output of the function f and let B be an arbitrary set of variables in $C(f)$ that does not include variables from X^{out}. Consider B as a guessed bit set and denote by β an arbitrary assignment for variables from B. Let $C(f,\gamma,\beta)$ denote the CNF resulting from the substitution of the assignments γ and β into $C(f)$.

Let A be some algorithm for solving SAT. Fix some positive number t. For each $\beta \in \{0,1\}^{|B|}$ build a CNF $C(f,\gamma,\beta)$ and apply the algorithm A to it, limiting the algorithm runtime to t. If A fails to solve the corresponding SAT instance in time t, then we terminate A and move to the next β. If for some $\beta \in \{0,1\}^{|B|}$ the algorithm A finds a satisfying assignment for $C(f,\gamma,\beta)$ in time $\leq t$, then thereby it will find $\alpha \in \{0,1\}^n : f(\alpha) = \gamma$.

It was shown in [3] that if α was randomly chosen in correspondence with a given on $\{0,1\}^n$ uniform distribution and if $\gamma : \gamma = f(\alpha)$ is known, then we can determine the probability of the following event: by applying the brute force strategy described above to the set $\{0,1\}^{|B|}$ we will find $\alpha : f(\alpha) = \gamma$. Denote this probability by ρ_B. It might be very small. Then we can repeat the strategy considering different outputs $\gamma_1, \ldots, \gamma_r$ of the function f (for many ciphers, it is enough to solve this problem at least for one such output to find the secret key). Then the probability of finding the preimage for at least one output $\gamma_1, \ldots, \gamma_r$ is

$$P_B^* = 1 - (1 - \rho_B)^r.$$

It is obvious that for a fixed $\rho_B > 0$ the probability P_B^* tends to 1 as r increases.

Note that the strategy described above takes time $2^s \cdot t$, $(s = |B|)$ to process one output γ_i, $i \in \{1, \ldots, r\}$, therefore, the upper bound for the total runtime of the corresponding attack is $2^s \cdot t \cdot r$. It was shown in [3] that if $r \approx \frac{3}{\rho_B}$, then $P_B^* > 0.95$, which means that the inversion of at least one of the r outputs of the function under consideration is an almost certain event. For this reason, it is desirable to find such B, for which the value

$$2^s \cdot t \cdot \frac{3}{\rho_B}$$

reaches the minimum on a set of possible alternatives of guessed bit sets. Unfortunately, if we want to calculate precisely the probability ρ_B, we have to search through the entire set $\{0,1\}^n$, which is infeasible. Therefore, the probability ρ_B takes the form of the expected value $E[\xi_B]$ of the random variable ξ_B of a special kind. Thus, it is required to minimize the function

$$2^s \cdot t \cdot \frac{3}{E[\xi_B]} \qquad (2)$$

over all possible sets B. The variable ξ_B is derived from simple probability experiments (see details in [3]), whereas we use the Monte Carlo method [14] to evaluate $E[\xi_B]$.

The paper [3] considers the problem of constructing an efficient SAT-based guess-and-determine attack in the context of the minimization problem of a black-box pseudo-Boolean function with the vector χ_B as the function's input. The unit components of χ_B select the set B in the set of variables found in $C(f)$. Then a probability experiment for calculating the function (2) is set up for this set B. The obtained number is an estimate of the time for the guess-and-determine attack, where B is used as a guessed bit set. The goal is to find B with the smallest value of the estimate for the function (2).

In [2,3], pseudo-Boolean optimization problems were solved by simple metaheuristics such as Simulated Annealing and Tabu Search. The main results of this paper deal with application of evolutionary algorithms to these problems.

3 Applying Evolutionary Computations to Construction of Guess-and-Determine Attacks

In this section we formulate the problem of constructing an efficient guess-and-determine attack as the pseudo-Boolean optimization problem. We also describe the basic techniques of evolutionary computing that we employ.

3.1 Construction of an Efficient Guess-and-Determine Attack as the Problem of Pseudo-Boolean Function Minimization

We consider the cryptanalysis problem as the problem of inversion of the function (1). Construct the CNF $C(f)$. Let X be a set of variables found in $C(f)$. Let $W = X \setminus X^{out}$ and let B be an arbitrary subset of W. We can present an arbitrary B with the help of the Boolean vector χ_B of length $q = |W|$. Ones in χ_B will indicate those variables from W that were included into B.

Following the ideas from [3], define the pseudo-Boolean function

$$\Phi : \{0,1\}^q \to \mathbf{R}. \tag{3}$$

The arbitrary vector $\chi_B \in \{0,1\}^q$ is an input of the function (3). We use this vector to build the set B, $B \subseteq W$.

Recall that X^{in}, $|X^{\mathrm{in}}| = n$ is the set formed by the variables from X corresponding to the input of the function (1). Define uniform distribution on $\{0,1\}^n$ and choose the corresponding Boolean vectors $\alpha_1, \ldots, \alpha_M$ ($\alpha_j \in \{0,1\}^n$, $j \in \{1, \ldots, M\}$). We will refer to this set of vectors as a random sample of size M. In view of the above, substitution of an arbitrary α_j into $C(f)$ and application of simple rules yields a derivation of assignments for all variables from X. Let γ_j be an assignment for variables from X^{out} obtained as a result of this derivation. By β_j we denote the assignment derived for variables from B, $j \in \{1, \ldots, M\}$.

Let us construct the CNFs $C(f, \beta_1, \gamma_1), \ldots, C(f, \beta_M, \gamma_M)$. Let each of these CNFs be an input of the SAT-solver A and set the solving time of each corresponding SAT instance as t. For each $j \in \{1, \ldots, M\}$ consider a random variable ξ_B^j, that evaluates to 1 if SAT instance for the CNF $C(f, \beta_j, \gamma_j)$ was solved in time $\leq t$, and $\xi_B^j = 0$ otherwise. The value of the function (3) at arbitrary point $\chi_B \in \{0,1\}^s$ is defined as follows:

$$\Phi(\chi_B) = 2^{|B|} \cdot t \cdot \frac{3M}{\sum_{j=1}^{M} \xi_B^j}, \tag{4}$$

where t and M are the parameters of this function.

In accordance with the Monte Carlo method, the bigger the size of the random sample M, the better approximation of (2) is given by the function (4).

Now consider the problem of finding the minimum of the function (4) over the Boolean hypercube $\{0,1\}^q$. To solve this problem we will use evolutionary algorithms.

3.2   Evolutionary Computation Techniques Used for Minimization of the Suggested Pseudo-Boolean Function

In this section we present some techniques that complement such strategies as the (1+1)-Evolutionary Algorithm [15] ((1+1)-EA) and one variant of the Genetic Algorithm (GA). We used these techniques to solve the minimization problem for functions of type (4) for several stream ciphers.

Basic Schemes of Evolutionary Computation. Algorithm 1, which is given below, is a common outline of an evolutionary algorithm for solving the minimization problem of an arbitrary function of type (4).

As an input, the algorithm takes the CNF $C(f)$, a random sample size M, the time limit t, the minimization strategy S, and the initial guessed bit set represented by the Boolean vector χ_{start}.

Algorithm 1. Evolutionary algorithm

Input: CNF formula $C(f)$, initial sample size M, time limit t, strategy S, initial guessed bits χ_{start}

1:   $P \leftarrow \text{INITPOPULATION}(S, \chi_{\text{start}})$
2:   $\langle \chi_{\text{best}}, v_{\text{best}} \rangle \leftarrow \langle \chi_{\text{start}}, \Phi(\chi_{\text{start}}, M, t) \rangle$
3:   $N_{\text{stag}} \leftarrow 0$ ▷ stagnation count
4:   **while not** STOPCONDITION( ) **do**
5:       $N_{\text{stag}} \leftarrow N_{\text{stag}} + 1$
6:       **for** χ **in** P **do**
7:           $v \leftarrow \Phi(\chi, M, t)$
8:           **if** $v < v_{\text{best}}$ **then**
9:              $\langle \chi_{\text{best}}, v_{\text{best}} \rangle \leftarrow \langle \chi, v \rangle$
10:             $N_{\text{stag}} \leftarrow 0$
11:          **end if**
12:          $M \leftarrow \text{SELECTM}()$
13:       **end for**
14:       **if** $N_{\text{stag}} < N_{\text{stag}}^{\max}$ **then**
15:          $P \leftarrow \text{GETNEXTPOPULATION}(S, P)$
16:       **else**
17:          $P \leftarrow \text{RESTART}(S, \chi_{\text{start}})$
18:          $N_{\text{stag}} \leftarrow 0$
19:       **end if**
20: **end while**
21: **return** $\langle \chi_{\text{best}}, v_{\text{best}} \rangle$

Each Boolean vector χ from $\{0, 1\}^q$ is considered as a population in the general concept of evolutionary computation. The value of the function Φ of type (4) at point χ is considered as a value of the fitness function for the corresponding population.

The function INITPOPULATION(S, χ_{start}) forms the initial population according to the input vector χ_{start} within the frames of the chosen evolutionary strategy S. The pair $\langle \chi_{\text{best}}, v_{\text{best}} \rangle$ corresponds to point in $\{0, 1\}^q$ with the current Best Known Value (BKV) of the function Φ.

The value of Φ at each specific point $\chi_B \in \{0, 1\}^q$ is computed using the scheme described in the previous section: construct the set B defined by χ, construct the random sample $\{\alpha_1, \ldots, \alpha_M\}$, find Boolean vectors $\gamma_1, \ldots, \gamma_M$ and $\beta_1, \ldots, \beta_M$ induced by this random sample, and finally build the SAT instances $C(f, \beta_1, \gamma_1), \ldots, C(f, \beta_M, \gamma_M)$. We use the SAT-solver A to solve all these SAT instances, each SAT-instance should be solved in time t. If the satisfying assignment is found or the time limit t is exceeded, the corresponding solving process is terminated. In the first case we set $\xi_B^j = 1$, in the second case $\xi_B^j = 0$ ($j = 1, \ldots, M$). Then we compute (4).

In Algorithm 1, the transition to the next population is performed via the function GETNEXTPOPULATION(S, P). The situation, when the algorithm fails to improve the current BKV during one iteration of the population change is called stagnation. If the number of stagnations exceeds some given limit $N_{\text{stag}}^{\max}$, then we perform a restart: a new starting population is formed using the function RESTART(S, χ_{start}). The algorithm stops if the given general limit (for example, 12 h of operation of a computing cluster) is exceeded.

Techniques of Improvement. Note that the value of the observed random variable becomes known only after algorithm A has run for some time (not greater than t). On the other hand, the greater M is, the more accurate estimate of the time of the guess-and-determine attack is given by the value of function (4). Accordingly, the computation time required to find the value of function (4) is critically dependent on the random sample size M. Therefore, the greater M is, the more computation is required to find the value of (4). When $M > 500$ we have to use a computing cluster to minimize (4). As we will see further from computational experiments, reduction of the random sample size allows significantly increasing the algorithm's speed (the speed correlated with the number of hypercube points in which the objective function value was calculated during optimization).

Next, we describe a special technique that helps significantly increase the number of points that the algorithm processes during a fixed time limit (function SELECTM). This technique is based on the dynamic change of the random sample size. However, before we proceed, we will focus on some important details.

Note that if we consider functions of type (4), we can significantly reduce the dimension of the search space by taking into account the features of the original cryptanalysis problem. It was shown in [2,3] that we can consider $\{0, 1\}^n$, $n = |X^{\text{in}}|$ as a hypercube over which we perform minimization. In other words, the set B can be searched for as some subset in the set of all variables of the secret key. This can explained by the fact that we have already mentioned above: substitution of arbitrary assignments of all variables from X^{in} into $C(f)$ derives the assignment for all variables from $X \setminus X^{\text{in}}$. The SAT-solver A performs this derivation very quickly (essentially, the running time is linearly dependent on $|C(f)|$, because in this case A does not have to solve a combinatorial problem). This happens because X^{in} is a Strong Unit Propagation Backdoor Set (SUPBS) [16]. Therefore, $\chi_{\text{start}} = 1^n$ (the Boolean vector that consists of n ones) can always be chosen as a starting population of our minimization procedure. This point corresponds to the situation $B_{\text{start}} = X^{\text{in}}$. Any subsequent point will define some subset of X^{in}.

Thus, values of the random variable ξ_B in point $\chi_{\text{start}} = 1^n$ are computed very quickly. This property holds for several subsequent points and can be observed in Fig. 2, where we present the results of two runs of the algorithm described above.

Since the time spent on solving a single SAT instance in the Random Sample during the Fast Descent phase is already small, the spectrum of the function (4) on such random samples contains close values. This means that there is no principal need to use random samples of big size. The general idea is to change M while the algorithm runs, depending on the "homogeneity" of the sample in terms of the values of the observed random variable. After a whole range of experiments,

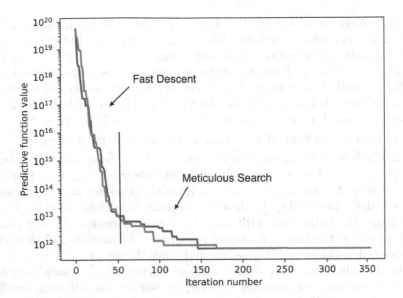

Fig. 2. Typical minimization of function (4). The process can be tentatively divided into two phases: Fast Descent and Meticulous Search.

we chose the following scheme: at the initial stage we use $M = 10$, then, passing through special checkpoints, M changes taking values $50, 100, 300, 500, 800$. The decision to change the sample size is made heuristically depending on the portion of $j \in \{1, \ldots, M\}$, for which $\xi_B^j = 1$ at the current value of M. General considerations here are as follows. Suppose that we consider two random samples R_1 and R_2 of size $M = 1000$ and that in 495 cases out of 1000 $\xi_B^j = 1$ for the sample R_1, and $\xi_B^j = 1$ for the sample R_2 in 510 cases out of 1000. The difference between these two values is $\approx 3\%$. Now let $M = 100$, and again we consider two random samples R_1, R_2. Suppose that $\xi_B^j = 1$ in 47 cases out of 100 for R_1 and $\xi_B^j = 1$ in 51 cases out of 100 for R_2. Here we have the difference $\approx 8\%$. Therefore, if in this situation we take $M = 100$, we will loose $\approx 5\%$ of accuracy, but we will spend 10 times less resources on computing the value of our function. Thus, in the initial search stage, when BKV often improves at each iteration, minor loss of accuracy is not an issue. In final iterations, when the algorithm takes too much time to improve some BKV, the accuracy of $E[\xi_B]$ is very important, which is why it is reasonable to considerably increase the sample size in later stages of the algorithm for minimization of (4).

When traversing the hypercube we use hash tables to store the passed points. If the mutation results in a point where the value of the function (4) was calculated earlier, then, after finding this point in the hash table, recalculation is not performed.

Finally, here we describe a special variant of the Genetic Algorithm (GA), which we used for minimization of (4). This algorithm employs the technique known as elitism. Our algorithm works with populations that consist of N

individuals. As in the case of $(1+1)$-EA, the GA starts with the point $\chi_{start} = 1^n$ for which it constructs N replicas. Each such replica is an individual. Let us describe an arbitrary iteration of the algorithm. Let $P_{curr} = \{I_1, \ldots, I_N\}$ and P_{next}, $|P_{next}| = N$ be the current and the new populations, respectively. Choose from P_{curr} L individuals with the best values of the function (4) and move them to P_{next} (elitism). Let $v_1, \ldots, v_N$ be the values of (4) for all individuals from P_{curr} and consider the set of numbers $U = \{u_1, \ldots, u_N\}$, where $u_j = \frac{1}{v_j}$. We associate with any individual $I_j \in P_{curr}$ a number p_j defined as $p_j = \frac{u_j}{\sum_{i=1}^{N} u_i}$. For individuals from P_{curr}, consider the set U with the probability distribution $X_U = \{p_1, \ldots, p_N\}$. Choose randomly H individuals from P_{curr} with respect to the distribution X_U and apply to each of them the standard mutation, flipping each bit with the probability $\frac{1}{n}$. Move the resulting individuals to P_{new}. Finally, choose from P_{curr} individuals with respect to the distribution X_U and perform over them one of the known crossover operations. Assume that G individuals were obtained as the result of crossover and move them to P_{new}.

Thus the individuals with smaller value of function (4) are more likely to be selected for mutation or crossover. Finally, we require the following condition to be satisfied: $L + H + G = N$. In our experiments, we used a variant of the described algorithm with $N = 10$, $L = 2$, $H = G = 4$.

4   Computational Experiments

We applied the algorithms and techniques described in the previous section to the cryptanalysis problems of keystream generators.

Stream cipher or keystream generator (see e.g. [17]) is a discrete function of kind (1), such that $m \gg n$. This function, taking an arbitrary n-bit word as input, generates a word of length m that behaves as a random sequence. The generator's input is called the *secret key*, the output is called the *keystream*. The practical implication of the keystream generator is very quick generation of a long keystream for a given random key. Short keys can be exchanged by participants via, for example, nonsymmetrical cryptography that provides substantial guarantees for resistance, but is extremely slow.

Suppose that for a common secret key α the participants **A** and **B** simultaneously generate the same keystream $\delta = f(\alpha)$. If **A** wants to send **B** a secret message $x \in \{0,1\}^m$, it creates a ciphertext $x \oplus \delta$ (a componentwise addition modulo 2), which is sent to the public channel. Upon receiving the ciphertext and knowing δ, **B** easily finds x. It often happens that a malefactor (or adversary) **M** knows a fragment of the message x – this can be, for example, some known proprietary information which is ciphered together with the secret message. This very system vulnerability for a long time had been present in the traffic encryption protocol of cellular telephony (see [18]). In this situation, **M** can use a ciphertext to find the fragment δ, denoted as γ, and then try to find α as an preimage of γ for a mapping f. To do that, **M** has to solve a system of algebraic equations that describes construction of γ from α, or he has to solve a corresponding SAT instance. For a correct definition of α, the length of the keystream fragment γ should not be smaller than the length of α (usually it is slightly larger). Such cryptanalysis of a keystream fragment in order to find a secret key is called the *known plaintext scenario* [17].

In view of the above, consider the inversion problem for the function $f : \{0,1\}^n \rightarrow \{0,1\}^k$, $k \approx n$, where for the known $\gamma \in Range\ f$ one needs to find such $\alpha \in \{0,1\}^n$ that $f(\alpha) = \gamma$. We analyzed several functions that can be used as keystream generators:

1. Stream ciphers of the family Trivium (Trivium-Toy 64, Trivium-Toy 96, Bivium).
2. Alternating Step Generator (ASG).

Below we give a brief description of these ciphers.

The Trivium stream cipher was proposed in [19]. Being one of the eSTREAM project winners, this cipher attracts a lot of attention from cryptanalysts. In several papers (see e.g. [20,21]) it was shown that there are guess-and-determine attacks for Trivium which are more efficient than brute force attacks in the context of the so-called state recovery problem.

The Bivium cipher is a weakened version of Trivium (it uses only two original registers out of three). This cipher, as well as Trivium, was described in [19], where the author states that it is of mainly a research interest. A series of attacks on Bivium (including the algebraic ones that use SAT) can be found in [21–25].

Papers [26,27] propose a general approach to the construction of Trivium-like ciphers with a smaller total size of registers which preserves the algebraic properties of the original Trivium. Below we follow [26] and refer to this family as Trivium-Toy. In particular, by Trivium-Toy L we denote a cipher from this family, in which L is a total size of state that should be recovered. Hereinafter we consider state recovery attacks on Trivium-Toy 96 (this cipher is described in [26]) and Trivium-Toy 64.

The Alternating Step Generator (ASG) was suggested in [28] and actually it is a common design of keystream generators that can be used for construction of stream ciphers with different lengths of the secret key. The most attractive property of ASG is ease of implementation and high speed of keystream generation. ASG was targeted by several attacks, the analysis of which is presented in [29]. Some of these attacks employ a significant amount of keystream. The original paper [28] describes an attack that tried all possible ways to fill the control register. Essentially, this is a guess-and-determine attack which uses the so-called Linear Consistency Test [30] as an algorithm for solving weakened systems of cryptanalysis equations. The paper [31] presents a SAT-based guess-and-determine attack for ASG based on building SAT Partitionings of hard variants of SAT described in [2].

We implemented the strategies (1+1)-EA and GA presented above and Tabu Search [32] in the same manner as it is described in [2,3]. The created program is an MPI application that uses the SAT solver ROKK [33] as a computational core.

Each experiment was conducted on 180 cores of a computing cluster equipped with Intel Xeon E5-2695 processors. The duration of one experiment was 12 h. In the minimization process of functions of the kind (4) we used a random sample of size $M = 500$. For each point taken as a solution we recalculated the value of the objective function using a sample of size $M = 10^4$. The resulting value was taken as final.

As test problems we considered cryptanalysis problems for keystream generators from the Trivium and the ASG families. The results of computational experiments are presented in Table 1. For a specific generator G notation G n/m means that one needs to find an n-bit secret key by analyzing m bits of a keystream. SAT instances encoding the corresponding cryptanalysis problems were constructed using the TRANSALG system [34] which translates algorithms for calculating discrete functions of the kind (1) into SAT.

Let us note that GA-elitism showed the best results among the considered strategies for minimization of functions of the kind (4), overtaking competitors in three test problems out of six.

Table 1. Experimental results for six cryptographic algorithms. The leftmost column contains the name of a keystream generator for which the cryptanalysis problem is considered. The remaining table is divided into three sections corresponding to strategies (1+1)-EA, GA and Tabu Search. The first column of each section contains the power of the guessed bit set which corresponds to the best guess-and-determine attack found. The second column contains a time estimation (in seconds) for the attack

	(1+1)-EA		GA		Tabu Search	
	Power of guessed bit set	G&D attack (seconds)	Power of guessed bits set	G&D attack (seconds)	Power of guessed bits set	G&D attack (seconds)
Trivium-Toy 64/75	21	**3.19e+07**	22	5.36e+07	17	4.30e+07
Trivium-Toy 96/100	33	1.28e+13	40	**2.09e+12**	34	3.14e+12
Bivium 177/200	32	2.60e+12	39	**1.49e+12**	40	4.29e+12
ASG 72/76	9	5604.8	8	6155.19	8	**5601.33**
ASG 96/112	13	6.76e+06	16	**3.72e+06**	14	3.95e+06
ASG 192/200	47	2.27e+18	44	2.84e+17	47	**1.14e+16**

5 Conclusion and Future Work

In the presented paper we applied evolutionary computation strategies to construct guess-and-determine attacks arising in algebraic cryptanalysis. Each of these attacks implies solving a family of algebraic equations over a finite field (usually, $GF(2)$). Each equation from such family is the result of substituting the values of some bits into a general equation describing how a considered cipher works. The substituted bits are called the guessed bits. To solve the equations we use SAT solvers, similar to a number of other works. It is clear that different sets of guessed bits correspond to guess-and-determine attacks with different runtime. We consider the problem of finding a set of guessed bits that yields an attack with the lowest runtime as the problem of minimization of a pseudo-Boolean function. To solve it we used two general strategies employed in evolutionary computation: (1+1)-EA and GA. The proposed algorithms were implemented in the form of a parallel MPI program for a computing cluster. Using these algorithms we constructed guess-and-determine attacks for several keystream ciphers. The obtained attacks are better than the ones constructed using the Tabu Search algorithm, but not dramatically better. However, from our point of view, we have only touched the potential of evolutionary computation in this area. In the nearest future we plan to significantly extend the spectrum of employed evolutionary computation techniques in application to problems of constructing Algebraic cryptanalysis attacks.

Acknowledgements. The authors would like to thank Daniil Chivilikhin, Maxim Buzdalov and anonymous reviewers for useful comments.

References

1. Bard, G.V.: Algebraic Cryptanalysis. Springer, New York (2009). https://doi.org/10.1007/978-0-387-88757-9
2. Semenov, A., Zaikin, O.: Algorithm for finding partitionings of hard variants of boolean satisfiability problem with application to inversion of some cryptographic functions. SpringerPlus 5(1), 554 (2016)
3. Semenov, A., Zaikin, O., Otpuschennikov, I., Kochemazov, S., Ignatiev, A.: On cryptographic attacks using backdoors for SAT. In: Proceedings of AAAI 2018, pp. 6641–6648 (2018)
4. Cook, S.A.: The complexity of theorem-proving procedures. In: Proceedings of the Third Annual ACM Symposium on Theory of Computing, pp. 151–158. ACM (1971)
5. Biere, A., Heule, M., van Maaren, H., Walsh, T. (eds.): Handbook of Satisfiability, Frontiers in Artificial Intelligence and Applications, vol. 185. IOS Press (2009)
6. Anderson, R.: A5 (was: hacking digital phones). Newsgroup Communication (1994). http://yarchive.net/phone/gsmcipher.html
7. Gendrullis, T., Novotný, M., Rupp, A.: A real-world attack breaking A5/1 within hours. In: Oswald, E., Rohatgi, P. (eds.) CHES 2008. LNCS, vol. 5154, pp. 266–282. Springer, Heidelberg (2008). https://doi.org/10.1007/978-3-540-85053-3_17
8. Courtois, N.T., Bard, G.V.: Algebraic cryptanalysis of the data encryption standard. In: Galbraith, S.D. (ed.) Cryptography and Coding 2007. LNCS, vol. 4887, pp. 152–169. Springer, Heidelberg (2007). https://doi.org/10.1007/978-3-540-77272-9_10
9. Courtois, N.T., Gawinecki, J.A., Song, G.: Contradiction immunity and guess-then-determine attacks on GOST. Tatra Mountains Math. Publ. 53, 65–79 (2012)
10. Semenov, A., Zaikin, O., Bespalov, D., Posypkin, M.: Parallel logical cryptanalysis of the generator A5/1 in BNB-grid system. In: Malyshkin, V. (ed.) PaCT 2011. LNCS, vol. 6873, pp. 473–483. Springer, Heidelberg (2011). https://doi.org/10.1007/978-3-642-23178-0_43

11. Marques-Silva, J., Lynce, I., Malik, S.: Conflict-driven clause learning SAT solvers. In: Frontiers in Artificial Intelligence and Applications, vol. 85, pp. 131–153 (2009)

12. Boros, E., Hammer, P.L.: Pseudo-Boolean optimization. Discrete Appl. Math. **123**(1–3), 155–225 (2002)

13. King, J.C.: Symbolic execution and program testing. Commun. ACM **19**(7), 385–394 (1976)

14. Metropolis, N., Ulam, S.: The Monte Carlo method. J. Am. Stat. Assoc. **44**(247), 335–341 (1949)

15. Rudolph, G.: Convergence Properties of Evolutionary Algorithms. Verlag Dr. Kovac, Hamburg (1997)

16. Williams, R., Gomes, C.P., Selman, B.: Backdoors to typical case complexity. In: IJCAI 2003, pp. 1173–1178 (2003)

17. Menezes, A.J., Vanstone, S.A., Oorschot, P.C.V.: Handbook of Applied Cryptography, 1st edn. CRC Press Inc., Boca Raton (1996)

18. Nohl, K.: Attacking Phone Privacy, pp. 1–6. Black Hat, Las Vegas (2010)

19. Cannière, C.: TRIVIUM: a stream cipher construction inspired by block cipher design principles. In: Katsikas, S.K., López, J., Backes, M., Gritzalis, S., Preneel, B. (eds.) ISC 2006. LNCS, vol. 4176, pp. 171–186. Springer, Heidelberg (2006). https://doi.org/10.1007/11836810_13

20. Raddum, H.: Cryptanalytic Results on Trivium. eSTREAM, ECRYPT Stream Cipher Project, Report 2006/039 (2006)

21. Maximov, A., Biryukov, A.: Two trivial attacks on TRIVIUM. In: Adams, C., Miri, A., Wiener, M. (eds.) SAC 2007. LNCS, vol. 4876, pp. 36–55. Springer, Heidelberg (2007). https://doi.org/10.1007/978-3-540-77360-3_3

22. Eibach, T., Pilz, E., Völkel, G.: Attacking Bivium using SAT solvers. In: Kleine Büning, H., Zhao, X. (eds.) SAT 2008. LNCS, vol. 4996, pp. 63–76. Springer, Heidelberg (2008). https://doi.org/10.1007/978-3-540-79719-7_7

23. Eibach, T., Völkel, G., Pilz, E.: Optimising Gröbner bases on Bivium. Math. Comput. Sci. **3**(2), 159–172 (2010)

24. Soos, M., Nohl, K., Castelluccia, C.: Extending SAT solvers to cryptographic problems. In: Kullmann, O. (ed.) SAT 2009. LNCS, vol. 5584, pp. 244–257. Springer, Heidelberg (2009). https://doi.org/10.1007/978-3-642-02777-2_24

25. Huang, Z., Lin, D.: Attacking Bivium and Trivium with the characteristic set method. In: Nitaj, A., Pointcheval, D. (eds.) AFRICACRYPT 2011. LNCS, vol. 6737, pp. 77–91. Springer, Heidelberg (2011). https://doi.org/10.1007/978-3-642-21969-6_5

26. Castro Lechtaler, A., Cipriano, M., García, E., Liporace, J., Maiorano, A., Malvacio, E.: Model design for a reduced variant of a Trivium type stream cipher. J. Comput. Sci. Technol. **14**(01), 55–58 (2014)

27. Teo, S.G., Wong, K.K.H., Bartlett, H., Simpson, L., Dawson, E.: Algebraic analysis of Trivium-like ciphers. In: Australasian Information Security Conference (ACSW-AISC 2014), vol. 149, pp. 77–81. Australian Computer Society (2014)

28. Günther, C.G.: Alternating step generators controlled by De Bruijn sequences. In: Chaum, D., Price, W.L. (eds.) EUROCRYPT 1987. LNCS, vol. 304, pp. 5–14. Springer, Heidelberg (1988). https://doi.org/10.1007/3-540-39118-5_2

29. Khazaei, S., Fischer, S., Meier, W.: Reduced complexity attacks on the alternating step generator. In: Adams, C., Miri, A., Wiener, M. (eds.) SAC 2007. LNCS, vol. 4876, pp. 1–16. Springer, Heidelberg (2007). https://doi.org/10.1007/978-3-540-77360-3_1

30. Zeng, K., Yang, C.H., Rao, T.R.N.: On the Linear Consistency Test (LCT) in cryptanalysis with applications. In: Brassard, G. (ed.) CRYPTO 1989. LNCS, vol. 435, pp. 164–174. Springer, New York (1990). https://doi.org/10.1007/0-387-34805-0_16

31. Zaikin, O., Kochemazov, S.: An improved SAT-based guess-and-determine attack on the alternating step generator. In: Nguyen, P., Zhou, J. (eds.) ISC 2017. LNCS, vol. 10599, pp. 21–38. Springer, Cham (2017). https://doi.org/10.1007/978-3-319-69659-1_2

32. Glover, F., Laguna, M.: Tabu Search. Kluwer Academic Publishers, Boston (1997)

33. Yasumoto, T., Okuwaga, T.: ROKK 1.0.1. In: Belov, A., Diepold, D., Heule, M., Järvisalo, M. (eds.) SAT Competition 2014, p. 70 (2014)

34. Otpuschennikov, I., Semenov, A., Gribanova, I., Zaikin, O., Kochemazov, S.: Encoding cryptographic functions to SAT using TRANSALG system. In: ECAI 2016. FAIA, vol. 285, pp. 1594–1595 (2016)

On the Use of Evolutionary Computation for In-Silico Medicine: Modelling Sepsis via Evolving Continuous Petri Nets

Ahmed Hallawa[1]([✉]), Elisabeth Zechendorf[2], Yi Song[1], Anke Schmeink[3],
Arne Peine[2], Lukas Marin[2], Gerd Ascheid[1], and Guido Dartmann[4]

[1] Chair for Integrated Signal Processing Systems, RWTH Aachen University,
52056 Aachen, Germany
{hallawa,song,ascheid}@ice.rwth-aachen.de
[2] Department of Intensive Care and Intermediate Care, University Hospital RWTH
Aachen, Aachen, Germany
{ezechendorf,apeine,lmartin}@ukaachen.de
[3] Research Area Information Theory and Systematic Design of Communication
Systems, RWTH Aachen University, 52056 Aachen, Germany
anke.schmeink@rwth-aachen.de
[4] Research Area Distributed System, Trier University of Applied Sciences,
Trier, Germany
g.dartmann@umwelt-campus.de

Abstract. Sepsis is one of the leading causes of death in Intensive Care
Units (ICU) world-wide. Continuous Petri Nets (CPNs) offer a promising solution in modelling its underlying complex pathophysiological processes. In this work, we propose a framework to evolve CPNs, i.e. evolve
its places, transitions, arc weights, topology, and kinetics. This facilitates modeling complex biological systems, including activated signalling
pathway in sepsis using limited experimental data. Inspired by Neuroevolution of Augmenting Topology (NEAT), which is adopted in Artificial
Neural Networks (ANNs), our framework includes a genotype to phenotype mapping based on the CPN incidence matrix, and a fitness function,
which considers both the behaviour of the evolving CPN and its emerging structural complexity. We tested our framework on ten different cases
with different complexity structures. In the worst case, results show the
NMSE less than 2% in the learning phase, and MSE of 13% in the validation phase. We applied our framework on real-world data from cell
culture experiments, representing a biological pathway in sepsis. Using
the output of these experiments, the proposed framework was able to
evolve a CPN to model this pathway with an MSE value of 10% in the
validation phase.

Keywords: Evolutionary computation · Continuous Petri Nets ·
Modelling sepsis · Intensive care medicine

© Springer Nature Switzerland AG 2019
P. Kaufmann and P. A. Castillo (Eds.): EvoApplications 2019, LNCS 11454, pp. 254–269, 2019.
https://doi.org/10.1007/978-3-030-16692-2_17

1   Introduction

One of the main causes of death in hospitals worldwide is sepsis [1]. In the United States, sepsis accounts for 20 billion US-Dollars, accounting for 5.2% of total US hospital costs [2]. Up to now, there is no causative therapeutic strategy available.

There are currently few new innovative approaches in the development of causative therapeutic strategies. In this regard, data indicates that synthetic antimicrobial peptides can modulate inflammation cascades by neutralizing multiple inflammatory mediators on the cell surface [3]. However, with the currently established drug development methodologies used in laboratories, the drug development process is both tedious in cost and time. As a result, a new approach is needed that integrates newly developed machine learning techniques and state-of-art modelling approaches. Thus, lowering the needed number of in-vivo and in-vitro experiments, which are complex, slow and expensive.

Continuous Petri Nets (CPNs) have a unique ability to model continuous dynamic processes, which can be described with the help of Ordinary Differential Equations (ODEs). This allows complex dynamic relationships to be graphically modelled, as CPNs in essence can be represented in a directed graph scheme. Modeling of biological system processes with the help of PNs has been studied in literature, however, most published approaches have the disadvantage that they can only model a small segment of a complex system. In particular, CPNs were used to model sepsis, but with the assumption that previous knowledge regarding its structure is available [4]. On the other hand, Evolutionary Algorithms (EAs) have been successfully used in learning artificial neural networks (ANN) in many applications such as breast cancer diagnosis [5], among so many other applications. This highlights the potential of using EA on CPNs, particularly as they provide a more effective modeling framework for the biochemical process that a ANN counterpart may lack [6].

In this work we present a framework that evolves a CPN that mimics the behaviour of a biological process such as the sepsis pathophysiological pathway with the help of in-vivo and in-vitro experiments. Thus, the framework answers: how many number of transitions needed in the CPN to model a process? how the places and transitions are connected? what are the weights of each edge? and finally, what are the kinetics of each transition?

2   Background

To introduce CPNs, it is important to highlight that each CPN has an equivalent representation in the form of a system of ODEs. This is why CPNs are suitable to be used to model applications in physics, biology and chemistry. Furthermore, each CPN has a graphical representation, e.g. Fig. 1. The basic structure of a CPN contains a set of places $\mathcal{P} = \{P_1 \ldots P_N\}$, and a set of transitions $\mathcal{T} = \{T_1 \ldots T_M\}$. Furthermore, $^\bullet\mathcal{P}(T)$ is the set of all incoming places of transition T, similarly, $^\bullet\mathcal{T}(P)$ and $\mathcal{T}^\bullet(P)$ are the incoming and outgoing of

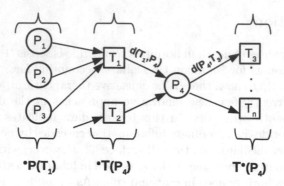

Fig. 1. Continuous Petri Net (CPN): An example

place P, respectively. Therefore, the equivalent system of ODEs for a CPN can be formalized as follows:

$$\frac{\partial m(p)}{\partial t} = \sum_{t \in {}^{\bullet}T(p)} \mathbf{d}(t,p)v(t) - \sum_{t \in T^{\bullet}(p)} \mathbf{d}(p,t)v(t) \qquad (1)$$

where $\frac{\partial m(p)}{\partial t}$ is the process modeled function of time, i.e. how different substrate change w.r.t. time, $m(p)$ stands for the tokens of place p in $\{P_1 \ldots P_N\}$, $\mathbf{d}$ is a firing rate function of place p and t, and $v(t)$ is the kinetics of transition. Therefore, the first summation term accumulates all weighted incoming tokens and the second sums all weighted outgoing tokens.

In addition, CPNs have been used to model sepsis. In Zechendorf et al. [4], researchers adopted EAs to learn the ODE coefficients of CPN to model sepsis. However, in their problem statement they assumed that the structure of the CPN is already known and as a result learned only its kinetics of the CPN. This limits the outcome of the modelling process as it minimizes the solution space, as a result, this was reflected in the fitness values presented in their work.

3   Methods

3.1   Genotype to Phenotype Mapping

Generally, an incidence matrix formalizes the relationship between nodes and edges in a graph [7]. Thus, if $\mathcal{G}$ is a graph with $V(\mathcal{G}) = \{v_1, \ldots, v_n\}$ and $E(\mathcal{G}) = \{e_1, \ldots, e_m\}$ are vertex and edge sets respectively, the corresponding incidence matrix $I(\mathcal{G})$ is an $n \times m$ matrix, whose row and column are indexed by $V(\mathcal{G})$ and $E(\mathcal{G})$ separately and, for example, can be represented as follows:

$$I(\mathcal{G}) = \begin{pmatrix} 1 & 1 & -1 & 0 & 0 & 0 \\ 1 & 0 & 0 & -1 & 0 & 0 \\ 0 & -1 & 0 & 0 & 1 & 0 \\ 0 & 0 & 1 & 0 & 0 & -1 \\ 0 & 0 & 0 & 1 & -1 & 1 \end{pmatrix}$$

where -1 represents the *ingoing* arc and $+1$ the *outgoing* arc, while 0 represents a no connection. This incidence matrix can be useful in forming a genotype for CPN. However, we can not adopt only an incidence matrix to represent fully a CPN, we still need to represent the weights and kinetics as well. However, incidence matrix can be used to represent topology and weights. Thus, instead of the row and column indexed by the vertex and edges in incidence matrix, in CPNs we can define the incidence matrix as $n \times m$ matrix, where n stands for the number of places, while m represents the number of transitions. Furthermore, the corresponding arc weights can be assigned on the corresponding position in the incidence matrix. For example, the arc weight which resides in the position (i, j) stands for the arc between place i and transition j and the arc weights value equals to the value at (i, j) in the CPN incidence matrix. Therefore, incidence matrix can offer an adequate representation of genotype-phenotype mapping.

However, one problem is that the number of edges are initially unknown, in fact, one of the requirements is to find it. Then we might face a problem in initializing the incidence matrix. To solve this problem, we use fixed-size genotype, which represent the maximum possible incidence matrix $(m \times n)$ size, and then we associate a mark with each CPN genotype indicating where its incidence matrix really ends, thus, the remaining positions are ignored. In other words, we allow dynamic size of the genotype, within a fixed maximum limit declared beforehand. Therefore, with this configuration we can use the incidence matrix to represent a CPN genotype. However, to facilitate adequate solution space exploration, we divided the genotype into three different parts: topology, arc weights, and kinetics.

Topology: Designing this part is done with the incidence matrix of CPNs in mind. We wanted to facilitate in the reproduction operators the possibility to explore the topology space. As a result, the topology gene is a string representation of the incidence, thus it is of length $n \times m$, where n and m are the number of places and transitions respectively. Moreover, three possibilities appear in the topology genotype: $+1$ stands for the arc directed from place to transition, -1 represents the inverse path from a transition to a place, while 0 represents no connection. Furthermore, for reproduction on the topology genotype, we adopted a uniform distribution, i.e. the crossover rules produce $+1$, -1 and 0 with equal probability as shown in Table 1. Details regarding the production rules is presented in Sect. 4.

Arc Weights: In CPNs the arc weights, as well as kinetics, are real values. Hence, the optimization problem of finding them is a continuous optimization problem. However, using EA, we need to discretize these values. We adopted a 4 bits representation to represent the weights, based on medical experts recommendation regarding the upper limits of these arc weights. In this regard, there are other methods than the binary coded decimal method we adopted, e.g. one's or two's complement method, or the gray code method, however the binary coded decimal is generally preferred [8]. Reproduction for the arc weight is conducted using the typical genetic algorithm (GA) operators: crossover and mutation on the binary representation of the real value.

Table 1. Topology reproduction rules with equal distribution

$Parent_1$	$Parent_2$	$Offspring$
1	1	1
-1	-1	-1
0	0	0
1	0	0
-1	0	0
0	1	1
0	-1	-1
1	-1	-1
-1	1	1

Kinetics: Similar to arc weights code, four binary bits represented the kinetics value, which varies from 0 to 15 (based on medical experts recommendations). Hence, assuming that place number is n, while m is the number of transitions, the genotype of kinetics code is an array which includes $m \times 4$ binary values. Similar to the arc weight, GA reproduction operators were adopted.

3.2 Fitness Function

The fitness function considered two aspects as suggested by [9]: behavioural and complexity. We set the behavioural fitness, which is defined based on the Normalized Mean Square Error (NMSE) between real and estimated data values as follows: Given objective data set, $X = \{x_1, \ldots, x_n\}$, and estimated values from the evolved CPN, $Y = \{y_1, \ldots, y_n\}$, the behavioural fitness is represented by the NMSE between x_i and y_i as follows:

$$\mathcal{E} = \frac{\frac{1}{n_t} \sum_{i=1}^{n_t} (x_i - y_i)^2}{E(X) \times E(Y)} \tag{2}$$

Where $E(X)$ and $E(Y)$ are the mean values of sets X and Y respectively, n_t is number of objective points in the training phase. Similarly, for the validation phase (after the CPN is evolved), we used the MSE:

$$MSE = \frac{1}{n_v} \sum_{i=1}^{n_v} (x_i - y_i)^2 \tag{3}$$

where n_v is the number of validation points. On the other hand, for complexity we calculated the number of points in the incidence matrix constructing the CPN and then divided this number by the maximum number of transitions in genotype and number of places. Therefore, the normalized complexity fitness factor is defined as follows:

$$\mathcal{C} = \frac{\sum_{i,j} |\mathbf{T}_r[i, j]|}{N_{place} \times \max_{tr}} \tag{4}$$

where $\sum_{i,j} |\mathbf{T}_r[i,j]|$ is the number of elements in the incidence matrix genome describing the CPN, N_{place} denotes the number of places while $\max_{tr}$ describes the maximum number of transitions in genotype. The overall fitness value $\mathcal{F}$ can be defined then as follows:

$$\mathcal{F} = \mathcal{E} + \lambda \times \mathcal{C} \tag{5}$$

Where λ is complexity coefficient, $\mathcal{E}$ is the NMSE and $\mathcal{C}$ is the normalized complexity. Therefore, the optimization problem is a minimization problem. It is worth mentioning that the complexity coefficient defines the weight of the complexity penalty in fitness and is considered a hyperparameter.

4   Results

In order to test our proposed framework, we applied it on two test case sets: A simple CPN set and a complex one, where each set consists of five randomly generated CPNs. The simple set has CPNs with maximum number of places and transition set as 10, and the complex set has at least 10 and maximum 20. The objective is to evolve a CPN that would match the behaviour of each generated CPN in both sets. Furthermore, each experiment was repeated 10 times to analyze the stochastic nature of the algorithm.

Table 2. Hyperparameter values used in the test cases

Hyperparameter	Simple CPN	Complex CPN	Real sepsis case
Population size	30	50	50
Mutation rate	0.02	0.03	0.03
Crossover rate	1	1	1
Complexity coefficient	0.1	0.1	0.1
No. of generations	100	300	100
No. of Min transition	2	2	2
No. of Max transition	6	20	20

Finally, we have tested the algorithm on a real sepsis case, where lab experiments where conducted and the framework was used to produce an evolved CPN, which replicates these values. Table 2 summarizes the hyperparameter values for each test case set. Table 4 summarizes the performance of the framework for each case.

However, before running test cases, we conducted an experiment using the complex CPN test set to check the effect of changing the distribution of the crossover rules. Therefore, we compared the performance of the evolutionary process in case the crossover rules where uniformly distributed with the non-uniformly distributed case. In case of uniform distribution, the output of the rules produce with equal

Table 3. Impact of different crossover rules on complex CPN test cases

Crossover rule	Case #	Fitness ($\mathcal{F}$) [%]	NMSE ($\mathcal{E}$) [%]	Complexity ($\mathcal{C}$) [%]	Transition [%]	Validation MSE [%]
Uniformly distributed across edge states	Case1	**2.75**	0.45	23.00	3	**1.99**
	Case2	**4.72**	2.72	20.0	4	**24.29**
	Case3	**2.91**	0.81	21.0	3	**6.60**
	Case4	4.22	1.42	28.0	4	**5.31**
	Case5	7.26	3.56	37.0	6	15.26
Non-uniformly distributed across edge states	Case1	5.5	2.16	40.0	4	2.51
	Case2	4.97	0.97	40.00	4	5.53
	Case3	3.5	1.50	20.00	2	13.93
	Case4	**4.08**	1.28	28.0	3	9.52
	Case5	**6.62**	2.62	40.00	4	**8.94**

probability one of the three possible edge states: An edge from a place to a transition (+1), an edge from a transition to a place (−1) and no edge at all (0). For the non-uniform case, this rule was not fulfilled. In Sect. 4, results show that uniformly distributed rules outperform non-uniformly distributed rules in three out of five complex CPNs test cases in the training phase, and four out of five in validation phase. Thus, we adopted the uniformly distributed rules (Table 3).

On the other hand, Table 4 summarize all results from the ten test cases. Furthermore, Figs. 2, 3, 4, 5, 6, 7, 8 and 9 present the detailed profile of cases 1 to 3 in the simple test case set and cases 1 to 4 in the complex test case set[1]. In each figure, parts a to f, shows the average fitness performance over 10 different runs and their respective variance, while part g, shows the difference between the behaviour difference between the real and evolved CPN for each substrate (places). Finally, Table 5 shows the experimental data conducted in-vitro by exposing cardiomyocytes to HS. In these experiments, HL-1 cells, murine cardiomyocytes, were exposed to 10 μg/ml HS for 16 h [10]. Protein was isolated using triton lyse buffer (300 mM NaCl, 20 mM TRIS (pH 7.4), 1% Triton-X100, 200 mM PMSF, 1 mM DTT, 2 mg/ml leupeptin, and 1 mg/ml pepstatin) [11,12]. The protein concentrations were determined by Bradford method. After protein separation by sodium dodecyl sulfate polyacrylamide gel electrophorese, proteins were transferred to an polyvinylidene difluoride membrane using the Trans-Blot Turbo Transfer System from Bio-Rad. Membrane was blocked with 5% bovine serum albumin for 1 h and incubated with an specifically antibody against poly-(ADP-ribose) polymerase (PARP) over night. Afterwards, the membrane was incubated with an secondary antibody and as previously described proteins were detected with ECL Prime Western Blotting Detection Reagent and LAS-4000-System [4]. Figure 9 shows the corresponding validation MSE.

---

[1] Due to page limit.

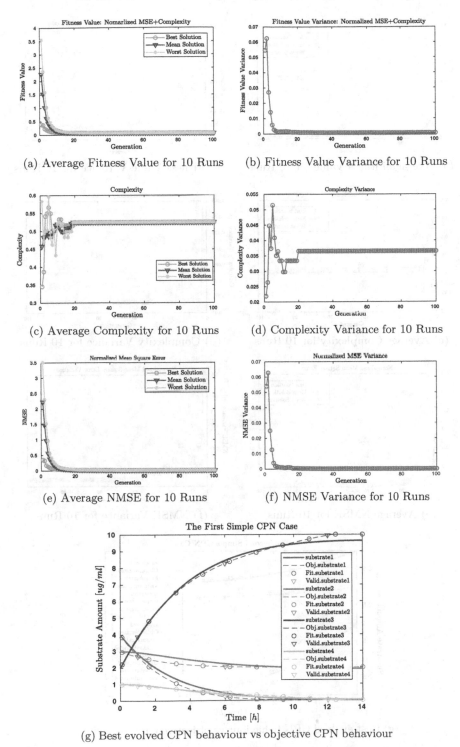

(a) Average Fitness Value for 10 Runs

(b) Fitness Value Variance for 10 Runs

(c) Average Complexity for 10 Runs

(d) Complexity Variance for 10 Runs

(e) Average NMSE for 10 Runs

(f) NMSE Variance for 10 Runs

(g) Best evolved CPN behaviour vs objective CPN behaviour

Fig. 2. CPN Simple Case 1

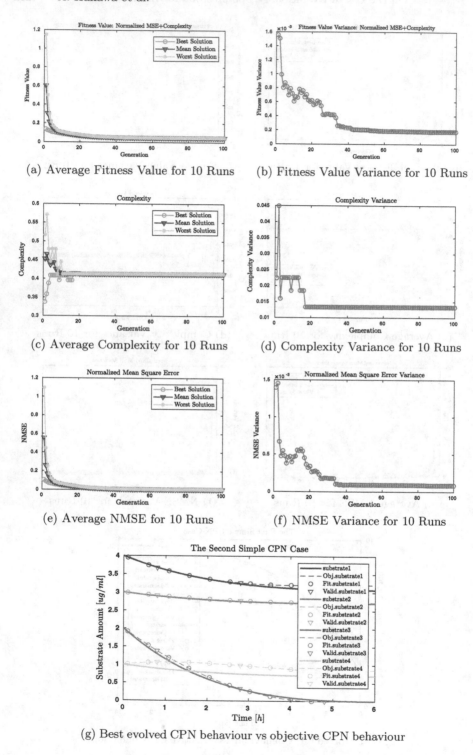

(a) Average Fitness Value for 10 Runs

(b) Fitness Value Variance for 10 Runs

(c) Average Complexity for 10 Runs

(d) Complexity Variance for 10 Runs

(e) Average NMSE for 10 Runs

(f) NMSE Variance for 10 Runs

(g) Best evolved CPN behaviour vs objective CPN behaviour

Fig. 3. CPN Simple Case 2

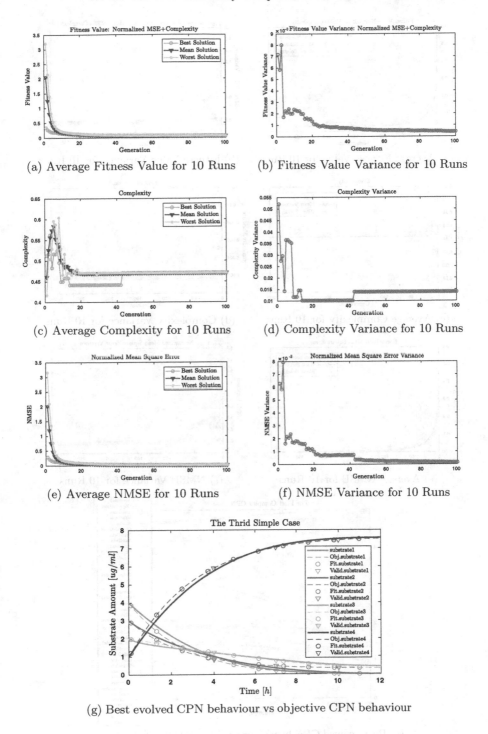

(a) Average Fitness Value for 10 Runs

(b) Fitness Value Variance for 10 Runs

(c) Average Complexity for 10 Runs

(d) Complexity Variance for 10 Runs

(e) Average NMSE for 10 Runs

(f) NMSE Variance for 10 Runs

(g) Best evolved CPN behaviour vs objective CPN behaviour

Fig. 4. CPN Simple Case 3

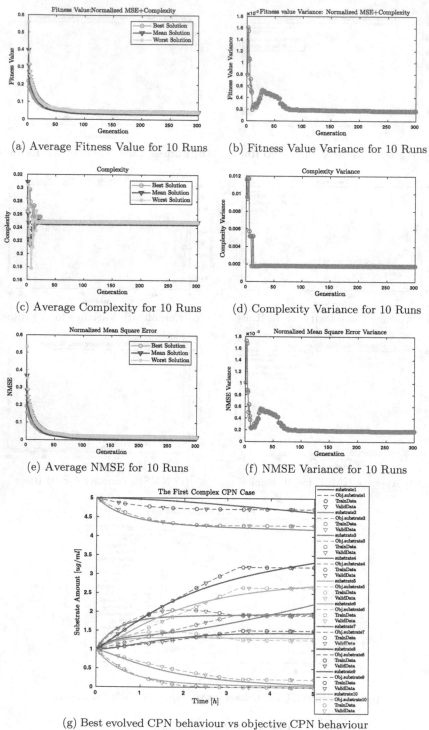

(a) Average Fitness Value for 10 Runs

(b) Fitness Value Variance for 10 Runs

(c) Average Complexity for 10 Runs

(d) Complexity Variance for 10 Runs

(e) Average NMSE for 10 Runs

(f) NMSE Variance for 10 Runs

(g) Best evolved CPN behaviour vs objective CPN behaviour

Fig. 5. CPN Complex Case 1

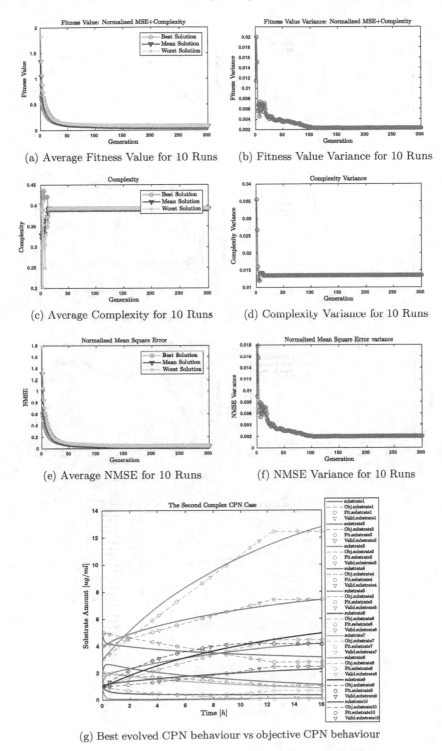

(a) Average Fitness Value for 10 Runs

(b) Fitness Value Variance for 10 Runs

(c) Average Complexity for 10 Runs

(d) Complexity Variance for 10 Runs

(e) Average NMSE for 10 Runs

(f) NMSE Variance for 10 Runs

(g) Best evolved CPN behaviour vs objective CPN behaviour

Fig. 6. CPN Complex Case 2

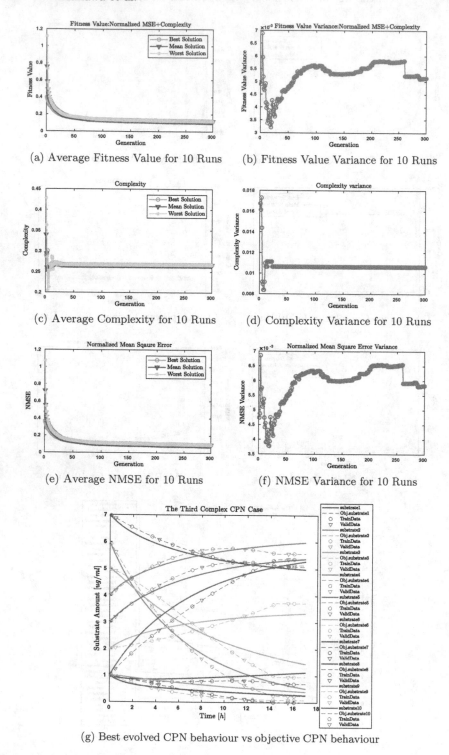

(a) Average Fitness Value for 10 Runs

(b) Fitness Value Variance for 10 Runs

(c) Average Complexity for 10 Runs

(d) Complexity Variance for 10 Runs

(e) Average NMSE for 10 Runs

(f) NMSE Variance for 10 Runs

(g) Best evolved CPN behaviour vs objective CPN behaviour

Fig. 7. CPN Complex Case 3

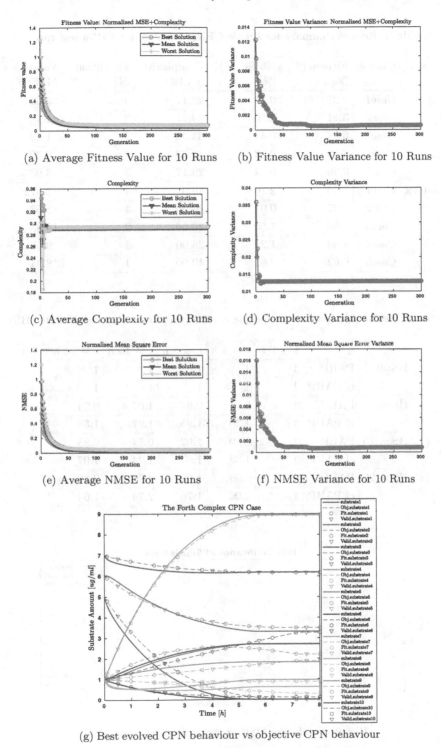

(a) Average Fitness Value for 10 Runs

(b) Fitness Value Variance for 10 Runs

(c) Average Complexity for 10 Runs

(d) Complexity Variance for 10 Runs

(e) Average NMSE for 10 Runs

(f) NMSE Variance for 10 Runs

(g) Best evolved CPN behaviour vs objective CPN behaviour

Fig. 8. CPN Complex Case 4

Table 4. Results summary for simple CPNs and complex CPNs test cases

Test set	Case #	Fitness ($\mathcal{F}$) [%]	NMSE ($\mathcal{E}$) [%]	Complexity ($\mathcal{C}$) [%]	Transition [%]	Validation MSE [%]
Simple	Case1	3.27	0.35	29.17	2	3.66
	Case2	3.21	0.30	29.17	2	1.14
	Case3	3.78	0.44	33.33	2	1.63
	Case4	3.46	0.13	33.33	2	0.20
	Case5	3.36	0.44	29.17	2	3.03
Complex	Case1	5.50	2.16	40.0	4	2.512
	Case2	4.97	0.97	40.00	4	5.53
	Case3	3.50	1.50	20.00	2	13.93
	Case4	4.08	1.28	28.00	3	9.52
	Case5	6.62	2.62	40.00	4	8.94

Table 5. Real sepsis test case: Exposing cardiomyocytes to Heparan Sulfate (HS)

		t = 0 h	t = 4 h	t = 8 h	t = 16 h	t = 24 h
HS=0	PARP	1	1	1	1	1
	cl PARP	1	1	1	1	1
HS=5	PARP	1	0.97	0.97	1.07	0.94
	cl PARP	1	1.29	1.95	1.97	1.58
HS = 10	PARP	1	0.89	1.02	0.74	0.88
	cl PARP	1	1.33	1.67	1.24	1.07
HS = 20	PARP	1	0.98	1.02	0.74	0.88
	cl PARP	1	2.06	1.76	2.74	1.64

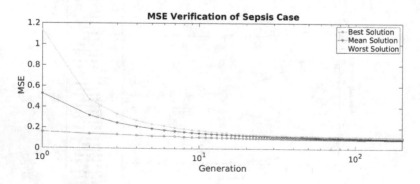

Fig. 9. Validation MSE of the real sepsis test case

5   Conclusions

In this work, we have introduced a new framework to evolve CPNs, which is suited to model continuous time varying biological processes. The work included a scheme for genotype to phenotype representation based on the incidence matrix. In addition, different crossover rules were tested, consequently, the uniformly distributed rules were adopted. For fitness function, both the behaviour of the evolved CPN and its complexity were considered. The framework was tested on ten different cases with different complexity structure and results show a validation MSE 13% in the worst case and less than 1% in the best case. Applied on real-world data from cell culture experiments, our framework was able to evolve a CPN to model this biological process with a verification MSE value of 10% from only 16 training points.

References

1. Komorowski, M., Celi, L.A., Badawi, O., Gordon, A.C., Faisal, A.A.: The artificial intelligence clinician learns optimal treatment strategies for sepsis in intensive care. Nat. Med. **24**, 1716 (2018)
2. Singer, M., et al.: The third international consensus definitions for sepsis and septic shock (sepsis-3). Jama **315**(8), 801–810 (2016)
3. Martin, L., van Meegern, A., Doemming, S., Schuerholz, T.: Antimicrobial peptides in human sepsis. Frontiers Immunol. **6**, 404 (2015)
4. Zechendorf, E., et al.: Heparan sulfate induces necroptosis in murine cardiomyocytes: a medical-in silico approach combining in vitro experiments and machine learning. Frontiers Immunol. **9**, 393 (2018)
5. Abbass, H.A.: An evolutionary artificial neural networks approach for breast cancer diagnosis. Artif. Intell. Med. **25**(3), 265–281 (2002)
6. Ahson, S.I.: Petri net models of fuzzy neural networks. IEEE Trans. Syst. Man Cybern. **25**(6), 926–932 (1995)
7. Bapat, R.B.: Incidence matrix. In: Graphs and Matrices, pp. 11–23. Springer, London (2010). https://doi.org/10.1007/978-1-84882-981-7_2
8. Brownlee, J.: Clever Algorithms: Nature-Inspired Programming Recipes, 1st edn. Lulu.com (2011)
9. Huizinga, J., Clune, J., Mouret, J.B.: Evolving neural networks that are both modular and regular: hyperneat plus the connection cost technique. In: Proceedings of the 2014 Annual Conference on Genetic and Evolutionary Computation, pp. 697–704. ACM (2014)
10. Claycomb, W.C., et al.: Hl-1 cells: a cardiac muscle cell line that contracts and retains phenotypic characteristics of the adult cardiomyocyte. Proc. Nat. Acad. Sci. **95**(6), 2979–2984 (1998)
11. Martin, L., et al.: Soluble heparan sulfate in serum of septic shock patients induces mitochondrial dysfunction in murine cardiomyocytes. Shock **44**(6), 569–577 (2015)
12. Martin, L., et al.: Peptide 19-2.5 inhibits heparan sulfate-triggered inflammation in murine cardiomyocytes stimulated with human sepsis serum. PLoS ONE **10**(5), e0127584 (2015)

A Cultural Algorithm for Determining Similarity Values Between Users in Recommender Systems

Kalyani Selvarajah[1]([⊠]) [iD], Ziad Kobti[1]([⊠]) [iD], and Mehdi Kargar[2]([⊠]) [iD]

[1] School of Computer Science, University of Windsor, Windsor, ON, Canada
{selva111,kobti}@uwindsor.ca
[2] Ted Rogers School of Information Technology Management, Ryerson University,
Toronto, ON, Canada
kargar@ryerson.ca

Abstract. Recommendation systems for online marketing often rely on users' ratings to evaluate the similarity between users in Collaborative Filtering (CF) recommender systems. This paper applies knowledge-based evolutionary optimization algorithms called Cultural Algorithms (CA) to evaluate the similarity between users. To deal with the sparsity of data, we combine CF with a trust network between users. The trust network is then clustered using Singular Value Decomposition (SVD) which helps to discover the top neighbors' trust value. By incorporating trust relationships with CF, we predict the rating by each user on a given item. This study uses the Epinions dataset in order to train and test the accuracy of the results of the approach. The results are then compared against those produced by Genetic Algorithms (GA), Cosine, and Pearson Correlation Coefficient (PCC) methods. The comparison of the results suggests that the proposed algorithm outperforms the other similarity functions.

Keywords: Cultural Algorithm · Recommender system ·
Collaborative Filtering · Trust aware

1 Introduction

With the increasing trend of e-commerce and online services, users are influenced by trusted groups of people who make recommendations on various items and purchase experiences. There is a growing trend for online retail businesses for instance to incorporate recommender systems (RS) in order to promote deals and assist consumers in matching their preferences given the massive amount of information. Recommendations typically speed up searches, make it easy for users to access their content of interest, and list similar contents of value that they might have not considered. RS continue to demonstrate their usefulness for both consumers and online service providers including e-commerce and social networks sites. Despite many advances, an important open problem in RS remains to improve the accuracy of predictions.

© Springer Nature Switzerland AG 2019
P. Kaufmann and P. A. Castillo (Eds.): EvoApplications 2019, LNCS 11454, pp. 270–283, 2019.
https://doi.org/10.1007/978-3-030-16692-2_18

Collaborative Filtering (CF) can be considered as either user-item filtering or item-item filtering. The foremost fundamental idea of user-item CF is that it utilizes the similarity between users. The similarity value is assigned as a weight, which is used to give priority to users with similar preferences. A user to item matrix is maintained to store the user ratings for a given item. Given the very large size of users and items, numerous items lack ratings. A worst case scenario when evaluating ratings from users for a given item is when the matrix lacks the rating data; given the large number of users and items, the probability of two random users rating the same item is extremely low due to the sparsity of data. Therefore, including the trust values of users in the system alleviates the mentioned problem as well as reduces the cold start issue [1]. Trust-based systems use a similar concept as user similarity. The difference is that, in trust-based systems, users assign a weight to other users based on how much they trust the other's rating on any given item. In this paper we combine trust-based systems with CF in order to address the shortcomings of the sparsity matrix.

The approach is to employ an Evolutionary Algorithm (EA) called Cultural Algorithm (CA) [2] which uses knowledge components in a belief space to guide the search in the population space. To our knowledge the CA was not previously used in CF based RS. The primary goal of this algorithm is to enhance the accuracy of the prediction by evaluating the similarity metric and to propose a new way for finding similarity values without using the standard similarity functions such as PCC, cosine similarity or Jaccard similarity functions. Since RS uses similarity values between −1 and 1 as a standard [3], we first attempt a possible solution for similarity metric with random values in the range of [−1..1] between any two users in the training dataset. If the similarity value takes negative values, it implies that those users are likely to have opposite opinions; otherwise, users' opinions are similar. The similarity metric which predicts the rating value approximately equals the actual rating value which will then be selected as the best similarity metric for the prediction.

The main contribution of this study is to employ a knowledge-based evolutionary algorithm, specifically the CA, for evaluating the similarity metric between each user's rating.

The rest of the paper is arranged as follows: The next section discusses the related work done on RS and CF and trust based systems. In Sect. 3, we present the proposed model using the CA. In Sect. 4, we present the experimental setup. Finally, Sect. 5 concludes the idea of this research paper with directions for future work.

2 Related Works

Many studies have been conducted in RS where most discussed the Content-Based approach, memory, modeled or hybrid based CF approaches. Moreover, some RS were developed using evolutionary algorithms. At the same time, trust has turned into an influential factor in online social systems. This section presents related work in CF and review the various functions for calculating Similarity metrics of CF and evolutionary algorithms in RS.

2.1   Collaborative Filtering and Trust Values

The CF is a popular approach in RS and applied in variety of applications. CF operates based on users' rating on given items. In other words, it basically (1) creates a user profile based on his or her rating history on various items, (2) creates the similarity metric among users based on rating information, (3) predict the rating of a specific user for a target item and recommend top rated related, or similar, items.

The authors of [4] proposed the first CF based RS. Following this, many rating based RS were implemented. Candillier et al. [5] reviewed the main CF filtering methods and compared their results, and highlighted the advantages and disadvantages of other approaches. Given a huge scale of recommendations, finding similar users had become a challenge [6] since some users refuse to rate items. As a solution, the authors of [1,6,7] proposed trust statement in the RS. Trust is considered as a prominent role in helping online users to gather reliable information despite the sparsity of ratings. A user's trust value can be considered as a reputation value which is explicitly provided by one user to another. The authors of [8] develop a framework with different trust propagation schemes in order to satisfy certain circumstances. However, the cold users, that is users who are new to the site and have no rating history yet, have limitation since they have to build a trust network before filtering. To alleviate this problem, the authors of [7] evaluated an implicit trust value between two users by averaging the prediction error of co-rated items between them. The authors [9] employed the same concept for calculating indirect trust value by applying a Mole-Trust approach in their paper leading a surprisingly enhanced accuracy of the results.

2.2   Similarity Metrics

Evaluating the similarity between users is the most significant technique for the CF based RS. A metric or a similarity measure decides the similarity of two user profiles or the similarity of two items. For this purpose, the ratings of all the items rated by the two users or the ratings of all users who have rated two items will be compared.

The authors of [1,7,9,10] used PCC to evaluate similarity and predicted the possible rating for an item to the particular user. Furthermore, the authors of [10] implemented the RS based on the concept of trust and distrust link, and by conducting several experiments, various social trusts are included to enhance the approach. Liu in [11] also used PCC for the similarity calculation in their proposed sequenced based trust model. They considered time and similarity as the principal factors in calculating trustworthiness of users.

2.3   Evolutionary Algorithms

The evolutionary algorithms used in RS helped to further enhance the accuracy. The Genetic Algorithm (GA) is a well-known evolutionary algorithm [12] to automate the global search and generate the preferred individuals representing the near-optimal solutions given a fitness evaluation function. The authors of [3] used GA in RS based on CF. They modeled a new similarity function to measure the similarity between users. Moreover, they used this weighted sum in GA for prediction of ratings. Similarly, in [13,14], GA was used to refine the similarity values. The authors of [14] Anand et al. [15] used GA to convert user-item space to user-feature space since the feature space was smaller than item space. The authors [16] proposed a GA-based hybrid approach which predicted the result based on the previous likes. Salehi et al. [17] also proposed hybrid based GA which considered the weight of implicit attribute of learning material as genes. In each iteration, the GA transforms to reach the optimum weight. Jia et al. [18] developed RS using a GA which selects weight and threshold values used in a similarity function to find neighbors in order to give the recommendation based on trust. Hence, the similarity and trust improved the efficiency and helped select the best neighbors. Velez-Langs et al. [19] proposed a different GA approach in RS learning capabilities and possibilities from the user's personal characteristics. It follows that the RS would recommend appropriate individuals. In addition, Kim et al. [20] utilized a GA to optimize K-means clustering seeds for the problem of understanding the characteristics of online customers' needs. To have a different choice of recommended items, the authors of [21] proposed an evolutionary approach called Invenire. This approach avoids the over-specialization problem.

3   Proposed Model

This section examines the well-ordered procedure of trust-aware collaborative filtering. We present the knowledge-based evolutionary model which is based on a CA to find the similarity between users.

3.1   Trust Evaluation

The trust network can be defined as a directed graph $G(U, E)$, where U represents the users and E represents the trust relationship between users. The system considers the direct trust matrix $T_{|U| \times |U|} = t_{(u,v)}$ where $t_{(u,v)} \in \{1, -1\}$ as an input and U is number of user. Here, if u trust v, $t_{(u,v)} = 1$ and if u distrust v, $t_{(u,v)} = -1$. The direct trust values are the values from one user to another user.

Motivated by the work of [22] on the signed network, we clustered the direct trust network as communities. The users in the same communities have similar opinion, whereas those in other communities have an opposite opinion. We choose Singular Value Decomposition (SVD) to cluster trust communities into the various groups [23]. Using SVD, we can decompose the direct trust matrix $T_{|U| \times |U|}$ into : $T_{|U| \times |U|} = S_{|U| \times r} \lambda_{r \times r} V_{r \times |U|}^T$ where columns of S are orthogonal to each other and, λ is diagonal matrix, rows of V are orthogonal to each other and r is the rank (we can set the value for the rank which normally is from 1 to 5). S and V represent the orthogonal eigenvectors associated with the r nonzero eigenvalues of TT^T and $T^T T$, respectively. Each portion of decomposed trust matrix S, λ and V have significant properties for our application. Based on the sign pattern of the values of rows of S and V, SVD clusters the users into different communities. The entries of λ are sorted as $\lambda_1 \geq \lambda_2 \geq \ldots \geq \lambda_r$, and help decide how the strength of trust each community has among others. For example, the SVD clusters the users based on the combination of sign value of S as shown in Fig. 1 when the number of users is 7 and the rank is 2 ($S_{7 \times 2}$).

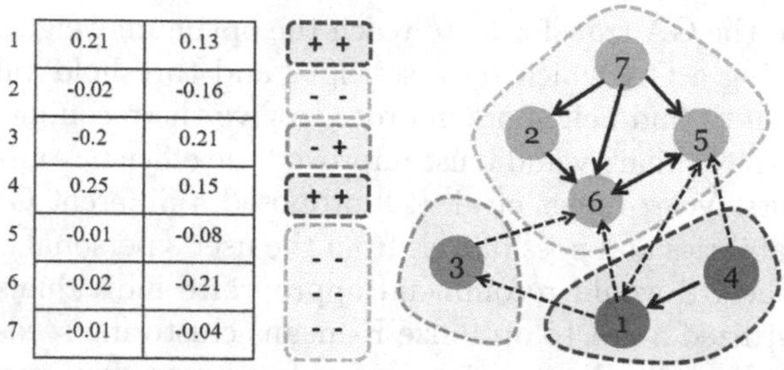

Fig. 1. The first 2 columns of matrix S are grouped based on a sign pattern then clusters the users of a trust network.

After the clustering, we collect the direct neighbors to have top trusted neighbors for each community network. Furthermore, to collect a sufficient number of neighbors, we consider both direct and indirect trust neighbors. We respectively assign the trust value to 1 and 0 for direct and indirect neighbors. If we reach the required number of direct neighbors up to the threshold K, we then stop the collection. If it is not reached, we will then explore the second step neighbors and continue the process until we collect the K number of neighbors. If we cannot reach the threshold value within 6 steps of propagation, we will stop exploring the neighbors based on the concept of the six degrees of separation theory. The modified implicit trust value of user u on user v is given in the following equation:

$$t_{u,v}^c = t_{u,v} \times dis_{u,v}^c \tag{1}$$

Where $t_{u,v}$ takes 0 or 1 from the Mole-Trust Algorithm [1] and $dis_{u,v}^c$ takes 1 if user v expertise level is greater than the average expertise level, else $\frac{1}{d}$ where $d = 6$ as per small world propagation theory. After that, the missing rating of an item i by user u using implicit trust value can be evaluated by the weighted average rating of u's trusted neighbors (TN_u) who rated the same item i as follows:

$$r_{u,v} = \frac{\sum_{v \in TN_u} t_{u,v} \cdot r_{v,i}}{\sum_{v \in TN_u} t_{u,v}} \tag{2}$$

Now we update the new rating values with the old values on all the users' profiles. This process helps us to fill all the missing ratings of an item i by user u in the past history as well.

3.2   Knowledge-Based Algorithm

Basically, CF uses users by items matrix $R_{|U|\times|I|}$ which is filled with rating values of users on various items. In the previous section we evaluated the new rating values of users based on the trust values. We use these updated rating values for the CA design. The CA in this study is used to calculate the similarity metric in order to enhance the result compared to the standard CF. Initially, the similarity between users will be randomly generated.

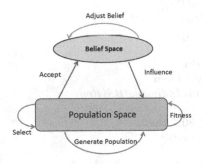

Fig. 2. Cultural Algorithm framework

As shown in Fig. 2, CA uses two phases of evolutionary inheritance system consisting of population and belief spaces [2]. Similar to other EAs such as GA, it begins with the initial random population of individuals. Then, the capability of the existing individuals can be evaluated using a fitness function. A set of individuals with better fitness are selected for further process. However with the CA, the selected group will be used for knowledge extraction in the belief space. Several types of knowledge are possible in the CA belief space such as situational and normative knowledge. The influence function, which extracts the knowledge from the belief space guides the generation of better individuals in each iteration, and thereby reduces the search domain while guiding the search direction. At the same time, some of the selected individuals perform crossover and mutation as in a typical GA. The acceptance function elevates the most fit individual knowledge into the belief space to be used later by the influence function. This process continues until it meets the termination condition. Finally, the individual with the best fitness value is considered as the solution to the problem.

The Algorithm in 1 shows our proposed method for finding the best similarity metric. We will review the main components of our model in the following section: representations, the fitness function, and the belief space structure.

Representations: The initial population has a set of individuals which represent possible solutions to the problem. An individual represents the similarity between the users. Generally, the similarity values in RS fall between $[-1, 1]$.

Algorithm 1. SimCA CF

Input: Rating matrix $|U|_n \times |I|_m$ of users on items
Output: Similarity Metric $n \times n$ where n is number of users

1: $n \leftarrow |population|$
2: $gs \leftarrow numberofiterations$
3: $t \leftarrow |selectedpopulation|$
4: $elite \leftarrow numberofelitcindividuals$
5: $Pop_{(1...n)} \leftarrow generaterandomindividuals$
6: **for** $i \leftarrow 1$ **to** gs **do**
7:      $MAE_{Pop} \leftarrow Fitness(Pop)$
8:      $(Pop) \leftarrow Sort_{(Fit)}Pop$
9:      $AVG_{MAE} \leftarrow Pop_{(1...t)}$
10:     $Pop_j \leftarrow Pop_{(1..elite)}$
11:     **for** $j \leftarrow elite$ **to** n **do**
12:         **if** random() $\leq 80\%$ **then**
13:             $Pop_j \leftarrow generateIndividualBS(Pop)$
14:         **else**
15:             **if** random() $\leq 80\%$ **then**
16:                 $Pop_j \leftarrow Crossover$
17:             **else**
18:                 $Pop_j \leftarrow Mutation$
19:             **end if**
20:         **end if**
21:     **end for**
22: **end for**
23: **return** Pop_1

Consequently, the individual can be generated initially with the random values between $[-1, 1]$ with the size of $|U|_m \times |U|_n \in \{x | x \in \mathbb{R}[-1, 1]x = 1 : m = n\}$, where $|U|$ is the number of users.

Fitness Function: The fitness function needs to decide the optimal similarity metric. Using individuals in the population space, we can predict the rating value of items by each user. The following well-known function will predict the rating of any user u on item i [9].

$$p_u^i = \overline{r}_u + \frac{\sum_{v=1}^{N_u}[sim(u,v) * (r_v^i - \overline{r}_v)]}{\sum_{v=1}^{N_u} sim(u,v)} \qquad (3)$$

Where, $\overline{r}_u$ is the average ratings of user u on items i, $sim(u,v)$ is similarity between users u and v, r_v^i is the rating of users v on item i and N_u is the set of users.

After we calculate the rating prediction based on the similarity values between users, we can evaluate the Root Mean Square Error (RMSE) which means the difference between the actual ratings and the predicted rating. We consider RMSE as our fitness function. If $RMSE$ is low, the predicted rating is closest to the actual rate value. The similarity metric using for this case would be the best individual [24]. As we mentioned before, the individual is stated as a 2D-Array with the size of the number of users.

$$RMSE = \sqrt{\frac{1}{N} \sum\nolimits_{i=1}^{N} (P_{U(i)} - R_{U(i)})^2} \qquad (4)$$

Belief Space: The main objective of the belief space is to exploit the extracted knowledge from selected individuals during the search in order to improve the performance of the CA. The best solution can be obtained from the combined elements of the best-selected individuals. In each iteration, the belief space updates the extracted knowledge to get the best solutions to be selected for the next generation.

Algorithm 2. Function: The best selected Individuals

1: generateIndividualsBS ← Function (Population)
2: $n \leftarrow |Population|$
3: $AVG_{RMSE} \leftarrow Average$
4: **for** $i \leftarrow 1$ to n **do**
5:     **if** $RMSE_{Pop_i} >= AVG_{RMSE}$ **then**
6:         $BS \leftarrow Pop_i$
7:     **end if**
8: **end for**
9: **return** BS

At this point we evaluate the average of selected individual solutions of the previous population and calculate the fitness (AVG_{RMSE}). The belief space will be updated in each iteration by the selected individual if its fitness (RMSE) is greater than the average fitness of best-selected individuals of the previous population as in the Algorithm 2. For example, if we select the best three individuals from the population based on their fitness value, the individual who has the least fitness value is selected as best among others. Then we calculate the average as shown in Fig. 3. The RMSE of this average similarity metric assigned as AVG_{RMSE}. It will help select the individuals for the next generation.

Fundamental Operators: Similar to the GA, a CA is also influenced by selection, crossover and mutation operators. The selection process is mainly based on fitness value. As we discussed above, the lower fitness value will give the best solution for the similarity metric in this study. Therefore, we choose the ranking method for the selection process. Since we consider the belief space as the main

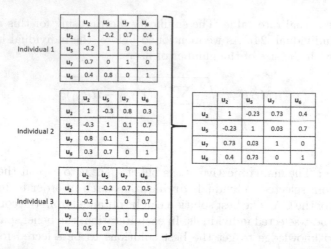

Fig. 3. Sample belief space with 3 individuals of 4 users in a community and the average of those three metrics.

component in our approach, the probability of using knowledge from the previous generation is 0.8. The other 0.2 is for crossover and mutation, 0.16 and 0.04 respectively.

Because we use a 2D-array to represent an individual, the crossover can be applied with either a vertical or horizontal sub-string method. We can process mutation by choosing random indexes of our 2D-arrays and replace them with random values between −1 and 1.

3.3 Primary Work-Flow of Proposed Model

The primary steps of the proposed model for the recommender system as depicted in Fig. 4 are given below:

1. Convert our data to sparse matrices: user-user trust and user-item rating.
2. Apply SVD to the user-user sparse matrix to cluster the users based on the sign.
3. Collect top-k neighbors for each users using the BFS algorithm & assign the trust value as 1 if they are directly connected & 0 if they are indirectly connected.
4. Modify the trust value using Eq. 1.
5. Combine the above modified trust matrix and user-item rating by using Eq. 2.
6. Generate the best similarity metric among users using CA.
7. Finally, for collaborative filtering, we generate the predicted rating of the user on any item.

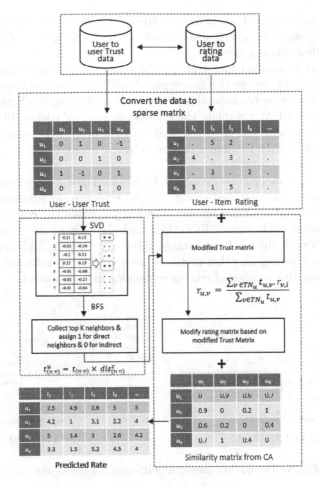

Fig. 4. Schematic view of the proposed model

4   Experiments

This section describes the performance evaluation of our model. We use Epinions[1] extended dataset, generated by [1] which is a consumer review website to collect the explicit trust value of users for various items such as books and movies. Users provide their opinion on items as reviews and ratings. Based on the review of other users they can provide their trust value as 1 for trusting and −1 for distrusting. This will help identify experts in the specific type of items. Other users can trust those reviews and ratings to decide on that specific item. Although the real extended dataset has a huge number of user reviews and trust ratings, due to the limited memory and processing speed, we collect users who have given trust rating for more than 10 users. The filtered data has 8658 users. At the same time, we collect rating

---

[1] http://www.trustlet.org/epinions.html.

items for the same users. The dataset is 23160 items with 377453 ratings. We use two evaluation methods: precision and recall, to check the quality of our proposed model with other existing methods.

Our dataset is divided into training data (i.e. 80%) and testing data (i.e. 20%). The result of our model is compared with the ones obtained from the other approaches using traditional metric on CF Recommender System: RMSE, precision, and recall. We evaluate the quality measure of cosine, PCC, and GA and compare them with our proposed model. In our proposed model, we generate a similarity metric with random numbers between −1 and 1 to create an initial population of 100.

We modified users' original ratings and generated new ratings based on users' trust values, named as modified ratings. So, we evaluate the precision and recall for both original and modified ratings. The Fig. 5a shows the comparison of the precision measure with cosine, PCC, GA and CA. Then, we examined those quality models with the modified ratings. Similarly, the Fig. 5b shows the recall quality measure among all other methods with CA. We conducted this experiment by varying the number of recommendations in each model.

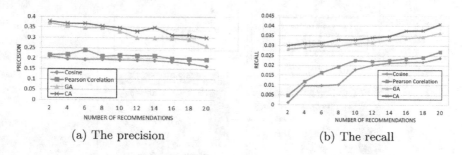

(a) The precision   (b) The recall

Fig. 5. The quality measures with original rating metric.

The Fig. 6a, b are the precision and recall quality measures for the modified rating metric respectively. Next, to measure the prediction performance of our proposed model, we evaluate the RMSE value for each case by varying the number of neighbors. The Fig. 7a illustrates the comparison of RMSE value with other models when we use explicit rating while Fig. 7b interprets when we use implicit rating. The Fig. 6b represents that the quality is enhanced in any number of recommendation and Fig. 6a proves that the quality measure is enhanced for any number of recommendations using our dataset.

We notice that the recalls of all the similarity measures increase with the increasing number of recommendations while the precision decreases. By considering the trust value in the new modified rating metric, the results show 12.37%, 62.7% and 69.6% improvement in prediction quality compared to GA, PCC and Cosine respectively. Figure 7a and b show that the accuracy improved relative to GA, PCC, Cosine and the standard rating which does not consider trust values.

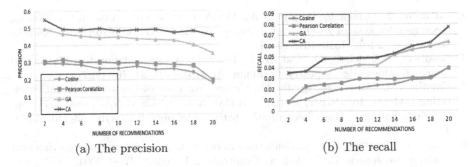

(a) The precision   (b) The recall

Fig. 6. The quality measures with original rating metric and trust between users.

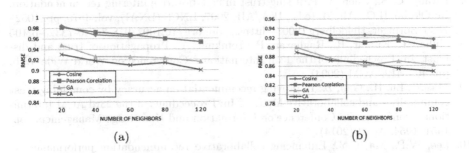

(a)   (b)

Fig. 7. The RMSE for original rating metric.

5   Conclusion

This study mainly focused on the evaluation of the similarity metric between users in CF and the importance of trust values between users. We proposed a new approach to evaluate the user similarity metric without using any standard similarity functions such as Cosine or PCC. We used Epinions dataset after pre-processing for the experimental setup. The empirical results from our proposed model show a significantly better result in both cases: original rating metric and modified rating metric (with trust values). At the same time, RMSE showed significant improvement with CA compared to the other methods. In future work, we would like to apply the CA to solve the over-specialization problem. Since it uses the various type of knowledge in the evolution process, CA would likely yield the best solution for this problem.

References

1. Massa, P., Avesani, P.: Trust-aware recommender systems. In: Proceedings of the 2007 ACM Conference on Recommender Systems, pp. 17–24. ACM (2007)
2. Reynolds, R.G.: An introduction to cultural algorithms. In: Proceedings of the Third Annual Conference on Evolutionary Programming, pp. 131–139. World Scientific (1994)

3. Bobadilla, J., Ortega, F., Hernando, A., Alcalá, J.: Improving collaborative filtering recommender system results and performance using genetic algorithms. Knowl.-Based Syst. **24**(8), 1310–1316 (2011)
4. Goldberg, D., Nichols, D., Oki, B.M., Terry, D.: Using collaborative filtering to weave an information tapestry. Commun. ACM **35**(12), 61–70 (1992)
5. Candillier, L., Meyer, F., Boullé, M.: Comparing state-of-the-art collaborative filtering systems. In: Perner, P. (ed.) MLDM 2007. LNCS (LNAI), vol. 4571, pp. 548–562. Springer, Heidelberg (2007). https://doi.org/10.1007/978-3-540-73499-4_41
6. Massa, P., Bhattacharjee, B.: Using trust in recommender systems: an experimental analysis. In: Jensen, C., Poslad, S., Dimitrakos, T. (eds.) iTrust 2004. LNCS, vol. 2995, pp. 221–235. Springer, Heidelberg (2004). https://doi.org/10.1007/978-3-540-24747-0_17
7. Hwang, C.-S., Chen, Y.-P.: Using trust in collaborative filtering recommendation. In: Okuno, H.G., Ali, M. (eds.) IEA/AIE 2007. LNCS (LNAI), vol. 4570, pp. 1052–1060. Springer, Heidelberg (2007). https://doi.org/10.1007/978-3-540-73325-6_105
8. Guha, R., Kumar, R., Raghavan, P., Tomkins, A.: Propagation of trust and distrust. In: Proceedings of the 13th International Conference on World Wide Web, pp. 403–412. ACM (2004)
9. Ma, X., Lu, H., Gan, Z.: Improving recommendation accuracy by combining trust communities and collaborative filtering. In: Proceedings of the 23rd ACM International Conference on Conference on Information and Knowledge Management, pp. 1951–1954. ACM (2014)
10. Lee, W.P., Ma, C.Y.: Enhancing collaborative recommendation performance by combining user preference and trust-distrust propagation in social networks. Knowl.-Based Syst. **106**, 125–134 (2016)
11. Liu, D.R., Lai, C.H., Chiu, H.: Sequence-based trust in collaborative filtering for document recommendation. Int. J. Hum. Comput. Stud. **69**(9), 587–601 (2011)
12. Goldberg, D.E., Holland, J.H.: Genetic algorithms and machine learning. Mach. Learn. **3**(2), 95–99 (1988)
13. Ar, Y., Bostanci, E.: A genetic algorithm solution to the collaborative filtering problem. Expert Syst.Appl. **61**, 122–128 (2016)
14. Alhijawi, B., Kilani, Y.: Using genetic algorithms for measuring the similarity values between users in collaborative filtering recommender systems. In: 2016 IEEE/ACIS 15th International Conference on Computer and Information Science (ICIS), pp. 1–6. IEEE (2016)
15. Anand, D.: Feature extraction for collaborative filtering: a genetic programming approach. Int J Comput Sci **9**, 348 (2012)
16. Gao, L., Li, C.: Hybrid personalized recommended model based on genetic algorithm. In: 2008 4th International Conference on Wireless Communications, Networking and Mobile Computing, WiCOM 2008, pp. 1–4. IEEE (2008)
17. Salehi, M., Pourzaferani, M., Razavi, S.A.: Hybrid attribute-based recommender system for learning material using genetic algorithm and a multidimensional information model. Egypt. Inf. J. **14**(1), 67–78 (2013)
18. Jia, Y.B., Ding, Q.Q., Liu, D.L., Zhang, J.F., Zhang, Y.L.: Collaborative filtering recommendation technology based on genetic algorithm. In: Applied Mechanics and Materials, vol. 599, pp. 1446–1452. Trans Tech Publ (2014)
19. Velez-Langs, O., De Antonio, A.: Learning user's characteristics in collaborative filtering through genetic algorithms: some new results. In: Jamshidi, M., Kreinovich, V., Kacprzyk, J. (eds.) Advance Trends in Soft Computing. SFSC, vol. 312, pp. 309–326. Springer, Cham (2014). https://doi.org/10.1007/978-3-319-03674-8_30

20. Kim, K.J., Ahn, H.: A recommender system using GA K-means clustering in an online shopping market. Expert. Syst. Appl. **34**(2), 1200–1209 (2008)
21. da Silva, E.Q., Camilo-Junior, C.G., Pascoal, L.M.L., Rosa, T.C.: An evolutionary approach for combining results of recommender systems techniques based on collaborative filtering. Expert Syst. Appl. **53**, 204–218 (2016)
22. Chiang, K.Y., Whang, J.J., Dhillon, I.S.: Scalable clustering of signed networks using balance normalized cut. In: Proceedings of the 21st ACM International Conference on Information and Knowledge Management, pp. 615–624. ACM (2012)
23. Golub, G.H., Reinsch, C.: Singular value decomposition and least squares solutions. Numerische mathematik **14**(5), 403–420 (1970)
24. Zadeh, P.M., Kobti, Z.: A multi-population cultural algorithm for community detection in social networks. Procedia Comput. Sci. **52**, 342–349 (2015)

Image and Signal Processing

Optimizing the C Index Using a Canonical Genetic Algorithm

Thomas A. Runkler[1]($\boxtimes$) and James C. Bezdek[2]

[1] Siemens AG, Corporate Technology, 80200 Munich, Germany
Thomas.Runkler@siemens.com
[2] University of Melbourne, Parkville, VIC 3010, Australia
jcbezdek@gmail.com

Abstract. Clustering is an important family of unsupervised machine learning methods. Cluster validity indices are widely used to assess the quality of obtained clustering results. The C index is one of the most popular cluster validity indices. This paper shows that the C index can be used not only to validate but also to actually find clusters. This leads to difficult discrete optimization problems which can be approximately solved by a canonical genetic algorithm. Numerical experiments compare this novel approach to the well-known c-means and single linkage clustering algorithms. For all five considered popular real-world benchmark data sets the proposed method yields a better C index than any of the other (pure) clustering methods.

Keywords: Cluster validity · c-means · Single linkage · C index · Genetic algorithm

1 Introduction

Clustering [1] is an unsupervised machine learning method that partitions a set of feature vectors

$$X = \{x_1, \dots, x_n\} \subset \Re^p \tag{1}$$

into $c \in \{2, \dots, n-1\}$ clusters. A cluster partition can be specified by an $n \times c$ binary partition matrix U with elements $u_{ik} \in \{0, 1\}$ so that

$$u_{ik} = \begin{cases} 1 & \text{if } x_k \in \text{cluster } i \\ 0 & \text{otherwise} \end{cases} \tag{2}$$

where

$$\sum_{k=1}^{n} u_{ik} > 0 \quad \text{for all } i = 1, \dots, c \tag{3}$$

$$\sum_{i=1}^{c} u_{ik} = 1 \quad \text{for all } k = 1, \dots, n \tag{4}$$

© Springer Nature Switzerland AG 2019
P. Kaufmann and P. A. Castillo (Eds.): EvoApplications 2019, LNCS 11454, pp. 287–298, 2019.
https://doi.org/10.1007/978-3-030-16692-2_19

Given c and n, the number of possible different partitions is [2]

$$\pi = \frac{1}{c!} \sum_{i=1}^{c} \binom{c}{i} (-1)^{c-i} i^n \approx \frac{c^n}{c!} \text{ for } c \ll n \tag{5}$$

Although the number π of different partitions may be extremely high for real world data sets, there exist many efficient algorithms to find cluster partitions. Two of the most popular clustering methods are *single linkage* (SL) [3] and *c-means* (CM) [4], and many variations and extensions of these have been proposed in the literature. Many stochastic clustering methods try to computationally mimic biological systems to find clusters in data, for example evolutionary algorithms (see [5] for a survey) or swarm-based methods such as ant algorithms [6,7], or particle swarm algorithms [8–11]. A drawback of these methods is that often nothing can be said about convergence or the properties of the obtained partitions. For a comprehensive survey of clustering algorithms we refer to [1]. Here, for simplicity we will restrict our comparisons to the popular single linkage (SL) and c-means (CM) methods.

Any clustering method will yield c clusters, whether or not the data X seem to contain c clusters, according to any model that defines the clusters computationally. Therefore, an important problem is the estimation of the number of (apparent) clusters in X. A common solution to this problem is based on *cluster validity*: For X, find a set $\mathcal{U}$ of *candidate partitions* U_c, $c = c_{\min}, \ldots, c_{\max}$, compute the validity of each of these partitions using a *cluster validity index* (CVI), and then pick the partition with the best validity, which is assumed to correspond with the apparently best choice for the number of clusters in X. Figure 1 illustrates the cluster validity procedure: Clustering may be done by optimizing an objective function, for example the c-means function J_{CM}, and validation is done by evaluating a cluster validity index CVI.

$$X, c \longrightarrow \boxed{J_{CM}} \longrightarrow \mathcal{U} \longrightarrow \boxed{CVI} \longrightarrow U_c$$

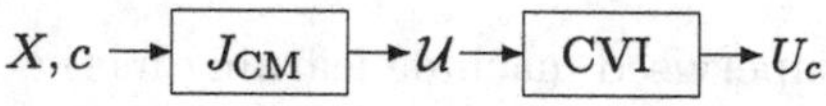

Fig. 1. Traditional clustering and cluster validity assessment: a clustering objective function (here: J_{CM}) is used to find cluster partitions $\mathcal{U}$, and a cluster validity index (CVI) selects the most valid partition U_c.

2   Cluster Validity: The C Index

One of the most popular cluster validity indices is the *C index* (CI) introduced by Dalrymple-Alford [12] and studied in more detail in [13,14]: We first compute the intra cluster sum of distances

$$\Gamma = \sum_{s=1}^{n-1} \sum_{t=s+1}^{n} \left(\sum_{i=1}^{c} u_{is} \cdot u_{it} \right) \cdot \|x_s - x_t\| \tag{6}$$

Cluster i has

$$n_i = \sum_{k=1}^{n} u_{ik} \tag{7}$$

points, so we have a total of

$$n_w = \sum_{i=1}^{c} \frac{n_i \cdot (n_i - 1)}{2} \tag{8}$$

intra cluster distances. We compute the vectors δ^+ and δ^- by sorting the distances between points $\|x_s - x_t\|$, $s = 1, \ldots, n - 1$, $t = s + 1, \ldots, n$, in ascending (δ^+) and descending (δ^-) order, and define

$$\Gamma_{\min} = \frac{1}{2} \sum_{i=1}^{n_w} \delta_i^+ \tag{9}$$

$$\Gamma_{\max} = \frac{1}{2} \sum_{i=1}^{n_w} \delta_i^- \tag{10}$$

Finally, we compute the C index (CI) as

$$J_{\text{CI}} = \frac{\Gamma - \Gamma_{\min}}{\Gamma_{\max} - \Gamma_{\min}} \tag{11}$$

According to the C index the best partition is associated with the lowest value of J_{CI}. A fuzzy extension of the C index was recently proposed by Runkler et al. [15]. Many other cluster validity indices have been proposed in the literature, for example the Dunn index [16,17], the Davies-Bouldin index [18], the silhouette index [19], just to mention a few. In a future project the authors are planning to investigate if the approach proposed in this paper can also be applied to these other cluster validity indices, and also for streaming data [20,21].

Optimizing the C index J_{CI} (when viewed as·a cluster-defining objective function) is a difficult discrete optimization problem. In this paper we apply a canonical genetic algorithm [22] to this optimization problem. We present an extensive study with five popular benchmark data sets where we compare our method against the SL and CM clustering algorithms, which are briefly reviewed in Sects. 3 and 4. Section 5 introduces a canonical genetic algorithm that (approximately) optimizes the C index (CI). Section 6 presents our experimental study. Section 7 summarizes our conclusions and gives an outline for planned future research.

3   Single Linkage Clustering

Single linkage (SL) [3] is defined by the following algorithm: Initialize a partition matrix by an $n \times n$ identity matrix $U_n = I_n$, so each feature vector is considered to be a singleton cluster. Then, for $t = 1, \ldots, n - c$, find the pair of clusters with the smallest inter cluster distance

$$d_{ij} = \min_{u_{il}=1, u_{jm}=1} \|x_l - x_m\|, \tag{12}$$

say $r, s \in \{1, \ldots, c\}$, where

$$d_{rs} = d_{\min} = \min_{i,j=1,\ldots,c;\, i \neq j} d_{ij}, \tag{13}$$

add a new row to U formed by merging (disjunction of) rows r and s, and delete the original rows r and s in U, so the number of rows in U is decreased by one in each step, and after step $t = n - c$, U has c rows. In (12) the *pairwise* distance between vectors x_l and x_m is shown as a norm. Any vector norm can be used; in the sequel, we use the Euclidean norm.

Given X and c, SL finds the partition U that minimizes the objective function

$$J_{\mathrm{SL}} = \max_{i=1,\ldots,c} \mathrm{diam}_i \tag{14}$$

with the diameter diam_i of cluster i defined as the minimum value of d so that all pairs of points in cluster i can be connected by a chain of links no greater than d, see page 199 of [23]. We compute the diameter of a cluster by applying the SL algorithm to the set of points in this cluster and obtain the diameter as the length of the last (maximum) link constructed by SL.

4   c-Means Clustering

The *c-means* (CM) [4] clustering model is defined by the objective function

$$J_{\mathrm{CM}}(U, V; X) = \sum_{i=1}^{c} \sum_{k=1}^{n} u_{ik} \cdot \|v_i - x_k\|^2 \tag{15}$$

with a set of cluster centers $V = \{v_1, \ldots, v_c\} \subset \Re^p$. Minimization of J_{CM} with respect to the two sets of variables U and V is an NP complete problem [24, 25]. Often good solutions can be found by randomly initializing U (or V) and then alternatingly computing [26] V and U (or U and V) using

$$v_i = \frac{\sum\limits_{k=1}^{n} u_{ik} \cdot x_k}{\sum\limits_{k=1}^{n} u_{ik}} \tag{16}$$

$$u_{ik} = \begin{cases} 1 & \text{if } \|v_i - x_k\| = \min\limits_{r \in 1,\ldots,c} \|v_r - x_k\| \\ 0 & \text{otherwise} \end{cases} \tag{17}$$

where ties may be handled by random assignment. Alternating optimization is terminated when successive estimates of U remain unchanged. The number of steps needed until termination depends on the data X and on the number of clusters c. Empirical studies indicate that the number of steps increases approximately linearly with c [27].

Alternating estimation of the CM objective function may or may not correspond to a global minimum of J_{CM}. To increase the probability of finding the global minimum, CM is often run multiple times with random initializations of U or V, and then the partition with the lowest J_{CM} is picked. Other approaches do not randomly initialize U or V but try to estimate the statistical distribution of the data in X to find better initializations of U or V, for example using the so-called *kmeans++* method [28].

Different kinds of stochastical methods have been proposed that try to find the global minimum of the CM objective function, for example simulated annealing [29], genetic algorithms [30], particle swarm optimization [31–34], ant colony optimization [35], wasp swarm optimization [36], or neural networks [37].

Approaches to approximate the global minimum of the CM objective function use continuous relaxations of the objective function in U, for example fuzzy c-means [38], possibilistic c-means [39], or noise clustering [40]. We remark that the global minimum of J_{CM} does not always agree with what humans might regard as the best clustering solution for a given data set. See [2] or [38] for an example.

5   A Canonical Genetic Algorithm for the C Index

Our goal is to find an $n \times c$ binary partition matrix U with elements $u_{ik} \in \{0,1\}$ that satisfies (3) and (4) that (globally) optimizes the C index (11). Here, to solve this optimization problem we use a canonical genetic algorithm as introduced in [22] that can be summarized as shown in Fig. 2.

```
randomly initialize population
determine fitness of each individual
perform selection
repeat
        perform crossover
        perform mutation
        determine fitness of each individual
        perform selection
until termination
```

Fig. 2. Algorithm of a canonical genetic algorithm (pseudo code).

We encode a partition U with the chromosome string $s = (s_1, \ldots, s_n)$ with $s_i \in \{1, \ldots, c\}$, $i = 1, \ldots, n$, so

$$u_{ik} = \begin{cases} 1 & \text{if } s_k = i \\ 0 & \text{otherwise} \end{cases} \tag{18}$$

for example

$$s = \boxed{1\;|\;3\;|\;1\;|\;2\;|\;\dots} \quad \longleftrightarrow \quad U = \begin{pmatrix} 1\;0\;1\;0\;\dots \\ 0\;0\;0\;1\;\dots \\ 0\;1\;0\;0\;\dots \end{pmatrix} \tag{19}$$

Any chromosome string will represent a partition matrix that satisfies (4) but not necessarily (3). If a chromosome violates (3), then it is deleted and not considered for generating offsprings. In the initialization step each element of the population is set to an n-dimensional vector where each element is randomly set to a number in $\{1, \dots, c\}$. The fitness of each individual is evaluated by generating a partition matrix U according to (18) and then computing the C index according to (11). In the selection step we consider all non-degenerate (3) individuals of the population for the generation of offsprings. We want to minimize not maximize J_{CI}, so each individual is chosen for crossover with the probability

$$p \sim \frac{e^{-J_{CI}} - e^{-J_{CImax}}}{e^{-J_{CImin}} - e^{-J_{CImax}}} \tag{20}$$

where J_{CImin} and J_{CImax} are the minimum and maximum values of J_{CI} in the current population, respectively, and all probabilities are normalized so that the sum of the probabilities for all individuals is equal to one. Crossover is applied to all individuals chosen in this way. In the crossover step for each pair of mating individuals we randomly choose a number $m \in \{1, \dots, n-1\}$ and construct an offspring by concatenating elements $1, \dots, m$ of the first individual and elements $m + 1, \dots, n$ of the second individual. In the mutation step we randomly pick a certain percentage of the n elements of each individual of the complete population and set each of these to a (different) random number in $\{1, \dots, c\}$. We terminate the algorithm after a fixed number of steps. In the following we will call this algorithm the *CIGA clustering algorithm*.

Throughout all experiments in this paper we will consider populations of 1000 individual chromosomes, mutation rate of 0.1%, and termination after 5000 steps. We will run this algorithm 100 times with different random initializations and report the best result, i.e. the partition that achieved the minimum value of the C index, and we will report the number of steps needed to get to this best result, so the number of steps will be ≤ 5000.

6   Experimental Studies

In our experimental study we consider the five real-world data sets available at the UCI machine learning repository (http://archive.ics.uci.edu/ml/) that have also been used in [15]: Iris, Wine, Seeds, Pima, and Vehicle. These data sets have appeared many times in the clustering literature; therefore, we show the main parameters of the data sets (Table 1) but do not provide further details here. Notice the extremely high numbers π of different partitions for the considered data sets.

Table 1. Parameters of the considered data sets: number of objects n, number of features p, number of classes/clusters c, and number of possible different partitions π.

	n	p	c	π
Iris	150	4	3	$6.2 \cdot 10^{70}$
Wine	178	13	3	$1.4 \cdot 10^{84}$
Seeds	210	7	3	$2.6 \cdot 10^{99}$
Pima	768	8	2	$7.8 \cdot 10^{230}$
Vehicle	846	18	4	$9.2 \cdot 10^{507}$

For each data set we randomly reorder the object indices so none of the algorithms can draw any cluster-relevant information from the indices (e.g. when neighboring elements in the chromosome are likely to belong to the same cluster). For each data set we fix c as the number of labeled classes in the data set, and we do not use the class labels in the algorithms.

There has been some discussion in the literature whether the Iris data set with 3 labeled classes contains 3 clusters or whether classes 2 and 3 belong together, so there are only 2 clusters. Note that the answer to the question "how many clusters are in Iris?" is entirely dependent on the model and algorithm chosen to look for them. Iris does not contain (natural) clusters—no data set does. So we run our experiments for the Iris data with both $c = 2$ (where class labels 3 are changed to 2) and $c = 3$.

For each of the data sets we run the SL, CM, and CIGA clustering algorithms. CM and CIGA are run 100 times with different random initializations. For given X and c, SL always deterministically produces the same result, so we do not have to run SL several times. CM is run until successive estimates of U remain unchanged, and CIGA is run for a fixed number of 5000 steps. For CIGA we consider the minimum J_{CI} of the 5000 steps and report the number of steps to find this minimum.

The left two columns of Fig. 3 show the histograms of the number of steps of the CM and CIGA algorithms (normalized, so the sum of the heights of all bars is 100%). Notice the different scaling of the horizontal axes for CM and CIGA. The average number of steps for CM increases with increasing size of the data sets, but it always needs less than 30 steps. For the Iris data set, CIGA needs less than 250 steps for $c = 2$ (row 1) but about $250, \ldots, 2500$ steps for $c = 3$ (row 2), sometimes even more. For the Wine and Seeds data sets, CIGA usually needs about $250, \ldots, 1250$ steps. For Pima, CIGA needs at least about 1500 and for Vehicle at least about 4000 steps. For Pima and Vehicle better results may be achieved if the maximum number of steps is increased but we had to restrict to 5000 steps because of run time. The experiments presented in this paper have been very time consuming. The total run time for all experiments on a 2.6 GHz Dual XEON workstation with 256 GB RAM was approximately 40 CPU weeks.

The right two columns of Fig. 3 show the normalized histograms of J_{CM} for the solutions obtained by the CM algorithm and of J_{CI} for the solutions obtained

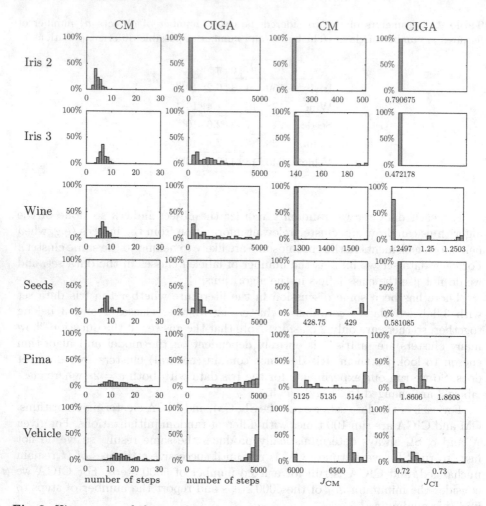

Fig. 3. Histograms of the number of steps (columns 1 and 2) and of the values of the objective functions (columns 3 and 4) for the CM (columns 1 and 3) and CIGA (columns 2 and 4) clustering algorithms.

by the CIGA algorithm. For Iris 2, Iris 3, and Seeds (rows 1, 2, and 4) CIGA always yields the same result, independent of initialization. All other rows show a certain distribution of the solutions found, and in the following we will discuss only the best of these solutions.

In Table 2 we consider four different partitions for each data set: the original class labels, the partition produced by SL, and the best partitions produced by CM and CIGA, so for each data set we see four rows and each of the rows corresponds to the (optimal) partition U according to the class labels, from SL, CM, and CIGA, respectively.

Table 2. Confusion matrices for the class labels and the SL, CM, and CIGA methods for the six different data sets. First four columns: Percentages of equivalent cluster assignments for each pair of clustering methods. Values on the diagonals are 100%. Off-diagonal percentages larger than 90% are shown in bold. Last three columns: Values of the J_{SL}, J_{CM}, and J_{CI} objective functions for the partitions according to the class labels and found by the SL, CM, and CIGA algorithms. Minima of each column are shown in bold.

	$\cap_{class}$	$\cap_{SL}$	$\cap_{CM}$	$\cap_{CIGA}$	J_{SL}	J_{CM}	J_{CI}
Iris 2							
Class	100%	**100%**	**100%**	**97.3333%**	**1.3892**	**220.8793**	0.81865
SL	**100%**	100%	**100%**	**97.3333%**	**1.3892**	**220.8793**	0.81865
CM	**100%**	**100%**	100%	**97.3333%**	**1.3892**	**220.8793**	0.81865
CIGA	**97.3333%**	**97.3333%**	**97.3333%**	100%	1.5534	230.0102	**0.79067**
Iris 3							
Class	100%	66%	85.3333%	82%	1.3892	165.4283	0.53283
SL	66%	100%	70%	69.3333%	**0.94322**	213.6504	0.80633
CM	85.3333%	70%	100%	**92.6667%**	1.3892	**138.8884**	0.47247
CIGA	82%	69.3333%	**92.6667%**	100%	1.3892	139.0418	**0.47218**
Wine							
Class	100%	37.6404%	**96.6292%**	**93.8202%**	4.7824	1292.6806	1.3021
SL	37.6404%	100%	35.9551%	35.3933%	**3.8495**	2198.9279	18.5903
CM	**96.6292%**	35.9551%	100%	**96.0674%**	4.5715	**1270.7491**	1.2576
CIGA	**93.8202%**	35.3933%	**96.0674%**	100%	4.7824	1279.7421	**1.2496**
Seeds							
Class	100%	34.7619%	**91.9048%**	**93.8095%**	1.6767	465.566	0.61912
SL	34.7619%	100%	34.2857%	34.2857%	**1.3066**	1422.0304	14.316
CM	**91.9048%**	34.2857%	100%	**98.0952%**	1.6767	**428.6082**	0.58135
CIGA	**93.8095%**	34.2857%	**98.0952%**	100%	1.6767	428.684	**0.58108**
Pima							
Class	100%	65.2344%	67.4479%	72.0052%	5.4204	5778.1891	2.2185
SL	65.2344%	100%	64.1927%	60.1563%	**3.4374**	6099.1163	164.2859
CM	67.4479%	64.1927%	100%	78.5156%	3.9746	**5122.0421**	2.0047
CIGA	72.0052%	60.1563%	78.5156%	100%	3.9746	5420.1357	**1.8605**
Vehicle							
Class	100%	26.2411%	35.9338%	35.2246%	9.8324	13318.408	1.2323
SL	26.2411%	100%	30.26%	30.6147%	**3.9827**	13807.5054	21.9063
CM	35.9338%	30.26%	100%	69.8582%	4.7206	**5973.6144**	0.79567
CIGA	35.2246%	30.6147%	69.8582%	100%	7.1918	7437.6411	**0.7432**

The first four columns show the percentages of matching cluster assignments for each pair of these partitions, where large off-diagonal values >90% are shown in bold. For Iris 2 we have the same partitions for SL and CM that correspond to the class labels (100%), and a very similar partition for CIGA, where only four data points are assigned differently, which corresponds to a matching percentage of $1 - 4/150 \approx 97.3333\%$. For Iris 3 the CM and CIGA results are very similar, for Wine and Seeds the class labels, CM, and CIGA. For Pima and Wine we have quite different partitions for all four methods.

The right three columns show the values of J_{SL}, J_{CM}, and J_{CI} for the (optimal) partitions from the class labels, SL, CM, and CIGA, where the minimum values for each J and for each data set are shown in bold. Again, for Iris 2 SL, CM, and CIGA yield the same result with the same values of the three objective functions, which is associated with the minimum of J_{SL} and J_{CM}. So, both SL and CM yield cluster partitions that exactly correspond to the original class labels. This may support the claim that the Iris data set contains two clusters. However, the C index of $J_{CI} \approx 0.81865$ corresponding to this partition is *not* optimal. CI yields a partition with a C index of only $J_{CI} \approx 0.79067$, which is about 3.4% lower. The C index obtained by CIGA is $J_{CI} \approx 0.79067$ for Iris 2, but only $J_{CI} \approx 0.47218$ for Iris 3, which supports the assumption that the Iris data set contains three but not two clusters. However, for Iris 3 the percentage of equivalent assignments between the class labels and CIGA is only 82%, i.e. only 82% of the original class labels match the optimal (in the sense of the C index) cluster assignments. In general, SL always yields the minimum of J_{SL}, CM yields the minimum of J_{CM}, and CIGA yields the minimum of J_{CI}. For each data set, CM yields a very good value of J_{CI}, but in all cases CIGA yields a partition with a smaller J_{CI} than all other considered methods (last column).

7 Conclusions

This paper has shown that cluster validity indices can not only be used for validating clusters but also to actually find clusters. This was specifically illustrated for the C index, one of the most popular cluster validity indices. Optimization of the C index is a difficult discrete optimization problem. We proposed to apply a canonical genetic algorithm to approximately optimize the C index, which we called the CIGA clustering algorithm.

An experimental study compared CIGA with the well-known single linkage (SL) and c-means (CM) clustering algorithms. In the experiments with five real world data sets (Iris, Wine, Seeds, Pima, Vehicle) the CIGA algorithm outperformed all other considered methods in terms of the C index. Our results therefore serve as valuable benchmarks for the clustering research community.

References

1. Jain, A.K., Dubes, R.C.: Algorithms for Clustering Data. Prentice Hall, Englewood Cliffs (1988)
2. Duda, R.O., Hart, P.E.: Pattern Classification and Scene Analysis. Wiley, New York (1973)
3. Sneath, P., Sokal, R.: Numerical Taxonomy. Freeman, San Francisco (1973)

4. MacQueen, J.: Some methods for classification and analysis of multivariate observations. In: Berkeley Symposium on Mathematical Statistics and Probability, vol. 14, pp. 281–297 (1967)
5. Hruschka, E.R., Campello, R.J.G.B., Freitas, A.A., de Carvalho, A.C.P.L.F.: A survey of evolutionary algorithms for clustering. IEEE Trans. Syst. Man Cybern. Part C (Appl. Rev.) **39**(2), 133–155 (2009)
6. Handl, J., Knowles, J., Dorigo, M.: Strategies for the increased robustness of ant-based clustering. In: Di Marzo Serugendo, G., Karageorgos, A., Rana, O.F., Zambonelli, F. (eds.) ESOA 2003. LNCS (LNAI), vol. 2977, pp. 90–104. Springer, Heidelberg (2004). https://doi.org/10.1007/978-3-540-24701-2_7
7. Kanade, P.M., Hall, L.O.: Fuzzy ants as a clustering concept. In: NAFIPS International Conference, Chicago, pp. 227–232, July 2003
8. Ji, C., Zhang, Y., Gao, S., Yuan, P., Li, Z.: Particle swarm optimization for mobile ad hoc networks clustering. In: International Conference on Networking, Sensing and Control, Taipeh, Taiwan, pp. 372–375, March 2004
9. Omran, M.G.H., Salman, A., Engelbrecht, A.P.: Dynamic clustering using particle swarm optimization with application in image segmentation. Pattern Anal. Appl. **8**(4), 332 (2006)
10. Tillett, J., Rao, R., Sahin, F., Rao, T.M.: Particle swarm optimization for the clustering of wireless sensors. Digit. Wirel. Commun. **5100**, 73–83 (2003)
11. Xiao, X., Dow, E.R., Eberhart, R., Miled, Z.B., Oppelt, R.J.: Gene clustering using self-organizing maps and particle swarm optimization. In: International Parallel and Distributed Processing Symposium, Nice, France, April 2003
12. Dalrymple-Alford, E.C.: Measurement of clustering in free recall. Psychol. Bull. **74**(1), 32 (1970)
13. Hubert, L., Arabie, P.: Comparing partitions. J. Classif. **2**(1), 193–218 (1985)
14. Hubert, L.J., Levin, J.R.: A general statistical framework for assessing categorical clustering in free recall. Psychol. Bull. **83**(6), 1072 (1976)
15. Bezdek, J.C., Moshtaghi, M., Runkler, T.A., Leckie, C.: The generalized C index for (internal) fuzzy cluster validity. IEEE Trans. Fuzzy Syst. **24**(6), 1500–1512 (2017)
16. Dunn, J.C.: Well-separated clusters and optimal fuzzy partitions. J. Cybern. **4**(1), 95–104 (1974)
17. Rathore, P., Ghafoori, Z., Bezdek, J.C., Palaniswami, M., Leckie, C.: Approximating Dunn's cluster validity indices for partitions of big data. IEEE Trans. Cybern. **9**, 1–13 (2019)
18. Davies, D.L., Bouldin, D.W.: A cluster separation measure. IEEE Trans. Pattern Anal. Mach. Intell. PAMI-1 **2**, 224–227 (1979)
19. Rousseeuw, P.J.: Silhouettes: a graphical aid to the interpretation and validation of cluster analysis. J. Comput. Appl. Math. **20**, 53–65 (1987)
20. Ibrahim, O.A., Keller, J.M., Bezdek, J.C.: Analysis of streaming clustering using an incremental validity index. In: IEEE International Conference on Fuzzy Systems, Rio de Janeiro, Brazil, pp. 1–8 (2018)
21. Moshtaghi, M., Bezdek, J.C., Erfani, S.M., Leckie, C., Bailey, J.: Online cluster validity indices for performance monitoring of streaming data clustering. Int. J. Intell. Syst. **34**, 541–563 (2019)
22. Holland, J.H.: Adaptation in Natural and Artificial Systems. The University of Michigan Press, Ann Arbor (1975)
23. Hartigan, J.A.: Clustering Algorithms. Wiley, New York (1975)
24. Aloise, D., Deshpande, A., Hansen, P., Popat, P.: NP-hardness of Euclidean sum-of-squares clustering. Mach. Learn. **75**(2), 245–248 (2009)

25. Brucker, P.: On the complexity of clustering problems. In: Henn, R., Korte, B., Oettli, W. (eds.) Optimization and Operations Research. LNE, vol. 157, pp. 45–54. Springer, Heidelberg (1978). https://doi.org/10.1007/978-3-642-95322-4_5
26. Bezdek, J.C., Hathaway, R.J.: Convergence of alternating optimization. Neural Parallel Sci. Comput. **11**(4), 351–368 (2003)
27. Runkler, T.A., Bezdek, J.C., Hall, L.O.: Clustering very large data sets: The complexity of the fuzzy c-means algorithm. In: European Symposium on Intelligent Technologies, Hybrid Systems and their implementation on Smart Adaptive Systems, Albufeira, Portugal, pp. 420–425, September 2002
28. Arthur, D., Vassilvitskii, S.: k-means++: The advantages of careful seeding. In: ACM-SIAM Symposium on Discrete Algorithms, pp. 1027–1035 (2007)
29. Rose, K., Gurewitz, E., Fox, G.: A deterministic annealing approach to clustering. Pattern Recogn. Lett. **11**(9), 589–594 (1990)
30. Krishna, K., Murty, M.N.: Genetic k-means algorithm. IEEE Trans. Syst. Man Cybern. Part B (Cybern.) **29**(3), 433–439 (1999)
31. Chen, C.Y., Ye, F.: Particle swarm optimization and its application to clustering analysis. In: International Conference on Networking, Sensing and Control, Taipeh, Taiwan, pp. 789–794, March 2004
32. Cui, X., Potok, T.E., Palathingal, P.: Document clustering using particle swarm optimization. In: IEEE Swarm Intelligence Symposium, Pasadena, pp. 185–191, June 2005
33. Runkler, T.A., Katz, C.: Fuzzy clustering by particle swarm optimization. In: IEEE International Conference on Fuzzy Systems, Vancouver, pp. 3065–3072, July 2006
34. van der Merwe, D.W., Engelbrecht, A.P.: Data clustering using particle swarm optimization. IEEE Congr. Evol. Comput. **1**, 215–220 (2003)
35. Runkler, T.A.: Ant colony optimization of clustering models. Int. J. Intell. Syst. **20**(12), 1233–1261 (2005)
36. Runkler, T.A.: Wasp swarm optimization of the c-means clustering model. Int. J. Intell. Syst. **23**(3), 269–285 (2008)
37. Tambouratzis, G., Tambouratzis, T., Tambouratzis, D.: Clustering with artificial neural networks and traditional techniques. Int. J. Intell. Syst. **18**(4), 405–428 (2003)
38. Bezdek, J.C.: Pattern Recognition with Fuzzy Objective Function Algorithms. Plenum Press, New York (1981)
39. Krishnapuram, R., Keller, J.M.: A possibilistic approach to clustering. IEEE Trans. Fuzzy Syst. **1**(2), 98–110 (1993)
40. Davé, R.N.: Characterization and detection of noise in clustering. Pattern Recogn. Lett. **12**, 657–664 (1991)

Memetic Evolution of Classification Ensembles

Szymon Piechaczek, Michal Kawulok[ID], and Jakub Nalepa[⊠][ID]

Silesian University of Technology, Gliwice, Poland
szymon.piechaczek@yahoo.com, {michal.kawulok,jakub.nalepa}@polsl.pl

Abstract. Creating classification ensembles may be perceived as a regularization technique which aims at improving the generalization capabilities of a classifier. In this paper, we introduce a multi-level memetic algorithm for evolving classification ensembles (they can be either homo- or heterogeneous). First, we evolve the content of such ensembles, and then we optimize the weights (both for the classifiers and for different classes) exploited while voting. The experimental study showed that our memetic algorithm retrieves high-quality heterogeneous ensembles, and can effectively deal with small training sets in multi-class classification.

Keywords: Ensemble classifier · Memetic algorithm · Classification

1 Introduction

Ensembles of classifiers can render better performance than a single classifier [1], and many methods for building such ensembles were proposed across various fields, including computer vision [2]. Multiple classifier systems (MCSs) can be categorized by the topology of an ensemble, the way the classifiers are selected, and how their decisions are combined [3]. Also, an MCS may be composed of homogeneous classifiers [4], trained on different datasets, different feature sets, or initialized differently, or heterogeneous ones [5] which combine various models.

Selecting the classifiers to an ensemble can be either static or dynamic. In the former approach, the competence regions of particular classifiers are determined during training [6], and in the latter, the competent classifiers are selected on the fly [7]. The dynamic techniques can be further split into dynamic classifier selection (DCS) [8], and dynamic ensemble selection (DES). The former picks a single classifier out of the ensemble, which is supposed to have the highest competence in a given region of the feature space. DES methods select a set of classifiers, whose decisions are fused. Such dynamic selection strategies include probabilistic models [9], dynamic overproduce-and-choose strategy [10], online pruning of base classifiers [11], and more. Ensemble classifiers also allow for combining multiple binary classifiers to solve multi-class problems—recently, Zhang et al. [12] showed that dynamic ensemble selection based on a competence measure can be successfully used for that purpose. Finally, DES was found very effective in dealing with the class imbalance problem [13].

© Springer Nature Switzerland AG 2019
P. Kaufmann and P. A. Castillo (Eds.): EvoApplications 2019, LNCS 11454, pp. 299–307, 2019.
https://doi.org/10.1007/978-3-030-16692-2_20

Evolutionary optimization has been used for building MCSs, both for DES, as well as for static ensembles, but this area has not been deeply explored yet. The task of fusing many classifiers (alongside the features and data) was treated as a multidimensional optimization problem in [14], solved with a genetic algorithm. In [15], multiobjective differential evolution was exploited to adjust weights in an MCS, whose classifiers were selected using static selection with majority voting. Also, it was proposed to use crowding to generate a pool of diverse (and heterogenous) classifiers, which finally form an ensemble. In [16], it was argued that the process of building MCSs can be inspired by biological mutualism.

In this paper, we follow this research pathway and propose a multi-level memetic algorithm (being a hybrid of genetic evolution enhanced with local-refinement procedures) for evolving different aspects of ensemble classifiers. In the first phase, we evolve the content of the ensembles (each containing $\mathcal{B}$ base classifiers selected from a pool of available pre-trained models). Then, the weights (utilized during the voting process) are subsequently optimized. Here, we optimize both the classifier weights (the answer of a "better" classifier should contribute more to the final class label elaborated by an ensemble), and the class weights (some classes, usually under-represented, may be more important to be spotted, hence enhancing their "importance" by increasing the corresponding weight should help in decreasing false negatives in this context). We validate our memetic algorithm using benchmark datasets, and apply it to evolve ensembles for hyperspectral image segmentation. The initial experiments shed light on the capabilities of our method, and showed it is able to deal with difficult (i.e., small and imbalanced) sets, and it elaborates high-quality classification engines.

Our multi-level memetic evolution of ensembles is introduced in Sect. 2. In Sect. 3, we discuss the experimental results. Section 4 concludes the paper.

2   Memetic Evolution of Classification Ensembles

We focus on the voting ensemble classifiers with hard voting—such classification engines undergo our memetic optimization. During prediction, each base learner votes for a class label of an incoming example with its given weight w_c. Then, depending on the predicted class, the vote is weighted again with the corresponding class weight w_l to extract the final label. The latter voting may be useful in specific applications; e.g., in medical image analysis, *not* spotting cancerous tissue in a medical scan is much less affordable than annotating healthy tissue as cancer, and cancerous examples almost always constitute the minority class. Hence, by increasing the cancerous class weight, we can potentially compensate for (extreme) imbalance of the training set, and help avoid false negatives by enhancing the *importance* of the corresponding class (however, more false positives may be introduced). In our **Multi-levEL Memetic Algorithm** for evolving **C**lassification ensembles (MELMAC), the ensemble evolution is divided into three consecutive phases which can be seen as the optimization levels (Fig. 1).

In phase (level) 1 (P1), we evolve the content of an ensemble of size $\mathcal{B}$ (being the number of base classifiers in an ensemble; $\mathcal{B}$ does *not* change during the

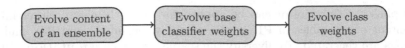

Fig. 1. In MELMAC, we evolve classification ensembles in three separate phases.

optimization). The initial population of N individuals which encode ensembles contains the base classifiers randomly drawn from the pool ($\mathcal{P}$) of pre-trained models[1] (in this phase, $w_c = w_l = 1$). Afterwards, the population of ensembles encompassing the base classifiers which were determined in P1 evolves to optimize the base classifier weights (P2, where $w_l = 1$), and the class weights (P3). In P2, each individual encodes $\mathcal{B}$ classifier weights, and in P3—$\mathcal{C}$ class weights, and their initial values are drawn randomly from the normal distribution $\mathcal{N}(0, 1)$.

The flowchart of a memetic evolution (which is generic to all of our optimization phases) is given in Fig. 2. To quantify the *fitness* η of individuals (chromosomes) in the population, we train the ensembles using a *training set* ($\mathbf{T}$), and classify the *validation set* ($\mathbf{V}$)—the classification accuracy obtained over $\mathbf{V}$ becomes the fitness of a corresponding chromosome (note that $\mathbf{T} \cap \mathbf{V} = \emptyset$). Importantly, the final generalization capabilities of the evolved ensembles is verified over the *test set* ($\mathbf{\Psi}$) which was not seen during the optimization ($\mathbf{T} \cap \mathbf{V} \cap \mathbf{\Psi} = \emptyset$).

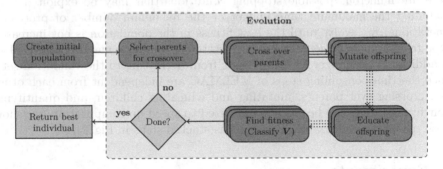

Fig. 2. Flowchart of an evolution exploited in each phase of MELMAC (the dotted education part is utilized in P2 and P3 only). The operations that are inherently parallelizable are rendered as stacked steps.

In all phases, we use the roulette **selection** to elaborate N parental pairs (the higher the fitness of an individual is, the larger the probability of selecting it for crossover becomes). Then, each pair of parents is **crossed over** using a single-point crossover (although the representations of individuals are different

---

[1] Note that the pool of base classifiers can be either homogeneous or heterogeneous, hence include various models trained over different training sets, and they can be parameterized differently.

for all phases, this operator can be applied for all of them—we either inherit the content of ensembles in P1, or the weight values in P2 and P3), and two offsprings are generated. To diversify the search, we execute **mutation**—in P1, we randomly select a base classifier (with probability $\mathcal{P}_m$) in an individual and replace it with a random model from a pool (which is not included in an ensemble already), whereas in P2 and P3, we apply uniform mutation to a randomly selected classifier/class weight (similarly, with probability $\mathcal{P}_m$).

In P2 and P3, we additionally utilize an **education** (local-search) operator to intensify the exploitation of the current part of a solution space (since this step is not performed in P1, we render it as dotted arrows in Fig. 2). Here, we analyze the neighborhood of a chromosome which is being educated—each weight is perturbed up to $\mathcal{E}$ times to explore its immediate neighborhood (of radius $\sigma_e \cdot \eta^2$) with probability $\mathcal{P}_e$, and the size of this neighborhood is dependent on the fitness score of the individual (the larger the fitness is, the bigger neighborhood is analyzed). If an educated individual is better-fitted than the original chromosome, it replaces it in the population. Finally, the next population of individuals includes N best chromosomes (from a set of $3N$ intermediate solutions, including those from the current generation alongside $2N$ generated children), therefore elitism is implicitly applied within each phase (i.e., the best individual always survives). After the full three-level optimization, we return the best individual (with the highest fitness value) obtained across all phases.

Our evolution progresses until the termination condition is met. Although there are numerous possible stopping conditions that may be exploited here (including the maximum time budget or the maximum number of processed generations), we evolve until the best fitness in the population is not increased for a given number of generations (this stopping condition may be perceived as an additional regularizer which prevents from overfitting to the validation set). Since the time-consuming steps of MELMAC are independent from each other (i.e., crossing over parents, mutating and educating children, and quantifying their fitness), they can be easily parallelized to speed up computation (we show those trivially parallelizable operations as stacked steps in Fig. 2).

3   Experiments

MELMAC was implemented in `Python 3.7` with an extensive usage of the `numpy` package, whereas `scikit-learn` was utilized to create and train the base classifiers. The parameters of MELMAC were kept constant across the entire study (they were tuned experimentally beforehand), and became $\mathcal{P}_m = 0.05$, $\mathcal{P}_e = 1.0$ (education is always applied), $\mathcal{E} = 5$ (the maximum number of education steps), $\sigma_e = 0.25$ (the neighborhood size), and the number of generations without the fitness improvement of the best chromosome after which the evolution was terminated was 10. The experiments were executed on a computer equipped with an Intel Core i7-3740QM 2.7 GHz (16 GB RAM) processor.

The experimental validation is split into two parts—(1) sensitivity analysis performed over the eight-class Mice Protein Expression dataset[2], and (2) application of our evolved ensembles for hyperspectral image (HSI) segmentation. In the latter experiment, we exploit three HSI benchmark multi-class datasets: Pavia University (340×610 image acquired using Reflective Optics System Imaging Spectrometer ROSIS sensor over the Pavia University in Lombardy, Italy; 103 hyperspectral bands, 9 classes of objects), Salinas Valley (217×512, NASA Airborne Visible/Infrared Imaging Spectrometer AVIRIS sensor over Salinas Valley in California, USA; 224 bands, 16 classes), and Indian Pines (145×145, AVIRIS over the North-Western Indiana, USA; 200 bands, 16 classes) scenes. For HSI, we do not perform any pre-processing, hence the feature vectors contain all reflectance information about the pixel, and their length corresponds to the number of bands (all datasets are very imbalanced; for details, see http:// www.ehu.eus/ccwintco/index.php/Hyperspectral_Remote_Sensing_Scenes).

For both experiments, the pool $\mathcal{P}$ of base classifiers consisted of 80 models (of different types and parameterizations; however trained over the same training sets). In $\mathcal{P}$, we included: Linear Discriminant Analysis with variable solvers, and with and without shrinkage to deal with small sets (if shrinkage is applied, its value is either automatically optimized or taken from the set $\{0.25, 0.5, 0.75, 1\}$), Linear Regression, k-Nearest Neighbors with variable $k \in \{5, 10, 15\}$ and weights (either uniform or relative to the distance between examples), Decision Trees with variable classification criteria (Gini or cross-entropy), splitting criteria (best or random), and the maximum depth $d \in \{10, 20, 30\}$, Support Vector Machines with variable $C \in \{0.001, 1, 100\}$, kernels (linear, radial-basis, polynomial, and sigmoid) and $\gamma \in \{0.1, 1, 10\}$ (note that γ is not used in all kernels), and Multi-Layer Perceptron with variable number of neurons n in the single hidden layer $n \in \{35, 75, 100\}$ and learning rate $\alpha \in \{0.0001, 0.001, 0.01\}$.

3.1 Experiment 1: Sensitivity Analysis

In this experiment, we verified the impact of the size of evolved ensembles $\mathcal{B}$ on their final classification capabilities (over the unseen test set Ψ). We focused on the Mice Protein Expression dataset which was divided into 10 non-overlapping folds (the overall number of examples in this dataset is 1081, and this set is imbalanced; here, we under-sampled the majority classes to balance the dataset, and the final number of examples analyzed in this experiment was 840, with 105 examples in each class). Then, a training set T consisted of 1 fold, whereas the examples from the remaining folds were split evenly into the validation V and test Ψ sets. For each split, the experiments were executed 10 times.

The experimental results reported in Table 1 show that increasing the number of base classifiers that are being evolved does not drastically affect the final classification accuracy (neither over the validation nor the test set—the differences are not statistically important according to the non-parametric Wilcoxon test at $p < 0.01$ for the majority of cases, excluding e.g., P3 for $\mathcal{B} = 20$ when compared

---

[2] Available at: https://www.kaggle.com/ruslankl/mice-protein-expression.

Table 1. The maximum (best across all runs) and average (across all runs) accuracy of the best evolved ensembles (and its standard deviation) for the validation and test sets after (a) P1, (b) P2, and (c) P3, alongside the average number of individuals successfully educated ($I_\mathcal{E}$) and mutated ($I_\mathcal{M}$), and the average number of generations g (with their average time g_t in sec.) before reaching convergence, averaged across all the folds for the Mice Protein Expression dataset.

Phase	$\mathrm{Acc}_{max}(V/\Psi)$	$\mathrm{Acc}_{avg}(V/\Psi)$	$\mathrm{Acc}_{\sigma}(V/\Psi)$	$I_\mathcal{E}$	$I_\mathcal{M}$	g	g_t
Number of base classifiers: $\mathcal{B} = 5$							
(a)	0.888/0.884	0.862/0.855	0.020/0.180	—	1.179	20.12	2.7
(b)	0.903/0.895	0.877/0.834	0.020/0.110	8.4	1.156	13.59	76.7
(c)	0.910/0.902	0.855/0.833	0.090/0.110	10.5	1.198	17.76	120.3
Number of base classifiers: $\mathcal{B} = 10$							
(a)	0.885/0.879	0.865/0.857	0.013/0.018	—	1.184	22.25	5.7
(b)	0.910/0.898	0.888/0.838	0.018/0.084	17.2	1.22	19.79	333.6
(c)	0.915/0.907	0.862/0.838	0.070/0.086	10.5	1.203	19.4	268.5
Number of base classifiers: $\mathcal{B} = 20$							
(a)	0.882/0.878	0.863/0.856	0.012/0.013	—	1.205	22.45	12.6
(b)	0.912/0.891	0.863/0.856	0.017/0.144	35.4	1.226	22.94	1490.6
(c)	0.908/0.898	0.816/0.785	0.131/0.146	10.9	1.187	18.92	598.5

with other $\mathcal{B}$'s); however, the stability of the models are slightly improved (standard deviation $\mathrm{Acc}_{\sigma}(V/\Psi)$ is decreased for larger ensembles). Interestingly, the overall accuracy may be lowered if the number of base classifiers is large (see the P3 results obtained for 20 base classifiers) likely due to premature convergence of the search (in a vaster solution space). Additionally, increasing $\mathcal{B}$ adversely affects the computation time (mainly because of the increased numbers of educations and mutations which require quantifying the fitness of offspring individuals). It is worth mentioning that an ensemble built with all models from $\mathcal{P}$ allowed for obtaining the average accuracy of 0.824 (with $w_c = w_l = 1$), whereas the average accuracy of the AdaBoost classifier was less than 0.3. On the other hand, three separate classifiers from $\mathcal{P}$ were able to outperform the evolved ensembles (i.e., Linear Discriminant Analysis with automatic shrinkage gave the average accuracy of 0.884, and linear and radial-basis support vector machines [with $C = 100$] resulted in the average accuracy of 0.863 and 0.86, respectively). It shows that MELMAC could benefit from better initialization of the population (these well-performing classifiers were not selected to the population in random initialization, and they were not introduced during mutation).

3.2   Experiment 2: Segmentation of Hyperspectral Images

In this experiment, we followed the training-test set splits given in [17]: the training set for Pavia University is balanced and contains 250 examples (pixels) for

Table 2. The maximum (best across all runs) and average (across all runs) accuracy of the best evolved ensembles (and its standard deviation) for the validation and test sets after (a) P1, (b) P2, and (c) P3, alongside the average number of individuals successfully educated ($I_\mathcal{E}$) and mutated ($I_\mathcal{M}$), and the average number of generations g (with their average time g_t in sec.) before reaching convergence, averaged across all the folds for the hyperspectral image datasets.

Phase	$Acc_{max}(V/\Psi)$	$Acc_{avg}(V/\Psi)$	$Acc_\sigma(V/\Psi)$	$I_\mathcal{E}$	$I_\mathcal{M}$	g	g_t
Pavia University							
(a)	0.833/0.823	0.830/0.792	0.009/0.019	—	1.3	16.5	19.644
(b)	0.837/0.822	0.836/0.724	0.001/0.175	8.255	1.2	12.3	468.149
(c)	0.770/0.822	0.769/0.724	0.002/0.175	11.924	1.2	14.9	885.258
Salinas Valley							
(a)	0.925/0.889	0.922/0.843	0.003/0.044	—	1.3	20	62.6
(b)	0.928/0.881	0.927/0.778	0.001/0.210	8.7	1.1	15.5	1022.9
(c)	0.846/0.881	0.844/0.778	0.004/0.210	21.5	1.1	20.7	3437.3
Indian Pines							
(a)	0.780/0.719	0.777/0.686	0.004/0.020		1.3	16.7	27.7
(b)	0.786/0.729	0.786/0.687	0.001/0.018	8.2	1.045	17.8	738.9
(c)	0.794/0.729	0.790/0.687	0.007/0.018	24.0	1.192	24.3	2503.7

each class, for Salinas Valley it is balanced too (300 examples per class), whereas it is imbalanced for Indian Pines (with the maximum number of examples in a class equal to 250), and the rest of HSI pixels become the test set. To create V, we randomly draw 30 and 80 examples from T for Pavia University and Salinas Valley, respectively, and 20% of training pixels from each class for Indian Pines.

The results gathered in Table 2 (here, we evolve ensembles of 5 base classifiers) show that the best scores are elaborated using ensembles with equal weights for classifiers and classes in most cases (for Indian Pines, the evolution of classifier weights in P2 slightly improved the generalization over Ψ). As in Experiment 1, the evolved ensembles outperformed those created using all base classifiers from the pool (they gave the average accuracy of 0.729, 0.823, and 0.616 for Pavia University, Salinas Valley, and Indian Pines, respectively; similarly, AdaBoost retrieved the classifiers which were not able to deal with such small training/validation sets and resulted in accuracy lower than 0.3 for all datasets). We can observe that due to a very small cardinality of V the evolved ensembles start to overfit (e.g., for Pavia University and Salinas Valley, when P1 is compared with P2). As in Experiment 1, there existed base classifiers which gave slightly better scores over Ψ, and were not included in the population, hence they could not boost the capabilities of our ensembles. Finally, the results elaborated using MELMAC are on par with those reported in [18] (obtained using convolutional neural networks). Therefore, including such deep models in $\mathcal{P}$ could help improve the overall accuracy of the evolved models even further.

4   Conclusions

In this paper, we introduced a multi-level memetic algorithm (MELMAC) for evolving classification ensembles (either homo- or heterogeneous). In MELMAC, the content of an ensemble (of a given size) is evolved in the first phase, whereas the weights applied for separate classifiers and classes are evolved afterwards. Our experimental study performed over benchmark datasets showed that MEL-MAC deals well with difficult (i.e., small and imbalanced) sets, and is able to retrieve high-quality ensembles. Our current work is focused on improving the generalization capabilities of evolved ensembles for small validation sets (to prevent from overfitting), and on employing deep models (with automatically learned data representations) into the pool of base classifiers. Finally, the initial assessment of base classifiers should help create better initial populations.

Acknowledgments. This work was supported by the National Science Centre, Poland, under Research Grant No. DEC-2017/25/B/ST6/00474, and JN was partially supported by the Silesian University of Technology under the Grant for young researchers (BKM-556/RAU2/2018).

References

1. Dietterich, T.G.: Ensemble methods in machine learning. In: Kittler, J., Roli, F. (eds.) MCS 2000. LNCS, vol. 1857, pp. 1–15. Springer, Heidelberg (2000). https://doi.org/10.1007/3-540-45014-9_1
2. Olague, G.: Evolutionary Computer Vision - The First Footprints. Natural Computing Series. Springer, Heidelberg (2016). https://doi.org/10.1007/978-3-662-43693-6
3. Woźniak, M., Graña, M., Corchado, E.: A survey of multiple classifier systems as hybrid systems. Inf. Fusion **16**, 3–17 (2014)
4. Claesen, M., De Smet, F., Suykens, J.A., De Moor, B.: Ensemblesvm: a library for ensemble learning using support vector machines. JMLR **15**(1), 141–145 (2014)
5. Nguyen, T.T., Nguyen, M.P., Pham, X.C., Liew, A.W.C.: Heterogeneous classifier ensemble with fuzzy rule-based meta learner. Inf. Sci. **422**, 144–160 (2018)
6. Kuncheva, L.I.: Clustering-and-selection model for classifier combination. In: 2000 Proceedings Fourth International Conference on Knowledge-Based Intelligent Engineering Systems and Allied Technologies, vol. 1, pp. 185–188. IEEE (2000)
7. Britto Jr., A.S., Sabourin, R., Oliveira, L.E.: Dynamic selection of classifiers—a comprehensive review. Pattern Recogn. **47**(11), 3665–3680 (2014)
8. Cruz, R.M., Sabourin, R., Cavalcanti, G.D.: Dynamic classifier selection: recent advances and perspectives. Inf. Fusion **41**, 195–216 (2018)
9. Woloszynski, T., Kurzynski, M.: A probabilistic model of classifier competence for dynamic ensemble selection. Pattern Recogn. **44**(10–11), 2656–2668 (2011)
10. Dos Santos, E.M., Sabourin, R., Maupin, P.: A dynamic overproduce-and-choose strategy for the selection of ensembles. Pattern Recogn. **41**(10), 2993–3009 (2008)
11. Oliveira, D.V., Cavalcanti, G.D., Sabourin, R.: Online pruning of base classifiers for dynamic ensemble selection. Pattern Recogn. **72**, 44–58 (2017)
12. Zhang, Z.L., Luo, X.G., García, S., Tang, J.F., Herrera, F.: Exploring the effectiveness of dynamic ensemble selection in the one-versus-one scheme. Knowl.-Based Syst. **125**, 53–63 (2017)

13. García, S., Zhang, Z.L., Altalhi, A., Alshomrani, S., Herrera, F.: Dynamic ensemble selection for multi-class imbalanced datasets. Inf. Sci. **445**, 22–37 (2018)
14. Gabrys, B., Ruta, D.: Genetic algorithms in classifier fusion. Appl. Soft Comput. **6**(4), 337–347 (2006)
15. Onan, A., Korukoğlu, S., Bulut, H.: A multiobjective weighted voting ensemble classifier based on differential evolution algorithm for text sentiment classification. Expert Syst. Appl. **62**, 1–16 (2016)
16. Lones, M.A., Lacy, S.E., Smith, S.L.: Evolving ensembles: what can we learn from biological mutualisms? In: Lones, M., Tyrrell, A., Smith, S., Fogel, G. (eds.) IPCAT 2015. LNCS, vol. 9303, pp. 52–60. Springer, Cham (2015). https://doi.org/10.1007/978-3-319-23108-2_5
17. Gao, Q., Lim, S., Jia, X.: Hyperspectral image classification using convolutional neural networks and multiple feature learning. Remote Sens. **10**(2), 299 (2018)
18. Ribalta, P., Marcinkiewicz, M., Nalepa, J.: Segmentation of hyperspectral images using quantized convolutional neural nets. In: Proceedings of IEEE DSD, pp. 260–267 (2018)

Genetic Programming for Feature Selection and Feature Combination in Salient Object Detection

Shima Afzali[✉], Harith Al-Sahaf, Bing Xue, Christopher Hollitt,
and Mengjie Zhang

School of Engineering and Computer Science,
Victoria University of Wellington, Wellington, New Zealand
{shima.afzali,harith.al-sahaf,bing.xue,
christopher.hollitt,mengjie.zhang}@ecs.vuw.ac.nz

Abstract. Salient Object Detection (SOD) aims to model human visual attention system to cope with the complex natural scene which contains various objects at different scales. Over the past two decades, a wide range of saliency features have been introduced in the SOD field, however feature selection has not been widely investigated for selecting informative, non-redundant, and complementary features from the existing features. In SOD, multi-level feature extraction and feature combination are two fundamental stages to compute the final saliency map. However, designing a good feature combination framework is a challenging task and requires domain-expert intervention. In this paper, we propose a genetic programming (GP) based method that is able to automatically select the complementary saliency features and generate mathematical function to combine those features. The performance of the proposed method is evaluated using four benchmark datasets and compared to nine state-of-the-art methods. The qualitative and quantitative results show that the proposed method significantly outperformed, or achieved comparable performance to, the competitor methods.

Keywords: Salient Object Detection · Genetic programming · Feature combination · Feature selection

1 Introduction

Human visual system can easily cope with the complex natural scene containing various objects at different scales using the visual attention mechanism. Salient Object Detection (SOD) aims to simulate the mentioned capability of the human visual system in prioritizing objects for high-level processing. SOD can be helpful to relieve complex vision problems such as scene understanding by detecting and segmenting salient objects [12]. Objects apparently catch more attention than background regions such as grass, sea, and sky. Therefore, if all generic objects can be detected in the first place, then scene understanding would be easily

© Springer Nature Switzerland AG 2019
P. Kaufmann and P. A. Castillo (Eds.): EvoApplications 2019, LNCS 11454, pp. 308–324, 2019.
https://doi.org/10.1007/978-3-030-16692-2_21

performed at the subsequent stages. SOD serves as an important pre-processing step for many tasks, such as image classification, image retargeting and object recognition [12]. Several applications benefit from saliency detection such as: visual tracking, image and video compression, content-based image retrieval, human-robot interaction, object recognition [12].

SOD methods are broadly classified into two groups, bottom-up and top-down methods [29]. In the bottom-up methods, multiple low-level features, such as intensity, color, and texture, are extracted from an image to compute the saliency values. Top-down methods are task-dependent and they usually utilize domain specific knowledge [22]. SOD methods generally comprise the following steps [26]: (a) extract/design some saliency features from the input image, (b) compute individual feature maps using biologically plausible filters such as Gabor or difference of Gaussian filters, and (c) combine these different features to generate the final saliency map.

To date, domain experts have designed different types of saliency features such as local contrast, global contrast, edge density, backgroundness, focuses, objectness, convexity, spatial distribution, and spareness [12]. Regarding the feature extraction stage, the design of new saliency features stimulates some possible questions. For example, whether the newly designed feature is informative for different image types (e.g., images with no much color variation, having cluttered background, having multiple objects, etc.), how does it effect other features, whether it is a duplicate feature (already exist) in the domain, which types of features it can complement, and how can we effectively use the new feature in different application scenarios. It is very difficult to provide definite answers for such questions. There are some potential ways to answer the aforementioned questions. One plausible way is to be a domain expert, this way suffers from some difficulties such as requiring domain knowledge of the task, domain experts are not always available and are very expensive to employ. Another way is developing a heuristic method which is very common in the literature [25,27]. However, it is becoming more and more challenging to design heuristic methods that are able to fully explore the potential of the existing features [12]. The next possible and suitable solution is to develop an automatic domain independent method. This method can widely explore characteristics of the existing and newly created features and find the relationship between them. In addition, it has the ability to select informative and non-redundant features that complement each other for different image types.

As mentioned before, the feature combination stage is one of the fundamental stages in SOD for generating the final saliency map [11]. In this regard, a few studies attempt to address the feature combination problem by finding the optimal values for the weights in the linear combination. For example, Liu et al. [23] employed the conditional random field (CRF) framework to learn the linear combination weights for different features. Afzali et al. [5] utilized particle swarm optimization (PSO) to learn a suitable weight vector for the saliency features and linearly combine the weighted features. Due to the highly nonlinearity of the visual attention mechanism, the above linear mapping might not perfectly capture the characteristics of feature combination of human visual system. Consequently, nonlinear methods are required to fuse saliency features to achieve higher performance on different image types. Moreover, in the majority of existing methods [17,22,29], saliency features and the combination stage have been manually designed by domain experts. In this scenario, the feature extraction and combination tasks are highly dependent on domain-knowledge and human intervention.

Genetic programming (GP) which is a well-known evolutionary computation (EC) technique has the ability to tackle image-related problems, such as region detection, feature extraction, feature selection, and classification, in a wide variety of applications [7]. GP can automatically generates solutions, it is problem-independent, and it has a flexible representation and global search ability. The tree-based representation of GP makes it a suitable tool to do feature manipulation such as feature construction and feature selection. GP has a good capability in creating different solutions which is not thought about by domain experts. Moreover, GP is well-known for being flexible due to the ability of evolving various models, e.g., linear and non-linear models, and operating on different types of data such as numerical and categorical data [18]. The aforementioned properties of GP motivates us to utilize it for feature selection and feature combination problems in SOD.

1.1 Goal

This paper aims at utilizing GP to automatically select features from different level and scales, and combine those selected features for the task of saliency object detection.

- Introduce a new automatic GP-based feature selection and feature combination method;
- Formulate an appropriate fitness function to evaluate GP solutions (programs);
- Evaluate the proposed method using dataset of varying difficulties to test the generalisability property of this method; and
- Compare the performance of the proposed method to that of nine handcrafted SOD methods to test whether those automatically evolved programs have the potential to achieve better or comparable performance to the domain-expert designed ones.

2 Background

2.1 Saliency Features

Jiang et al. [17] developed a discriminative regional feature integration (DRFI) approach which is a supervised SOD method. In DRFI, 93 features have been introduced, which contains three types of regional saliency descriptor including regional contrast, regional backgroundness, regional property. Regional contrast descriptor is a 29-dimensional feature vector contains color and texture features.

Color features are extracted from three different color spaces including RGB (red, green, and blue), HSV hue, saturation, and value), and $L^*a^*b^*$. For texture feature extraction, they used local binary patterns (LBP) feature [14] and responses of Leung-Malik (LM) filter bank [17]. Regional backgroundness descriptor that is a 29-dimensional feature vector extracted by computing the difference between each region and a pseudo-background region as a reference. Finally, regional property descriptor that is 35-dimensional feature vector computed by considering the generic properties of a region such as appearance and geometric features. Hence, a 93-dimensional $(2 \times 29 + 35)$ feature vector is obtained.

2.2 Related Work

Over the past two decades [16], an extremely rich set of saliency detection methods have been developed. Most previous methods have focused on designing various hand-crafted features to find objects with visual features different than those from the background. As a pioneer, Itti et al. [16] proposed bottom-up visual saliency method using center-surround differences across multi-scale image features. Prior efforts employed simple features such as color and grayscale, edges, or texture, as well as more complex features such as objectness, focusness and backgroundness. The reader shall refer to the survey paper by Borji et al. [12] for more details on traditional saliency features.

In addition to the feature extraction, many SOD methods attempt to design different feature combination frameworks. Zhu et al. [30] used an optimization-based framework to combine multiple foreground and background features as well as the smoothness terms to automatically infer the optimal saliency values. Zhou et al. [29] developed a SOD method by integrating compactness and local contrast features. [29] addressed the drawback of combining global contrast and compactness features by considering local contrast, since local contrast can detect some salient regions ignored by compactness. In [29], a diffusion process based on manifold ranking is employed to propagate saliency information. Lin et al. [22] introduced a method to predict salient object by extracting multiple features such as local contrast, global contrast, and background contrast in different feature extraction scales, e.g., pixel-level, region-level and object-level. In [22], the authors manually designed a framework to integrate the features, e.g., background priors, refined global contrast, and local contrast. Liu et al. [23] employed the conditional random field to learn an optimal linear combination of of local, regional, and global features. Jiang et al. [17] used random forest to the fusion weights of feature maps. In [17], they used three types of regional saliency features including contrast, backgroundness, and property to 93-dimensional feature vector for each region. The majority of the aforementioned studies suffers from requiring domain-knowledge and human intervention in designing a good feature combination method. Hence, recent studies attempt to address the mentioned problem by developing automatic approaches.

Compared with traditional methods that use hand-crafted features, convolutional neural network (CNN) based methods that adaptively extract high-level semantic information from raw images have shown impressive results in predicting saliency maps [28]. Lee et al. [20] considered both hand-crafted features and high-level features extracted from CNNs. To combine the features together, they designed a unified fully connected neural network to compute saliency maps. Recently, Hou et al. [15] developed a CNN method which combines both low-level and high-level features from different scales. Although CNN based developments have achieved high performance in the SOD domain in recent years, the top CNN methods require nontrivial steps such as generating object proposals, applying post-processing, enforcing smoothness through the use of superpixels or defining complex network architectures [24].

Among EC techniques, GP has the ability to solve various complex problems in many research areas such as feature extraction, classification, and object detection [10]. Lensen et al. [21] developed a GP approach to automatically select regions, extract histogram of oriented gradients (HOG) features and perform binary classification on a given image. Al-Sahaf et al. [8] showed that GP has the ability to automatically extract features, perform feature selection and image classification. Later on, Al-Sahaf et al. [9], used multitree GP representation to automatically evolve image descriptors. Unlike existing hand-crafted image descriptors, [9] automatically extracts feature vectors. Afzali et al. [4] utilized GP to construct informative foreground and background features from the extracted saliency features, then combined the constructed two features employing a spatial blending phase. However, this method was not fully automatic and required human intervention in the feature combination stage. Ain et al. [6] developed GP-based method to do feature selection and feature construction for skin cancer image classification. The authors claimed that the GP selected and constructed features helped the classification algorithms to produce effective solutions for the real-world skin cancer detection problem. In the mentioned studies, GP made the proposed approaches free from any requirement for human intervention or domain-specific knowledge. Apart from the existing GP methods, this study investigates using GP for feature selection and feature combination in SOD field.

3   The Proposed Method

This section describes the proposed GP based method for automatically feature selection and feature combination (GPFSFC) for saliency detection. The overall structure is depicted in Fig. 1. For the training stage, first, different image segmentation-levels are computed for each image in the training set, then saliency features are extracted from the segmented images. Second, the saliency feature set and ground truth are fed into GP. Third, the GP process generates and evaluates the GP programs. Finally, the GP process results 50 evolved individuals. For the validation stage, after completing the segmentation and feature extraction parts, the saliency feature set and ground truth of the validation images are used to select the best individual from the evolved GP individuals. For the test stage, for a given test image, similar to the training and validation stages, multi-level image segmentation and feature extraction are computed. Then, the saliency map of the image is produced by employing the selected GP individual.

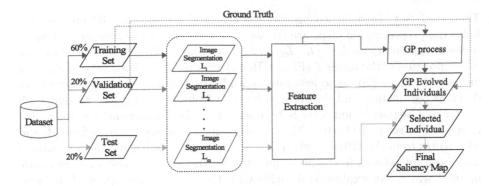

Fig. 1. The overall algorithm of GPFSFC.

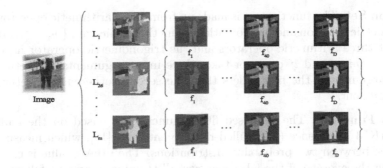

Fig. 2. Feature extraction from different segmentation levels.

3.1 Feature Selection and Feature Combination by GP

GP evolves the initial population using the ramped-half-and-half method. Each individual in the population takes saliency features and constants as terminals and combines them using the operations from the function set. The goodness of each individual is evaluated by a fitness function (more details below). The fitness value of each individual is computed by taking the average of the performance over all training images. In the subsequent generation, a population of new individuals are generated by applying the different genetic operators such as crossover, mutation and elitism on some individuals selected from the current population. This process continues until the maximum number of generations is reached, and the best evolved program is then returned which is a mathematical function of different selected operations and features. The terminal set, function set and fitness function are presented and discussed in detail in the following sections.

Terminal Set. In order to provide the terminal set for the GP process, the following preprocessing steps are employed. Firstly, for a given image I, a set of m-level segmentations $L = \{L_1, L_2, ..., L_m\}$ is computed, each segmentation is a decomposition of the image I (Fig. 2). Here, the graph-based image segmentation method [13] is employed to generate multiple segmentations using m groups of different parameters. In this study, m is set to 48 by following [17]. Second, each region of a segmentation level is represented by D-dimensional feature vector (Fig. 2). D is set to 103 $(10 + 93)$ by collecting 10 saliency features from [4] and 93 features from the DRFI method [17]. Since a large number of saliency features have been introduced in the literature, it is worthwhile to develop an automatic method which can explore and tackle with the large search space of different features. In this study, we provided a wide range of features to investigate how GP can cope with. Figure 2 demonstrates visualizations of different segmentation levels and saliency features belong to each level.

Function Set. The function set is made up from three arithmetic operators, one trigonometric function and one conditional function, which are $\{+, -, \times, sin, if\}$. The first three arithmetic operators and the trigonometric operator have their regular meaning, and if operator takes three input arguments and returns the second argument if the first is less than the second; otherwise, it returns the third argument.

Fitness Function. The proposed fitness function is based on the Kullback-Leibler (KL) divergence (also called relative entropy) [19], which measures the difference between two probability distributions. The fitness value is computed by taking the average of the KL value over all training images as follow

$$Fitness = \frac{1}{n} \sum_{i=1}^{n} KL(I_i, G_i) \tag{1}$$

where, n is the number of images, I_i and G_i are the ith image (the output saliency image computed using GP) and its corresponding ground truth. We utilize the KL divergence to see whether the output of GP is closer to the ground truth. Before applying KL divergence, we apply softmax on the output of GP and the ground truth to compute the probability of each one for the KL divergence. The KL divergence is defined as

$$KL(p, q) = \sum_{r \in R} p(r) \frac{\ln p(r)}{q(r)} \tag{2}$$

where r is a region from the region vector R. p and q are two probability distributions of the ground truth and the output of GP, respectively.

4 Experiment Design

4.1 Benchmark Datasets

In this paper, we evaluate the performance of the proposed method using four benchmark datasets in SOD. We choose these datasets based on the following four criteria: (1) being widely-used, (2) containing both large and small number of images, and (3) having different biases (e.g. number of salient objects, image clutter, center-bias). We split each dataset into three parts: a training set (60%), a validation set (20%) and a test set (20%).

The first dataset is single-object (*SED1*), which is a subset of the SED dataset [12]. The SED1 dataset includes 100 images containing only one salient object in each image. Pixel-wise ground truth annotations for the salient objects in SED1 are provided. Here, we employed the SED1 dataset, since we only considered single salient object images.

The second dataset is ASD, which is a subset of the MSRA10K dataset [23]. The MSRA10K dataset provides bounding boxes manually drawn around salient regions by nine users. However, a bounding box-based ground truth is far from being accurate. Thus, [23] created an accurate object-contour based ground truth dataset of 1000 images. Each image is manually segmented into foreground and background. Most images have only one salient object and strong contrast between objects and backgrounds.

The third dataset is ECSSD dataset which contains 1000 semantically meaningful but structurally complex images [12]. In contrast to some simple datasets such as MSRA, in which background structures are simple and smooth, the ECSSD dataset contains more complex image types. Ground truth masks are provided by 5 subjects.

The fourth dataset is PASCAL which contains 850 images [11]. The reason to use PASCAL-S datasets was to assess performance of the proposed method over scenes with multiple objects with high background clutter.

In this study, the raw images and respective ground truth are re-sized to 200×200 pixels to leverage computational efficiency.

Following [12], we utilized three universally-agreed, standard, and easy-to-compute evaluation criteria, which are precision-recall (PR) curve, receiver operating characteristic (ROC) curve and F-measure to evaluate the different SOD methods. We used these criteria to compute the quantitative performance of each SOD method.

The PR curve is obtained by binarizing the saliency map using a number of thresholds ranging from 0 to 255, as in [25]. To compare the quality of the different saliency maps, the threshold is varied from 0 to 255, and the precision and recall at each value of the threshold are computed by comparing the binary mask and the corresponding ground truth. Then a precision-recall curve is plotted using the sequence of precision-recall pairs. Precision corresponds to the fraction of the pixels correctly labeled against the total number of pixels assigned salient, whereas recall is the fraction of the pixels correctly labeled in relation to the number of ground truth salient pixels.

$$Precision = \frac{TP}{TP + FP} \qquad (3)$$

$$Recall = \frac{TP}{TP + FN} \qquad (4)$$

where TP (true positive) is the number of foreground pixels that are correctly detected as foreground, FP (false positive) is the number of background pixels that are incorrectly detected as foreground, and FN (false negative) is the number of foreground pixels that are incorrectly detected as background.

The ROC curve can also be generated based on the true positive rates (TPR) and false positive rates (FPR) obtained during the calculation of the PR curve. TPR and FPR can be computed as

$$TPR = \frac{TP}{TP + FN} \qquad (5)$$

$$FPR = \frac{FP}{FP + TN} \qquad (6)$$

An image dependent adaptive threshold T_a proposed by Achanta et al. [1] is used to binarize the saliency map ($\mathbf{S}$). T_a is computed as twice as the mean saliency of $\mathbf{S}$.

$$T_a = \frac{2}{W \times H} \sum_{x=1}^{W} \sum_{y=1}^{H} \mathbf{S}(x, y) \qquad (7)$$

where W and H are the width and height of the saliency map $\mathbf{S}$, respectively, and $\mathbf{S}(x, y)$ is the saliency value of the pixel at position (x, y).

Often, neither precision nor recall can fully evaluate the quality of a saliency map. To this end, the F-measure (F_β) (Eq. 8) is used as the weighted harmonic mean of precision and recall with a non-negative weight β^2.

$$F_\beta = \frac{(1 + \beta^2) Precision \times Recall}{\beta^2 Precision + Recall} \qquad (8)$$

As suggested in many salient object detection works [1,25], we set β^2 to 0.3, to weight precision more. The reason is because recall rate is not as important as precision [23]. For instance, 100% recall can be easily achieved by setting the whole map to be foreground.

4.2   Parameter Tuning

GP has a number of parameters which can be altered for a given problem. In this study, the initial population is created by ramped half-and-half method. The population size is restricted to 300 individuals, since the computational costs of dealing with images is high. The minimum tree depth is 2 and the maximum

Table 1. GP parameters.

Parameter	Value	Parameter	Value
Population size	300	Initial population	Half-and-half
Generations	50	Tree depth	2–10
Mutation rate	0.40	Selection type	Tournament
Crossover rate	0.60	Tournament size	7

depth is 10. The evolutionary process is terminated when the maximum number of 50 generations is reached. The evolutionary process is independently executed 30 times using different random seed values and the average performance is reported. Here, the mutation and crossover rates are set to 40% and 60%, respectively, based on the fact that a higher mutation rate could produce better training performance by allowing a wider exploration of the search space. The best evolved program is kept to prevent the performance of the subsequent generation from degrading. The tournament selection method is used for selecting individuals for the mating process and the tournament size is set to 7. Table 1 gives a summary for the GP parameters.

4.3   Methods for Comparison

The proposed method is compared to nine state-of-the-art methods, five methods including DRFI, GS, GMR, SF, and RBD are selected from [12], and four other methods including MSSS [2], wPSO [5], GPFBC [4], and FBC [3].

5   Results and Discussions

5.1   Quantitative Comparisons

In Fig. 3(a) and (b), precision-recall and ROC curves of GPFSFC is comparable with DRFI and outperforms other methods on the SED dataset. Figure 3(c) shows that GPFSFC has similar average precision and F-measure results to DRFI, but higher average recall than DRFI. On the SED dataset, DRFI and GPFSFC have generally good results regarding the average precision, recall, and F-measure among the other SOD methods.

As it can be seen in Fig. 4(a) and (b), GPFSFC is performing as the second best method after DRFI regarding the precision-recall and ROC curves on the ASD dataset. In Fig. 4(c), although GPFSFC has slightly lower average recall than DRFI, RBD, GS, and FBC on the ASD dataset, GPFSFC has the highest average precision and F-measure among all the other methods. Regarding precision-recall and ROC curves in Fig. 5(a) and (b), GPFSFC is the third good performing method among the nine state-of-the-art methods, where DCNN outperforming all the methods, although it could not perform as good as GPFSFC and DRFI on the SED and ASD datasets. In Fig. 6 similar to Fig. 5, GPFSFC shows the good results after DCNN and DRFI on the PASCAL dataset.

Considering quantitative results, although GPFSFC loses its performance to DCNN on the ECSSD and PASCAL datasets, it shows good performance comparing to DCNN on the SED and ASD datasets. This limitation is mainly due to lack of high-level features in the feature set of GPFSFC, while DCNN benefits of combining both low-level and high-level features. Extracting more powerful feature representations and training the method with complex scenes can be used to deal with the challenging datasets.

To compare GPFSFC with DRFI, as DRFI employed an ensemble learning containing a large number of decision trees to predict the saliency value of the regions, it can generally generate more accurate results than the automatically evolved GP programs which is only one independent tree in this experiment.

5.2   Qualitative Comparisons

The qualitative comparisons of GPFSFC and nine other state-of-the-art methods are illustrated in Figs. 7 and 8. For the qualitative comparison, multiple representative images are selected from different datasets which incorporate a variety of difficult circumstances, including complex scenes, salient objects with center bias, salient objects with different sizes, low contrast between foreground and background. Figure 7 presents some challenging cases where GPFSFC can successfully highlight the salient object and suppress the background. For example, the first row is a complex image where the building is not homogeneous, it has reflection on the water and complex background. However, GPFSFC and DRFI can deal with it, while other methods such as DCNN, RBD, GMR, GS, SF, and MSSS performs poorly to detect the object. DRFI is good in detecting object, but it wrongly highlights some part of the background regions.

In 4th row, GPFSFC can completely suppress background where the background is cluttered. This is due to the advantage of selecting informative background features by the evolved GP program. As can be seen in the 5th image, it has non-homogeneous foreground and complex background, GPFSFC shows good performance, while the other nine methods are struggling in both highlighting the object and completely suppressing the background. In the 6th row, GPFSFC properly covers the foreground object, although the color contrast between the object and background is low. Choosing informative contrast, backgroundness, and property (appearance and geometric) features and combining them using suitable mathematical operations is the key point for having good performance on the aforementioned images.

Figure 7 shows that DCNN fails to completely detect salient object when the image has complex background and low contrast between the foreground and background. One potential reason can be lack of segment-level information like prior knowledge on segment level.

Although both GPFSFC and DRFI show good performance in different scenarios, these methods suffer in some challenging cases such as images in Fig. 8. GPFSFC fails to completely identify the foreground object in all three images and wrongly highlights the background in the 2nd and 3rd images. This problem is due to the lack of the high-level knowledge and enough training samples to learn different scenarios and object types.

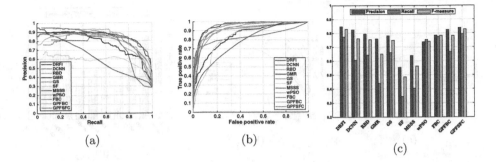

(a)   (b)

(c)

Fig. 3. Performance of GPFSFC compared to nine other methods on **SED1**.

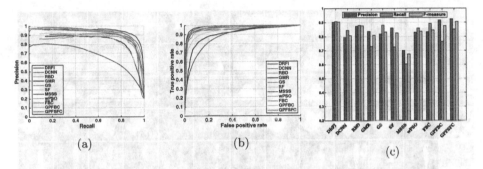

(a)   (b)

(c)

Fig. 4. Performance of GPFSFC compared to nine other methods on **ASD**.

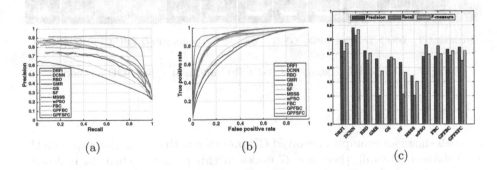

(a)   (b)

(c)

Fig. 5. Performance of GPFSFC compared to nine other methods on **ECSSD**.

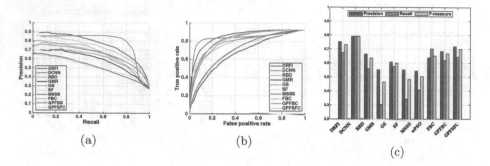

(a)   (b)   (c)

Fig. 6. Performance of GPFSFC compared to nine other methods on **PASCAL**.

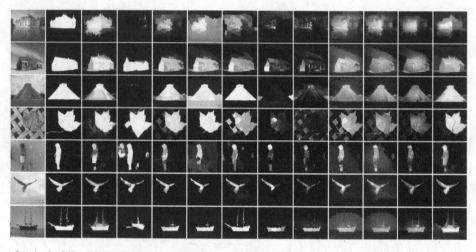

Original GT DRFI DCNN RBD GMR GS SF MSSS wPSO FBC GPFBC GPFSFC

Fig. 7. Some visual examples of tGPFSFC and nine other SOD methods.

5.3 Further Analysis

Figure 9 shows an example of evolved GP program with high performance on the ASD dataset. Overall, there are 27 nodes in this program where 14 nodes are leaves and the other 13 are functions. The description of the selected features by GP represented in Table 2. As it can be seen in Fig. 9, five regional background features $\{f_{32}, f_{34}, f_{35}, f_{36}, f_{40}\}$ are selected to suppress background regions. Three regional property features $\{f_{65}, f_{66}, f_{71}\}$ are selected to consider the generic properties of regions. Finally, three contrast features $\{f_0, f_{94}, f_{96}\}$ are chosen to capture the color differences space (changes), as a region is likely thought to be salient if it is different from the other regions. This GP program only chooses 11 features from 103 features and decrease the dimensionality nearly nine times. The GP process considers complementary characteristic of

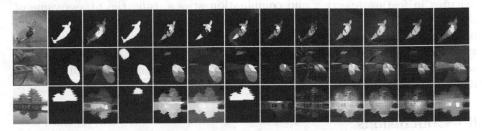

Original GT DRFI DCNN RBD GMR GS SF MSSS wPSO FBC GPFBC GPFSFC

Fig. 8. Some visual examples of GPFSFC and nine other SOD methods.

Table 2. Selected feature's description.

Feature	Description	Feature type
f_0	Average R value [17]	Regional contrast
f_{32}	Average B value [17]	Regional backgroundness
f_{34}	Average H value [17]	Regional backgroundness
f_{35}	Average S value [17]	Regional backgroundness
f_{36}	Average V value [17]	Regional backgroundness
f_{40}	Average b^* value [17]	Regional backgroundness
f_{65}	Normalized perimeter [17]	Regional property
f_{66}	Aspect ratio of the bounding box [17]	Regional property
f_{71}	Variances of the a^* value [17]	Regional property
f_{94}	Local contrast [23]	
f_{96}	Global contrast [23]	

×
−
if
f40
f65
+
0.26
f66
+
+
×
f66
+
f36
f35
f71
+
+
×
f34
f94
×
+
f36
f32
f0
f96

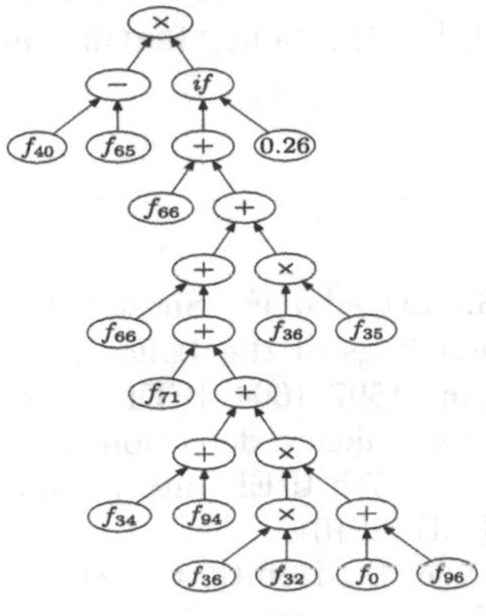

Fig. 9. Sample program evolved by GP.

features in feature selection and combination stages using the fitness value of the evolved GP programs. Figure 9 demonstrates that the evolved GP program or mathematical expression is a non-linear function. Furthermore, it represents a good example to present how the combination of different types of features such as color, backgroundness, appearance and geometric is important in properly detecting salient object.

6   Conclusions

In this study, GP has been successfully employed to automatically select and combine saliency features to produce the final saliency map. The proposed method can easily incorporate any additional features and select the features complement each other. It makes no assumption of linear superposition or equal weights of features and it does not require domain-expert. GPFSFC has the ability to tackle a wide range of saliency features from different segmentation levels and explore various mathematical expressions for the feature combination stage. The saliency features by themselves are not that accurate and sufficient to properly detect the salient object and suppress background. Therefore, a good feature selection and combination method plays an important role in achieving high performance. The quantitative and qualitative results reveal that GPFSFC can effectively choose the features which complement each other and have good effect on each other, thus, the final combination of those features results in a good saliency map. In this paper, although GPFSFC was slightly worse on the ECSSD and PASCAL datasets, it showed promising results by outperforming one of the well-know and recent CNN methods (DCNN) on two datasets, i.e., SED and ASD.

For future work, as GP showed promising results for feature selection, feature construction, and feature combination processes, it is worth studying the use of GP for automatically extracting saliency features from the row images and investigate whether GP is successful in generating new features that are as informative as the hand-crafted features. Moreover, we would like to explore GP for extracting semantic (high-level) saliency features and provide further analysis on the extracted features.

References

1. Achanta, R., Hemami, S., Estrada, F., Susstrunk, S.: Frequency-tuned salient region detection. In: Proceedings of the IEEE Conference on Computer Vision and Pattern Recognition, pp. 1597–1604. IEEE (2009)
2. Achanta, R., Süsstrunk, S.: Saliency detection using maximum symmetric surround. In: Proceedings of the 17th IEEE International Conference on Image Processing, pp. 2653–2656. IEEE (2010)
3. Afzali, S., Al-Sahaf, H., Xue, B., Hollitt, C., Zhang, M.: Foreground and background feature fusion using a convex hull based center prior for salient object detection. In: Proceedings of the 33rd International Conference on Image and Vision Computing New Zealand, pp. 1–6. Springer (2018)

4. Afzali, S., Al-Sahaf, H., Xue, B., Hollitt, C., Zhang, M.: A genetic programming approach for constructing foreground and background saliency features for salient object detection. In: Mitrovic, T., Xue, B., Li, X. (eds.) AI 2018. LNCS (LNAI), vol. 11320, pp. 209–215. Springer, Cham (2018). https://doi.org/10.1007/978-3-030-03991-2_21

5. Afzali, S., Xue, B., Al-Sahaf, H., Zhang, M.: A supervised feature weighting method for salient object detection using particle swarm optimization. In: Proceedings of the IEEE Symposium Series on Computational Intelligence, pp. 1–8. IEEE (2017)

6. Ain, Q.U., Xue, B., Al-Sahaf, H., Zhang, M.: Genetic programming for feature selection and feature construction in skin cancer image classification. In: Geng, X., Kang, B.-H. (eds.) PRICAI 2018, Part I. LNCS (LNAI), vol. 11012, pp. 732–745. Springer, Cham (2018). https://doi.org/10.1007/978-3-319-97304-3_56

7. Al-Sahaf, H., Al-Sahaf, A., Xue, B., Johnston, M., Zhang, M.: Automatically evolving rotation-invariant texture image descriptors by genetic programming. IEEE Trans. Evol. Comput. 21(1), 83–101 (2017)

8. Al-Sahaf, H., Song, A., Neshatian, K., Zhang, M.: Two-tier genetic programming: towards raw pixel-based image classification. Expert Syst. Appl. 39(16), 12291–12301 (2012)

9. Al-Sahaf, H., Xue, B., Zhang, M.: A multitree genetic programming representation for automatically evolving texture image descriptors. In: Shi, Y., et al. (eds.) SEAL 2017. LNCS, vol. 10593, pp. 499–511. Springer, Cham (2017). https://doi.org/10.1007/978-3-319-68759-9_41

10. Al-Sahaf, H., Zhang, M., Al-Sahaf, A., Johnston, M.: Keypoints detection and feature extraction: a dynamic genetic programming approach for evolving rotation-invariant texture image descriptors. IEEE Trans. Evol. Comput. 21(6), 825–844 (2017)

11. Borji, A., Cheng, M.M., Jiang, H., Li, J.: Salient object detection: a benchmark. IEEE Trans. Image Process. 24(12), 5706–5722 (2015)

12. Borji, A., Cheng, M., Jiang, H., Li, J.: Salient object detection: a survey. CoRR abs/1411.5878 (2014)

13. Felzenszwalb, P.F., Huttenlocher, D.P.: Efficient graph-based image segmentation. Int. J. Comput. Vis. 59(2), 167–181 (2004)

14. Heikkilä, M., Pietikäinen, M., Schmid, C.: Description of interest regions with local binary patterns. Pattern Recogn. 42(3), 425–436 (2009)

15. Hou, Q., Cheng, M.M., Hu, X., Borji, A., Tu, Z., Torr, P.: Deeply supervised salient object detection with short connections. In: Proceedings of the IEEE Conference on Computer Vision and Pattern Recognition, pp. 5300–5309. IEEE (2017)

16. Itti, L., Koch, C., Niebur, E.: A model of saliency-based visual attention for rapid scene analysis. IEEE Trans. Pattern Anal. Mach. Intell. 20(11), 1254–1259 (1998)

17. Jiang, H., Wang, J., Yuan, Z., Wu, Y., Zheng, N., Li, S.: Salient object detection: a discriminative regional feature integration approach. In: Proceedings of the IEEE Conference on Computer Vision and Pattern Recognition, pp. 2083–2090 (2013)

18. Koza, J.R.: Genetic Programming: On the Programming of Computers by Means of Natural Selection, vol. 1. MIT press, Cambridge (1992)

19. Lavrenko, V., Allan, J., DeGuzman, E., LaFlamme, D., Pollard, V., Thomas, S.: Relevance models for topic detection and tracking. In: Proceedings of the Second International Conference on Human Language Technology Research, pp. 115–121. Morgan Kaufmann Publishers Inc. (2002)

20. Lee, G., Tai, Y.W., Kim, J.: Deep saliency with encoded low level distance map and high level features. In: Proceedings of the IEEE Conference on Computer Vision and Pattern Recognition, pp. 660–668 (2016)

21. Lensen, A., Al-Sahaf, H., Zhang, M., Xue, B.: Genetic programming for region detection, feature extraction, feature construction and classification in image data. In: Heywood, M.I., McDermott, J., Castelli, M., Costa, E., Sim, K. (eds.) EuroGP 2016. LNCS, vol. 9594, pp. 51–67. Springer, Cham (2016). https://doi.org/10.1007/978-3-319-30668-1_4

22. Lin, M., Zhang, C., Chen, Z.: Predicting salient object via multi-level features. Neurocomputing **205**, 301–310 (2016)

23. Liu, T., et al.: Learning to detect a salient object. IEEE Trans. Pattern Anal. Mach. Intell. **33**(2), 353–367 (2011)

24. Luo, Z., Mishra, A.K., Achkar, A., Eichel, J.A., Li, S., Jodoin, P.M.: Non-local deep features for salient object detection. In: Proceedings of the IEEE Conference on Computer Vision and Pattern Recognition. vol. 2, p. 7 (2017)

25. Perazzi, F., Krähenbühl, P., Pritch, Y., Hornung, A.: Saliency filters: contrast based filtering for salient region detection. In: Proceedings of the IEEE Conference on Computer Vision and Pattern Recognition, pp. 733–740. IEEE (2012)

26. Song, H., Liu, Z., Du, H., Sun, G., Le Meur, O., Ren, T.: Depth-aware salient object detection and segmentation via multiscale discriminative saliency fusion and bootstrap learning. IEEE Trans. Image Process. **26**(9), 4204–4216 (2017)

27. Yang, C., Zhang, L., Lu, H., Ruan, X., Yang, M.H.: Saliency detection via graph-based manifold ranking. In: Proceedings of the IEEE Conference on Computer Vision and Pattern Recognition, pp. 3166–3173. IEEE (2013)

28. Zhang, P., Wang, D., Lu, H., Wang, H., Ruan, X.: Amulet: aggregating multi-level convolutional features for salient object detection. In: Proceedings of the IEEE International Conference on Computer Vision, pp. 202–211 (2017)

29. Zhou, L., Yang, Z., Yuan, Q., Zhou, Z., Hu, D.: Salient region detection via integrating diffusion-based compactness and local contrast. IEEE Trans. Image Process. **24**(11), 3308–3320 (2015)

30. Zhu, W., Liang, S., Wei, Y., Sun, J.: Saliency optimization from robust background detection. In: Proceedings of the IEEE Conference on Computer Vision and Pattern Recognition, pp. 2814–2821. IEEE (2014)

Variable-Length Representation for EC-Based Feature Selection in High-Dimensional Data

N. D. Cilia, C. De Stefano, F. Fontanella[✉], and A. Scotto di Freca

Dipartimento di Ingegneria Elettrica e dell'Informazione (DIEI),
Università di Cassino e del Lazio meridionale,
Via G. Di Biasio, 43, 03043, Cassino, FR, Italy
{nicoledalia.cilia,destefano,fontanella,a.scotto}@unicas.it

Abstract. Feature selection is a challenging problem, especially when hundreds or thousands of features are involved. Evolutionary Computation based techniques and in particular genetic algorithms, because of their ability to explore large and complex search spaces, have proven to be effective in solving such kind of problems. Though genetic algorithms binary strings provide a natural way to represent feature subsets, several different representation schemes have been proposed to improve the performance, with most of them needing to a priori set the number of features. In this paper, we propose a novel variable length representation, in which feature subsets are represented by lists of integers. We also devised a crossover operator to cope with the variable length representation. The proposed approach has been tested on several datasets and the results compared with those achieved by a standard genetic algorithm. Results of comparisons demonstrated the effectiveness of the proposed approach in improving the performance obtainable with a standard genetic algorithm when thousand of features are involved.

Keywords: Feature selection · Evolutionary algorithms ·
Variable length representation

1 Introduction

In the last few years, machine learning applications have seen a strong growth of data represented by a large number of features, with some, or even most of them, be redundant or irrelevant. The presence of these features may be harmful to the classifier training process of any machine learning application [1–3]. In fact, this kind of features, typically slow down the learning process, makes the learnt model more complex, and, eventually, cause a deterioration of the classification performance. In order to identify and eliminate these features, it is possible to use feature selection techniques. These techniques aim to find the smallest subset of features providing the most discriminative power and imply the definition of a search procedure and an evaluation function. The first, to effectively explore

© Springer Nature Switzerland AG 2019
P. Kaufmann and P. A. Castillo (Eds.): EvoApplications 2019, LNCS 11454, pp. 325–340, 2019.
https://doi.org/10.1007/978-3-030-16692-2_22

the search space made of all the possible subsets of the whole set of features, whereas the second implements the criteria chosen to evaluate the discriminative power of these subsets.

Evaluation functions can be divided into two broad classes *filter* and *wrapper* [4] Wrapper approaches use a classification algorithm as an evaluation function. The goodness of a given subset is measured in terms of classification performance; this leads to high computational costs when a large number of evaluations is required, especially when large datasets are involved. Filter evaluation functions, instead, are independent of any classification algorithm and, in most cases, are computationally less expensive and more general than wrapper algorithms.

As concerns the search strategies, given a measure, the optimal subset can be found by exhaustively evaluating all the possible solutions. Unfortunately, this strategy is impracticable in most situations, because the total number of possible solutions is 2^N, for a dataset with N features. For this reason, many search techniques have been applied to feature selection, such as complete search, greedy search and heuristic search [4,5]. Since these algorithms do not take into account complex interactions among the features, most of these methods suffer from stagnation in local optima and/or high computational cost.

Evolutionary computation (EC) techniques have recently received much attention from the feature selection community as they are well-known for their global search ability [6–8]. Moreover, EC techniques do not need domain knowledge and do not make any assumptions about the search space, such as whether it is linearly or non-linearly separable, and differentiable [9–11]. Among the EC based approaches, Genetic Algorithms (GAs) have been widely used. GA binary vectors provide a natural and straightforward representation for feature subsets: the value 1 or 0 of the chromosome i-th element indicates whether the i-th feature is included or not (see Fig. 1). This allows GA-based algorithms to be used for feature selection problem without any modification [12–14]. However, the chromosome length is equal to the number of available features. As consequence, when this number grows, the GA evolution becomes inefficient. To cope with this problem, have been proposed several enhancements. In [15], the authors proposed a binary vector to represent each chromosome, where a predefined small number of binary bits n_b are converted to an integer number i, indicating that the i-th feature is selected. Therefore, the length of the genotype was determined by multiplying n_b and the desired number of features. This representation reduced the dimensionality of the GA search space, which resulted in better performance than the traditional representation on high-dimensional datasets. Chen et al. [16] also developed a binary representation, which included two parts, wherein the first part was converted to an integer representing the number of features to be selected while the second showed which features were selected. Jeong et al. [17] proposed a new representation to further reduce the dimensionality, where the length of the chromosome was equal to the number of desired features. The values in chromosomes indicated the indices of the included features. When the index of a feature(s) appeared multiple times, a partial SFFS

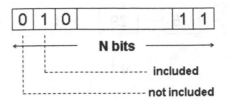

Fig. 1. The layout of the proposed system.

operator was applied to choose alternative features to avoid duplication. The main limitation of the just mentioned approaches is that the number of features needs to be predefined, which might not be the optimal size. To address this limitation, Yahya et al. [18] proposed a variable length representation, whereby each chromosome showed the selected features only and different chromosomes may have different lengths. New genetic operators were accordingly developed to cope with the variable length representation. However, the performance of the proposed algorithm was only tested on a single problem, for cues phrase selection and neither compared with other GAs-based methods.

In this paper, we present a novel variable-length representation scheme for encoding subsets of features and a set of genetic operators. In particular, a feature subset is represented by an ordered list of integers, where the value contained in the i-th element of the list represents the index of the i-th feature of the subset and the length of the list coincide with the cardinality of the subset. As concerns the genetic operators, we defined a crossover operator that implements the recombination of the parent chromosomes and that can produce offspring with a length different from that of the parents. We also defined two mutation operators. The first has been devised in such a way that it always produces a valid offspring, i.e. its chromosome does not contain any double feature index. The second mutation operator, instead, is simpler than the first one and may produce invalid offspring that needs to be repaired, by removing any possible double index. As for the fitness function, we adopted a filter one which uses correlation to evaluate feature subset quality [19]. The effectiveness of the proposed approach has been tested on six different datasets publicly available, whose total number of features ranges from thirty to about two thousand. In a first set of experiments, we compared the two mutation operators and found that the simpler one, performs better. Then, we compared the performance of our approach with that achieved by a standard GA. In this case, the experimental results proved, as expected, that when the number of features is low (less than one thousand) the two algorithms have similar performance, whereas when the number of features is higher, the proposed approach outperforms the standard GA. Finally, we also compared the results achieved by our approach with those of two state-of-the-art algorithms for feature selection. The comparison was performed in terms of the classification performance, obtained by using decision tree classifiers. Also in this case the results confirmed the effectiveness of the proposed approach when a huge number of features is involved.

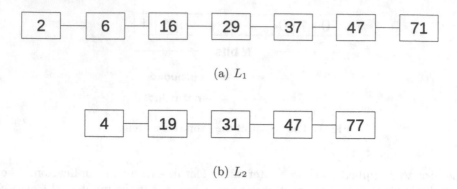

(a) L_1

(b) L_2

Fig. 2. Two examples of the representation scheme adopted, with $N = 100$. (a) the list L_1 represents the subset of features $S_1 = \{2, 6, 16, 29, 37, 47, 71\}$. (b) the list L_2 represents the subset of features $S_2 = \{4, 19, 31, 47, 77\}$

The remainder of the paper is organized as follows: in Sect. 2 the representation scheme proposed and the related operators are described, Sect. 3 illustrates the evolutionary algorithm implemented, including the fitness function. In Sect. 4 the experimental results are detailed. Finally, Sect. 5 is devoted to the conclusions.

2   The Representation Scheme and the Operators

As mentioned in the Introduction, GA binary vectors provide a natural representation for the feature selection problem, where the chromosome length is equal to the number N of available features and the cardinality of the subset represented is given by the number of 1's in the chromosome. On the other hand, the choice to represent also the information about the missing features (the 0's in the vector) makes the evolution inefficient when N is huge. Thus, it is needed a more efficient representation scheme, in which only the features actually included in the solution (subset) are represented. For this reason, we have chosen a direct representation scheme, in which a feature subset is encoded by a sorted list of integers and the length of the list represents the cardinality of the subset encoded. It is worth noting that, though the ordering may require repairing the chromosome after the application of the operators, it allows the implementation of more effective operators. As concerns the length of the list, it can be modified by the application of the crossover operator.

Let N be the number of available features and S a subset of features of cardinality N_S: $S = \{f_1, f_2, \ldots, f_{N_S}\}$, with $f_i < f_{i+1}$ and $1 \leq f_i \geq f_{i+1}$. In our approach S is represented by an ordered list of integers L, where the i-th element of the list contains the value f_i and the cardinality of S is intrinsically represented by the length of L. Two examples of representation are shown in Fig. 2.

2.1   Crossover

The crossover operator allows us to generate two new individuals by swapping parts of the lists (chromosomes) of the parents given in input to the operator. Let us assume that two lists L_1 and L_2, having length N_1 and N_2 respectively, have been selected. Without loosing generality, let us also assume that $N_1 \leq N_2$. The crossover operator produces two new lists M_1 and M_2, whose size vary in the interval $[N_1, N_2]$, by carrying out the following steps:

1. Choose randomly a number t_1 in the interval $[1, N_1 - 1]$. Split L_1 into two sublists L_1' and L_1'';
2. Choose randomly a number t_2 in the interval $[t_1, t_1 + (N_2 - N_1)]$. Split L_2 into two sublists L_2' and L_2'';
3. Set $M_1 = L_1'$;
4. Set $M_2 = L_2'$;
5. Append L_2'' to M_1;
6. Append L_1'' to M_2;
7. Reorder M_1 and M_2;
8. Remove double indices from M_1 and M_2;

The constraint on the value of t_2 has been introduced to avoid the generation of offspring exhibiting very different lengths between them. Such constraint forces crossover to generate new individuals whose length is included in the interval $[N_1, N_2]$. Nonetheless, in order to improve the exploration ability of the evolutionary algorithm, we have relaxed the above constraint by extending the range in which the length of the offspring may vary. In particular, we have introduced a tolerance θ such that the value of t_2 may be chosen in the interval $[t_1 - \theta, t_1 + (N_2 - N_1) + \theta]$, so allowing the generation of individuals whose minimum and maximum size respectively is $N_1 - \theta$ and $N_2 + \theta$. In the experiments reported in Sect. 4, the value of the tolerance θ has been fixed to 1. An example of application of the crossover operator is shown in Fig. 3.

2.2   The Mutation Operators

We have defined two mutation operators, named *mut1* and *mut2*. Both operators scan the list making up the chromosome and modify its i–th element e_i according to a given probability p_m. The operator mut1, always generates valid offspring, because it replaces the value contained in e_i with one included in the interval delimited by its previous and subsequent elements. The operator mut2, instead, is simpler and may require the reordering of the list or the removal of double indices.

Given a list L of size n_l, the operator mut1 modifies L through the following steps:

if *flip*(p_m) **then**
   Choose randomly a value v in the interval $[1, e_i[$;
   Substitute e_1 with v;

end if {scan the list until element}
for $i \leftarrow 2$ to $(n_l - 1)$ **do**
  if $flip(p_m)$ **then**
    Choose randomly a value v in the interval $]e_{(i-1)}, e_{(i+1)}[$;
    Substitute e_i with v;
  end if
  if $flip(p_m)$ **then**
    Choose randomly a value v in the interval $[e_i, N[$;
    Substitute e_{n_l} with v;
  end if
end for

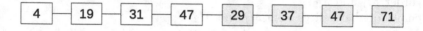

(a) The final M_1 offspring.

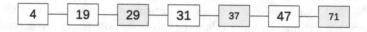

(b) M_2 before the reordering and removal of double indices.

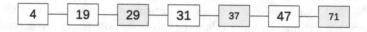

(c) The final M_2 offspring.

Fig. 3. An example of application of the crossover operator to the chromosomes shown in Fig. 2. To highlight the swapped parts of L_1 and L_2, the elements of L_1 are shaded. Note that M_1 does need any reordering and/or removal of double indices, whereas M_2 must be reordered and the double index 47 removed.

Note that in the procedure above, the first and last element are processed separately because they don't have respectively the previous and subsequent element. Moreover, since the mutation interval of each element is delimited by the values of its previous and subsequent element, each new mutated element does not alter the list ordering and cannot be already present in the list.

As concerns the second mutation operator mut2, given a list L of size n_l, it modifies L through the following steps:

for $i \leftarrow 1$ to n_l **do**
  if $flip(p_m)$ **then**
    Choose randomly a value v in the interval $[1, N]$;
    Substitute e_i with v;
  end if
end for
Reorder L ;
Remove double indices from L;

3   The Evolutionary Algorithm

The system presented here has been implemented by using a generational evolutionary algorithm. The algorithm starts by generating a population of P individuals, randomly generated. The procedure for the random generation of these individuals takes in input three terms: the number of available features N and the minimum and maximum chromosome length l_{min} and l_{max}, with the last two values expressed as a fraction of N. The procedure starts randomly generating a value l in the interval $[l_{min}, l_{max}]$. Then l values are randomly picked up from the set $\{1, \ldots, N\}$, without replacement, and added to the list making up the chromosome of the individual. Finally, the list is sorted, in ascending order. In the experiments reported in Sect. 4, the values of l_{min} and l_{max} have been set to 0.1 and 0.5, respectively.

Afterwards, the fitness of the generated individuals is evaluated according to the formula in (4). After this preliminary evaluation phase, a new population is generated by selecting $P/2$ couples of individuals using the tournament method. Then the genetic operators detailed in the previous section are applied. First, the one point crossover operator is applied to each of the selected couples, according to a given probability factor p_c. Afterwards, one the two mutation operators is applied with a probability p_m. The value of p_m has been set to $1/l_c$, where l_c is the chromosome length. This probability value allows, on average, the modification of only one chromosome element. This value has been suggested in [20] as the optimal mutation rate below the error threshold of replication. Finally, these individuals are added to the new population. The process just described is repeated for N_g generations. The basic outline of the second module algorithm is schematically illustrated in Fig. 4.

3.1   The Fitness Function

In order to introduce the subset evaluation function adopted, let us briefly recall the well known information-theory concept of entropy. Given a discrete variable X, which can assume the values $\{x_1, x_2, \ldots, x_n\}$, its entropy $H(X)$ is defined as:

$$H(X) = -\sum_{i=1}^{n} p(x_i) \log_2 p(x_i) \tag{1}$$

where $p(x_i)$ is the probability mass function of the value x_i. The quantity $H(X)$ represents an estimate of the uncertainty of the random variable X. The concept of entropy can be used to define the conditional entropy of two random variables X and Y taking the values x_i and y_j respectively, as:

$$H(X|Y) = -\sum_{i,j} p(x_i, y_j) \log \frac{p(y_j)}{p(x_i, y_j)} \tag{2}$$

where $p(x_i, y_j)$ is the joint probability that at same time $X = x_i$ and $Y = y_j$. The quantity in (2) represents the amount of randomness in the random variable X when the value of Y is known.

As fitness function, we chose a filter one, called CFS (Correlation-based Feature Selection) [19], which uses a correlation based heuristic to evaluate feature subset quality. Given two features X and Y, their correlation r_{XY} is computed as follows[1]:

$$r_{XY} = 2.0 \cdot \frac{H(X) + H(Y) - H(X, Y)}{H(X) + H(Y)} \tag{3}$$

Where the symbol H denotes the entropy function. The function r_{XY} takes into account the usefulness of the single features for predicting class labels along with the level of inter-correlation among them. The idea behind this approach is that good subsets contain features highly correlated with the class and uncorrelated with each other.

Given a feature selection problem in which the patterns are represented by means of a set Y of N features, the CFS function computes the merit of the generic subset $X \subseteq Y$, made of k features, as follows:

$$f_{CFS}(X) = \frac{k \cdot \overline{r_{cf}}}{\sqrt{k + k \cdot (k-1) \cdot \overline{r_{ff}}}} \tag{4}$$

where $\overline{r_{cf}}$ is the average feature-class correlation, and $\overline{r_{ff}}$ is the average feature-feature correlation. Note that the numerator estimates the discriminative power of the features in X, whereas the denominator assesses the redundancy among them. The CFS function allows irrelevant and redundant features to be discarded. The former because they are poor in discriminating the different classes at the hand; the latter because they are highly correlated with one or more of the other features. In contrast to previously presented approaches [21,22], this fitness function is able to automatically find the number of features and does not need the setting of any parameter.

Table 1. The values of the parameters used in the experiments. Note that p_m depends on the chromosome length l_c.

Parameter	Symbol	Value
Population size	$\mathcal{P}$	100
Crossover probability	p_c	0.6
Mutation probability	p_m	$1/l_c$
Number of generations	N_g	1000

4   Experimental Results

The proposed approach was tested on six, publicly available, datasets: *Colon, Musk, Ovarian, Satd, Spam* and *Ucihar*. The characteristics of the datasets are

---

[1] Note that the same holds also for the feature-class correlation.

```
begin
    randomly initialize a population of P individuals;
    evaluate the fitness of each individual;
    g = 0;
    while g < N_g do
        for i = 0 to P/2 do
            select a couple of individuals;
            replicate the selected individuals;
            if flip(p_c) then
                apply the crossover operator;
            end if
            perform the mutation on the offspring;
        end for
        evaluate the fitness of each individual;
        replace the old population with the new one;
        g = g + 1;
    end while
end
```

Fig. 4. The outline of the evolutionary algorithm implemented. The function $\text{flip}(p)$ returns the value 1 with probability p and the value 0 with a probability $(1 - p)$.

Table 2. The datasets used in the experiments.

Datasets	Attributes	Samples	Classes
Colon	2000	62	2
Musk	166	476	14
Ovarian	2191	216	2
Satd	36	6435	6
Spam	57	3037	2
Ucihar	561	2948	6

summarized in Table 2. They present different characteristics as regards the number of attributes, the number of classes (two or multiple classes problems) and the number of samples. For each dataset, we performed thirty runs. At the end of every run, the feature subset encoded by the individual with the best fitness was stored as the solution provided by that run. As for the parameters of the evolutionary algorithm outlined in Fig. 4, we performed some preliminary trials to set them. These parameters were used for all the experiments described below and are reported in Table 1.

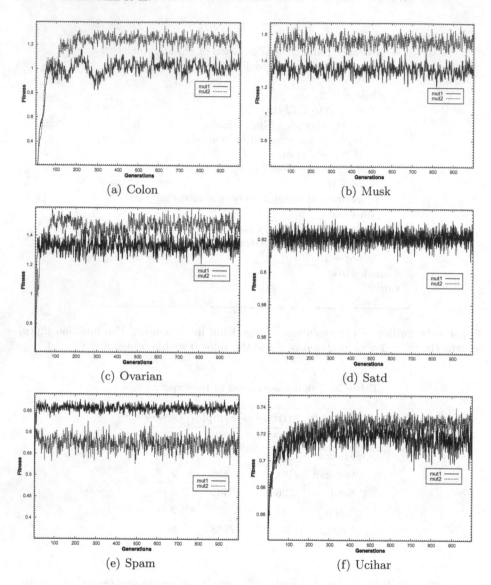

Fig. 5. Comparison results between the mut1 and mut2 operators.

In order to test the effectiveness of the proposed approach, we performed several experiments. In the first set of experiments, we compared the performance of the two mutation operators devised. The comparison was performed in terms of the population average fitness during the run that achieved the best fitness. The results are shown in Fig. 5. From the plots, it can be seen that, except for Spam, the mutation operator mut2 performs better than mut1. In fact, the average fitness of the population is higher when is applied the mut2 operator.

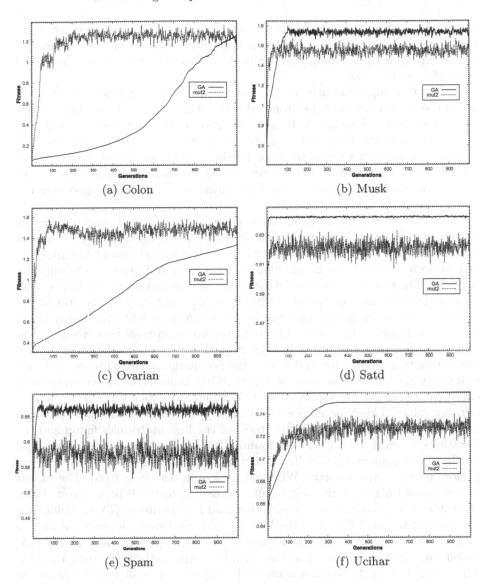

Fig. 6. Comparison results between the GA and mut2 operator.

This suggests that mut2 has a better exploration ability than mut1. This is due to the fact that, given a list L (chromosome) of length l_c representing the subset $S = \{f_1, f_2, \ldots, f_{l_c}\}$, mut2 can change L by replacing each element e_i of L by a value in the interval $[1, N]$, whereas mut1 can replace each e_i only with a value in the interval $[e_{i-1}, e_{i+1}]$. This choice, although allowing the generation of offspring always valid, seems to limit the exploration ability of the operator. Moreover, as concerns Ucihar, even if the differences are less evident, they tend to increase in the final part of the evolution. Finally, for Satd, instead, there is no performance difference. Most probably, this is due to the low number of feature involved (fifty-seven), that limits the dimension of the search space and make the search "easier".

After the comparison between the two mutation operators devised, we compared the performances of the proposed approach, by using the operator mut2, with those achieved by a GA. In order to have a fair comparison, it was performed by using the same fitness function and the same set of parameters. The comparison was performed in terms of: (i) the population average fitness during the run that achieved the best fitness, among the thirty runs performed (see Fig. 6); (ii) the fitness values of the best individuals found, averaged over the thirty runs performed (see Table 3).

Note that in the plots in Fig. 6, the curve of the proposed approach is the dotted one and is referred as "mut2" in the legend. From the figure, it can be noted that for all the datasets with less than one thousand features, the GA performs better than mut2, whereas for Colon and Ovarian the opposite happens. The plots of these two datasets show a common trend: the GA reaches average fitness values comparable with those of mut2 only at the end of the evolution. Moreover, as concerns Ovarian these values are still significantly below than those achieved by mut2. This confirms our hypothesis that when a huge number of features is involved, the GA evolution becomes inefficient. It is also interesting to note that, for all datasets, the population average fitness values for mut2 vary much more than those of the GA. This confirms the better exploration capability of mut2 with respect to the GA, which exhibits a very low variability of these values. This hypothesis is confirmed by the comparison results reported in Table 3. The Table compares the fitness of the best individuals found and the iterations at which they have been found. The reported results were computed on the thirty runs performed. To statistically validate the obtained results, we performed the non-parametric Wilcoxon rank-sum test ($\alpha = 0.05$). The fitness values in bold highlight the best results, according to the Wilcoxon test. From the table, as expected, it can be seen that mut2 outperforms GA on Colon and Ovarian, in terms of fitness of the best individuals found. However, as for the other datasets, mut2 is still able to achieve the same performances as the GA, despite its population has an average fitness significantly lower than the GA. This result is a further confirmation of the exploration ability of the proposed representation scheme.

Finally, we tested the devised approach in terms of classification performance. We used decision trees (DT in the following) as classifiers, trained by using the C4.5 algorithm. To this aim, at the end of each of the thirty runs performed, the feature subset encoded by the individual with the best fitness was used to build a DT. The classification performances of the classifiers built have been obtained by using the 10-fold cross-validation approach. The results reported in the following were obtained averaging the performance of the 30 DTs built.

Table 3. Comparison results between our approach, by using the mut2 operator, and the GA.

Dataset	Mut2				GA			
	Fitness		#iterations		Fitness		#iterations	
	avg	std	avg	std	avg	std	avg	std
Colon	**1.63**	**0.008**	601	140	1.26	0.027	997	1.2
Musk	1.94	0.001	258	120	1.94	0.004	220	150
Ovarian	**1.87**	**0.01**	49.5	23.4	1.31	0.01	990	1.3
Satd	0.645	0.00	33.10	12.10	0.645	0.00	16	3.4
Spam	0.722	0.00	10.60	3.75	0.722	0.00	86.7	62.1
Ucihar	0.751	0.0001	836	152	0.751	0.0001	613	200

Table 4. Comparison results in terms of classification performance.

Dataset	Our approach		mRMR		FCBF	
	RR	NF	RR	NF	RR	NF
Colon	**90.98**	**5.15**	75.96	22.00	89.52	18.00
Musk	78.15	6.00	79.90	7.00	80.21	11.00
Ovarian	**93.56**	**7.65**	74.25	13.00	88.42	18.00
Satd	83.95	20.00	85.04	6.00	**86.09**	**4.00**
Spam	**93.02**	**3.00**	92.05	7.00	92.19	17.00
Ucihar	89.68	52.10	69.37	23.00	**93.02**	**32.00**

We compared the achieved results with those obtained by the following two state-of-the-art algorithms for feature selection:

- *Fast Correlation-Based Filter.* This algorithm uses the concept of "predominant correlation" to select good features that are strongly relevant to the class concept and are not redundant. In order to identify non-linear correlations, the adopted measure is based on the well-known information-theoretical concept of entropy, which measures the uncertainty of a random variable. The algorithm consists essentially of two steps: in the first step the features are ranked according to their correlation to the class; in the second step, the ordered list is further processed to remove redundant features. Further details of the algorithm can be found in [23]. It will be denoted as *FCBF* in the following.
- *minimum Redundancy Maximum Relevance.* This approach finds the best feature subset made up of the features that are highly correlated with the class concept (max. relevance) and minimally correlated each other (minimum redundancy). As correlation measure, this approach uses the mutual information criterion for the discrete variables and the F-test for the continuous ones. Further details of the algorithm can be found in [24]. It will be denoted as *mRMR* in the following.

Since the above methods are deterministic, they generate a single feature subset. In order to perform a fair comparison with the proposed approach, for each dataset, 30 DT's have been learned by using the 10-fold cross-validation technique, with different initial seeds. The results reported in the following have been obtained by averaging the performance of the 30 DT's learned. Table 4 reports the obtained results in terms of recognition rate (RR) and number of features (NF). Also in this case, we performed the Wilcoxon test mentioned above to statistically validate the results. For each dataset, the recognition rate in bold highlights the best result, according to the Wilcoxon test. From the table, it can be seen that, on Colon and Ovarian, the results achieved by our approach are better than those of FCBF and mRMR, both in terms of RR and NF. Thus, this result confirms the effectiveness of the presented technique in finding good solutions in huge search spaces. Moreover, also on Spam, our approach outperforms FCBF and mRMR, by selecting only three features. This result suggests that, in some cases, even for "smaller" search spaces, the proposed approach allows finding better solutions.

5   Conclusions

We presented a novel variable-length representation to be used in EC-based algorithms for feature selection. We also presented a set of genetic operators. In the proposed approach, a feature subset is represented by an ordered list of integers. As concerns the genetic operators, we devised a crossover operator and two mutation operators. The first recombines the individuals and evolves variable-length offspring, whereas the second explore the search space made of all the possible feature subsets. As for the fitness function, we adopted a filter one.

We tested the proposed approach on six datasets and performed three sets of experiments. In the first set, we compared the performance of the two mutation operators devised, whereas in the second one we compared the performance of our approach with that of a standard GA. Finally, we compared our results with those of two state-of-the-art algorithms for feature selection, namely the Fast Correlation-Based Filter and the minimum Redundancy Maximum Relevance. The experimental results confirmed that, for problems with a huge number of features, the proposed approach outperforms both the standard GA and the two algorithms taken into account for the comparison.

Future works will investigate different feature evaluation functions, both filter and wrapper. Moreover, system performance will be evaluated also for different classification schemes.

References

1. Cordella, L.P., De Stefano, C., Fontanella, F., Scotto di Freca, A.: A weighted majority vote strategy using bayesian networks. In: Petrosino, A. (ed.) ICIAP 2013 Part II. LNCS, vol. 8157, pp. 219–228. Springer, Heidelberg (2013). https://doi.org/10.1007/978-3-642-41184-7_23

2. De Stefano, C., Fontanella, F., Folino, G., di Freca, A.S.: A Bayesian approach for combining ensembles of GP classifiers. In: Sansone, C., Kittler, J., Roli, F. (eds.) MCS 2011. LNCS, vol. 6713, pp. 26–35. Springer, Heidelberg (2011). https://doi. org/10.1007/978-3-642-21557-5_5

3. De Stefano, C., Fontanella, F., Scotto Di Freca, A.: A novel Naive Bayes voting strategy for combining classifiers. In: 2012 International Conference on Frontiers in Handwriting Recognition, pp. 467–472, September 2012

4. Dash, M., Liu, H.: Feature selection for classification. Intel. Data Anal. 1(1–4), 131–156 (1997)

5. Xue, B., Zhang, M., Browne, W.N., Yao, X.: A survey on evolutionary computation approaches to feature selection. IEEE Trans. Evol. Comput. 20(4), 606–626 (2016)

6. Bevilacqua, V., Mastronardi, G., Piscopo, G.: Evolutionary approach to inverse planning in coplanar radiotherapy. Image Vis. Comput. 25(2), 196–203 (2007)

7. Menolascina, F., Tommasi, S., Paradiso, A., Cortellino, M., Bevilacqua, V., Mastronardi, G.: Novel data mining techniques in acgh based breast cancer subtypes profiling: the biological perspective. In: 2007 IEEE Symposium on Computational Intelligence and Bioinformatics and Computational Biology, pp. 9–16, April 2007

8. Menolascina, F., et al.: Developing optimal input design strategies in cancer systems biology with applications to microfluidic device engineering. BMC Bioinform. 10(12), October 2009

9. Bevilacqua, V., Costantino, N., Dotoli, M., Falagario, M., Sciancalepore, F.: Strategic design and multi-objective optimisation of distribution networks based on genetic algorithms. Int. J. Comput. Integr. Manuf. 25(12), 1139–1150 (2012)

10. Bevilacqua, V., Pacelli, V., Saladino, S.: A novel multi objective genetic algorithm for the portfolio optimization. In: Huang, D.-S., Gan, Y., Bevilacqua, V., Figueroa, J.C. (eds.) ICIC 2011. LNCS, vol. 6838, pp. 186–193. Springer, Heidelberg (2011). https://doi.org/10.1007/978-3-642-24728-6_25

11. Bevilacqua, V., Brunetti, A., Triggiani, M., Magaletti, D., Telegrafo, M., Moschetta, M.: An optimized feed-forward artificial neural network topology to support radiologists in breast lesions classification. In: Proceedings of the 2016 on Genetic and Evolutionary Computation Conference Companion, GECCO 2016 Companion, pp. 1385–1392. ACM, New York, NY, USA (2016)

12. Cilia, N.D., De Stefano, C., Fontanella, F., Scotto di Freca, A.: A ranking-based feature selection approach for handwritten character recognition. Pattern Recogn. Lett. 121, 77–86 (2018)

13. De Stefano, C., Fontanella, F., Marrocco, C.: A GA-based feature selection algorithm for remote sensing images. In: Giacobini, M., et al. (eds.) EvoWorkshops 2008. LNCS, vol. 4974, pp. 285–294. Springer, Heidelberg (2008). https://doi.org/10.1007/978-3-540-78761-7_29

14. De Stefano, C., Fontanella, F., Marrocco, C., Scotto di Freca, A.: A GA-based feature selection approach with an application to handwritten character recognition. Pattern Recogn. Lett. 35, 130–141 (2014)

15. Hong, J.H., Cho, S.B.: Efficient huge-scale feature selection with speciated genetic algorithm. Pattern Recogn. Lett. 27(2), 143–150 (2006)

16. Chen, T.C., Hsieh, Y.C., You, P.S., Lee, Y.C.: Feature selection and classification by using grid computing based evolutionary approach for the microarray data. In: 2010 3rd International Conference on Computer Science and Information Technology, vol. 9, pp. 85–89, July 2010

17. Jeong, Y.S., Shin, K.S., Jeong, M.K.: An evolutionary algorithm with the partial sequential forward floating search mutation for large-scale feature selection problems. J. Oper. Res. Soc. 66(4), 529–538 (2015)

18. Yahya, A.A., Osman, A., Ramli, A.R., Balola, A.: Feature selection for high dimensional data: an evolutionary filter approach. J. Comput. Sci. **7**, 800–820 (2011)
19. Hall, M.A.: Correlation-based feature selection for discrete and numeric class machine learning. In: Proceedings of the Seventeenth International Conference on Machine Learning, pp. 359–366. Morgan Kaufmann Publishers Inc., San Francisco, CA, USA (2000)
20. Ochoa, G.: Error thresholds in genetic algorithms. Evol. Comput. **14**(2), 157–182 (2006)
21. Huang, J., Cai, Y., Xu, X.: A hybrid genetic algorithm for feature selection wrapper based on mutual information. Pattern Recogn. Lett. **28**(13), 1825–1844 (2007)
22. Oreski, S., Oreski, G.: Genetic algorithm-based heuristic for feature selection in credit risk assessment. Expert Syst. Appl. **41**(4, Part 2), 2052–2064 (2014)
23. Yu, L., Liu, H.: Feature selection for high-dimensional data: A fast correlation-based filter solution. In: Proceedings of the Twentieth International Conference on International Conference on Machine Learning, ICML2003, pp. 856–863. AAAI Press (2003)
24. Peng, H., Long, F., Ding, C.: Feature selection based on mutual information criteria of max-dependency, max-relevance, and min-redundancy. IEEE Trans. Pattern Anal. Mach. Intell. **27**(8), 1226–1238 (2005)

Life Sciences

A Knowledge Based Differential Evolution Algorithm for Protein Structure Prediction

Pedro H. Narloch[iD] and Márcio Dorn[✉][iD]

Institute of Informatics, Federal University of Rio Grande do Sul,
Porto Alegre, Brazil
mdorn@inf.ufrgs.br

Abstract. Three-dimensional protein structure prediction is an open-challenging problem in Structural Bioinformatics and classified as an NP-complete problem in computational complexity theory. As exact algorithms cannot solve this type of problem, metaheuristics became useful strategies to find solutions in viable computational time. In this way, we analyze four standard mutation mechanisms present in Differential Evolution algorithms using the Angle Probability List as a source of information to predict tertiary protein structures, something not explored yet with Differential Evolution. As the balance between diversification and intensification is an essential fact during the optimization process, we also analyzed how the Angle Probability List might influence the algorithm behavior, something not investigated in other algorithms. Our tests reinforce that the use of structural data is a crucial factor to reach better results. Furthermore, combining experimental data in the optimization process can help the algorithm to avoid premature convergence, maintaining population diversity during the whole process and, consequently, reaching better conformational results.

Keywords: Protein structure prediction · Differential Evolution · Structural Bioinformatics

1 Introduction

Proteins are essential molecules for every living organism due to biological functions they provide [1]. Their biological functions are directly related with a stable three-dimensional (3D) structure, called as protein's tertiary structure. As their structures being so important, if a protein folds unexpectedly, it can be harmful to the biological system. In light of these facts, the determination of the three-dimensional protein structure is vital for the understanding of how life goes on. However, the experimental determination of these structures made by Nuclear Magnetic Resonance (NMR) and X-ray crystallography are not cheap, neither straightforward. Hence, computational methods could be interesting to reduce these costs and shorten the difference between known sequences and

© Springer Nature Switzerland AG 2019
P. Kaufmann and P. A. Castillo (Eds.): EvoApplications 2019, LNCS 11454, pp. 343–359, 2019.
https://doi.org/10.1007/978-3-030-16692-2_23

already determined structures. In this way, the Protein Structure Prediction (PSP) problem became a critic and challenging problems in Structural Bioinformatics [2].

There are different manners to computationally approach the PSP problem, each one varying the degree of freedom. Due to these large number of possible conformations a protein can fold, the PSP is considered, according the computational complexity theory, an NP-complete problem [3]. Since exact algorithms are inefficient for these class of problems, bio-inspired algorithm became interesting, although they do not guarantee an optimum solution. In this way, some works have aggregated domain-based knowledge to boost metaheuristics and get better and native-like protein structures. Besides these improvements and the significant number of researches done [4], the PSP still an open, challenging problem.

Among different evolutionary algorithms, the Differential Evolution algorithm (DE) has been showing good results not only in important competitions [5], but also in PSP applications [6,7]. Due to these facts, in this paper, our objective is to analyze four classical DE mutation mechanisms with domain-based knowledge provided by an *Angle Probability List* (APL) [8], using the preference of amino acids in the population initialization procedure. Besides this source of knowledge has shown promising results in other evolutionary algorithms, it has not be used with the DE algorithm yet. Moreover, a populational diversity metric [9] is employed to verify if the APL might influence the balance between DE intensification and diversification capacities, something not observed yet (besides its importance). The next sections in this paper are organized as follows. Section 2 presents a literature review of the problem, the classical DE algorithm, and related works. The method, metrics, and tools are described in Sect. 3. In Sect. 4 the results obtained by the different approaches are discussed. Finally, the conclusions and future works are given in Sect. 5.

2 Preliminaries

2.1 Proteins, Structure and Representation

From a structural perspective, a protein is an ordered linear chain of building blocks known as amino acids. The thermodynamic hypothesis created by Anfinsen [10] states that the protein's conformation is given by the lowest free energy of the entire system. In this way, it is possible to assume that a protein folds into its tertiary structure purely by its amino acid sequence (primary structure) and environment conditions. Over the last years, different computational strategies were applied to the PSP problem in order to achieve the minimum of its free energy. As proteins are complex molecules, the computational representation of them is not a trivial task. Thus, there were proposed different manners to describe a protein and to simulate factors which contribute to the folding process. The most real computational representation includes all atoms in the system, where each atom has its atomic coordinates in a three-dimensional space. However, as this type of representation is very similar to real proteins, it becomes computationally expensive due to the significant number of atoms it can assume. A less expensive, but equally important representation, is the dihedral angles representation where each amino acid has a determined number of angles to be set. This representation overcomes the all-atom problem, and it maintains the protein characteristics.

A polypeptide (or protein) is composed by a set of amino acids chained by a chemical bond called peptide bond. All amino acids found in proteins have the same backbone structure, composed by an amino-group (N), a central carbon atom called by alpha-carbon (C_α), four hydrogens (H) and a carboxyl-group (C). What differs each amino acid is their side-chain composition which can vary in 20 different types. The peptide bond, responsible for bonding two amino acids, is formed by the C-N interaction, forming the ω angle which tends to be planar. There are two other angles which are free to rotate in the space: the ϕ angle which rotates around N-C_α and the ψ angle which rotates around the C_α-C bond, varying from $-180°$ to $+180°$. The number of side-chain angles (χ) varies according to each amino acid ranging from 0 to 4 angles. The set of consecutive torsion angles represent the internal rotations of a polypeptide main chain. A single polypeptide may contain multiple secondary structures. α-helix and β-sheet are the most stable secondary structures and they can be considered as the principal elements present in 3D structures of proteins. There is another type of regular secondary structure, known as β-turn, that does not occur so frequently as α-helices and β- strands. The β-turn structure is a set of short segments and are often connect two β-strands.

There are different functions which calculate the protein free energy according to its computational representation. The *Rosetta energy function* [11] is one well-known all-atom high-resolution strategy used in different high-performance predictors [12]. Nowadays, there are more than 18 energy terms that compose the *Rosetta energy function*, and most of them are composed by knowledge-based potentials. The final energy is the sum of all these terms organized in five classes as shown in Eq. 1.

$$E_{Rosetta} = \begin{cases} E_{physics-based} + E_{inter-electrostatic} \\ +E_{H-bonds} + E_{knowledge-based} + E_{AA} \end{cases} \quad (1)$$

where $E_{physics-based}$ calculates the 6–12 Lennard-Jones interactions and Solvatation potential approximation, $E_{inter-electrostatic}$ stands for inter-atomic electrostatic interactions and $E_{H-bonds}$ hydrogen-bond potentials. In $E_{knowledge-based}$ the terms are combined with knowledge-based potentials while the free energy of amino acids in the unfolded state is in E_{AA} term.

2.2 Angle Probability List - APL

Given the complexity related to the prediction problem, it is reasonable to enhance the search strategy with already known structural data from a determined database. As discussed in [13], residues can assume different torsion angles values accordingly to the secondary structure they might assume, thus being valuable information which should be used in order to reduce the search space and improve the search capabilities. With this in mind, an *Angle Probability List* (APL) is proposed by [8] based on the conformational preferences of amino acids according to its secondary structure using high-quality information from the Protein Data Bank (PDB) [14]. To compose the knowledge-database, a set of 11,130 structures with resolution $\leq 2.5\text{Å}$ stored in PDB until December 2014 were selected. An APL is built from a histogram matrix ($H_{aa,ss}$) of $[-180, 180] \times [-180, 180]$ for each amino acid residue (aa) and secondary structure (ss). It is possible to generate different combinations of amino acids considering degrees of the neighborhood from the reference one. As promising results were obtained using the APL, a web tool called NIAS[1] (*Neighbors Influence of Amino acids and Secondary structures*) is available to compute APLs [15]. Figure 1 illustrates the conformational preference of three amino acid residues in coil secondary structure: Glycine (GLY), Asparagine (ASP) and Proline (PRO).

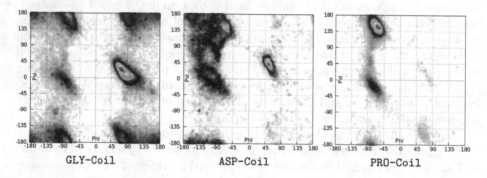

GLY-Coil   ASP-Coil   PRO-Coil

Fig. 1. Example of APL's for an amino acid sequence "GNP" with secondary structure "CCC". The dark red color marks the densest regions of the Ramachandran plot. The boldface letters represent the reference amino acids and their SS(Color figure online).

2.3   Differential Evolution

With the intention to handle nonlinear and multimodal cost functions, Storn and Price developed in 1997 one of the best optimization algorithms since then: the *Differential Evolution* (DE) [16]. The DE is a population-based evolutionary algorithm composed of four steps: initialization, mutation, crossover, and selection. Initially, a population of NP solution vectors with D dimensions is randomly generated. During the optimization, the algorithm iterates a defined number of generations over the mutation, crossover, and selection process. In the mutation step, for each target vector x_i^g, a mutant vector v_i^{g+1} is generated according to a

---

[1] http://sbcb.inf.ufrgs.br/nias.

mutation strategy. One of the initial formulations proposed by Storn and Price is called $DE/rand/1$ (Eq. 2).

$$DE/rand/1 : v_i^{g+1} = x_{r1}^g + F \cdot (x_{r2}^g - x_{r3}^g) \tag{2}$$

where g represents the generation, x_{r_n} a random solution from the current population, and $F > 0$ a parameter for scaling the difference between vectors. All selected vectors are mutually exclusive and different from the target vector.

During the mutation process, a crossover mechanism selects dimensions from the mutant vector which are mixed by the employed mutation strategy, creating a trial vector u_i^g. Generally, DE applications use a binomial crossover scheme. In this case, the dimension is mutated whenever a randomly generated number is less than the crossover rate (CR) parameter. The binomial crossover scheme is expressed by Eq. 3.

$$u_{i,d}^g = \begin{cases} v_{i,d}^g & \text{if } d = d_{rand} \text{ or rand } [0,1] \leq CR, \\ x_{i,d}^g & \text{otherwise} \end{cases} \tag{3}$$

where d_{rand} is any random dimension to guarantee at least one modification and $rand$ an uniform random number between 0 and 1.

After mutation and crossover stages, the trial vector is passed by a score function to evaluate the new solution. In this stage, the selection mechanism act to determines whether the target or the trial vector will compose the population in the next generation (iteration). It is possible to describe the selection operator as shown in Eq. 4.

$$x_i^{g+1} = \begin{cases} u^g & \text{if } f(u^g) \leq f(x_i^g), \\ x_i^g & \text{otherwise} \end{cases} \tag{4}$$

where $f(x)$ is the score function to be minimized. In this way, the solution with the best fitness value will be part of the offspring, and consequently, part of the new population in the next generation. Algorithm 1 presents the classical Differential Evolution scheme for a minimization problem.

Besides the straightforward implementation of classical DE, as shown in Algorithm 1, with few parameters to be selected, it has been getting high ranks in different optimization competitions in a wide variety of objective functions when compared with other evolutionary computation techniques [5].

2.4   Related Works

Metaheuristics are broadly used in hard optimization problems due to their capacity to reach feasible results since exact algorithms cannot handle NP-Complete problems [17]. Some of the well-known methods are Genetic Algorithms (GA), Simulated Annealing (SA), Differential Evolution (DE), Artificial Bee Colony (ABC), Particle Swarm Optimization (PSO), Ant Colony Optimization (ACO) and others [17]. As the protein structure prediction be considered an NP-Complete problem [3], different methods were applied to predict protein tertiary structure.

Algorithm 1. Classical Differential Evolution

Data: NP, F and CR
Result: The best individual in population
Generate initial population with NP individuals
while $g \leq$ *number of generations* **do**
    for *each i individual in population* **do**
        select three random individuals (x_{r1}, x_{r2}, x_{r3})
        $d_{rand} \leftarrow$ select a random dimension to mutate
        for *each d dimension* **do**
            if $d = d_{rand}$ **or** $random \leq CR$ **then**
                | $u_{i,d} \leftarrow x_{r1,d}^{g} + F \cdot (x_{r2,d}^{g} - x_{r3,d}^{g})$
            else
                | $u_{i,d} \leftarrow x_{i,d}$
            end
        end
        if $u_{i,fitness} \leq x_{i,fitness}$ **then**
            | add u_i in the offspring
        else
            | add x_i in the offspring
        end
    end
    population $\leftarrow$ offspring
    $g \leftarrow g + 1$
end

As bio-inspired methods could be easily modified, many variations of the same base-algorithm compose different approaches to the problem. In [6] the DE algorithm was tested using two mechanisms, known as generation gap and Gaussian mutation, in order to maintain the populational diversity during the optimization process. Another version of DE is proposed in [18] where the optimization process is divided in four slices and each part a different mutation mechanism is used. A multi-objective formulation is combined with the self-adaptive DE in [19] where the energy function is divided by bonded and non-bonded terms. It is important to notice that none of these works have used APL information to enhance their search algorithms and all of them have competitive results despite the number of objectives. In [7] a self-adaptive differential evolution (SADE) is combined with a library of amino acid segments based on different motifs present in proteins. Reported results have shown that DE is a good meta-heuristic to find plausible proteins conformation and it can be enhanced with domain-based knowledge.

Another well known evolutionary algorithm used to predict tertiary structures is the GA and its different versions. In [20] a knowledge-based Genetic Algorithm was proposed with the intention to reduce the search space. The used information uses torsion angles intervals for amino acids based on previous occurrences in experimentally determined proteins, a very similar approach provided by APL. Another knowledge-based GA was proposed in the APL paper [8] and compared with a PSO algorithm using the same source of information. The results obtained by this work showed that algorithms enhanced with APL get better energy values and conformational similarity when compared with the experimentally known structure. Among different algorithms found in literature, only two of them [6,18] in some way monitored the populational diversity generated by different mechanisms. Also, is noteworthy that when methods use some problem-domain information, the search algorithm reaches better values of energy and more similar structural conformation. In this way, this paper test four different DE mutation mechanisms with APL and analyze if there is an impact on the algorithm behavior.

3   Materials and Methods

Tertiary protein structure prediction is not an easy task to do, and there are different ways to approach the problem. In light of the conclusions presented by the Anfinsen's thermodynamic hypothesis [10], it is possible to declare that proteins reach stability with the minimum free energy. Considering the challenges presented by the problem, there are three essential components to create a PSP solver: (a) a way to computationally represents the protein structure; (b) an energy function to evaluate the protein, and (c) an algorithm to explore the search space in order to find a solution with the minimum possible energy [4].

Scoring Function:
In order to create a solver for the PSP problem, we used a *Python-based* interface [21] to interact with a state-of-art molecular modeling suite known as *Rosetta* [11]. With this interface, it is possible to calculate the free energy of each possible protein using the *score3 energy function*[2]. The *score3 energy function* uses a centroid-based representation for the side chains of each amino acid, reducing not only the computational cost for energy calculation but also the representation vector used in the search algorithm. Hence, the problem dimensionality is 2N, where N is the length of the primary structure. Moreover, a new term is added to the fitness function in order to benefit well-formed secondary structures. To identify the formed secondary structures, an implementation of DSSP [22] by *PyRosetta* is used during the optimization process. In this way, every time that search algorithm finds a solution which the secondary structure matches with the one given as an input, a reinforcement score is assigned to the *score3* value. On the other hand, if the perturbation made by the algorithm does not correctly find the secondary structure, a punishment is ascribed to the solution. Equation 5 formulates the fitness function used in this work.

$$E_{total} = E_{score3} + E_{SecondaryStructure} \qquad (5)$$

It is important to mention that choosing an energy function to guide a search algorithm in PSP task is not a trivial effort. The rules that govern the biochemical processes and relations are only partially known, which makes it harder to design efficient computational strategies for these situations. There is not a function which correctly describes the potential energy of a real system, which implies that different energy functions could lead to different final structural results. Although the energy function is the fitness function which guides our search

---

[2]   https://www.rosettacommons.org.

algorithm, the *Root Mean Square Deviation* (RMSD) might be considered as well. The RMSD is a measurement which compares the distance (in angstroms) among atoms in two structures. In our case, we use it to compare a final solution with the already known structure. Equation 6 presents the $RMSD_\alpha$, where only C_α atoms are compared.

$$RMSD(a, b) = \sqrt{\frac{\sum_{i=1}^{n} |r_{ai} - r_{bi}|^2}{n}} \qquad (6)$$

where r_{ai} and r_{bi} are the ith atoms in a group of n atoms from structures a and b. The closer RMSD is from 0Å more similar are the structures.

Search Strategy: Over the years, different approaches were proposed to predict tertiary protein structures. However, a different approach was proposed in [8] where conformational preferences of amino acid residues can be used to improve the search algorithm, leading to better values of energy and structural quality. The *Angle Probability List* (Sect. 2.2) has been tested with a *Biased Random-Key Genetic Algorithm*, *Particle Swarm Optimization* [8], and a memetic algorithm [2]. With this in mind, our approach combines the data found with APL and four different mutation strategies in the *Differential Evolution* algorithm using the *PyRosetta score3* energy function. The mutation mechanisms are listed in Table 1. As used in other works, the information will be used only in the initialization of the population, where random individuals are generated based on the APL.

Table 1. Classical mutation strategies in DE.

Approach	Equation
$DE_{best/1/bin}$	$v_i^{g+1} = x_{best}^g + F \cdot (x_{r2}^g - x_{r3}^g)$
$DE_{rand/1/bin}$	$v_i^{g+1} = x_{r1}^g + F \cdot (x_{r2}^g - x_{r3}^g)$
$DE_{curr-to-rand}$	$v_i^{g+1} = x_i^g + F1 \cdot (x_{r1}^g - x_i^g) + F2 \cdot (x_{r2}^g - x_{r3}^g)$
$DE_{curr-to-best}$	$v_i^{g+1} = x_i^g + F1 \cdot (x_{best}^g - x_i^g) + F2 \cdot (x_{r2}^g - x_{r3}^g)$

Furthermore, a populational diversity measure (Eq. 7) is applied to verify if the usage of APL information has some impact in the diversity during the optimization process, something not analyzed in any work. This metric was proposed in [9], and it can be used in continuous-domain problems.

$$GDM = \frac{\sum_{i=1}^{N-1} ln \left(1 + \min_{j[i+1,N]} \frac{1}{D} \sqrt{\sum_{k=1}^{D} (x_{i,k} - x_{j,k})^2} \right)}{NMDF} \qquad (7)$$

where D represents the dimensionality of the solution vector, N is related to the population size and x the individual (or solution vector). The NMDF is a normalization factor which corresponds to the maximum diversity value so far.

The population diversity metric starts with the value 1, meaning maximum diversity and as this index tends to 0 means that individuals are getting closer (without considering the fitness function). This type of measurement is helpful in the understanding the algorithm convergence behavior. It also helps to identify if the algorithm is getting trapped in local optima, which leads to premature convergence.

4   Experiments and Analysis

In order to compare the four different mutation strategies and APL usage, we have chosen ten proteins to compose our experiment. In Table 2 all proteins are organized by alphabetical order with their size (amino acid quantity) and secondary structure types. These proteins were selected based on literature works [7,8]. For our simulation we used the same DE parameters (listed in Table 3) based on literature works [6,18,19] with 1 million of fitness evaluations which corresponds to 10 thousand generations.

Table 2. Target protein sequences.

PDB ID	Size	Secondary Structure Content
1AB1	46	One sheet/Two helices
1ACW	29	One sheet/Two helices
1CRN	46	One sheet/Two helices
1ENH	54	Three helices
1ROP	63	Two helices
1UTG	70	Five helices
1ZDD	35	Two helices
2MR9	44	Three helices
2MTW	20	One helix
2P81	44	Two helices

Table 3. DE parameters.

Parameter	Value	Description
NP	100	Population size
CR	1	Crossover factor
F	0.5	Mutation factor

With the intention to keep a fair comparison among different versions, each mutation mechanism started with the same initial population. In this way, it is possible to ensure that none of the mechanism was benefited by randomness when created the initial population. For each protein, in each mutation mechanism, thirty experiments were done in the same environment. The results obtained are listed in Table 4, organized by protein and mutation mechanism.

Adopting the Angle Probability List: Analyzing the general energy results listed in Table 4, one can observe that in all predicted proteins, the lowest energy found is always obtained by a mechanism that used the APL as a source of information. Nonetheless, some solutions found by approaches that not used the APL

Table 4. Results obtained for target proteins using different mutation mechanisms.

PDB	Strategy	Energy	
		With APL	Without APL
1AB1	$DE_{rand/1/bin}$	$-98.00(-75.48 \pm 9.54)$	$-129.64(53.73 \pm 118.64)$
	$DE_{best/1/bin}$	$-152.24(-95.32 \pm 18.48)$	$-61.21(11.44 \pm 40.25)$
	$DE_{curr-to-rand}$	$-169.14(-109.07 \pm 17.29)$	$-161.30(-103.47 \pm 15.76)$
	$DE_{curr-to-best}$	$-158.14(-122.57 \pm 15.64)$	$-112.02(-54.29 \pm 29.50)$
1ACW	$DE_{rand/1/bin}$	$-148.22(-25.17 \pm 41.97)$	$-56.26(3.36 \pm 29.10)$
	$DE_{best/1/bin}$	$-133.85(-88.22 \pm 39.33)$	$-31.23(50.65 \pm 38.53)$
	$DE_{curr-to-rand}$	$-135.75(-63.13 \pm 24.92)$	$-151.10(-95.40 \pm 33.96)$
	$DE_{curr-to-best}$	$-160.84(-111.85 \pm 26.69)$	$-116.13(-32.59 \pm 50.66)$
1CRN	$DE_{rand/1/bin}$	$-95.03(-72.76 \pm 6.13)$	$-114.07(103.40 \pm 91.24)$
	$DE_{best/1/bin}$	$-136.18(-93.92 \pm 16.06)$	$-97.59(18.49 \pm 62.02)$
	$DE_{curr-to-rand}$	$-188.41(-113.55 \pm 23.59)$	$-145.69(-103.30 \pm 15.22)$
	$DE_{curr-to-best}$	$-173.95(-129.20 \pm 23.17)$	$-119.92(-58.94 \pm 25.79)$
1ENH	$DE_{rand/1/bin}$	$-343.13(-334.83 \pm 3.08)$	$-303.81(110.16 \pm 191.09)$
	$DE_{best/1/bin}$	$-364.38(-348.84 \pm 7.92)$	$-170.09(-65.82 \pm 61.22)$
	$DE_{curr-to-rand}$	$-376.11(-363.21 \pm 10.90)$	$-330.66(-275.93 \pm 32.43)$
	$DE_{curr-to-best}$	$-368.94(-359.37 \pm 5.06)$	$-263.78(-185.41 \pm 31.73)$
1ROP	$DE_{rand/1/bin}$	$-498.18(-485.32 \pm 6.59)$	$-290.06(122.87 \pm 237.18)$
	$DE_{best/1/bin}$	$-471.52(-458.66 \pm 6.13)$	$-224.12(-46.93 \pm 102.87)$
	$DE_{curr-to-rand}$	$-484.88(-475.80 \pm 3.14)$	$-415.73(-331.45 \pm 36.06)$
	$DE_{curr-to-best}$	$-477.11(-468.65 \pm 4.64)$	$-308.32(-195.35 \pm 64.94)$
1UTG	$DE_{rand/1/bin}$	$-514.55(-487.69 \pm 10.24)$	$-406.34(276.78 \pm 248.75)$
	$DE_{best/1/bin}$	$-516.13(-497.01 \pm 9.29)$	$-208.48(4.16 \pm 81.38)$
	$DE_{curr-to-rand}$	$-545.70(-533.13 \pm 8.03)$	$-381.62(-299.04 \pm 53.70)$
	$DE_{curr-to-best}$	$-536.09(-515.88 \pm 9.49)$	$-313.49(-183.90 \pm 73.30)$
1ZDD	$DE_{rand/1/bin}$	$-233.00(-225.00 \pm 3.78)$	$-241.13(-80.29 \pm 131.59)$
	$DE_{best/1/bin}$	$-232.28(-225.54 \pm 3.66)$	$-164.58(-79.73 \pm 51.30)$
	$DE_{curr-to-rand}$	$-245.71(-236.38 \pm 4.22)$	$-231.66(-216.58 \pm 11.51)$
	$DE_{curr-to-best}$	$-240.61(-231.89 \pm 4.05)$	$-226.97(-185.55 \pm 29.04)$
2MR9	$DE_{rand/1/bin}$	$-287.20(-264.20 \pm 11.33)$	$-241.53(-22.16 \pm 166.52)$
	$DE_{best/1/bin}$	$-282.84(-270.72 \pm 6.96)$	$-153.54(-70.01 \pm 36.29)$
	$DE_{curr-to-rand}$	$-296.22(-289.38 \pm 3.28)$	$-269.38(-230.85 \pm 16.97)$
	$DE_{curr-to-best}$	$-290.33(-283.44 \pm 4.76)$	$-218.43(-157.42 \pm 25.07)$
2MTW	$DE_{rand/1/bin}$	$-109.56(-102.87 \pm 3.45)$	$-107.41(-100.68 \pm 5.22)$
	$DE_{best/1/bin}$	$-95.02(-90.62 \pm 2.12)$	$-97.16(-65.23 \pm 20.75)$
	$DE_{curr-to-rand}$	$-104.58(-98.74 \pm 2.88)$	$-103.79(-100.00 \pm 1.99)$
	$DE_{curr-to-best}$	$-101.91(-94.70 \pm 2.53)$	$-100.69(-92.48 \pm 8.22)$
2P81	$DE_{rand/1/bin}$	$-249.80(-236.02 \pm 5.37)$	$-260.32(22.46 \pm 129.73)$
	$DE_{best/1/bin}$	$-252.24(-242.28 \pm 4.69)$	$-164.87(-92.15 \pm 45.06)$
	$DE_{curr-to-rand}$	$-266.89(-252.41 \pm 10.84)$	$-251.31(-227.03 \pm 14.42)$
	$DE_{curr-to-best}$	$-257.72(-251.24 \pm 3.70)$	$-237.72(-184.42 \pm 31.72)$

could be considered outliers, e.g., the energy obtained by DE$_{rand/1/bin}$ for protein 1AB1. This emphasizes that APL information is an important factor that might be present in PSP solvers since the use of this database enhance the algorithm and help the search mechanism to find solutions with lower energy values. As shown in Fig. 2, it is possible to notice that approaches which used a more significant amount of individuals to compose the new solution have maintained higher levels of diversity during the optimization process and, consequently, better energy values ($DE_{curr-to-rand}$ and $DE_{curr-to-best}$). This behavior can be related to the better populational quality generated by APL, leading to a better combination among different individuals thus, avoiding the premature convergence. In DE$_{rand/1/bin}$ and $DE_{best/1/bin}$ this behavior was not observed, showing that using three individuals to compose a new one might be influenced by selective pressure, where local optima influenced the algorithm's evolution contributing to the premature convergence. This pattern was observed in all cases. Therefore, only one convergence comparison is plotted as an example.

Among the for methods that used APL as a source of information, the methods DE$_{curr-to-rand}$ and DE$_{curr-to-best}$ achieved better energy results. A post-hoc non parametric test (Dunn's Test) was used to verifies the null hypothesis (Table 5), comparing all four methods that used APL. The better energy values obtained can be related to the diversity maintenance during the whole optimization process as shown in Fig. 2. Moreover, presented results keep showing a decreasing tendency, even in the last generation, showing that it is possible to reach better energy values if the optimization had continued. In contrast, the $DE_{best/1/bin}$ had a premature convergence, which made the evolution impossible to happen. Finally, with the results obtained and discussed in this section, it is possible to ensure that APL information does guide the search algorithm to better results in three factors: energy, RMSD and diversity indexes (depending on the employed strategy). In general, the $DE_{curr-to-rand}$ showed to be better than other three mutation strategies, or at least equivalent, not only in energy values but also in diversity maintenance. In Table 6 the minimum energy solutions (blue) are compared with the experimental structures (red) found in PDB. It is possible to notice that similar secondary structures are found in all cases besides the alignment in some cases. Achieved results are comparable in terms of folding organization with state-of-the-art prediction methods, corroborating the effectiveness of our proposal.

Besides the DE$_{curr-to-rand}$ showed to be better, or at least equivalent, than other methods in energy terms, it is not the case when the RMSD value of the minimum energy found is compared with the e.g. DE$_{best/1/bin}$. As energy functions are approximations methods to computationally evaluate the potential energy of a protein, and the search space has multimodal characteristics, it is expected that proteins with lower energy values can have bigger RMSD results, meaning that the global optimum was not found, since the minimum potential energy might describes the native conformation of a protein (0 Angstrom in comparison with the experimental data).

Table 5. Dunn's Multiple Comparison Test for all versions which used APL as source of information. Values with values down to 0.05 means that there is statistical significance between a pair of methods.

Protein		$DE_{rand/1/bin}$	$DE_{best/1/bin}$	$DE_{curr-to-rand}$
1AB1	$DE_{best/1/bin}$	0.00	–	–
	$DE_{curr-to-rand}$	0.00	0.00	–
	$DE_{curr-to-best}$	0.00	1.00	0.00
1ACW	$DE_{best/1/bin}$	0.00	–	–
	$DE_{curr-to-rand}$	0.04	0.03	–
	$DE_{curr-to-best}$	0.00	0.00	0.00
1CRN	$DE_{best/1/bin}$	0.00	–	–
	$DE_{curr-to-rand}$	0.00	0.01	–
	$DE_{curr-to-best}$	0.00	0.00	0.14
1ENH	$DE_{best/1/bin}$	0.00	–	–
	$DE_{curr-to-rand}$	0.00	0.00	–
	$DE_{curr-to-best}$	0.00	0.00	0.71
1ROP	$DE_{best/1/bin}$	0.00	–	–
	$DE_{curr-to-rand}$	0.00	0.00	–
	$DE_{curr-to-best}$	0.00	0.00	0.00
1UTG	$DE_{best/1/bin}$	0.12	–	–
	$DE_{curr-to-rand}$	0.00	0.00	–
	$DE_{curr-to-best}$	0.00	0.00	0.00
1ZDD	$DE_{best/1/bin}$	1.00	–	–
	$DE_{curr-to-rand}$	0.00	0.00	–
	$DE_{curr-to-best}$	0.00	0.00	0.02
2MR9	$DE_{best/1/bin}$	0.78	–	–
	$DE_{curr-to-rand}$	0.00	0.00	–
	$DE_{curr-to-best}$	0.00	0.00	0.03
2MTW	$DE_{best/1/bin}$	0.00	–	–
	$DE_{curr-to-rand}$	0.40	0.00	–
	$DE_{curr-to-best}$	0.00	0.00	0.00
2P81	$DE_{best/1/bin}$	0.01	–	–
	$DE_{curr-to-rand}$	0.00	0.00	–
	$DE_{curr-to-best}$	0.00	0.00	1.00

Table 6. Cartoon representation of experimental structures (red) compared with lowest energy solutions (blue) found by each mutation mechanism.

PDB	$DE_{rand/1/bin}$	$DE_{best/1/bin}$	$DE_{curr-to-rand}$	$DE_{curr-to-best}$
1AB1	7.42Å	3.59Å	4.58Å	9.53Å
1ACW	4.87Å	5.53Å	3.72Å	3.57Å
1CRN	7.40Å	6.15Å	2.40Å	6.00Å
1ENH	6.24Å	2.94Å	4.99Å	5.86Å
1ROP	8.54Å	1.90Å	3.98Å	3.43Å

(continued)

Table 6. (continued)

PDB	$DE_{rand/1/bin}$	$DE_{best/1/bin}$	$DE_{curr-to-rand}$	$DE_{curr-to-best}$
1UTG	5.03Å	3.90Å	3.94Å	11.69Å
1ZDD	2.78Å	1.27Å	2.48Å	2.65Å
2MR9	4.11Å	3.14Å	3.27Å	3.26Å
2MTW	5.90Å	5.43Å	6.84Å	5.08Å
2P81	8.11Å	1.56Å	5.58Å	2.37Å

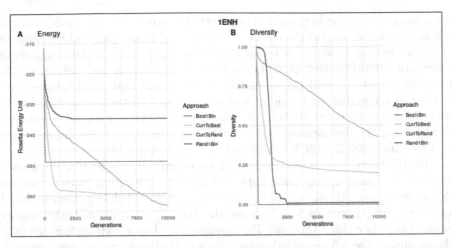

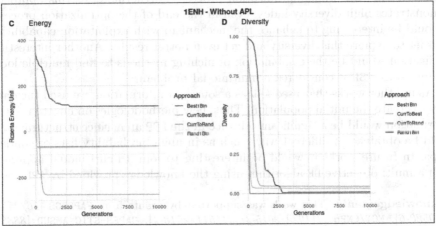

Fig. 2. PDB ID 1ENH convergence of energy and diversity for all mutation mechanisms with and without APL. **A** and **B** plots are the mean results obtained using APL whereas **C** and **D** were obtained without APL information.

5 Conclusion

Meaningful progress in protein structure prediction area has happened in the last decade. However, it is still necessary the development of methods which enhance search algorithms with well structured biological data. In this paper we have tested one valuable source of information called APL with the DE algorithm in four different mutation mechanisms, comparing the DE behavior with and without APL. In order to compare these different versions, ten proteins with different secondary structures were predicted using score3 energy function provided by *PyRosetta package*. Results showed that the combination of well-structured data with the differential evolution algorithm achieve better conformational results in comparison with the same algorithm without experimental data, even if the data is used only to create the initial population. The overall contributions of our work are the following: (a) the use of computational techniques and concepts to develop a new algorithm for a relevant biological problem; (b) the analysis of conformational preferences of amino acid residues in proteins and its use to 3D protein structure prediction methods. We observed that when we associate the type of an amino acid residue and secondary structure, it is possible to obtain valuable information about the preferences of this amino acid residues; and finally (c) the development and evaluation of different DE versions to search the three-dimensional protein conformational space using APL.

Aside the better results obtained by all approaches that used APL, the knowledge database used in this work improved the DE exploration capacity in two different versions: $DE_{curr-to-rand}$ and $DE_{curr-to-best}$, helping them to avoid premature convergence. In 8 of 10 cases, the $DE_{curr-to-rand}$ got better energy results than other three versions. This version also has shown higher diversity indexes during the whole optimization process. Furthermore, as this approach demonstrates high diversity index, even in the end of the optimization process, it would be interesting to enhance the mechanism with exploitation capabilities in order to explore this diversity and get even better results. Another interesting application would be the combination of niching methods to find multiple local minima since PSP is considered a multimodal problem.

As in other works that used APL as a source of information, we used the data only to create the initial population. Thus, new methodologies on how to use the information would be a significant advance in the PSP area, since structural data could be obtained in different ways such as in angle probability lists or contact maps. In future works it would be interesting to join a multi modal approach with a multi objective DE algorithm using the knowledge provided by APL.

Acknowledgements. This work was supported by grants from FAPERGS [16/2551-0000520-6], MCT/CNPq [311022/2015-4; 311611/2018-4], CAPES-STIC AMSUD [88887.135130/2017-01] - Brazil, Alexander von Humboldt-Stiftung (AvH) [BRA 1190826 HFST CAPES-P] - Germany. This study was financed in part by the Coordenacão de Aperfeiçoamento de Pessoal de Nível Superior - Brazil (CAPES) - Finance Code 001.

References

1. Walsh, G.: Proteins: Biochemistry and Biotechnology. Wiley, Hoboken (2014)
2. Corrêa, L.d.L, Borguesan, B., Krause, M.J., Dorn, M.: Three-dimensional protein structure prediction based on memetic algorithms. Comput. Oper. Res. **91**, 160–177 (2018)
3. Guyeux, C., Côté, N.M.L., Bahi, J.M., Bienie, W.: Is protein folding problem really a NP-complete one? First investigations. J. Bioinf. Comput. Biol. **12**, 1350017-1–1350017-24 (2014)
4. Dorn, M., E Silva, M.B., Buriol, L.S., Lamb, L.C.: Three-dimensional protein structure prediction: methods and computational strategies. Comput. Biol. Chem. **53**, 251–276 (2014)
5. Das, S., Mullick, S.S., Suganthan, P.N.: Recent advances in differential evolution-an updated survey. Swarm Evol. Comput. **27**, 1–30 (2016)

6. Narloch, P.H., Parpinelli, R.S.: Diversification strategies in differential evolution algorithm to solve the protein structure prediction problem. In: Madureira, A.M., Abraham, A., Gamboa, D., Novais, P. (eds.) ISDA 2016. AISC, vol. 557, pp. 125–134. Springer, Cham (2017). https://doi.org/10.1007/978-3-319-53480-0_13

7. Oliveira, M., Borguesan, B., Dorn, M.: SADE-SPL: a self-adapting differential evolution algorithm with a loop structure pattern library for the PSP problem. In: IEEE Congress on Evolutionary Computation, pp. 1095–1102 (2017)

8. Borguesan, B., E Silva, M.B., Grisci, B., Inostroza-Ponta, M., Dorn, M.: APL: an angle probability list to improve knowledge-based metaheuristics for the three-dimensional protein structure prediction. Comput. Biol. Chem. **59**, 142–157 (2015)

9. Corriveau, G., Guilbault, R., Tahan, A., Sabourin, R.: Review of phenotypic diversity formulations for diagnostic tool. Appl. Soft Comput. J. **13**, 9–26 (2013)

10. Anfinsen, C.B.: Principles that govern the folding of protein chains. Science **181**, 223–230 (1973)

11. Rohl, C.A., Strauss, C.E., Misura, K.M., Baker, D.: Protein structure prediction using Rosetta. Methods Enzymol. **383**, 66–93 (2004)

12. Tai, C.H., Bai, H., Taylor, T.J., Lee, B.: Assessment of template-free modeling in CASP10 and ROLL. Proteins Struct. Funct. Bioinf. **82**, 57–83 (2014)

13. Ligabue-Braun, R., Borguesan, B., Verli, H., Krause, M.J., Dorn, M.: Everyone is a protagonist: residue conformational preferences in high-resolution protein structures. J. Comput. Biol **25**, 451–465 (2017)

14. Berman, H.M., et al.: The protein data bank. Nucleic Acids Res. **28**, 235–242 (2000)

15. Borguesan, B., Inostroza-Ponta, M., Dorn, M.: NIAS-Server: neighbors influence of amino acids and secondary structures in proteins. J. Comput. Biol. **24**, 255–265 (2017)

16. Storn, R., Price, K.: Differential evolution - a simple and efficient heuristic for global optimization over continuous spaces. J. Global Optim. **11**, 341–359 (1997)

17. Du, K.-L., Swamy, M.N.S.: Search and Optimization by Metaheuristics. Techniques and Algorithms Inspired by Nature. Springer, Cham (2016). https://doi.org/10.1007/978-3-319-41192-7

18. Narloch, P., Parpinelli, R.: The protein structure prediction problem approached by a cascade differential evolution algorithm using ROSETTA. In: Proceedings-2017 Brazilian Conference on Intelligent Systems, BRACIS 2017 (2018)

19. Venske, S.M., Gonçalves, R.A., Benelli, E.M., Delgado, M.R.: ADEMO/D: an adaptive differential evolution for protein structure prediction problem. Expert Syst. Appl. **56**, 209–226 (2016)

20. Dorn, M., Inostroza-Ponta, M., Buriol, L.S., Verli, H.: A knowledge-based genetic algorithm to predict three-dimensional structures of polypeptides. In: IEEE Congress on Evolutionary Computation, pp. 1233–1240 (2013)

21. Chaudhury, S., Lyskov, S., Gray, J.J.: PyRosetta: a script-based interface for implementing molecular modeling algorithms using Rosetta. Bioinformatics **26**, 689–691 (2010)

22. Kabsch, W., Sander, C.: Dictionary of protein secondary structure: Pattern recognition of hydrogen-bonded and geometrical features. Biopolymers **22**, 2577–2637 (1983)

A Biased Random Key Genetic Algorithm with Local Search Chains for Molecular Docking

Pablo F. Leonhart[ID] and Márcio Dorn[✉][ID]

Institute of Informatics, Federal University of Rio Grande do Sul, Porto Alegre, Brazil
{pfleonhart,mdorn}@inf.ufrgs.br

Abstract. Molecular Docking is an essential tool in drug discovery. The procedure for finding the best energy affinity between ligand-receptor molecules is a computationally expensive optimization process because of the roughness of the search space and the thousands of possible conformations of ligand. In this way, besides a realistic energy function to evaluate possible solutions, a robust search method must be applied to avoid local minimums. Recently, many algorithms have been proposed to solve the docking problem, mainly based on *Evolutionary Strategies*. However, the question remained unsolved and its needed the development of new and efficient techniques. In this paper, we developed a *Biased Random Key Genetic Algorithm*, as global search procedure, hybridized with three variations of *Hill-climbing* and a *Simulated Annealing* version, as local search strategies. To evaluate the receptor-ligand binding affinity we used the *Rosetta* scoring function. The proposed approaches have been tested on a benchmark of protein-ligand complexes and compared to existing tools AUTODOCK VINA, DOCKTHOR, and jMETAL. A statistical test was performed on the results, and shown that the application of local search methods provides better solutions for the molecular docking problem.

Keywords: Structural Bioinformatics · Molecular Docking · Memetic Algorithms · Local Search Chains

1 Introduction

Automated prediction of molecular interactions is an essential step in rational drug design. Molecular Docking (MD) methods are used to predict the protein-ligand interactions and to aid in selecting potential molecules as part of a virtual screening of large databases [1]. A MD simulation begins with the molecules unbound, and then an algorithm tests many ligand conformations and orientations to find the best receptor-ligand binding affinity [2]. Because of the infinite number of possible conformations that a ligand can assume, considering the translation and rotation of the molecule, besides the rotation of their bonds, the MD problem has a high computational complexity [3]. MD has been classified in computational complexity as a NP-Hard problem [4].

© Springer Nature Switzerland AG 2019
P. Kaufmann and P. A. Castillo (Eds.): EvoApplications 2019, LNCS 11454, pp. 360–376, 2019.
https://doi.org/10.1007/978-3-030-16692-2_24

Some aspects must be considered in the development of a molecular docking strategy. The scoring function (i) to evaluate the binding energy interaction has to be realistic enough to assign the most favorable scores to the experimentally determined complex [5]. Also, the strategy adopted to find solutions in the conformational search space (ii) must be robust to escape from local minimums. Another aspect is the discretization of the search space (ii) that works together with the algorithm in a more realistic simulation. Nowadays, many algorithms have been developed to solve the molecular docking problem. The most common approaches involve *Genetic Algorithms* (GA) [1], *Differential Evolution* (DE) [6] and other *Evolutionary Strategies* (ES) [7]. Hybrid (or memetic) methods also have been tackled presenting local procedures such as *Simulated Annealing* (SA) in combination with global search strategies (GA, DEE, ES, etc). Although the recent advances into the development of robust MD methods, the problem remain unsolved and the development of new and robust approaches is still needed.

In this paper, we present an approach inspired in *Biased Random-Key Genetic Algorithm* (BRKGA) [8] (*as a global search strategy*) hybridized with three variations of *Hill-climbing* and *Simulated Annealing* methods (*as local search strategies*). The binding site of the receptor was discretized in sub-cubes according to the strategy originally proposed by Leonhart et al. [9]. This representation's schema allows the BRKGA to explore the receptor-ligand conformational search space better, and the local search methods perform the exploitation process with the aim to escape from local minima. The high-resolution score function of *Rosetta* [10], was used to compute the binding energy of ligand and receptor molecules. The main contributions of this work are: (i) development and assessment of a hybrid (memetic) evolutionary algorithm to determine the preferred orientation of one molecule (ligand) to a protein (receptor); (ii) evaluation of different local search strategies combined with a global search strategy for exploring the ligand-receptor conformational space. The remainder of the article is organized as follows: Sect. 2 presents some fundamental concepts of the molecular docking problem and a general overview of metaheuristics applied to the MD problem. Section 3 describes the proposed method. Section 4 shows the achieved results and the methodology adopted to evaluate the algorithms. Section 5 concludes the paper and points out directions for further research.

2   Preliminaries

In Structural Bioinformatics there is a wide range of unanswered biological questions, which can be answered neither by experiments nor by classical modeling and simulation approaches. Nature-inspired metaheuristic algorithms have become powerful and popular in many of these problems. In this paper, we investigate the combination of an algorithm inspired in *Biased Random Key Genetic Algorithm* (BRKGA) with different local search procedures for the MD problem for the Molecular Docking problem. The BRKGA brings the concept of *Genetic Algorithms* and castes to keep the population diversified and able to evolute. The local search heuristics play a significant role in the exploitation process to find good solutions. Our proposal also incorporates a discretization scheme with the aim to keep the population diverse and guarantee a better exploration of the search space by using local and global solution competition to leave local minimums. According with Brooijmans et al. [5] there are three essential aspects to consider in a docking method: (i) the representation of the structure and search space's discretization; (ii) an energy function to evaluate the conformation between molecules; and (iii) a search method to find the best binding energy of each complex. In the following, we discuss each one of these components.

Receptor-Ligand Representation: One vital factor to be considered in Molecular Docking methods is the flexibility of the receptor and ligand molecules [11]. During the docking process, the topology of these structures is shaped so that the chemical bond occurs in the most stable and possible form. In the most traditional approach, the receptor is treated as a rigid structure along the docking process. The ligand is mantained flexible due its internal dihedral angles. To represent the interaction between both molecules we must consider three aspects about the ligand conformation: (i) the translation of the body in Cartesian coordinate space; (ii) the orientation, described by four quaternion values; (iii) the ligand flexibilities, represented by the free torsion's rotation, also known as dihedral angles. These degrees represent the rotations around single bonds which are considered the freedom's level in molecules. The number of these angles varying by the size and chemical structure of the ligand. A high number of degrees of freedom significantly increases the problem complexity. In the proposed method, each solution is encoded by three values referent to the ligand translation, the four values following corresponds to the ligand orientation, and the remaining n values to the dihedral angles. The translation has the range values defined by the size of the search area, which corresponds to the active site of the receptor. In the case of the rotation and torsion angles, the representation is in radians and encoded in the range of $[-\pi, \pi]$. Figure 1 schematizes the solution encoding adopted in this work.

Scoring Function: The scoring function is important for the accuracy of a docking algorithm. Unfortunately, its complexity can largely increase the runtime. As a minimal requirement, the energy function used for docking must account for hydrophilic and hydrophobic surface complementarity. In this paper, we adopt the *Rosetta* energy function and weights established by Gray et al. [10]. The authors attempted to assemble a free energy model that can best discriminate docking decoys by capturing *van der* Waals interactions, solvation, hydrogen bonding, and electrostatics in addition to local internal energies such as torsion angle strains. Equation 1 show the score function used in this work. The binding affinity of two molecules is measured by a linear combination of an attractive *van der* Waals score (S_{atr}), a repulsive *van der* Waals score (S_{rep}), an implicit solvation score (S_{sol}), a surface area-based solvation term (S_{sasa}), a hydrogen bonding score (S_{hb}), a rotamer probability term (S_{dun}), a residue-residue pair probability term (S_{pair}), and simple electrostatic terms divided into short-range and long-range attractive and repulsive components ($S_{elec}^{sr-rep}, S_{elec}^{sr-atr}, S_{elec}^{lr-rep}, S_{elec}^{lr-atr}$). We use the

PyRosetta toolkit [12] to compute the binding affinity between a protein and a ligand molecule.

$$S = w_{atr}S_{atr} + w_{rep}S_{rep} + w_{sol}S_{sol} + w_{sasa}S_{sasa} + w_{hb}S_{hb} + w_{dun}S_{dun}$$
$$+ w_{pair}S_{pair} + w_{elec}{}^{sr-rep}S_{elec}{}^{sr-rep} + w_{elec}{}^{sr-atr}S_{elec}{}^{sr-atr} \quad (1)$$
$$+ w_{elec}{}^{lr-rep}S_{elec}{}^{lr-rep} + w_{elec}{}^{lr-atr}S_{elec}{}^{lr-atr}$$

Metaheuristics for the Molecular Docking Problem: Most of the existing challenging optimization problems in Structural Bioinformatics cannot be optimally solved by any known computational method due to the high dimensionality and complexity of the search space. To overcome these issues, metaheuristics techniques are being applied in an attempt to find near-optimal solutions to these problems. Recently, many metaheuristics have been developed for the MD problem [13,14]. The most common search techniques utilized in MD are *Genetic Algorithms* (GA) [1], *Differential Evolution* (DE) [6], and *Particle Swarm Optimization* (PSO) [7]. These methods were applied to minimize the receptor-ligand binding energy by changing the ligand's orientation. In Rosin et al. [15] the authors present a GA hybridized with a local search strategy to better explore the binding conformational search space. They adopted *Simulated Annealing* and the *Solis-Wets* as LS strategies. Tagle et al. [16] tested three variations of *Simulated Annealing* as LS strategy in a *Memetic Algorithm*. There are other studies reporting the combination of global/local search methods: [5,17,18]. In this paper, we investigate the use of different local search strategies combined with a *Biased Random Key Genetic Algorithm*.

3   The Proposed Method

In this section, we describe the proposed method for the MD problem. The search strategy based on the *Biased Random Key Genetic Algorithm* (BRKGA) [8] is explained. The approach to discretize the search space, adopting a treatment to the problem's multimodality and avoid local minimums is also described. We present the continuous local search (LS) strategies used in this work. They are composed by three variations of *Hill-climbing* [19,20], a well-known basic local search: *Best Improvement*, *First Improvement*, and *Stochastic Hill Descent*, and the *Simulated Annealing* method [21,22].

Biased Random Key Genetic Algorithm: *Genetic Algorithms* (GA) were firstly described in the 1960s [23]. GA is a metaheuristic based on a population of solutions that runs for many iterations (generations) until a defined stop criterion is satisfied. The first step is to create random individuals called chromosomes to form the population. A chromosome is composed of genes, which represents each variable coding the solution to an optimization problem. In each iteration of the algorithm, selection's operators chose individuals to suffer the crossover operation, as well as the mutation process is applied to change some individuals' gene, then the new individuals generated will compose the next generation [24].

The *Biased Random Key Genetic Algorithm* (BRKGA) combines the concepts of GA and random keys to present another way to select individuals to crossover operation and an encoding method to represent the individuals by real values $[0, 1]$ [8, 25]. Nevertheless, in our method, we directly use real-coded values to represent each gene, instead of values between 0 and 1. The population is organized in *castes* (*elite* and *non-elite*), according with the fitness of each individual. The crossover operates over an individual from each set, choosing every gene by a probabilistic parameter. This parameter is skewed to prefer more genes from the elite because they are the best solutions. Figure 1 illustrates the structure of a BRKGA. After the generation of the initial population, the value of objective function indicates the order of individuals, and the castes are defined. The process to build the next generation occurs firstly by copying the p_e chromosomes that form the *elite* caste. The mutation process that is responsible for insert new individuals in the population represents p_m from the total of individuals. The crossover step generates the remaining of new individuals. In this process, one chromosome from each caste is selected, and then a random scheme is applied to select with a probability of ρ_e the genes from the *elite* instead of genes from the *non-elite*. In this way, the size of individuals generated in crossover is equal to $p - p_e - p_m$. According to [26] these are the recommended parameter values for the BRKGA: $p = a.n$, where p is the population size, $1 <= a \in \mathbb{R}$, and n is the chromosome size; $0.10p \leq p_e \leq 0.25p$ (elite population size); $0.10p \leq p_m \leq 0.30p$ (mutant population size); and $0.5p \leq \rho_e \leq 0.80p$ (elite allele inheritance probability).

Search Space Discretization: The molecular docking problem needs a defined search space containing the binding site of the receptor. This area can be represented as a cube showed in Fig. 1. The search space is discretized dividing the area into smaller cubes [9]. The cube's volume is represented by three variables: Δx, Δy, and Δz, corresponding to each axis. With the information of central point and the volume of the area, we can create smaller cubes by dividing the search space. Equation 2 computes the size of the central cube in the search space.

$$
\begin{aligned}
V_{c1} = {} & [|p_{cx} + \frac{\Delta x}{s}| + |p_{cx} - \frac{\Delta x}{s}|] \\
& \times [|p_{cy} + \frac{\Delta y}{s}| + |p_{cy} - \frac{\Delta y}{s}|] \\
& \times [|p_{cz} + \frac{\Delta z}{s}| + |p_{cz} - \frac{\Delta z}{s}|]
\end{aligned}
\tag{2}
$$

in which s indicates how many small cubes will be divided the search space. The values of p_{cx}, p_{cy}, and p_{cz} represents the central point of the cube. From this variables is possible to define the coordinates of each small cube. In the Fig. 1 is defined $s = 6$, where 27 smaller cubes were created in the search area.

The idea of the division into smaller cubes is to explore the ligand-receptor binding search space in a better way. The initialization of the population by the BRKGA is responsible for creating individuals in every cube in a distributed way. During the algorithm iterations, the diversity is maintained by forcing some individuals to keep in their cubes. Operations like crossover, mutation and the local search applications can move an individual from one cube to another. Then the core of the algorithm needs to control when and which individual can migrate between the cubes. The number of solutions in every small area depends on the population size p and the percentual of individuals non-migrants p_{nm} in each one.

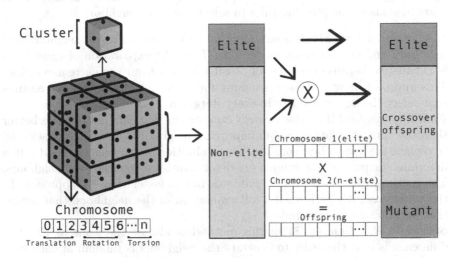

Fig. 1. Discretization of the search space, clustering, and representation of chromosomes. Also, the population scheme of BRKGA and the process of updating solutions by copying the elite set, and applying crossover and mutation operators.

This discretization approach allows the use of clustering techniques to group similar individuals (in terms of structural similarity) in a sub-cube. In this way, there is an internal competition of each solution by cube and a global competition between all individuals from search space. The crossover operation can generate an offspring from a different cube position from his parents, or the same if them belongs to the same cube. The mutation process is divided into two steps: (i) first, to fill the eventually empty cubes due to crossover or local search processes; (ii) after filling, the new individuals can be created in any one area of search space. Also, the application of local search methods can modify the cube of the solution subjected to it. The way that these local searches operate with the BRKGA in this search space discretization is described in the following.

Hill-climbing Algorithm: The *Hill Climbing algorithm* (HC), also known as descent, or iterative improvement, is an oldest and simplest local search metaheuristic [19, 20]. The procedure consists in starts with a given initial solution, and at each iteration, the method swaps the current candidate by the best neighbor found that improves the objective function's value. The search stops when all neighbors are worse than the current solution, what means that reaches a local optimum.

Some problems have representations with large neighborhoods, then to speedup the search is required to adopt a strategy to restrict the candidate solutions like a subset of the search space. This algorithm is a type of monotonous local search because it allows only modifications that improve the objective function. Thus, there are variations of the method according to the order in which the neighboring solutions are generated (deterministic or stochastic), and the selection's strategy of the solution. In our approach, we adopt three variants of HC, that are also known as pivoting rules to select the best neighbor:

- *Best improvement* (BI): is a strategy that evaluates the whole neighborhood in a fully and deterministic mode. Therefore, the exploration of candidates is exhaustive because explores all possible moves from the current solution. This approach may be time-consuming for large neighborhoods but ensures that select the best candidate in every iteration.
- *First Improvement* (FI): this strategy chooses the first candidate that is better than the current solution. So, an improving neighbor is immediately selected to replace the known best solution. The evaluation of the neighborhood is in a deterministic manner following a pre-defined order to generate the neighbors. The method visits part of all candidates but is faster than BI approach. In the worst case, it's performed a full evaluation of the neighbors, that means that no improves were found.
- *Stochastic Hill Descent* (SHD): this method is almost the same at FI. The difference is that the order to generate the neighbors is random at each iteration of the search process. Thus, the strategy ensures that all regions of a solution's neighbors have equal conditions to be selected.

Simulated Annealing (SA): Is a stochastic method that enables, in some conditions, the acceptance of worst solutions [21,22]. The aim is to avoid local optimums and then delay the search convergence. From the initial solution, the SA generates a random neighbor at each iteration. The candidate R is always accepted if better than the current solution S. But, if it's worst, there is a rule that allows the solution according to a probability:

$$\Delta E = \frac{Quality(R) - Quality(S)}{t}$$
$$P(t, R, S) = e^{\Delta E} \tag{3}$$

where $t >= 0$ and ΔE represents the quality difference between R and S. According to the current temperature and the degradation of the function objective the candidate get the probability to be accepted. If R is so much worse than S, the fraction is larger, then the probability it's almost 0. If R is close to S, then the probability is close to 1, ensuring that R could be selected. The temperature factor has an important role too. If t is close to 0, then the probability is also close to 0. Otherwise, if t is high, the probability is close to 1. Thus, the idea is that in the beginning, the method works like a random walk, accepting solutions regardless of how good it's. When t decreases, the probability also decreases and then the procedure acts like a HC algorithm.

BRKGA with LS Chains: The concept of *Local Search Chains* was proposed by [27]. The idea is that an individual utilized in a LS call may later become the initial state of a subsequent LS application. Thus, the final strategy parameter values achieved by the solution will form the initial values of the new LS call. Some particular individuals are subject to a limited amount of LS but in general, the application of this chaining process, throughout the BRKGA evolution, allows the LS operators to be extended in specific promising search zones. According to [28] two aspects must be taken into account the management of LS chains. The number of LS intensity (i) must be fixed, called LS intensity stretch (I_{str}). Thus, every LS application has the same number of function evaluations. Also, at the end of the LS operation, the parameters that guided the method to get the current state of the solution are stored (ii). So, when the individual is selected again to be improved, the initial values for the LS parameters will be loaded from the previous execution (if it exists). For instance, in the HC methods are saved the last neighbor generated in every variant, and for the FI and SHD algorithms, is also kept the previous neighborhood's order employed. However, the SA procedure has no chaining application because it's applied all I_{str} evaluations to a unique solution in each LS call. In the LS procedure, there is an important parameter that defines the neighborhood exploration, the search radius perturbation. This one indicates the value to be added or subtracted from each gene to produce the neighbors. As seen, each algorithm has a specific way to generate the neighborhood. The variations it's in the order of gene visitation and the direction, i.e., increasing or decreasing each position by the radius value. This parameter is the same for all positions of the vector, which include translation, rotation, and internal torsion angles. Thus, a complete search in the neighborhood, like performs Best Improvement search, evaluates $2 * n$ neighbors, where n is the number of genes in the individual representation.

Our approach combining BRKGA and HC methods use the concepts of LS chains. Algorithm 1 shows the pseudo-code of the proposed method. The input values P, P_e, P_m, n, ρ_e, L, I_{ls} represent the population size, the elite size, the mutant population size, the number of positions of the random key vector, the crossover probability, the number of clusters, and the minimum individuals to run a LS call, respectively. The step that defines the set of potential individuals to be improved by LS follows two rules: (i) is selected the best individuals from the most populated cubes that weren't exhaustively improved by the LS; (ii) if the size of this set is less than I_{ls}, are selected the best individuals from the whole population to fill. The output is the optimized solution X. The algorithm loops until the number of energy evaluations be satisfied.

4   Experiments and Results

The BRKGA and LS methods have been implemented in the *Python language*. They were tested with a benchmark composed of 16 complexes having receptor and ligand molecules. These structures were previously generated and classified in [29]. The selection of them was motivated because they are difficult docking problems containing a wide range of ligands sizes. The instances were obtained from the PDB [30]. The experiments have two objectives: (i) identify and evaluate

Algorithm 1. Pseudocode of the developed BRKGA-LS algorithm.

Input: P, P_e, P_m, n, ρ_e, L, I_{ls}
Output: Best Solution X

1: $P \leftarrow$ initialize with n vectors of real-coded values
2: **while** not reach the maximum of energy evaluations **do**
3:     Cluster solutions in L clusters
4:     Evaluate energy of each solution in P
5:     Divide P in P_e and $P_{\overline{e}}$
6:     Copy elite set to the next generation: $P^+ \leftarrow P_e$
7:     $P^+ \leftarrow P^+ \bigcup crossover()$
8:     Cluster P
9:     $P^+ \leftarrow P^+ \bigcup mutation()$
10: **end while**
11: **return** best solution X

Crossover operation

12: **for all** $i \leftarrow 1$ to $|P| - |P_e| - |P_m|$ **do**
13:     Select random parent a from P_e; Select random parent b from $P_{\overline{e}}$;
14:     **for all** $j \leftarrow 1$ to n **do**
15:         Randomize boolean variable B with probability of ρ of resulting **True**
16:         **if** $B == True$ **then**
17:             $c[j] \leftarrow a[j]$
18:         **else**
19:             $c[j] \leftarrow b[j]$
20:         **end if**
21:     **end for**
22:     $evaluateSolution(c)$
23: **end for**

Mutation process

24: **while** exists empty cubes **do**
25:     $evaluateSolution$(newIndividual)   ▷ According to empty cube label
26: **end while**
27: Fill the remaining set space with solutions in random cubes

---

Solution evalutation

28: $fitness = calculateEnergy$(solution)
29: **if** currentEvaluations is multiple of **LS** intensity **then**
30:     $prepareLocalSearch()$
31: **end if**
32: **return** $fitness$

Preparation of Local search

33: Make a list of potential candidates to apply **LS**   ▷ According to I_{ls} value
34: **while** not enough evaluations in the I_{str} **do**
35:     $runLocalSearch()$ ▷ Iterating over the previous list
36: **end while**

Local search step

37: Load solution configuration   ▷ Keep the previous conditions
38: Perform the **LS** procedure
39: Save solution configuration   ▷ Allows make the LS chains

the most promising local search method working together with the BRKGA for the MD problem; (ii) evaluate the behavior of our two bests approaches with other algorithms in the art state.

Benchmark: A set of 16 structures based on the HIV-protease receptor were selected to perform the tests on the proposed methods. In the proteins based on HIV-protease the peptidomimetic inhibitor it's involved in a tunnel-shaped active site [29]. The set of molecules was downloaded from the PDB database and divided into four groups, according to ligand size. The instances are categorized and identified by the PDB code and the range of crystallographic resolution in Ångströms (Å) as follow: *small* (1AAQ, 1B6L, 1HEG, 1KZK, 1.09 − 2.50); *medium* (1B6J, 1HEF, 1K6P, 1MUI, 1.85 − 2.80); *large* (1HIV, 1HPX, 1VIK, 9HVP, 2.00 − 2.80) size inhibitors, as well as *cyclic urea* (1AJX, 1BV9, 1G2K, 1HVH, 1.80 − 2.00) inhibitors. Each complex was equally prepared for all methods applied:

– *Preparing the structures:* small molecules such as solvent molecules, noninteracting ions, and water were removed from the X-ray crystallographic structures using PYMOL [31].
– *Definition of torsional angles:* AutoDockTools was used to generate PDBQT files from the ligand and macromolecules, and also get the ligand root atom. This conversion scripts add partial charges and hydrogens to structures. The algorithms will optimize the torsional degrees of freedom presents in this ligand file. The maximum number of torsions is set to be 10, but can be less according to the ligand structure. The selected torsions are that allowed the fewest number of atoms to move (keeping freeze the molecule core) [29].
– *File conversions:* the *Open Babel tool* [32] was used to convert PDB files in *.mdl* format. After, tools of the *PyRosetta toolkit* were used to generate the necessary files to apply its scoring function. We produce *.params* files from the previous *.mdl* files created, and build a unique PDB structure with the macro and ligand molecules. Then, it's possible to load the complexes and evaluate the energy score with the PyRosetta. However, our algorithms are responsible by producing the changes in the ligand conformation and repass the coordinates to Rosetta interface.
– *Docking environment:* from the ligand root atom get with AutoDockTools we define the center and dimensions of the search space. The binding site size was set in 11 Å for each axis, and the grid spacing defined as 0.375 Å. Since we have the box size, the initial ligand conformation and the position in the active site are randomized for each algorithm run.
– *Running algorithms:* the *Rosetta* energy function was applied to calculate the energy score. Considering the use of a stochastic method, we execute 31 independent runs per instance with a stop criterion of 1,000,000 energy evaluations. So, we are acquiring statistical confidence in the present results.

Parametrization: Initially, we ran the BRKGA for each instance and save the first population and initial structure of the ligand in each execution, considering that in each of them we randomize the shape of ligand to take no advantage. In reason to have the same conditions to analyze the results between different

methods, we apply this same configuration of the population and initial ligand
to all algorithms.

The parameter settings are summarized in Table 1. For the BRKGA algorithm,
we have adopted the recommended values for the parameters according to [8].
The population size is 150 individuals, in which 20% refers to elite set, 30% to
mutation set, and the elite allele inheritance probability on the random choice is
0.5–0.7, as described in Sect. 3. Also, it's important to emphasize that we have
adopted real values to encode the genes. The local search parameters were defined
as 0.5 to radius perturbation in each gene of the individual, the LS intensity
stretch is equal to 500, and the local/global search ratio is 0.5, according to
were adopted in [27,28]. SA has another parameters referent to the minimum
and maximum temperature, 2,5 and 25,000 respectively.

Table 1. Parameter settings.

BRKGA parameters		Local search parameters	
Population size	150 individuals	Candidate selection	Best individuals from most populated cubes
Max. evaluations	1,000,000	Radius perturbation	0.50
p_e	0.20	I_{str}	500
p_m	0.30	$r_{L/G}$	0.50
ρ_e	0.50–0.70	**Simulated Annealing parameters**	
p_{nm}	0.30	Minimum temperature	2.50
		Maximum temperature	25,000

Achieved Results:

We divide the tests into two stages: (i) applying all algorithm combinations
over half of the instances with 11 runs; (ii) executing the best methods with all
benchmark and 31 runs. In the first step we are interested in discovering if LS
improves the results of BRKGA in the docking problem, and if yes, which one of
them. In the second experiment stage, we run the best algorithms of global plus
local search and compare with the pure BRKGA, and with other methods.

Achieved results were evaluated according to its energy in *kcal/mol*, and
with the *Root Mean Square Deviation* (RMSD) which consists in comparing the
structure of the ligand generated by the algorithm with the crystallographic
structure of the molecule. In Table 2 is presented the best energy found and
the corresponding RMSD, as well the average and standard deviation of each one
obtained by the best solutions generated in each configuration. Considering just
the best energy values, we have an equal distribution between the methods.
However, when looking for the average, there are two approaches in detach, the
SA that has best values in 25% of instances, and the SHD which is the best in
the remaining 75%. Considering the execution time, comparing with the BRKGA,

Table 2. Achieved results from the comparison of the BRKGA with all local search methods. Third and fourth columns contains the lowest energy (*kcal/mol*) and its RMSD values for each method. Fifth and sixth columns show the average energy and RMSD, respectively, and standard deviation for 11 runs of the algorithms. Cells shaded in gray highlight the best solution obtained, and cells shaded in blue, the best solutions average, for each instance.

ID	Method	Best solution Energy	RMSD	11 runs average Energy	RMSD	ID	Best solution Energy	RMSD	11 runs average Energy	RMSD
1AJX	BRKGA	-240.67	12.01	-238.65 ± 1.87	12.36 ± 0.47	1HPX	-348.37	5.85	-347.43 ± 1.34	5.59 ± 0.69
	BI	-250.80	3.10	-242.29 ± 3.91	10.13 ± 3.91		-356.47	2.06	-349.02 ± 3.63	5.29 ± 1.48
	FI	-249.35	1.38	-241.47 ± 2.69	11.34 ± 3.20		-355.67	2.86	-349.24 ± 3.06	5.10 ± 1.04
	SHD	-250.51	0.98	-243.95 ± 4.10	8.30 ± 4.68		-357.34	3.15	-350.88 ± 3.87	4.58 ± 1.46
	SA	-250.62	1.22	-243.22 ± 4.24	9.18 ± 4.90		-356.79	2.97	-351.73 ± 3.92	4.55 ± 1.60
1B6J	BRKGA	328.69	1.11	-316.67 ± 8.55	5.50 ± 2.34	1K6P	-381.24	6.12	-373.70 ± 5.27	7.43 ± 1.62
	BI	-324.98	1.42	-320.42 ± 1.92	6.89 ± 2.19		-382.64	5.40	-380.03 ± 1.36	5.77 ± 2.23
	FI	-328.66	1.12	-322.14 ± 3.48	4.94 ± 2.44		-384.01	6.43	-380.32 ± 2.57	6.02 ± 1.45
	SHD	-327.06	2.08	-322.18 ± 3.40	4.87 ± 2.80		-383.26	5.50	-380.66 ± 1.73	5.54 ± 1.51
	SA	-328.27	1.18	-321.34 ± 3.02	5.96 ± 2.25		-383.08	5.36	-380.59 ± 1.85	6.64 ± 1.22
1G2K	BRKGA	-455.77	1.34	-442.09 ± 12.37	4.49 ± 2.70	1KZK	-440.32	1.91	-422.67 ± 13.09	5.77 ± 2.19
	BI	-456.73	0.84	-443.77 ± 8.81	4.76 ± 2.32		-440.88	0.34	-436.38 ± 4.98	3.63 ± 2.30
	FI	-456.20	2.89	-447.06 ± 8.12	4.68 ± 2.64		-440.99	0.84	-433.40 ± 7.15	4.21 ± 2.38
	SHD	-456.71	0.88	-449.73 ± 7.50	3.64 ± 2.53		-440.92	0.84	-438.90 ± 2.34	2.13 ± 1.50
	SA	-456.00	0.88	-451.66 ± 6.00	3.24 ± 1.95		-440.87	1.56	-434.32 ± 5.78	4.63 ± 2.26
1HEG	BRKGA	358.26	8.62	361.53 ± 1.45	9.09 ± 1.73	1VIK	175.03	2.56	236.20 ± 77.73	6.30 ± 3.16
	BI	358.42	8.73	359.60 ± 0.01	8.13 ± 1.54		175.46	3.57	192.35 ± 10.86	6.30 ± 3.28
	FI	357.99	6.79	359.64 ± 1.19	8.36 ± 2.25		175.00	2.19	194.60 ± 21.46	4.66 ± 3.21
	SHD	358.35	5.72	359.11 ± 0.51	7.63 ± 1.20		175.45	2.11	183.06 ± 12.51	3.27 ± 2.26
	SA	358.38	8.43	359.73 ± 1.32	8.44 ± 2.50		174.37	2.55	193.10 ± 23.09	4.70 ± 2.55

the *Best Improvement* increases in 50%, while *First Improvement* and *Stochastic Hill Descent* in almost 100%, and *Simulated Annealing* in only 7%.

In the second test stage we compare BRKGA against SA and SHD. Achieved results were also compared with AUTODOCK VINA [33], DOCKTHOR [34], and jMETAL [35], a multi-objective docking approach. Table 3 summarizes the achieved results. The energy values are compared only between our proposed methods because of the particular function energy adopted which differs from the others methods. The comparison of RMSD was done considering all approaches. The results show that the SHD method is better in 94% of complexes when comparing the average energy and average RMSD between the proposed approaches. Nevertheless, when comparing for the best average RMSD against all state of the art methods, our algorithm is better only to 1AAQ complex, and DOCKTHOR the best in 81%.

To analyze the significant difference between the methods, we performed the *Dunn test* [36], a nonparametric test for multiple comparisons procedure. Table 4 presents the significant levels of energy and RMSD. The cells highlighted, above the diagonal, shows the values for energy comparisons and the remaining cells for RMSD comparisons. Using a significance of $\alpha < 0.05$, we find that when we compared BRKGA with the SHD, there is a significant difference in 93% of instances regarding energy, and 75% in RMSD. Comparing BRKGA with SA, there is difference of energy in 87% and RMSD in 56%. However, between the two LS methods there is no significant difference, only in energy for the 1AAQ case.

Table 3. Achived Results from BRKGA, *Stochastic Hill Descent, Simulated Annealing,* DOCKTHOR, AUTODOCK VINA, and jMETAL. Third and fourth columns contains the lowest energy (*kcal/mol*) and its RMSD Å values for each method. Fifth and sixth columns show the average energy and RMSD, respectively, and standard deviation for 31 runs of the algorithms. Cells shaded in gray highlight the best solution obtained in every instance (considering only our approaches), and cells shaded in blue, the best solutions average when comparing all methods.

ID	Method	Best solution Energy	RMSD	31 runs average Energy	RMSD
1AAQ	BRKGA	36.70	1.14	42.75 ± 5.89	3.12 ± 2.47
	SHD	36.88	1.13	37.94 ± 0.47	1.09 ± 0.14
	SA	37.04	1.14	39.12 ± 2.68	1.59 ± 1.51
	VINA	3.93	9.75	9.07 ± 5.05	9.06 ± 0.86
	DOCKTHOR	3.52	12.15	5.12 ± 1.41	8.89 ± 5.11
	JMETAL	-16.00	1.87	-12.56 ± 3.48	2.11 ± 1.15
1AJX	BRKGA	-246.85	4.40	-239.68 ± 1.87	12.24 ± 1.50
	SHD	-250.51	0.98	-243.68 ± 4.39	8.59 ± 4.75
	SA	-250.62	1.22	-242.21 ± 3.84	10.32 ± 4.22
	VINA	-10.74	1.52	-9.82 ± 0.42	6.21 ± 2.82
	DOCKTHOR	48.68	0.85	49.33 ± 2.30	0.88 ± 0.09
	JMETAL	-18.00	4.03	-12.84 ± 4.35	3.88 ± 1.64
1B6J	BRKGA	-328.69	1.11	-316.83 ± 7.49	5.85 ± 2.33
	SHD	-328.70	1.09	-322.42 ± 3.26	5.06 ± 2.76
	SA	-328.27	1.18	-320.81 ± 3.99	5.94 ± 2.22
	VINA	-9.12	2.89	-2.39 ± 3.64	6.92 ± 3.08
	DOCKTHOR	38.78	0.61	38.96 ± 0.11	0.61 ± 0.06
	JMETAL	-18.00	3.47	-11.45 ± 6.79	2.49 ± 0.71
1B6L	BRKGA	-317.46	3.09	-312.62 ± 3.11	6.64 ± 1.94
	SHD	-319.21	2.80	-316.64 ± 1.83	3.54 ± 1.99
	SA	-318.57	2.96	-315.72 ± 2.34	4.42 ± 2.41
	VINA	-12.71	0.89	-12.02 ± 1.29	2.22 ± 3.07
	DOCKTHOR	30.51	0.42	30.72 ± 0.13	0.43 ± 0.02
	JMETAL	-16.00	2.70	-13.14 ± 3.68	2.17 ± 0.63
1BV9	BRKGA	-80.07	0.74	-16.33 ± 44.55	7.27 ± 2.86
	SHD	-80.64	0.67	-57.17 ± 18.63	6.83 ± 4.84
	SA	-80.41	0.70	-53.44 ± 22.76	6.90 ± 4.59
	VINA	14.56	5.78	20.65 ± 2.64	8.41 ± 1.49
	DOCKTHOR	55.69	0.91	56.59 ± 1.04	0.95 ± 0.05
	JMETAL	-20.00	1.99	-14.61 ± 5.64	2.45 ± 1.31
1G2K	BRKGA	-455.77	1.34	-436.89 ± 9.44	6.04 ± 2.05
	SHD	-456.71	0.88	-452.56 ± 5.60	2.83 ± 2.28
	SA	-456.62	0.78	-450.82 ± 6.38	3.43 ± 2.36
	VINA	-10.01	4.41	-9.34 ± 0.86	6.05 ± 1.77
	DOCKTHOR	15.46	0.36	16.10 ± 1.49	0.43 ± 0.17
	JMETAL	-20.00	2.73	-12.84 ± 5.31	3.01 ± 1.69
1HEF	BRKGA	175.78	13.48	176.69 ± 0.57	12.25 ± 0.56
	SHD	173.69	11.43	175.26 ± 1.00	11.79 ± 0.67
	SA	173.69	11.36	175.46 ± 0.87	12.00 ± 0.80
	VINA	-1.79	9.36	1.42 ± 2.26	8.77 ± 0.56
	DOCKTHOR	67.68	1.79	68.07 ± 0.36	1.85 ± 0.14
	JMETAL	-22.00	6.60	-10.74 ± 6.25	5.81 ± 1.67
1HEG	BRKGA	357.76	6.91	361.20 ± 1.45	8.67 ± 1.68
	SHD	357.18	6.45	359.00 ± 0.59	7.53 ± 1.36
	SA	356.76	5.56	359.11 ± 1.32	7.77 ± 1.82
	VINA	-5.85	5.50	-5.51 ± 0.26	6.00 ± 0.98
	DOCKTHOR	58.37	4.10	60.50 ± 1.91	3.98 ± 1.51
	JMETAL	-14.50	5.03	-9.15 ± 3.90	5.24 ± 1.73

ID	Method	Best solution Energy	RMSD	31 runs average Energy	RMSD
1HIV	BRKGA	-164.16	2.92	-150.52 ± 3.10	6.65 ± 1.11
	SHD	-164.94	3.41	-163.67 ± 0.82	2.28 ± 0.55
	SA	-165.15	3.39	-163.19 ± 2.67	2.65 ± 1.04
	VINA	-0.29	7.49	12.01 ± 18.70	8.09 ± 0.92
	DOCKTHOR	55.13	0.28	55.32 ± 0.14	0.29 ± 0.04
	JMETAL	-32.00	7.93	-18.14 ± 3.68	2.58 ± 1.45
1HPX	BRKGA	-348.40	5.28	-346.81 ± 1.59	5.87 ± 1.04
	SHD	-357.34	3.15	-352.56 ± 3.99	4.04 ± 1.36
	SA	-356.79	2.97	-351.63 ± 3.73	4.66 ± 1.49
	VINA	-9.08	5.74	-6.52 ± 2.18	6.18 ± 1.03
	DOCKTHOR	88.64	5.84	96.05 ± 9.69	7.07 ± 2.11
	JMETAL	-18.00	4.61	-11.73 ± 4.77	3.84 ± 0.94
1HVH	BRKGA	428.01	8.94	434.16 ± 2.60	10.33 ± 2.22
	SHD	427.29	8.38	429.98 ± 2.45	8.65 ± 1.62
	SA	427.00	8.42	430.81 ± 2.26	8.66 ± 1.80
	VINA	-8.65	7.34	-7.04 ± 1.12	5.98 ± 1.71
	DOCKTHOR	127.18	8.34	130.22 ± 2.76	6.10 ± 1.58
	JMETAL	-18.00	3.06	-12.45 ± 4.42	2.89 ± 0.52
1K6P	BRKGA	-382.38	4.42	-375.89 ± 4.85	6.88 ± 1.76
	SHD	-383.78	0.83	-368.36 ± 67.28	5.39 ± 1.96
	SA	-383.08	5.36	-380.13 ± 1.85	6.29 ± 1.56
	VINA	-5.16	5.22	-0.53 ± 3.20	7.45 ± 2.09
	DOCKTHOR	143.30	1.76	150.40 ± 6.76	2.08 ± 1.89
	JMETAL	-20.00	3.20	-14.69 ± 4.77	3.20 ± 0.85
1KZK	BRKGA	-440.32	1.91	-422.16 ± 17.60	5.68 ± 2.14
	SHD	-441.16	1.54	-438.45 ± 2.99	2.83 ± 1.97
	SA	-441.03	1.55	-435.52 ± 5.89	3.94 ± 2.35
	VINA	-9.85	2.37	-8.05 ± 0.83	5.73 ± 2.57
	DOCKTHOR	27.61	0.79	27.79 ± 0.16	0.75 ± 0.13
	JMETAL	-24.00	9.02	-10.44 ± 12.79	7.95 ± 1.63
1MUI	BRKGA	-36.39	1.83	-26.57 ± 6.68	5.85 ± 3.02
	SHD	-36.79	1.82	-34.99 ± 1.86	1.72 ± 0.68
	SA	-36.57	1.93	-33.67 ± 2.64	2.53 ± 2.21
	VINA	-8.22	6.68	-6.29 ± 1.15	7.48 ± 1.09
	DOCKTHOR	17.06	0.79	17.42 ± 0.15	0.48 ± 0.20
	JMETAL	-20.00	4.65	-13.04 ± 4.16	2.57 ± 0.98
1VIK	BRKGA	174.70	2.18	228.25 ± 50.90	6.85 ± 2.82
	SHD	173.46	2.20	186.06 ± 16.52	3.82 ± 2.81
	SA	173.35	2.16	187.99 ± 18.04	4.59 ± 2.93
	VINA	96.97	11.19	140.08 ± 29.70	9.18 ± 2.02
	DOCKTHOR	165.54	1.39	267.46 ± 123.25	2.97 ± 2.60
	JMETAL	-40.00	6.65	-19.12 ± 14.44	4.01 ± 0.98
9HVP	BRKGA	359.99	1.09	372.04 ± 14.29	4.98 ± 3.39
	SHD	360.26	1.46	362.91 ± 4.20	3.11 ± 3.44
	SA	360.09	1.05	362.94 ± 4.75	2.50 ± 2.46
	VINA	-3.74	10.09	0.50 ± 4.03	9.39 ± 0.95
	DOCKTHOR	25.96	1.19	26.04 ± 0.05	1.21 ± 0.05
	JMETAL	-22.00	4.17	-14.89 ± 4.13	2.72 ± 1.01

Table 4. The table entries above the diagonal show the p-values for energy comparisons, and the cells below for the RMSD values. In both, the significance level is $p < 0.05$.

ID	Method	BRKGA	SHD	SA	ID	BRKGA	SHD	SA	ID	BRKGA	SHD	SA	ID	BRKGA	SHD	SA
1AAQ	BRKGA	—	0.00	0.07	1BV9	—	0.00	0.00	1HIV	—	0.00	0.00	1KZK	—	0.00	0.00
	SHD	0.00	—	0.03		1.00	—	1.00		0.00	—	1.00		0.00	—	0.21
	SA	0.00	1.00	—		1.00	1.00	—		0.00	0.69	—		0.02	0.16	—
1AJX	BRKGA	—	0.00	0.00	1G2K	—	0.00	0.00	1HPX	—	0.00	0.00	1MUI	—	0.00	0.00
	SHD	0.00	—	0.64		0.00	—	0.34		0.00	—	1.00		0.00	—	0.20
	SA	0.20	0.45	—		0.00	0.79	—		0.02	0.38	—		0.00	1.00	—
1B6J	BRKGA	—	0.00	0.07	1HEF	—	0.00	0.00	1HVH	—	0.00	0.00	1VIK	—	0.00	0.00
	SHD	0.74	—	0.58		0.01	—	1.00		0.01	—	0.59		0.00	—	1.00
	SA	1.00	0.87	—		0.14	0.87	—		0.00	1.00	—		0.03	0.47	—
1B6L	BRKGA	—	0.00	0.00	1HEG	—	0.00	0.00	1K6P	—	0.00	0.00	9HVP	—	0.08	0.01
	SHD	0.00	—	0.54		0.06	—	1.00		0.03	—	0.82		0.20	—	1.00
	SA	0.00	1.00	—		0.14	1.00	—		0.50	0.74	—		0.06	1.00	—

5 Conclusion and Future Work

The performed statistical test shows that the energy values obtained have a difference between BRKGA and SHD and SA. We highlight that only our approaches use the same energy function. The comparisons of RMSD also present significant differences in most instances for all methods. Although, SHD has presented better average results against SA, there is no significant difference for RMSD, and an unique case for energy (1AAQ). However, the difference between the approaches with local search and the BRKGA is significant for almost all instances in energy and RMSD. These results confirm that the application of LS brings improves to the search for good solutions in the conformational space.

Analyzing both local searches applied, we conclude that the SHD algorithm showed better results than another two variations of *Hill-climbing* because of its random characteristic in visit the neighborhood in the search process. This behavior avoids cyclic search and allows that every gene can be improved. The *Simulated Annealing* has a process similar to HC in acceptance of solutions but allows worst configurations with a certain probability. This characteristic and the random behavior at generating neighbors in the search process, make the method competitive with the *Stochastic Hill Descent*.

This work brings contributions to the use of hybrid (memetic) computational techniques and concepts to develop a promising method for the molecular docking problem. The combination of methods to global and local search can lead to efficient applications in many domains, especially in the MD problem. Also, this works carries exciting research topics, with a gamma of applications in computational biology and bioinformatics. There is a gap to apply other methods as global procedures, such as GA, DE, PSO, and GRASP, as well as algorithms to work as the local procedure, like *Solis and Wets* and *Nelder-Mead*.

Another recent and attractive approach is to develop an auto-adaptive algorithm. In this case of a hybrid (memetic) solution, we could insert the concept of auto-adaptation in the choice of which method applies as local search, i.e., use different local search methods during the same execution. This can be justified by our experiment, that reveals no much significant difference between SHD and

SA algorithms. Then, in future works, the combination of more than one method as LS can reveal better solutions to molecular docking problem.

Acknowledgements. This work was supported by grants from FAPERGS [16/2551-0000520-6], MCT/CNPq [311022/2015-4; 311611/2018-4], CAPES-STIC AMSUD [88887. 135130/2017-01] - Brazil, Alexander von Humboldt-Stiftung (AvH) [BRA 1190826 IIFST CAPES-P] - Germany. This study was financed in part by the Coordenação de Aperfei çoamento de Pessoal de Nível Superior - Brazil (CAPES) - Finance Code 001.

References

1. López-Camacho, E., Godoy, M.J.G., Nebro, A.J., Aldana-Montes, J.F.: jMetalCpp: optimizing molecular docking problems with a C++ metaheuristic framework. Bioinformatics **20**, 437–438 (2013)
2. García-Godoy, M.J., López-Camacho, E., García-Nieto, J., Nebro, A.J., Aldana-Montes, J.F.: Solving molecular docking problems with multi-objective metaheuristics. Molecules **20**(6), 10154–10183 (2015)
3. Stockwell, G.R., Thornton, J.M.: Conformational diversity of ligands bound to proteins. J. Mol. Biol. **356**(4), 928–944 (2006)
4. Sadjad, B., Zsoldos, Z.: Toward a robust search method for the protein-drug docking problem. IEEE/ACM Trans. Comput. Biol. Bioinf. **8**, 1120–1133 (2011)
5. Brooijmans, N., Kuntz, I.D.: Molecular recognition and docking algorithms. Annu. Rev. Biophys. Biomol. Struct. **32**(1), 335–373 (2003)
6. Kukkonen, S., Lampinen, J.: GDE3: the third evolution step of generalized differential evolution. In: IEEE CEC, pp. 443–450 (2005)
7. Nebro, A., Durillo, J., García-Nieto, J., Coello, C., Luna, F., Alba, E.: SMPSO: a new PSO-based metaheuristic for multi-objective optimization. In: 2009 IEEE Symposium on Computational Intelligence in Multicriteria Decision-Making, pp. 66–73 (2009)
8. Gonçalves, J.F., Resende, M.G.C.: Biased random-key genetic algorithms for combinatorial optimization. J. Heuristics **17**(5), 487–525 (2011)
9. Leonhart, P.F., Spieler, E., Ligabue-Braun, R., Dorn, M.: A biased random key genetic algorithm for the protein–ligand docking problem. Soft Comput. 1–22 (2018)
10. Gray, J.J., et al.: Protein-protein docking with simultaneous optimization of rigid-body displacement and side-chain conformations. J. Mol. Biol. **331**(1), 281–299 (2003)
11. Andrusier, N., Mashiach, E., Nussinov, R., Wolfson, H.: Principles of flexible protein-protein docking. Proteins **73**(2), 271–289 (2008)
12. Chaudhury, S., Lyskov, S., Gray, J.J.: PyRosetta: a script-based interface for implementing molecular modeling algorithms using Rosetta. Bioinformatics **26**, 689–691 (2010)
13. Huang, S.Y., Zou, X.: Advances and challenges in protein-ligand docking. Int. J. Mol. Sci. **11**(8), 3016 (2010)
14. Lameijer, E.W., Back, T., Kok, J.N., Ijzerman, A.D.P.: Evolutionary algorithms in drug design. Nat. Comput. **4**, 177–243 (2005)

15. Rosin, C.D., Halliday, R.S., Hart, W.E., Belew, R.K.: A comparison of global and local search methods in drug docking. In: Proceedings of the Seventh International Conference on Genetic Algorithms, pp. 221–228. Morgan Kaufmann (1997)
16. Ruiz-Tagle, B., Villalobos-Cid, M., Dorn, M., Inostroza-Ponta, M.: Evaluating the use of local search strategies for a memetic algorithm for the protein-ligand docking problem. In: 2017 36th International Conference of the Chilean Computer Science Society (SCCC), pp. 1–12, October 2017
17. Halperin, I., Ma, B., Wolfson, H., Nussinov, R.: Principles of docking: an overview of search algorithms and a guide to scoring functions. Proteins Struct. Funct. Bioinf. **47**(4), 409–443 (2002)
18. Taylor, R., Jewsbury, P., Essex, J.: A review of protein-small molecule docking methods. J. Comput. Aided Mol. Des. **16**(3), 151–166 (2002)
19. Papadimitriou, C.H., Steiglitz, K.: Combinatorial Optimization: Algorithms and Complexity. Courier Corporation, North Chelmsford (1998)
20. Aarts, E., Lenstra, J.K. (eds.): Local Search in Combinatorial Optimization, 1st edn. Wiley, New York (1997)
21. Kirkpatrick, S., Gelatt, C.D., Vecchi, M.P.: Optimization by simulated annealing. Science **220**(4598), 671–680 (1983)
22. Černý, V.: Thermodynamical approach to the traveling salesman problem: an efficient simulation algorithm. J. Optim. Theory Appl. **45**(1), 41–51 (1985)
23. Mitchell, M.: An Introduction to Genetic Algorithms. MIT Press, Cambridge (1998)
24. Holland, J.H.: Adaptation in Natural and Artificial Systems: An Introductory Analysis with Applications to Biology, Control and Artificial Intelligence. MIT Press, Cambridge (1992)
25. Gonçalves, J.F., de Almeida, J.R.: A hybrid genetic algorithm for assembly line balancing. J. Heuristics **8**(6), 629–642 (2002)
26. Goulart, N., de Souza, S.R., Dias, L.G.S., Noronha, T.F.: Biased random-key genetic algorithm for fiber installation in optical network optimization. In: 2011 IEEE Congress of Evolutionary Computation (CEC), pp. 2267–2271, June 2011
27. Molina, D., Lozano, M., Herrera, F.: Memetic algorithm with local search chaining for continuous optimization problems: a scalability test. In: ISDA 2009–9th International Conference on Intelligent Systems Design and Applications, pp. 1068–1073 (2009)
28. Molina, D., Lozano, M., Sánchez, A.M., Herrera, F.: Memetic algorithms based on local search chains for large scale continuous optimisation problems: MA-SSW-Chains. Soft Comput. **15**(11), 2201–2220 (2011)
29. Morris, G.M., Huey, R., Lindstrom, W., Sanner, M.F., Belew, R.K., Goodsell, D.S., Olson, A.J.: Autodock4 and autodocktools4: automated docking with selective receptor flexibility. J. Comput. Chem. **30**(16), 2785–2791 (2009)
30. Berman, H.M., Westbrook, J., Feng, Z., Gilliland, G., Bhat, T.N., Weissig, H., Shindyalov, I.N., Bourne, P.E.: The protein data bank. Nucleic Acids Res. **28**(1), 235–242 (2000)
31. Schrödinger, LLC: The AxPyMOL molecular graphics plugin for Microsoft PowerPoint, version 1.8, November 2015
32. O'Boyle, N.M., Banck, M., James, C.A., Morley, C., Vandermeersch, T., Hutchison, G.R.: Open babel: an open chemical toolbox. J. Cheminf. **3**(1), 1–14 (2011)

33. Trott, O., Olson, A.J.: AutoDock Vina: improving the speed and accuracy of docking with a new scoring function, efficient optimization, and multithreading. J. Comput. Chem. **31**(2), 455–461 (2010)
34. de Magalhães, C.S., Almeida, D.M., Barbosa, H.J.C., Dardenne, L.E.: A dynamic niching genetic algorithm strategy for docking highly flexible ligands. Inf. Sci. **289**, 206–224 (2014)
35. Durillo, J., Nebro, A.: jMetal: a Java framework for multi-objective optimization. Adv. Eng. Softw. **42**, 760–771 (2011)
36. Dunn, O.J.: Multiple comparisons using rank sums. Technometrics **6**(3), 241–252 (1964)

Self-sustainability Challenges of Plants Colonization Strategies in Virtual 3D Environments

Kevin Godin-Dubois(✉), Sylvain Cussat-Blanc, and Yves Duthen

University of Toulouse, IRIT - CNRS UMR 5505, 2 rue du Doyen Gabriel Marty,
31042 Toulouse, France
{kevin.dubois,sylvain.cussat-blanc,yves.duthen}@irit.fr

Abstract. The Biosphere is a bountiful source of inspiration for the biologically inclined scientist, though one may be seized by the twists and turns of its complexity. Artificial Life emerged from the conundrum of condensing this overwhelming intricacy into a tractable volume of data.

To tackle the distant challenge of studying the long-term dynamics of artificial ecosystems, we focused in this work our efforts on plant-plant interactions in a simplified 3D setting. Through an extension of K. Sims' directed graphs, we devised a polyvalent genotype for artificial plants development. These individuals compete and collaborate with one another in a shared plot of earth subjected to dynamically changing environmental conditions. We illustrate and analyze how the use of multi-objective fitnesses generated a panel of diverse morphologies and strategies. Furthermore, we identify two driving forces of the emerge of self-reproduction and investigate their effect on self-sustainability.

Keywords: Artificial plants · Ecosystems · Autonomous reproduction · Self-sustainability

1 Introduction

Natural ecosystems are staggering by virtue of their intricate interactions networks and seemingly endless detail levels. Adaptation, co-evolution and arming races are a small subset of the myriad evolutionary strategies observed in the world. In order to translate this onto an artificial medium, one has to address three key issues. The first is finding functional, adaptive and yet computationally tractable representations for both the genomic and phenotypic aspects of the 'creatures'. Then, genetic variation and discrimination methods, the keystones of Darwin's natural selection process, ought to be devised. Ultimately, environmental pressure, whether coming from the inanimate (abiotic) or living (biotic) surroundings, should be strong enough so that diversity can emerge without excessively narrowing the viable behaviors range.

© Springer Nature Switzerland AG 2019
P. Kaufmann and P. A. Castillo (Eds.): EvoApplications 2019, LNCS 11454, pp. 377–392, 2019.
https://doi.org/10.1007/978-3-030-16692-2_25

A number of models have been devised to tackle the problem of artificial morphogenesis starting with the L-Systems which successfully generated plant morphologies in both 2-D [1] and 3-D environments [2], as well as mobile creatures [3]. In a similar fashion, the directed graphs from [4], define distinct organs and the relationship between them, allowing for very compact encoding of motile morphologies. Closer to biology is the use of Genetic Regulatory Networks to control cellular building blocks. By using such a finer-grained representation it was shown possible to evolve cell clusters towards specific shapes [5] or generate creatures with self-organized organs in non-trivial environments [6].

With rules as simple as the one in the Game of Life [7], elaborate ecosystems emerged with no reliance on evolutionary techniques. Other contributions simulated self-reproducing computer programs into a 2D memory grid where the organisms had their duplication mechanism embedded into their life-cycle [8]. Further complexification of the phenotypic space led to ecosystems such as [9,10] where the environment was a 2D continuous grid and the 'animals' had to manage their reproduction cycle by actively searching for mates in their surroundings.

In case of co-evolution or competition (e.g. [11]), there is a biotic component to the ecosystem in the sense that from the viewpoint of an individual, every other entity is a hard-to-control part of its environment. When considering the abiotic, i.e. non-living, component of earth's ecosystems, it has been shown that, with only water availability and temperatures, it is possible to model most of the biodiversity observed in nature [12]. The same holds true for artificial simulations where the impact of environmental factors can be a driving force for the speciation process [2]. Additionally, dynamically changing the local constraints pushes individuals out of local minimums and promotes adaptability [13].

Our objective is to design a virtual ecosystem which tackles all three points by modeling both plants and animals in a shared environment thus inducing complex interactions and survival/reproduction strategies starting, in this paper, with the growth of vegetals. To this end we first detail, in Sect. 2, the use of both an extended version of the directed graphs (hereafter named 'Graphtals') from [4], which produces functional morphologies in the face of unknown (a)biotic constraints, and a novel reproduction scheme with self-controlled speciation capabilities. We then move on to exploring, in Sect. 3, various outcomes stemming from this implementation in terms of growth and adaptation strategies as well as the impact of the evolution process itself before highlighting the necessary extensions to this model.

2   Self-reproducing Vegetals

Starting from the breakdown of the various components of an individual plants genome (Sect. 2.1), we expand upon how organs interact with one another to promote the whole organisms well-being and survival capacities (Sect. 2.2). A detailed presentation of a speciation-oriented approach to self-controlled, self-reproduction follows in Sect. 2.3. We conclude this overview of the model by the external components of the system, i.e. the environment (Sect. 2.4) and the englobing ecosystem (Sect. 2.5).

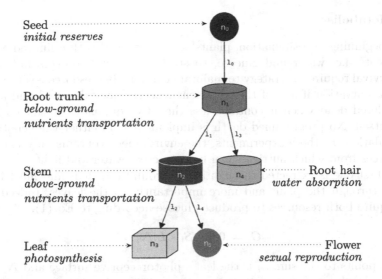

Fig. 1. Base graphtal in all following experiments

2.1 Graphtal Fields

As in [14], our genotype is composed of both structural (Fig. 1) and behavioral instructions. Each node n_l describes an abstract sub-organ and is composed of an id and a set of parameters n_i^d specifying its skill which in turn defines its shape (sphere, box, cylinder). Furthermore, it also codes for the initial dimensions, density, color and maximal growth factor of the corresponding organ.

Behavior is controlled by two tuples $A, S \in [0, 1]^E$ with E the number of elements (limited to water and glucose in this work). A models an organ's balance between production and consumption: a value of 0 (resp. 1) indicates a source (resp. sink) for this specific element. S enables quiescent behavior by imposing a threshold below which no growth or budding actions can be performed.

A link $l_i : n_1 \rightarrow n_2$ expresses a *growth relationship* from n_1 (the parent node noted l^i) to n_2 (the 'child' node noted l^o) and, also, contains a set of parameters noted l^d describing the position on the parent's surface and the relative orientation. A special field r is used to regulate recurrent connections by imposing a maximal depth ($r > 0$) or indicate a terminal node ($r = -1$).

To easily code for the highly regular structures observed in nature, each link can select a repetition pattern from:

- *none*: no repetition
- *radial*(V, N): N-1 copies evenly rotated around V
- *random*(N, S): N-1 copies randomly placed

Ultimately, a handful of plant-wide parameters are stored separately. Namely the growth speed and maximal size factor, the individual's sex and two sets of parameters dW and $\{\mu, \sigma^i, \sigma^o\}$ that will be detailed more thoroughly in following sections.

2.2   Metabolism

At the beginning of a simulation, plants' seeds are filled with a limited amount of nutrients, i.e. water and glucose, to start the growth process but longer-term survival requires a strategy to maintain comfortable resource levels. Organs die from starvation if any of their nutrient reserves are exhausted and a plant is considered dead when it contains less than two organs as it can no longer sustain itself. No programmed death is implemented in either the genotype or the simulation. In these experiments, the environment contains only two types of resources from which nutrients can be extracted: water and light.

The first one is extracted through below-ground root hair and must be dispatched through the plant and more importantly to the leaves. Indeed these later require both resources to produce glucose according to Eq. (1).

$$G_i = .025 * S_i * \textbf{L}.\textbf{N}_i \tag{1}$$

where $\textbf{L}$ points to the sun, S_i is the leaf's photoreceptive surface and $\textbf{N}_i$ the z axis in its local coordinates system.

Resources distribution is implemented through a decentralized mechanism which follows the gradient of nutrient. Every simulation step, organs share a portion of their reserves according to the transport Eq. (2) with $e \in g, w$ a nutrient.

$$d^e_{A \to B} = \frac{k^e_{AB} * stored(A) * needs(B)}{\sum\limits_{o \in C(A)} needs(o)} \tag{2}$$

where $needs(x)$ (resp $stored(x)$) is the need (resp. stored amount) in nutrients in an organ x, $C(x)$ the set of organs connected to x and A, B two organs so that $0 \leq needs(A) < needs(B) \leq 1 \wedge B \in C(A)$. The k^e_{AB} term is a refractory coefficient dependant on the types of A and B.

2.3   Autonomous Reproduction

One of the most powerful tools available to Life is its ability to adapt through the process of natural selection. Over the course of history numerous propagation scheme have been developed. In this work, we chose to focus on sexual reproduction because of its greater degree of interactions and inter-species diversity.

To this end, we improved upon the previous work in [14] by including genomic components devoted to reproduction: sex, compatibility metrics $\{\mu, \sigma_i, \sigma_o\}$ and sexual organs. These interact with one another according to Algorithm 1. Given $G_1 = (N_1, L_1)$ and $G_2 = (N_2, L_2)$, two genomes, the alignment procedure creates three subsets:

- $M_a = \{\{l_1, l_2\}, l^{id}_1 = l^{id}_2 \wedge l^i_1 = l^i_2 \wedge l^o_1 = l^o_2\}$
- $M_{ia} = \{\{l_1, l_2\}, l^{id}_1 = l^{id}_2 \wedge (l^i_1 \neq l^i_2 \vee l^o_1 \neq l^o_2)\}$
- $M_i = \{l_1 \in L_1, \nexists l_2 \in L_2, l^{id}_1 = l^{id}_2\} \cup \{l_2 \in L_2, \nexists l_1 \in L_1, l^{id}_2 = l^{id}_1\}$

```
Data: P, set of plants
M ← {p ∈ P/p is male};
for m ∈ M do
    Gₘ ← genotype(m);
    for sₘ, stamen ∈ m do
        f ← random female, with pistils, in range of m;
        G_f ← genotype(f);
        A ← align(Gₘ, G_f);
        C_mf ← compatibility(A, G_f);
        if random toss with probability C_mf then
            delete sₘ;
            p_f ← random pistil from f;
            p_f ← Fruit(Mutate(Cross(A, Gₘ, G_f)));
        end
    end
end
```

Algorithm 1. Mating process

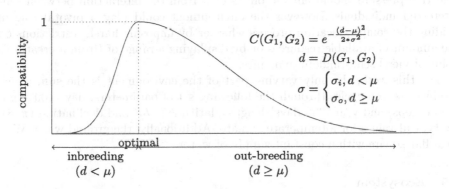

$$C(G_1, G_2) = e^{-\frac{(d-\mu)^2}{2*\sigma^2}}$$

$$d = D(G_1, G_2)$$

$$\sigma = \begin{cases} \sigma_i, d < \mu \\ \sigma_o, d \geq \mu \end{cases}$$

optimal

inbreeding $(d < \mu)$ out-breeding $(d \geq \mu)$

Fig. 2. Genetic compatibility function

which highlight the structural similarities between both individuals. The genetic distance d is then computed based on the average field-by-field difference of comparable node and link data, i.e. those in M_a and M_{ia}. Links in M_i are given the maximal distance of 1. The total sum is given by Eq. (3).

$$d(l_1, l_2) = \frac{1}{3}(d(l_1^i, l_2^i) + d(l_1^o, l_2^o) + d(l_1^d, l_2^d))$$

$$D(G_1, G_2) = |M_i| + \sum_{\{l_1, l_2\} \in M_a \cup M_{ia}} d(l_1, l_2) \tag{3}$$

This crossover operator differs from those commonly found in the literature [4,6,15] on three points: (1) it can fail early on, (2) is biased by the *female* genome and (3) has low resistance to large structural differences. The rationale behind point 3 is that, instead of devising a robust operator that can produce a somewhat viable offspring from two completely unrelated individuals, a minimalist alignment procedure is better suited to sexual reproduction of same species creatures in which the population is mostly homogeneous. Indeed, point 1 guarantees that the more both genomes are different the less likely it is that crossing will be attempted at all. The decision of aborting or proceeding with the reproduction is left to the female individual as, in this sexual scheme, it will have to provide the fruit in resources.

Having the compatibility function embedded into the genome gives each species a segregation scheme which is of utmost importance as the number of nodes/links is not fixed, thus requiring the optimal genetic distance to be adapted as evolution goes. Furthermore having both in-/out-breeding coefficients allow for the specification of the search spaces with adaptive plants accepting a broader range of incoming genetic material while more conservative ones could instead focus on controlled inbreeding to solidify their alleles.

2.4   Environment

Selective pressure should emerge on its own from the interaction between self-interested individuals. However the environment could play a pivotal role in guiding the complexification process whether by imposing harsh restrictions on the amount of available resources or by displaying a range of (from a creature's point of view) semi-random dynamics.

For this work, the only varying part of the environment is the sun, which produces season cycles through the following set of parameters: day (100 simulation steps) and year (300 days) lengths, latitude ($\pi/4$) and declination ($\pi/8$), i.e. typical values for a temperate climate. Additionally, the ground was a W = 10 m flat square with a constant supply of water.

2.5   Ecosystem

A complete ecosystem is composed of both a description of the environment (currently not evolved) and a set of plants 'templates'. Each of these latter represents whole species whose strategies for survival will be pitted against one another in their shared piece of earth.

The procedure to translate these templates into a densely populated ecosystem is straightforward. First the environment is divided into as many cells as the requested number of plants (100 in this experiment) and the largest seed size is tested against half the cell size. If this fails, the whole ecosystem is deemed non-viable and the simulation is aborted, thus preventing plants from having too large initial reserves. Otherwise, each cell is subdivided once more in four and a plant is placed in a single subcell with a random genome from the set of templates and a random vertical rotation. This leaves enough room for autonomous reproduction to place offspring even when the initial population has not entirely died out.

In the current experimental settings, the number of plants is set to 100 and only one species is considered. Furthermore, every random number used during the simulation (plants position, rotations, iterations, etc.) is generated from a fixed seed provided by the genotype (not evolved but randomly set).

3 Colonization Dynamics

3.1 Evolution Protocol

This work comes within the scope of studying long-term evolutionary trends in elaborate 3D ecosystems. However, evolving, from scratch, such systems with a non-trivial degree of complexity would require a prohibitive amount of computational resources. Stemming from this intent, the following experiment was designed to generate usable individuals to seed an environment with. Viable plants would thus have to develop strategies to both survive and reproduce so that their genetic material does not die off.

The evolution protocol relied on evolution programming where plants' genomes underwent single point, equiprobable, mutation on all of the fields mentioned in Sect. 2.1. Evaluating a genotype implies populating an empty environment as described in Sect. 2.5 and then stepping back for a maximum of $N = 60000$ simulation steps (2 simulated years) to see whether autonomous dynamics would emerge.

In order to limit the search space to the genetic fields of the plants, the environment was kept constant in all runs. As we aimed for both efficiency and diversity, we devised a range of fitness functions F_* as described below.

$$\nu = \frac{1}{NP}$$

$$F_b = \nu \sum_{t \in N} \sum_{p \in P} biomass(p, t)$$

$$F_p = \nu \sum_{t \in N} \sum_{p \in P} production(t, p)$$

$$F_c = \frac{\nu}{W^2} \sum_{t \in N} surface(t)$$

$$F_a = \nu \sum_{p \in P} lifespan(p) * 2^{-\alpha_p}$$

where $surface(t)$ corresponds to the total surface covered by plants at time t and α_p is designed to provide a smoother gradient towards reproduction. Plainly put, these aim at producing plants which are: (F_b) large, (F_p) many-leaved, (F_c) wide, (F_a) fast reproducers. Given that every fitness is likely to be exploited into non-desired behaviors, a fifth one F_m is introduced that evaluates genomes on all four criteria at the same time.

Furthermore, in order to prevent local optimum a novelty metric is used as proposed in [16]. An individual's 'footprint', i.e. its synthetic behavioral description, is (F_a, F_b, F_p, R, G, S), with R the number of successful autonomous reproductions, G the number of autonomous generations and S the seed size.

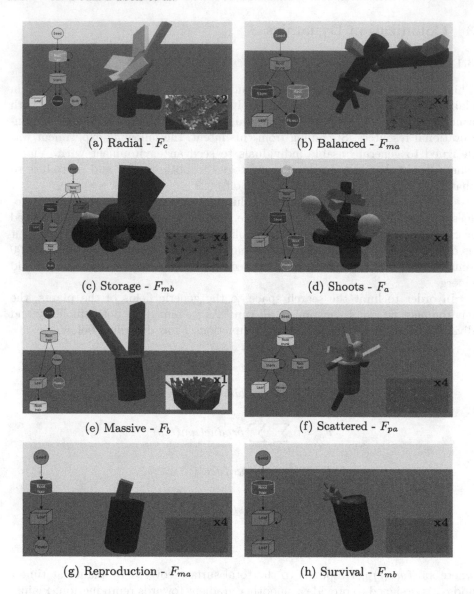

(a) Radial - F_c

(b) Balanced - F_{ma}

(c) Storage - F_{mb}

(d) Shoots - F_a

(e) Massive - F_b

(f) Scattered - F_{pa}

(g) Reproduction - F_{ma}

(h) Survival - F_{mb}

Fig. 3. Examples of the morphologies developed. (a) to (g) are at the 20th day and (h) is at the 30th. From left to right: Genome, Single individual, Ecosystem. The fitness that produced this individual is indicated in the caption with F_{ma} indicating the age criterion of the multi-objective fitness F_m. Videos of these individuals' full ecosystem can be seen on https://vimeo.com/album/5075632.

An autonomous *reproduction*, in this context, is defined as two individuals (m, p) embedded in a simulation deciding on generating an offspring through the process described in Algorithm 1. The autonomous *generation* g_c of such an offspring is $g_c = max(g_m, g_p) + 1$, where g_m (resp. g_p) is the autonomous *generation* of the mother m (resp. father p).

In each scenario, plants are evaluated on two to five criteria using a tournament selection where 3 participants are randomly selected from the population and compete on a random objective as described in [6].

Ten runs per fitness were dispatched on a cluster of Bi-Intel(r) IVYBRIDGE 2,8 Ghz 10-cores and were re-launched as soon as they completed an evolution (250 generations) with a maximal, total, duration of five hours.

3.2 Morphologies

While three out the five fitnesses performed an average of two evolutions in the given time frame ($F_a = 2.4$, $F_c = 2.5$, $F_m = 1.8$), the remaining two behaved very differently: while F_b produced 8.4 'champions' per run, F_p did not manage to bring a single one to the 250 generations threshold (min $= 93$, max $= 199$). This can be explained by observing the evaluation times of those final individuals which range from 12 ms up to 10+ minutes.

In order to gain a further grasp on the situation, we manually examined the phenotype of the 40 best champions (out of a total of 215) that is 5 for every single objective fitness and another 5 for each criterion in F_m. Summarized in Fig. 3 are the morphologies of the most interesting creatures.

As one can see these evolutions produced very different strategies to cope with the environment and their respective fitness. Variation in the sun's position and the plants' relative orientation led to either having large leaves so that production is maximized during short favorable moments Fig. (3a,c,e), or numerous, evenly spread, leaves so that sunlight can be efficiently gathered throughout the day by different parts of the plant Fig. (3d,f,h).

Root morphology was not thoroughly investigated, due to the uniform water distribution exerting only very limited evolutionary pressure, and most individual manage with a simple root trunk connected to a handful of capillary tubes Fig. (3b,d,f). Some even went as far as to completely forsake the former Fig. (3a,e,g).

As the autonomous reproduction process starts from flowers, their growth is of utmost importance for a species' permanence. All plants except two from Fig. 3 actually generate at least one such organ, though only Fig. 3g,e,a manage to bring them to maturity.

3.3 Strategies

From these morphologies and their associated dynamics graphs, we can extract three main strategies: quiescence, expansion and reproduction as illustrated in Fig. 4. The first one (in red) is quite straightforward in its survival method. One can see on the graphs that after a short burst of activity, early on in the simulation, this type of individuals goes into a quiescent state, keeping its metabolic value in a comfortable range so that most plants make it to the end of simulation. The expansionist (in blue) however, adopts a radically different approach: instead it tries to reach as fast as it can a mature state which can, depending on

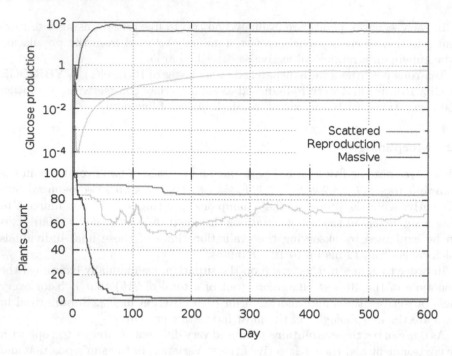

Fig. 4. Typical examples of the three strategies' dynamics. Individuals are taken from Fig. 3 with the red, green and blue curves corresponding to Fig. 3f,g and e (Color figure online)

the plant, take up to a full year. This allows the ecosystem to compensate for the extremely high mortality rate: in the example depicted, 96% of the population dies in the first hundred days. Finally, the reproduction strategy (in green) relies on having the smallest possible morphology, i.e. a small seed and a single root hair directly connected to the leaf. Resources are mostly directed towards producing mature fruits as quickly as can be, thus maintaining a population in a safe range ($[60, 80]$ in this case).

It is interesting to compare these behaviors with those obtained in [2] where varying environmental factors led to the emerge of the CSR triangle [17]. Indeed, a plant population under favorable conditions should evolve towards individual competition while with decreasing resources availability a slower, more conservative, metabolism is expected. If exposed to recurrent, localized uncontrolled deaths, the ruderals would thrive with their fast life cycle and colonization approach.

The fact that all three strategies emerged within identical environments shows that, on the one hand, the genetic search space is large enough to contain very distinct viable genomes even before being subjected to an evolutionary process. On the other hand, it also warms about a possibly *too* large search space with functional genotypes separated by wide unfit combinations.

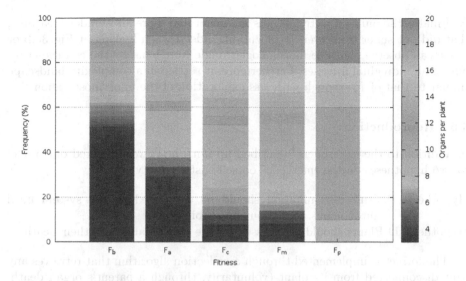

Fig. 5. Repartition of average organ count per fitness across all runs.

3.4 Influence of Evaluation Criteria

We now turn our attention to the evolution procedure itself and more specifically the contribution of our fitness functions set. The diversity of criteria used induced a similar amount of variability in the obtained genomes as one can see the range of morphologies and behavior obtained. All fitnesses, however, did not perform equally both in terms of complexity (see Fig. 5) and relevance. Indeed, while F_b produced plants that could grow at a sustained pace, they proved quite simplistic, morphologically speaking, with almost 60% of the champions growing less than five organs. Given that every random graphtal starts with this specific amount, it shows that evolution discovered that the bigger one wants to grow, the smaller the genotype.

A similar trend can be observed in F_a, though with a slight offset caused by the necessity of having sexual organs. The global strategy for this fitness is as described in details for individual Fig. 3g: small genome, small plant, fast mating. This tendency is reversed in F_p, with no instance in the 'Minimalist' section of the phenotypic space. Indeed, as glucose production requires both efficiently positioned leaves and sufficient water uptake, evolution favored genotypes with repetitive structure. On the downside, this also led to extremely long evaluation times which prevent all runs to reach 250 generations with no overwhelming advantages over the competing fitnesses.

The coverage-oriented evolutions performed by F_c led to a more balanced distribution of organ count between minimalism and over-complexification. While this is the less biologically inspired criterion, it proved more robust to being exploited by the evolutionary algorithm and, paradoxically, brought more life-like individuals about, such as Fig. 3a.

Finally, the multi-objective fitness F_m generated more all-rounder creatures, that did not suffer from over-optimization. Indeed when looking at Fig. 3c,b or f, one can observe plausible morphologies made functional by the contradictory pull of all individual fitnesses. Furthermore, it settled in a complexity landscape similar to that of F_c, though with less exploration of the uppermost region.

3.5   Reproduction

In addition to the experiment described up until now, we subjected our model to two hypotheses on the emergence on self-sustainability:

Hypothesis **F** Fruit dissemination should be supported by the system until autonomous abscission can stabilise.
Hypothesis **D** Plants should be stressed by the unavoidability of their deaths.

The former is implemented through a collection algorithm that retrieves any fruit disconnected from its plant (voluntarily, through a parent's organ death, etc) and proceeds to its dissemination through the usual algorithm. The latter is emulated by extending the simulation duration s_d when in presence of self-reproductive behavior. That is, for a number g of autonomous generations, the allotted number of years is $s_d = min(10, max(g, 1) * 2)$. This allows genomes exhibiting self-reproduction capabilities to reach much higher fitness values thus increasing their chance of producing offspring in the next generation of the genetic algorithm.

We tested all four combinations with none (f,d), one (f,D and F,d) and both (F,D) of the hypotheses active in the runs to see how this would influence the capacity to develop self-reproduction (Fig. 6).

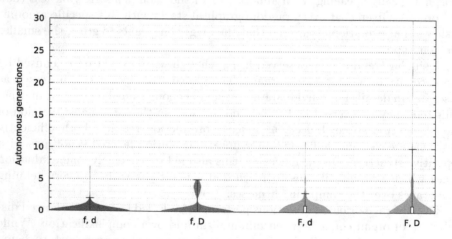

Fig. 6. Violin plots of the number of autonomous generation per run type (see the text for details). The colored area displays the kernel density overlayed with a boxplot with whiskers spanning the 95% confidence interval.

The initial conditions (f,d) proved quite detrimental for the emerge of self-sustainability: indeed, aside from a single outlier reaching 7 autonomous generations, 93% of the individuals obtained do not reproduce at all. Only introducing adaptive simulation duration (f,D) reduces this number to 84% and creates a secondary behavioral cluster around the 4th generation. This dynamic is inverted when only collecting immature fruits with only 54% of non-reproducing runs and 38% reaching the first generation. Finally when both features [F,D] are enabled the threshold of less than 50% of infertile individuals is crossed, albeit slightly, and peak performance is at absolute maximal across all alternatives (19th, 22nd and 30th autonomous generations obtained in the 10 years allotted time-frame).

Based on the capacity to produce self-reproducing plants and the results of T-Test evaluations we can surmise that run (f,d) is outperformed by every other alternative (p-value <= 0.001). Additionally, while no significant differences were detected between runs (f,D) and (F,D) the last one (F,D) shows better results (p-value < 0.01) than both of them.

In order to further understand the dynamics behind these differences in self-reproductive behavior, we investigated what 'checkpoint' individuals tended to stop at (Fig. 7). The different categories are

– None: No sexual organs were produced by the plants
– Flowers: Some were produced but never fecundated
– Fruits: Seeds were produced but never planted
– Repro.: Self-reproduction occurred.

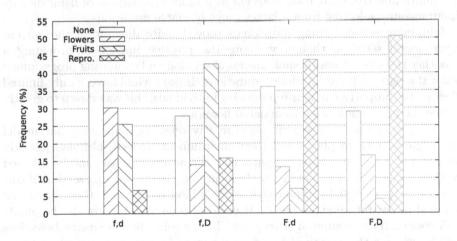

Fig. 7. Repartition of checkpoints for autonomous reproduction. Colors are the same as Fig. 6, for details of the checkpoints see text.

At first glance, we can note different locations for the point(s)-of-failure depending on which Hypothesis was enabled. While run (f,d) seems to struggle at every checkpoint, loosing almost a third of its population each time, run (f,D) understood the importance of producing fruits but rarely found how to disseminate them into the world. Both (F,d) and (F,D) show no particular problem on this point due to the algorithm taking the lead when necessary and, instead, are clustered between individuals that do not attempt to reproduce and those that succeed.

In the end, both hypotheses are verified by this additional runs as, on the one hand, emphasizing the need to circumvent individual death promotes self-replication while, on the other hand, providing a fallback mechanism, until self-controlled abscission can stabilise, leads to fruit generation.

4   Conclusion

In this work, we devised a complex genetic encoding for plant morphologies derived from K. Sims' seminal work. The intricate mapping into a phenotype is highly dependent on the environmental conditions of both the abiotic (light, water) and biotic (competition) components. A self-reproduction scheme, based on our bail-out crossover algorithm, was introduced to generate self-sustaining ecosystems.

The ensuing experiment highlighted the difficulty of developing a balanced strategy between survival and reproduction with individuals mostly falling into either extremes. Furthermore, we observed that the size of the genotypic space was large enough to, in itself, create a speciation phenomena solely based on the starting point of an evolutionary run. We additionally noticed the positive impact of a multi-objective evaluation criterion to promote robustness of demeanor by simultaneously selecting from a larger range of viable mutations.

Obtaining self-sustaining individuals proved quite difficult in the current experimental settings, though we note the positive impact of providing a smoother transition towards spontaneous abscission and emulating programmed death through adaptive simulation durations. Indeed while having only limited effect on the proportion of non-reproducing individuals, this was shown to greatly reduce the production of un-fecundated flowers.

Multiple avenues of research are currently open for improving our model with regards both to plants and their environment. Indeed, although already quite complex, our graphtals currently lack the means to sense, and be affected by, the temperature at ground level. As mentioned in the opening section of this paper, it is an essential component in earth's biodiversity and, thus, should be included in our future experiments. It could also be used to provide the plants with some form of temporal perception. This would allow adaptive behaviors such as delaying the growth of a seed until a more favorable season or restarting leaf production comes 'spring'.

The environment itself was kept mostly static for this experiment with the exception of the sun. There is ongoing development to add more diversity both topological, e.g. changing the altitude of different patches, and temporal, e.g. varying the hygrometry. These changes would induce stress in the ecosystem, thus promoting adaptation and diversity. By evolving these environments and selecting for those with the most stimulating patterns, we could obtain simulations with highly adapted yet clustered populations.

A further step in this direction would bring about co-evolution and competition by increasing the number of species used to seed the environment. We could then observe the richness of interaction between multiple strands of individuals, both plants and animal, on a long time-scale (as introduced in https://vimeo.com/godinduboisalife/futureworks). The use of graphtal to grow plants being straightforward to extend to animals, as this technique is extensively used in the literature.

Acknowledgments. This work was performed using HPC resources from CALMIP (Grant P16043) and the Bullet Physics SDK http://bulletphysics.org.

References

1. Prusinkiewicz, P., Lindenmayer, A.: The Algorithmic Beauty of Plants. The Virtual Laboratory, vol. 54. Springer, New York (1990). https://doi.org/10.1007/978-1-4613-8476-2
2. Bornhofen, S., Barot, S., Lattaud, C.: The evolution of CSR life-history strategies in a plant model with explicit physiology and architecture. Ecol. Model. **222**(1), 1–10 (2011)
3. Hornby, G.S., Pollack, J.B.: Evolving L-systems to generate virtual creatures. Comput. Graphics (Pergamon) **25**(6), 1041–1048 (2001)
4. Sims, K.: Evolving 3D morphology and behavior by competition. Artif. Life **1**(4), 353–372 (1994)
5. Joachimczak, M., Wróbel, B.: Evo-devo in silico-a model of a gene network regulating multicellular development in 3D space with artificial physics. In: Alife, pp. 297–304 (2008)
6. Disset, J., Cussat-Blanc, S., Duthen, Y.: Evolved development strategies of artificial multicellular organisms. In: 15th International Symposium on the Synthesis and Simulation of Living Systems (ALIFE XV 2016), pp. 1–8. The MIT Press, Cancun (2016). https://hal.archives-ouvertes.fr/hal-01511892
7. Gardner, M.: Mathematical games: the fantastic combinations of John Conway's new solitaire game "life". Sci. Am. **223**, 120–123 (1970)
8. Adami, C., Brown, T., Kellogg, W.K.: Evolutionary learning in the 2D artificial life system Avida. In: Artificial Life IV. MIT Press (1994)
9. Metivier, M., Lattaud, C., Heudin, J.c.: A stress-based speciation model in LifeDrop characters. In: Artificial Life (2002). http://www.alife.org/alife8/proceedings/sub265.pdf
10. Ventrella, J.: GenePool: exploring the interaction between natural selection and sexual selection. In: Adamatzky, A., Komosinski, M. (eds.) Artificial Life Models in Software, pp. 81–96. Springer, London (2005). https://doi.org/10.1007/1-84628-214-4_4
11. Miconi, T.: Evosphere: evolutionary dynamics in a population of fighting virtual creatures. In: 2008 IEEE Congress on Evolutionary Computation, CEC 2008, pp. 3066–3073 (2008)
12. Woodward, F.I., Williams, B.G.: Climate and plant distribution at global and local scales. Vegetatio **69**(1–3), 189–197 (1987)

13. Canino-koning, R., Wiser, M.J., Ofria, C.: The evolution of evolvability: changing environments promote rapid adaptation in digital organisms. In: Proceedings of the European Conference on Artificial Life, pp. 268–275 (2016). https://www.mitpressjournals.org/doi/abs/10.1162/ecal_a_0047

14. Dubois, K., Cussat-Blanc, S., Duthen, Y.: Towards an artificial polytrophic ecosystem. In: Morphogenetic Engineering Workshop, at the European Conference on Artificial Life (ECAL) 4 September 2017 (2017). http://doursat.free.fr/docs/MEW2017_program_and_abstracts.pdf#page=4

15. Bonfim, D.M., de Castro, L.N.: FranksTree: a genetic programming approach to evolve derived bracketed L-systems. In: Wang, L., Chen, K., Ong, Y.S. (eds.) Advances in Natural Computation, pp. 1275–1278. Springer, Heidelberg (2005). https://doi.org/10.1007/11539087_168

16. Lehman, J., Stanley, K.O.: Exploiting open-endedness to solve problems through the search for novelty. In: Artificial Life (2008). http://eplex.cs.ucf.edu/papers/lehman_alife08.pdf

17. Grime, J.P.: Evidence for the existence of three primary strategies in plants and its relevance to ecological and evolutionary theory. Am. Nat. **111**(982), 1169–1194 (1977)

Networks and Distributed Systems

Early Detection of Botnet Activities
Using Grammatical Evolution

Selim Yilmaz[✉] and Sevil Sen

WISE Lab., Department of Computer Engineering, Hacettepe University,
Ankara, Turkey
{selimy,ssen}@cs.hacettepe.edu.tr

Abstract. There have been numerous studies proposed for detecting
botnets in the literature. However, it is still a challenging issue as most of
the proposed systems are unable to detect botnets in their early stage and
they cannot perform satisfying performance on new forms of botnets. In
this study, we propose an evolutionary computation-based approach that
relies on grammatical evolution to generate a botnet detection algorithm
automatically. The performance of the proposed flow-based detection
system reveals that it detects botnets accurately in their very early stage
and performs better than most of the existing methods.

Keywords: Botnet · Flow-based detection ·
Evolutionary computation · Grammatical evolution

1 Introduction

Botnet is a number of compromised devices called as *bots or zombies* which
are controlled through a special Command and Control (C&C) channel by
an intruder node known as *botmaster*. By taking advantage of computational
resources of bots, botnets are used for performing several distributed illegal
activities such as spamming, distributed denial-of-service attacks. The increasing size of botnets have now reached to an unprecedented level such that more
than 80% of the Internet traffic is propagating through botnets [1]. Distributing
such bots are cheap, however they could have drastic effects on the economy. It
is reported that malware distribution has caused a damage of 13.2 billion to 67.2
billion USD to the global market within only two years in the past [1]. Hence,
developing robust and effective detection systems towards the different forms of
botnets has become a must.

In this study, we proposed an evolutionary computation-based botnet detection system for early detection of botnets. Grammatical Evolution (GE) is
employed due to its capability in generating interpretable computer programs
for security experts. As inputs to GE, the TCP/UDP packet flows have been
preferred to the packet payloads due to allowing us to inspect encrypted or
obfuscated network traffic. Since most of the network traffic today is encrypted,

© Springer Nature Switzerland AG 2019
P. Kaufmann and P. A. Castillo (Eds.): EvoApplications 2019, LNCS 11454, pp. 395–404, 2019.
https://doi.org/10.1007/978-3-030-16692-2_26

monitoring information in the packet headers for a possible suspicious activity has become the only way. Furthermore, since the packet headers correspond to only a small fraction of the whole network traffic, it reduces the computational overhead. In addition, working on the smaller time windows rather than all individual flows enables our system to detect bot activity before the attack phase. To the best of the authors' knowledge, this is the first study that explores the use of grammatical evolution for detecting botnets at the early stage and that uses the most recent dataset in the literature for identifying new botnet types.

The rest of the paper is organized as follows: Sect. 2 summarizes the related approaches in the literature. Section 3 explains grammatical evolution algorithm in detail. The proposed framework is outlined in Sect. 4. The experimental setup and results are provided in Sect. 5. Section 6 concludes the paper.

2   Related Work

Machine learning-based techniques have already been proposed to detect different types of botnets (i.e., IRC, HTTP, and P2P) within the recent years. However, most of these approaches employ deep packet inspection, which cannot analyze encrypted traffic. Given that the majority of today's network traffic is encrypted, researchers have moved their focus on developing flow-based detection systems as complementary to such systems recently. In this section, such flow-based detection approaches are reviewed.

Huseynov et al. [2] proposed a method based on the similarity of communication patterns between botmasters and bots. It relies on a semi supervised method where only 10% of all traffic was labeled. Firstly, similar activities are clustered by using ant colony optimization and then, clusters are identified based on their similarities with the labeled data. Narang et al. [3] applied different feature selection methods for identification of botnet activities. The authors concluded that built time considerably increases when the full feature set is applied.

Kirubavathi et al. [4] proposed a method for detecting HTTP-based botnets. Therefore, the inputs to the proposed neural network consist of only TCP related features. They extended their study in [5] to study other types of communication schemes (IRC & P2P) and to investigate other classification methods (i.e., J48, Naïve Bayes, SVM). In another study based on neural networks, Nogueira et al. [6] proposed a framework called botnet security system (BoNeSSy) where historical profiling provided by each application were used to train the network.

A behavioral-based P2P botnet identification system was proposed by Saad et al. [7]. They regarded network traffic as data stream where bots tend to behave differently over time. The authors also studied the ability of five different learning algorithms for online botnet detection. Although the performance of these algorithms was promising, they were not enough to satisfy all the requirements of an online detection system. Another behavioral-based botnet detection method by Wang et al. [8] identifies malicious domain names and IP addresses by using fuzzy pattern recognition technique.

Livadas et al. [9] proposed an IRC-based botnet detection system of two stages. The first stage filters out non-chat flows. In the second stage, real chat flows are distinguished from botnet flows using J48, Naïve Bayes, and Bayesian network classifiers. Fedynyshyn et al. [10] proposed a host-based approach to detect botnet traffic by also employing random forest and J48 classification algorithms.

3   Grammatical Evolution

Grammatical evolution is an evolutionary computation technique inspired by the biological process of generating a protein from the genetic material of an organism. It first generates a certain number of individuals which represent candidate solutions for the targeted problem and then, examines the 'fitness' of each individual and finally creates new individuals by applying genetic operators (such as crossover, mutation, and, etc.). These are the steps of a single evolution step. The best solution is yielded by GE at the end of the evolution process.

GE is capable of generating a solution (program) in any language. The grammar of this language is expressed in a notation named Backus-Naur Form (BNF). BNF grammars consists of **terminals**, which are the end-items in the language and **nonterminals**, which can be expanded by the terminals or nonterminals. A grammar is represented by a four-tuple {T,N,P,S}. T represents the terminal set, N represents the nonterminal set, P is production rule set comprising of a number of grammar rules, and S is a start symbol indicating the entry-point of the grammar. To generate a program, GE makes use of a 8-bit binary string genomes (called *codons*) that are assigned to every individual and mapped to a sequence of integer values. The mapping process in GE results in a higher genetic diversity in the population than the other evolutionary-based approaches, which is the main advantage of GE [11].

To elaborate the process of automatic program generation, one-step evolution of GE on a symbolic regression problem is explained. Let the problem to be examined be $f(X) = X^4 + X^3 + X^2 + X$, and the BNF notation for the grammar be:

(1)	`<expr>::= <expr><op><expr>`	(0)
	`| (<expr><op><expr>)`	(1)
	`| <pre-op>(<expr>)`	(2)
	`| (<var>)`	(3)
(2)	`<op>::= +`	(0)
	`| -`	(1)
	`| /`	(2)
	`| *`	(3)
(3)	`<pre-op>::= Sin`	(0)
	`| Cos`	(1)
	`| Exp`	(2)
	`| Log`	(3)
(4)	`<var>::= X`	(0)
	`| 1.0`	(1)

The symbols enclosed by brackets (<>) are the nonterminals and others are the terminals, starting symbol (S) is <expr>. Suppose the genome of an individual is:

220	35	84	42	251	15	47	66

The first codon value (220 in this example) is used to expand the first nonterminal item in the language, which is initially indicated by S. Every encountered nonterminal in the BNF is replaced with the following rule:

$$rule = (codon\ integer\ value)\ MOD\ (\#\ of\ rules\ of\ nonterminal)$$

The first nonterminal here is <expr> and the codon value is 220 which selects rule 0 (220 MOD 4 = 0) and <expr> is replaced by <expr><op><expr>. The left-most nonterminal item is still <expr> but the codon value is now 35 which results in a selection rule of 3 (35 MOD 4 = 3). Now the language has become (<var>)<op><expr>. This process continues until all nonterminal items are replaced to terminals. In the case where the last codon value is reached but there still remains nonterminals in the language, the codon pointer switches back to the first codon value that is 220 – this process is called *wrapping*. The final language generated according to the genome and the grammar is '(X)/(1.0)'. Please refer to [12] for the detailed description of the algorithm.

4   The Proposed Method

The proposed approach comprises of three consecutive phases. The first phase involves obtaining traffic flows in different time windows from a real-world dataset (in *.pcap* format) in order to determine the optimal window size for obtaining high

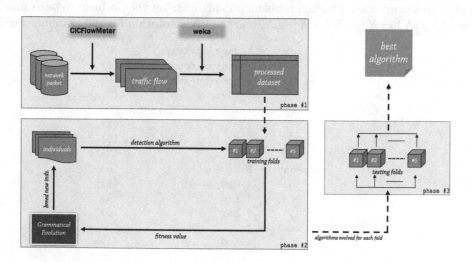

Fig. 1. General framework of the proposed botnet detection system

detection accuracy in botnet detection. Moreover, a number of distinctive features of network flows are extracted for botnet detection. In the second phase, grammatical evolution is employed to evolve a detecting algorithm automatically by using these features. The final phase evaluates the performance of all evolved detection algorithms. The general framework of our approach has been demonstrated in Fig. 1. Each phase is explained in detail in the following two subsections.

4.1 Flow Generation and Feature Extraction

A flow is a collection of packets having the same five-tuple that is source IP, destination IP, source port, destination port, and protocol. These features can be extracted directly from the TCP/UDP headers. However an extra computation is required to extract additional statistical features from these packets such as flow duration. CICFlowMeter [13], a Java-based bidirectional flow generator and analyzer tool integrating jNetPcap [14] library, has been used to generate traffic flows and to extract all traffic-related features from these flows. This tool supports 76 traffic-related statistical features calculated separately for the forward and backward directions. The identifying attributes like 'flow ID' have been discarded from the feature set as they might not be good indicators especially when the training and testing data come from different network domains. The tool terminates a flow depending on the flow timeout or the FIN flag. In this study, the flow timeout value is set to 30 min. It labels a flow *bot* or *benign* considering the IPv4 addresses of the sender or the receiver. The more information about the feature set can be found in [13]. In addition to these features, in this study, bidirectional initial packet sizes are also extracted from the flow as they are known to be very effective in identifying botnet activity.

Individual flows are split into different time windows. In order to obtain an appropriate window-length, we have conducted a pre-experiment, in which every individual flow has been windowed from 120 s to 360 s in the multiples of 60 s. The results showing the effect of the window's size on botnet detection are discussed in Sect. 5.2. In the training, the optimal window size of this pre-experiment is used. Moreover, as the feature values are in a high scale, all features' values are mapped to a range from 0 to 1.

4.2 Evolution of Botnet Detection Algorithm

In this study, GE has been explored for the evolution of an algorithm which detects botnets by inspecting the most discriminative features of the flow. Every individual in GE represents a candidate detection algorithm and is evaluated depending on its detection ability which has been measured by the following 'fitness' equation:

$$fitness = TP - (\omega \times FP) \tag{1}$$

where TP and FP are the true positive and false positive, respectively, and ω (=3, empirically found) is a constant factor which increases the magnitude of FP. The idea behind why ω constant has been introduced into the equation is to avoid over-stimulation to the botnet patterns as the dataset is highly imbalanced (with extracted 2,390,624 benign and 79,428 botnet flows). We have employed 5 fold cross-validation and thus the whole dataset is divided into 5 folds such that 4 of them have been used for training and the remainder has been used for the testing of the evolved algorithms. Due to the stochastic nature of evolutionary-based algorithms, we have run GE 10 times per a fold (50 runs in total). Java-based evolutionary computation toolkit (ECJ) [15] has been used for the GE implementation. All the GE parameters used during the training have been provided in Table 1. The parameters not listed here are the default parameters of ECJ.

The BNF grammar for the problem is provided in Fig. 2. Traffic-related features generated through CICFlowMeter and random values (**rnd**) generated from a uniform distribution between 0 and 1 are employed as operands. A number of arithmetic and relational operations as well as mathematical functions are defined as operators. The detection performance of all detection algorithms has been evaluated on the folds spared for testing, then the best algorithm has been determined, which corresponds to the final phase of our methodology. The details regarding this phase is outlined in Sect. 5.2.

Table 1. GE Parameters

Parameters	Value
Functions	$+, -, *, /$, sin, cos, log, ln, sqrt, abs, exp, ceil, floor, max, min, pow, mod, $<, \leq, >, \geq, ==, !=$, and, or
Terminals	rnd(0,1), features in [13]
Population Size	50 (number of elite individuals = 5)
Crossover probability	0.9
Mutation probability	0.1
Selection strategy	Tournament selection (Tournament size: 7)
Generations	1000

```
(1)    <algorithm>  ::= if(<cond>) { alert(); }
(2)    <cond>       ::= <cond><set-op><cond> | <expr><rel-op><expr>
(3)    <expr>       ::= <expr><op><expr> | (<expr><op><expr>)
                        | <pre-op>(<expr>) | <var>
(4)    <op>         ::= + | - | / | *
(5)    <pre-op>     ::= sin | cos | log | ln | sqrt | abs | exp | ceil
                        | floor | max | min | pow | modulus
(6)    <rel-op>     ::= < | ≤ | > | ≥ | == | !=
(7)    <set-op>     ::= and | or
(8)    <var>        ::= rnd | feature set in [13].
```

Fig. 2. BNF grammar of the botnet detection problem.

5   Experiments

5.1   Datasets

We have incorporated two datasets. The first dataset is the well-known Information Security and Object Technology (ISOT) dataset. It contains traffics belonging to the P2P Storm and Waledac botnets captured from 2007 to 2009. Benign traffics, on the other hand, were recorded from the everyday traffics such as HTTP web browsing, gaming packets, torrent packets like Azureus. The second dataset is CICIDS2017. It was obtained from the traffic generated between July 3 to July 7 in 2017, where different attacks were implemented for each day. We have used only botnet (Ares) and benign packets of this dataset captured on July 7. Please refer to [7,16] for the detailed description of the ISOT and CICIDS2017 datasets, respectively.

5.2   Results

As discussed earlier, we have conducted a pre-experiment in order to investigate the effect of the window size on botnet detection. GE has been run 5 times under the same settings given in Table 1 but with 500 generations. The results show that the algorithm detects bot activity with a higher performance as the window size increases up to 240 s (see Fig. 3). Hence, the optimal window size is set to be 240 s for the evolution of botnet detection algorithm in this study.

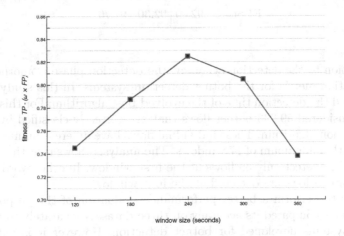

Fig. 3. Effect of the window size on botnet detection

The best, worst, and the mean statistical findings of the best detection algorithm generated from every testing fold have been provided in Table 2. As it is shown in the table, the best algorithm has shown very similar performances on different test settings. The best performance suggests that the evolved algorithm has been able to detect bot traffic with an accuracy of 92.92%. The higher rate of precision value shows that the algorithm has successfully detected bot activities with rare *false alarms*, which is known to be crucial for such an imbalanced traffic.

The evolved algorithm that gives the best accuracy is given in Algorithm 1. GE is known with its ability in generating interpretable programs. Although the best evolved program is readable, it is not easily interpretable since we have employed a large number of functions and features (see Fig. 2). In addition to that, we have not limited the length of the generated program. We can conclude from the algorithm that, out of the 78 features, only 31 features have been enough to distinguish the bot traffic from the benign traffic. With 12 source originated and 11 destination originated features used in the detection algorithm, it can be deduced that forward and backward related features have an equal effect on the botnet detection.

Table 2. Statistical performance analysis of the best evolved algorithm

	Performances (%)		
	Best	Mean	Worst
TP	88.77	88.10	87.17
FP	2.54	2.80	2.98
Precision	97.19	96.93	96.75
Accuracy	92.92	92.65	92.17
F1 Score	92.60	92.30	91.76

In addition to the detection accuracy, the detection phase is another important evaluation metric for the botnet detection systems. In this study, we have also analyzed the detection time of the evolved best algorithm. For this purpose, we have considered 59,933 botnet flows that are correctly classified by our proposed detection algorithm. These individual flows have different numbers of time windows with a maximum of 17 windows. The analysis shows that the algorithm have missed to detect only 69 flows at the first window. In other words, 99.88% of all individual flows are detected at the first window.

In order to perform a better performance evaluation of the proposed algorithm, we have compared its accuracy on three datasets separately to the state-of-the-art systems developed for botnet detection. However it is not easy to conduct a fair comparison due to the different network environments, bot binaries, and etc. [17]. The comparative results (see Table 3) show that the proposed method performs best on the ISOT dataset. In addition, the proposed algorithm has performed better than most of the existing methods.

Table 3. Comparison results

Study	Dataset	C& C structures	Accuracy (%)
Huseynov et al. [2]	ISOT	P2P	72.15
Nogueira et al. [6]	Generated traces	IRC, P2P, HTTP	87.56
Saad et al. [7]	ISOT	P2P	89.00
Huseynov et al. [2]	ISOT	P2P	89.85
Livadas et al. [9]	*Dartmouth* trace [18]	IRC	92.00
Narang et al. [3]	ISOT & Generated trace	P2P	92.55
Fedynyshyn et al. [10]	Generated traces	IRC, P2P, HTTP	92.90
Present study	ISOT & CICIDS2017	P2P	92.92
Wang et al. [8]	Generated traces	IRC, P2P, HTTP	95.00
Present study	CICIDS2017	P2P	95.09
Present study	ISOT	P2P	96.24
Kirubavathi and Anitha [4]	Generated traces	HTTP	99.02
Kirubavathi and Anitha [5]	ISOT & Others*	IRC, P2P, HTTP	99.14

* They merged 11 datasets including various types of bots like Skynet, Rbot and etc.

6 Conclusion

In this study, an evolutionary computation-based botnet detection system is proposed to address the aforementioned issues raised by existing techniques. GE is employed to generate a detection algorithm in a readable form to distinguish botnet traffic from normal traffic. P2P-based botnets, which are the latest and the most challenging type of botnets currently available, are focused in this study. To the best of our knowledge, it is the first study that explores the use of GE for detecting botnet flows. The results show that the proposed method has achieved a very high detection accuracy and performed better than most of the detection methods. In addition, it has accurately detected 99.88% of botnet traffics in their first window which proves the early detection capability of our approach. Therefore, we can easily conclude that grammatical evolution is very promising and applicable for botnet detection.

References

1. Karim, A., Salleh, R.B., Shiraz, M., Shah, S.A.A., Awan, I., Anuar, N.B.: Botnet detection techniques: review, future trends, and issues. J. Zhejiang Univ. Sci. C **15**(11), 943–983 (2014)
2. Huseynov, K., Kim, K., Yoo, P.D.: Semi-supervised botnet detection using ant colony clustering. In: Proceedings of Symposium on Cryptography and Information Security (SCIS), pp. 1–7 (2014)

3. Narang, P., Reddy, J.M., Hota, C.: Feature selection for detection of peer-to-peer botnet traffic. In: Proceedings of the 6th ACM India Computing Convention, Compute 2013, pp. 16:1–16:9. ACM, New York (2013)

4. Kirubavathi Venkatesh, G., Anitha Nadarajan, R.: HTTP botnet detection using adaptive learning rate multilayer feed-forward neural network. In: Askoxylakis, I., Pöhls, H.C., Posegga, J. (eds.) WISTP 2012. LNCS, vol. 7322, pp. 38–48. Springer, Heidelberg (2012). https://doi.org/10.1007/978-3-642-30955-7_5

5. Kirubavathi, G., Anitha, R.: Botnet detection via mining of traffic flow characteristics. Comput. Electr. Eng. **50**, 91–101 (2016)

6. Nogueira, A., Salvador, P., Blessa, F.: A botnet detection system based on neural networks. In: 2010 Fifth International Conference on Digital Telecommunications, pp. 57–62, June 2010

7. Saad, S., et al.: Detecting P2P botnets through network behavior analysis and machine learning. In: 2011 Ninth Annual International Conference on Privacy, Security and Trust, pp. 174–180, July 2011

8. Wang, K., Huang, C.Y., Lin, S.J., Lin, Y.D.: A fuzzy pattern-based filtering algorithm for botnet detection. Comput. Netw. **55**(15), 3275–3286 (2011)

9. Livadas, C., Walsh, R., Lapsley, D., Strayer, W.T.: Using machine learning techniques to identify botnet traffic. In: Proceedings. 2006 31st IEEE Conference on Local Computer Networks, pp. 967–974, November 2006

10. Fedynyshyn, G., Chuah, M.C., Tan, G.: Detection and classification of different botnet C&C channels. In: Calero, J.M.A., Yang, L.T., Mármol, F.G., García Villalba, L.J., Li, A.X., Wang, Y. (eds.) ATC 2011. LNCS, vol. 6906, pp. 228–242. (2011). https://doi.org/10.1007/978-3-642-23496-5_17

11. O'Neill, M., Ryan, C.: Grammatical evolution. IEEE Trans. Evol. Comput. **5**(4), 349–358 (2001)

12. Ryan, C., Collins, J.J., Neill, M.O.: Grammatical evolution: evolving programs for an arbitrary language. In: Banzhaf, W., Poli, R., Schoenauer, M., Fogarty, T.C. (eds.) EuroGP 1998. LNCS, vol. 1391, pp. 83–96. Springer, Heidelberg (1998). https://doi.org/10.1007/BFb0055930

13. CICFlowMeter: Network Traffic Flow Analyzer. http://netflowmeter.ca/netflowmeter.html. Accessed 25 Nov 2018

14. jnetpcap. http://jnetpcap.com. Accessed 01 July 2018

15. ECJ: A java-based evolutionary computation research system (2017). https://www.cs.gmu.edu/eclab/projects/ecj/

16. Sharafaldin, I., Lashkari, A.H., Ghorbani, A.A.: Toward generating a new intrusion detection dataset and intrusion traffic characterization. In: Proceedings of the 4th International Conference on Information Systems Security and Privacy. SciTePress - Science and Technology Publications (2018)

17. Lu, W., Rammidi, G., Ghorbani, A.A.: Clustering botnet communication traffic based on n-gram feature selection. Comput. Commun. **34**(3), 502–514 (2011). Special Issue of Computer Communications on Information and Future Communication Security

18. Henderson, T., Kotz, D., Abyzov, I.: The changing usage of a mature campus-wide wireless network. In: Proceedings of the 10th Annual International Conference on Mobile Computing and Networking, MobiCom 2004, pp. 187–201. ACM, New York (2004)

Exploring Concurrent and Stateless Evolutionary Algorithms

Juan J. Merelo[1(✉)], J. L. J. Laredo[2], Pedro A. Castillo[1],
José-Mario García-Valdez[3], and Sergio Rojas-Galeano[4]

[1] Universidad de Granada/CITIC, Granada, Spain
{jmerelo,pacv}@ugr.es
[2] RI2C-LITIS, Université du Havre Normandie, Le Havre, France
juanlu.jimenez@univ-lehavre.fr
[3] Instituto Tecnológico de Tijuana, Calzada Tecnológico s/n, Tijuana, Mexico
mario@tectijuana.edu.mx
[4] School of Engineering, Universidad Distrital Francisco José de Caldas,
Bogotá, Colombia
srojas@udistrital.edu.co

Abstract. Creating a concurrent and stateless version of an evolutionary algorithm implies changes in its algorithmic model. From the performance point of view, the main challenge is to balance computation with communication, but from the evolutionary point of view another challenge is to keep diversity high so that the algorithm is not stuck in local minima. In a concurrent setting, we will have to find the right balance so that improvements in both facets do not cancel out. In this paper we address such an issue, by exploring the space of parameters of a population based concurrent evolutionary algorithm that yields to find out the best combination for a particular problem.

Keywords: Concurrent algorithms · Distributed computing ·
Stateless algorithms · Algorithm implementation ·
Performance evaluation

1 Introduction

Concurrent programming adds a layer of abstraction on the machinery of processors and operating systems to offer a high-level interface that enables the user to program code that might be executed in parallel either in threads or in different processes [1].

Different languages offer different facilities for concurrency at the primitive level, and mainly differ on how they deal with shared state, that is, variables that are accessed from several processes. Actor-based concurrency [2] eliminates shared state by introducing a series of *actors* that store state and can change it; on the other hand, channel based concurrency follows the *communicating sequential processes* methodology [3], which is effectively stateless, with different processes reacting to channel input without changing state, and writing to these channels.

© Springer Nature Switzerland AG 2019
P. Kaufmann and P. A. Castillo (Eds.): EvoApplications 2019, LNCS 11454, pp. 405–412, 2019.
https://doi.org/10.1007/978-3-030-16692-2_27

This kind of concurrency is the one implemented by many modern languages such as Go or Perl 6 [4]. However, the statelessness of the implementation requests for a change in the implementation of any algorithm. In particular, evolutionary algorithms need to migrate its computation model to an architecture that creates and processes streams of data using functions that do not change state.

Despite the emphasis on hardware-based techniques such as cloud computing or GPGPU, there are not many papers [5] dealing with creating concurrent evolutionary algorithms that work in a single computing node or that extend seamlessly from single to many computers.

For instance, the EvAg model [6] is a locally concurrent and globally parallel evolutionary algorithm that leaves the management of the different agents (i.e. threads) to the underlying platform scheduler and displays an interesting feature: the model is able to scale seamlessly and take full advantage of CPU threads. In a first attempt to measure the scalability of the approach experiments were conducted in [7] for a single and a dual-core processor showing that, for cost functions passing some milliseconds of computing time, the model was able to achieve near linear speed-ups. This study was later on extended in [8] by scaling up the experimentation to up to 188 parallel machines. The reported speed-up was ×960 which is beyond the linear ×188 that could be expected if local concurrency were not taken into account.

The aforementioned algorithm used a protocol that worked asynchronously, leveraging its peer-to-peer capabilities; in general the design of concurrent EAs has to take into account the communication/synchronization between processes, which nowadays will be mainly threads. Although the paper above was original in its approach, other authors targeted explicitly multi-core architectures, such as Tagawa [9] which used shared memory and a clever mechanism to avoid deadlock. Other authors [10] actually use a message-based architecture based in the concurrent functional language Erlang, which separates GA populations as different processes, although all communication takes place with a common central thread.

In our previous papers [11,12], we presented the proof of concept and initial results with this kind of stateless evolutionary algorithms, implemented in the Perl 6 language. These evolutionary algorithms use a single channel where entire populations are sent. The (stateless) functions read a single population from the channel, run an evolutionary algorithm for a fixed number of generations, which we call the *generation gap* or simply *gap*, and send the population in the final generation back to the channel. Several populations are created initially, and a concurrent *mixer* is run which takes populations in couples, mixes them leaving only a single population with the best individuals selected from the two merged populations. This *gap* is then conceptually, if not functionally, similar to the *time to migration* in parallel evolutionary algorithms (with which concurrent evolutionary algorithms show a big resemblance).

We did some initial exploration of the parameter space in [13]. In these initial explorations we realized that the parameters we used had an influence at the algorithmic level, but also at the implementation level, changing the wallclock performance of the algorithm.

In this paper we will explore the parameter space systematically looking particularly at two parameters that have a very important influence on performance: population size and generation gap. Our intention is to create a rule of thumb for setting them in this kind of concurrent model, so that they are able to achieve the best performance.

We will present the experimental setup next.

2 Experimental Setup and Results

Initially, we will work with two threads, one for mixing the populations and another for evolution. This setup, while being concurrent, allows us to focus more on the basic features of the algorithm, which is what we aim exploring in this paper.

What we want to find out in these set of experiments is what is the generation gap that gives the best performance in terms of raw time to find a solution, as well as the best number of evaluations per second. In order to do that, we prepared an experiment using the OneMax function with 64 bits, a concurrent evolutionary algorithm such as the one described in [11], which is based in the free Perl 6 evolutionary algorithm library `Algorithm::Evolutionary::Simple`, and run the experiments in a machine with Ubuntu 18.04, an AMD Ryzen 7 2700X Eight-Core Processor at 2195 MHz. The Rakudo version was 6.d,

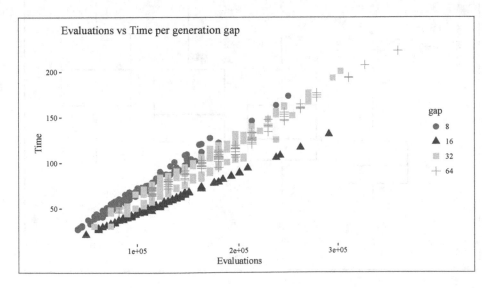

Fig. 1. Number of evaluations vs. time needed for different generation gaps (8 to 64)

which had recently been released with many improvements to the concurrent core of the language. All processing scripts as well as data obtained in the experiments are available, under a free license, from our GitHub repository.

We used a population size of 256, as well as generation gaps increasing from 8 to 64. Many experiments were run for every configuration, up to 150 in some cases. We logged the upper bound of the number of evaluations needed (by multiplying the number of messages by the number of generations and number of individuals evaluated; this means that this number will be an upper bound, and not the exact number of evaluations until a solution is reached). We will first look at the general picture by plotting the wallclock time in seconds (measured by taking the time of the starting of the algorithm and the last message and subtracting the latter from the former) vs the number of evaluations that have been performed. The result is shown in Fig. 1. Experiments with different generation gaps are shown with different colors (where available) and shapes,

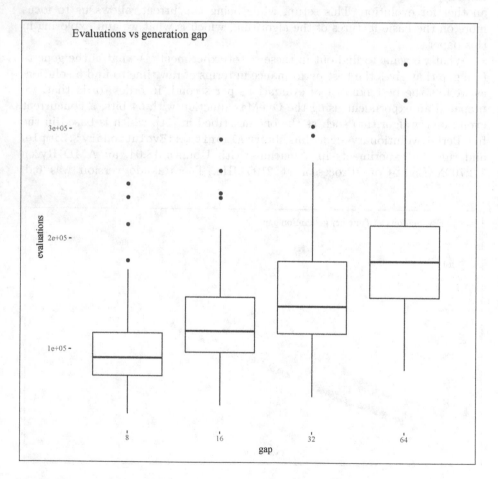

Fig. 2. Number of evaluations needed for different generation gaps (8 to 64)

and they spread in an angle which is roughly bracketed by the experiments with a generation gap of 8, which need the most time for the same number of evaluations, and the experiments with a gap of 16, which usually need the least. The experiments with gaps = 32 or 64 are somewhere in between.

In that same chart it can also be observed that the number of evaluations needed to find the 64 bit OneMax solution is quite different. We make a boxplot of the number of evaluations vs the generation gap in Fig. 2. This figure shows an increasing number of evaluations per gap size. Differences are significant between every generation gap and the next. This increasing number of evaluations per generation gap is probably due to the fact that the increasing number of isolated generations makes the population lose diversity, making finding the solution increasingly difficult. This is the same effect observed in parallel algorithms, as reported in [14], so it is not unexpected. What is unexpected is the combination of generation gap size and the concurrent algorithm, since it is impossible to know in advance what is the optimal computation to communication balance.

We plot the number of evaluations per second in Fig. 3. These show a big difference for a generation gap of 16, with a number of evaluations which is almost 50% higher than for the rest of the generation gaps, where the difference is not so high.

The number of evaluation per second does not follow a clear trend. It falls and remains flat for a generation gap higher than 16; it is also slightly higher than for the minimum generation gap that has been evaluated, 8. This generation gap, however, presents also the lowest number of evaluations to solution, which means that, on average, the solution will be found faster with a generation gap of 8 or 16. This is shown in Fig. 4.

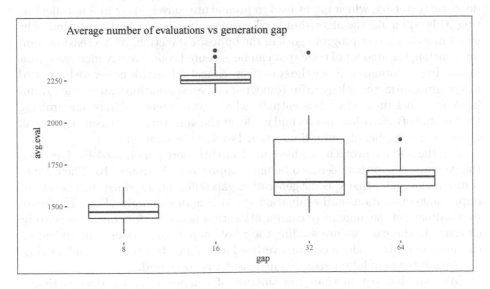

Fig. 3. Average number of evaluations per second for different generation gaps (8 to 64)

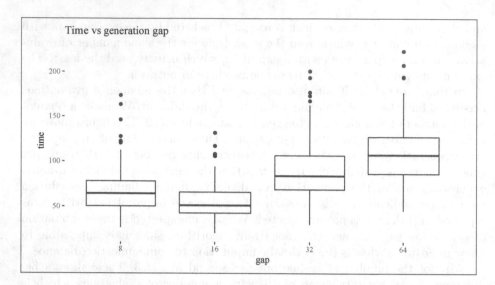

Fig. 4. Boxplot of the average wallclock time for different generation gaps (8 to 64)

3 Conclusions and Discussion

In this paper we have set out to explore the interaction between the generation gap and the performance measures in a concurrent and stateless evolutionary algorithm. From the point of view of the algorithm, increasing the generation gap favors exploitation over exploration, which might be a plus in some problems, but also decreases diversity, which might lead to premature convergence; in a parallel setting, this will make the algorithm need more evaluations to find a solution. The effect in a concurrent program goes in the opposite direction: by decreasing communication, the amount of code that can be executed concurrently increases, thus improving performance. Since the two effects cancel out, in this paper we have used a experimental methodology to find out what is the combination that is able to minimize wallclock time, which is eventually what we are interested in by maximizing the number of evaluations per second while, at the same time, increasing by a small quantity the number of evaluations needed to find the solution.

For the specific problem we have used in this short paper, a 64-bit Onemax, the generation gap that seems to be more appropriate is around 16. The time to communication for that specific generation gap is around 2 s, since 16 generations imply 4096 evaluations and evaluation speed is approximately 2K/s. This gives us a ballpark of the amount of computation that is needed for concurrency to be efficient. In this case, we are sending the whole population to the communication channel, and this implies a certain overhead in reading, transmitting and writing. Increasing the population size also increases that overhead.

We can thus reason than the amount of computation, for this particular hardware, should be in the order of 2 s, so that it effectively overcomes the amount of communication needed. This amount could be played out in different way, for instance by increasing the population; if the evaluation function takes more time, different combinations should be tested so that no message is sent unless that particular amount of time is reached.

With these conclusions in mind, we can set out to work with other parameters, such as population size or number of initial populations, so that the loss of diversity for bigger population sizes can be overcome. Also we have to somehow address the problem of the message size during communication to and from the channel; in this respect, we are planning to resort to a compact representation of the population, obtained either by a statistical distribution model or a standard compression algorithm. This is left as an avenue for future work.

Acknowledgements. This paper has been supported in part by projects TIN2014-56494-C4-3-P s (Spanish Ministry of Economy and Competitiveness), DeepBio (TIN2017-85727-C4-2-P) and AMED (co-funded by European Regional Development Fund and the region Normandy). I would like to express my gratefulness to the users in the #perl6 IRC channel, specially Elizabeth Mattijsen, Timo Paulsen and Zoffix Znet, who helped us with the adventure of programming efficient concurrent evolutionary algorithms.

References

1. Andrews, G.R.: Concurrent Programming: Principles and Practice. Benjamin/Cummings Publishing Company, San Francisco (1991)
2. Schippers, H., Van Cutsem, T., Marr, S., Haupt, M., Hirschfeld, R.: Towards an actor-based concurrent machine model. In: Proceedings of the 4th workshop on the Implementation, Compilation, Optimization of Object-Oriented Languages and Programming Systems, pp. 4–9. ACM (2009)
3. Hoare, C.A.R.: Communicating sequential processes. Commun. ACM **21**(8), 666–677 (1978). https://doi.org/10.1145/359576.359585
4. Lenz, M.: Perl 6 Fundamentals. Apress, Berkeley (2017). https://doi.org/10.1007/978-1-4842-2899-9
5. Li, X., Liu, K., Ma, L., Li, H.: A concurrent-hybrid evolutionary algorithms with multi-child differential evolution and guotao algorithm based on cultural algorithm framework. In: Cai, Z., Hu, C., Kang, Z., Liu, Y. (eds.) ISICA 2010. LNCS, vol. 6382, pp. 123–133. Springer, Heidelberg (2010). https://doi.org/10.1007/978-3-642-16493-4_13
6. Jiménez-Laredo, J.L., Eiben, A.E., van Steen, M., Merelo-Guervós, J.J.: EvAg: a scalable peer-to-peer evolutionary algorithm. Genet. Program. Evolvable Mach. **11**(2), 227–246 (2010)
7. Laredo, J., Castillo, P., Mora, A., Merelo, J.: Exploring population structures for locally concurrent and massively parallel evolutionary algorithms. In: WCCI 2008 Proceedings, pp. 2610–2617. IEEE Press (2008). http://atc.ugr.es/I+D+i/congresos/2008/CEC_2008_2610.pdf
8. Laredo, J.L.J., Bouvry, P., Mostaghim, S., Merelo-Guervós, J.-J.: Validating a peer-to-peer evolutionary algorithm. In: Di Chio, C., et al. (eds.) EvoApplications 2012. LNCS, vol. 7248, pp. 436–445. Springer, Heidelberg (2012). https://doi.org/10.1007/978-3-642-29178-4_44

9. Tagawa, K.: Concurrent differential evolution based on generational model for multi-core CPUs. In: Bui, L.T., Ong, Y.S., Hoai, N.X., Ishibuchi, H., Suganthan, P.N. (eds.) SEAL 2012. LNCS, vol. 7673, pp. 12–21. Springer, Heidelberg (2012). https://doi.org/10.1007/978-3-642-34859-4_2
10. Kerdprasop, K., Kerdprasop, N.: Concurrent computation for genetic algorithms. In: Proceedings of the 1st International Conference on Software Technology, pp. 79–84 (2012)
11. Merelo, J.J., García-Valdez, J.M.: Mapping evolutionary algorithms to a reactive, stateless architecture: using a modern concurrent language. In: Proceedings of the Genetic and Evolutionary Computation Conference Companion, GECCO 2018, pp. 1870–1877. ACM, New York (2018). https://doi.org/10.1145/3205651.3208317
12. García-Valdez, J.M., Merelo-Guervós, J.J.: A modern, event-based architecture for distributed evolutionary algorithms. In: Proceedings of the Genetic and Evolutionary Computation Conference Companion, GECCO 2018, pp. 233–234. ACM, New York (2018). https://doi.org/10.1145/3205651.3205719
13. Merelo, J.J., García-Valdez, J.-M.: Going stateless in concurrent evolutionary algorithms. In: Figueroa-García, J.C., López-Santana, E.R., Rodriguez-Molano, J.I. (eds.) WEA 2018. CCIS, vol. 915, pp. 17–29. Springer, Cham (2018). https://doi.org/10.1007/978-3-030-00350-0_2
14. Cantú-Paz, E.: Migration policies and takeover times in genetic algorithms. In: Proceedings of the 1st Annual Conference on Genetic and Evolutionary Computation, GECCO 1999, vol. 1, 775 p. Morgan Kaufmann Publishers Inc., San Francisco (1999). http://dl.acm.org/citation.cfm?id=2933923.2934003

Evolving Trust Formula to Evaluate Data Trustworthiness in VANETs Using Genetic Programming

Mehmet Aslan[✉][iD] and Sevil Sen[iD]

WISE Lab., Department of Computer Engineering,
Hacettepe University, Ankara, Turkey
{mehmetaslan,ssen}@cs.hacettepe.edu.tr

Abstract. Vehicular Ad Hoc Networks (VANETs) provide traffic safety, improve traffic efficiency and present infotainment by sending messages about events on the road. Trust is widely used to distinguish genuine messages from fake ones. However, trust management in VANETs is a challenging area due to their dynamically changing and decentralized topology. In this study, a genetic programming based trust management model for VANETs is proposed to properly evaluate trustworthiness of data about events. A large number of features is introduced in order to take into account VANETs' complex characteristics. Simulations with bogus information attack scenarios show that the proposed trust model considerably increase the security of the network.

Keywords: Evolutionary computation · Genetic programming ·
Trust management · Data trust ·
Vehicular Ad Hoc Networks (VANETs)

1 Introduction

Vehicles equipped with smart modules such as Wi-Fi, GPS and computing power can communicate with other vehicles on the road and form a mobile, decentralized, structureless wireless network called Vehicular Ad Hoc Networks (VANETs). Vehicles in VANETs share information to indicate an accident, traffic congestion or another event on a road by sending new event messages or forwarding existing ones to other vehicles. VANET applications use this information to provide traffic safety, improve traffic efficiency and present infotainment.

Complex characteristics of VANETs such as being structureless, decentralized and mobile cause some security challenges. Any vehicle can enter to and exit from a VANET without any control or permission of an authority and start sending or forwarding messages about events occurred on this dynamic environment. This openness makes VANETs vulnerable to several attacks such as bogus information [1]. In this attack scenario, attackers could send false messages about events and also send messages about fake events as if they exist. Vehicles must

© Springer Nature Switzerland AG 2019
P. Kaufmann and P. A. Castillo (Eds.): EvoApplications 2019, LNCS 11454, pp. 413–429, 2019.
https://doi.org/10.1007/978-3-030-16692-2_28

distinguish bogus information to achieve reliable data transfer within network and maintain traffic safety and efficiency. Trust management model is widely used as a solution against such attacks.

Dynamically changing, decentralized and self-organized topology of ad hoc networks make trust management an optimization problem. Thanks to similar nature of biological systems, nature-inspired optimization algorithms are being used to address some problems in ad hoc networks [2]. A taxonomy of nature-inspired algorithms that used to solve problems in ad hoc networks is given as online/offline techniques, centralized/decentralized systems and proposals using local/global knowledge in [3]. Despite of many aspects of ad hoc networks addressed by evolutionary algorithms, there is a lack of studies that use bio-inspired algorithms to bring a solution to trust management in ad hoc networks.

In this research, a genetic programming based trust management system is proposed to properly evaluate trustworthiness of VANET application data against bogus information attacks. The proposed system uses much more trust evidence than other studies to satisfy the requirements of trust management in VANETs. It selects the most appropriate trust evidences as features to make the right decision about the received event messages and their sender vehicles by evolving a trust calculation formula. The simulation results show that the evolved formula prevents propagation of bogus messages successfully.

The rest of this paper is organized as follows: studies on trust management for VANETs and evolutionary computation (EC) techniques for ad hoc networks are reviewed in Sect. 2. The network model is explained in Sect. 3 and the proposed trust management method is described in Sect. 4. Section 5 gives details of the experiments and presents analysis of the simulation results. Finally, conclusions are drawn and future work direction is presented in Sect. 6.

2   Related Work

Previous studies are classified into two categories based on their main focus and relevance to this research. Trust management systems for VANETs found in the literature are reviewed in Sect. 2.1. Section 2.2 highlights evolutionary computation algorithms used to solve different problems in ad hoc networks.

2.1   Trust Management for VANETs

Trust management has many aspects that should be taken into account to establish a proper trust based framework for both VANETs and other ad hoc networks. These aspects are called trust management components and are defined as trust properties, trust management properties, trust metrics and attacks to trust model in several surveys [4–8]. Dynamicity, incomplete/partially transitivity and context-dependency are described in [4,6] and subjectivity and asymmetry are also described in [4] as trust properties. Nonetheless, none of the proposed approaches for VANETs covers all trust properties [4].

In highly dynamic and distributed environments such as VANETs, trust management should be fully decentralized [8]. It is described as one of the most important trust management properties, since a centralized authority cannot be assumed to be exist for trust computation in VANETs [4]. Because of the possibility of interactions with the same vehicle might be low in a fast and dynamic VANET environment, vehicles cannot wait until direct interactions reach a threshold [8]. Another property that should be considered is capturing dynamicity of VANET environment to calculate the trust based on the current situation using event/task type, location and time information [8]. Moreover, the possibility of uncooperative vehicles to enter VANETs freely should also be taken into account in developing a trust management model [4,8].

Decentralized trust models in VANETs that are based on past interactions and environmental information to take dynamic infrastructure of VANETs into consideration are grouped into three categories: entity-oriented trust models, data-oriented trust models, and hybrid trust models [6,8]. Entity-oriented trust model is the traditional way for trust computing that is proposed for many ad hoc networks including VANETs and MANETs (mobile ad hoc networks). It only considers the trustworthiness of nodes in the network and does not compute different trust values for different messages sent from the same nodes. Calculating only the trustworthiness of messages sent from nodes without considering trust values of the nodes themselves is called data-oriented trust model. Hybrid trust models evaluate both entity and data trust. They use trust value of an entity as a parameter in addition to other trust evidences to evaluate trust values of messages sent from it and also update the entity trust value according to calculated data trust value to maintain a trust relationship based on past interactions.

Wei and Chen [9] propose a hybrid trust model to evaluate the trustworthiness of an event message using beacon trust, event trust and reputation trust values of a vehicle in VANETs. It employs both beacon messages and event messages in order to calculate the trust value, and also update the reputation trust value of vehicles by using the latest event's trust value. Event messages are forwarded either to support or to deny opinion according to a trust threshold in this model. They simulate the model with scenarios including both alteration attacks and bogus information attacks and evaluate the model using F_1 [10] measure. However, they only consider a vector of position, velocity and direction values of vehicle and similarity between event location and estimated location of vehicle as trust evidence with a threshold for distance between receiver and sender and a threshold for time delay between event message time and current time.

Yao et al. [11] proposed an entity-oriented trust model and a data-oriented trust model, however they did not integrate these two models. Even though they use trust value of vehicles in VANETs as a parameter of data-oriented trust model, they do not update the trust value of vehicle using the trust value of data sent from it. They take into account different event types and different vehicle types by assigning weights to them and, introduce a weighted version of successful data forwarding rate using event weight called malicious tendency. This value and vehicle type are then used to calculate trust values of vehicles in the entity-oriented trust model. In order to calculate the data trust, in addition to the trust value of the sender vehicle, they use the distance between the event position and the sender vehicle's position and, the difference between the time of event occurrence and the time of event message. They focus on secure routing in the simulations of the proposed entity-oriented trust model and, use black hole attack and selective forwarding attack scenarios as well as a network scenario without attacks. Network based metrics are used to evaluate the entity-oriented trust model and an analysis is made for the data-oriented trust model.

To sum up, studies that focus on decentralized trust models in VANETs either take into account very limited trust evidence or do not attach much importance to hybrid trust models. In this paper, we propose a hybrid trust model that mainly aims to evaluate data trustworthiness by using a large number of items of trust evidence that is gathered from the network. Trust values of entities are also calculated based on data trust values of messages sent by these entities.

2.2   Evolutionary Computation Techniques for Ad Hoc Networks

Nature-inspired algorithms developed for solving different problems in ad hoc networks are classified according to their execution mode, information requirement and executing platform in [3]. Firstly, algorithms are classified as online and offline techniques. Secondly, requirement of information about network is considered and algorithms are classified as global knowledge if they need the whole network information and local knowledge if the nodes only use information gathered by themselves. Lastly, optimization algorithms that are run on a central unit are classified as centralized system and optimization algorithms that are run on each node of the network locally are classified as decentralized system. Authors also classified existing studies based on this taxonomy but they did not mention any research about trust management in ad hoc networks. Most of the bio-inspired algorithms used in ad hoc networks are mainly based on two categories, one is centralized and offline with global knowledge and the other is decentralized and online with local knowledge. The latter is more appropriate for trust management in VANETs as each vehicle must evaluate trust values by using only its own local information while moving online on the network.

A recent survey reviews the applications of evolutionary algorithms that is proposed to solve optimization problems in mobile ad hoc networks in the literature [12]. The survey focuses on MANETs, VANETs and DTNs (delay tolerant networks) and divided the reviewed studies into five categories: topology management, broadcasting algorithms, routing protocols, mobility models and data dissemination. Another survey focuses on the applications of evolutionary computation methods for cybersecurity in MANETs [13]. This survey covers evolutionary algorithms (EA), swarm intelligence (SI), artificial immune systems (AIS) and evolutionary games (EG) and classifies these algorithms based on attack types that they counteract and defense mechanisms implemented by them including node trust and reputation systems. It is shown that most of the proposals in the literature is based on EG [13]. The only application of the EA method to trust and reputation systems is proposed for peer-to-peer networks [14]. To sum up, as far as we know the current study is the first application of evolutionary computation techniques to the trust management problem in VANETs.

3   The Network Model

Since there is no well-accepted standard for VANETs yet, an application layer protocol that the proposed trust model is built on is introduced and explained in this section.

3.1   Basic Assumptions

Ad hoc networks are formed by nodes that participate into the network dynamically and contribute to the network communication by behaving both as nodes and as routers. In terms of VANETs, these nodes are vehicles that move at different speeds and generally arrive at different destinations. These nodes encounter other vehicles in the traffic network and make communication with them on the move. Vehicles generally communicate with each other for just a short time, and then never see each other again, which makes the safely communication harder for such dynamic networks. Unfortunately, there is no standard about communication model in VANETs yet, so researchers have been developing new communication models. In the following, some assumptions about vehicles in order to propose a communication model are introduced.

All vehicles have all required devices to communicate with other vehicles over wireless links and form VANET. They could send messages about themselves and events on the road to other vehicles within their communication range. Vehicles have also a unit for calculating trust levels of other vehicles and their messages. Identities and types of all vehicles are also assumed to be controlled and signed by the authorities and this information cannot be changed by vehicles itself.

3.2   Application Model

Many applications running on VANETs mainly focus on sharing information about events that vehicles come across. Vehicles send messages to others while moving on the road to communicate and improve safety and efficiency of the traffic. They mainly send two types of message: beacon and event messages. Events can be considered situations occurred on traffic or road that is worth to share information about them such as a traffic accident, a traffic jam, a toll road or another.

Beacon Messages are periodically sent messages without an observation of an event. Vehicles send beacon messages at every second to their neighbour nodes that are in their direct communication range. This message shows that the sender vehicle of this message is in the traffic network and moving. The beacon message includes current position and velocity data of vehicle at the time of sending this message in addition to unique identifier and type of the vehicle as shown in Table 1.

Table 1. Beacon message format

Vehicle identifier	Vehicle type	Message time	Vehicle position	Vehicle velocity

Event Messages are sent by vehicles only when an event is observed. Events that occur in traffic can be categorized into three groups: safety events, efficiency events and infotainment events. Messages about safety events are the most critical type, since it aims to increase traffic safety in critical events such as traffic accident, wet/icy road. Efficiency event messages are sent to establish an efficient traffic network in the case of events such as traffic congestion, road maintenance, closed road. Infotainment event messages carry some information about the facilities nearby such as toll road, scenic area, restaurant, parking/petrol station. Event messages include event type, event description and event position besides the fields exist in beacon messages as shown in Table 2.

Table 2. Event message format

Vehicle identifier	Vehicle type	Message time	Vehicle position	Vehicle velocity	Event type	Event description	Event position

3.3   Bogus Information Attack

Suitable security solutions are needed for VANETs in order to overcome vulnerabilities caused by allowing any vehicle to enter to the network such as selfish vehicles, misbehaving ones or even attackers. Selfish vehicles use the network for their own intent. They collect all information from other vehicles but do not send any or send very limited data to them. Their main motivation is using their own resources for only themselves and not being helpful for other vehicles in the network. Misbehaving vehicles could have some malfunctioned device or could be captured by an attacker and send false information unintentionally. Vehicles that aim to damage the network deliberately are called attackers.

In this study, attackers carry out bogus information attacks in order to harm the network. In this attack scenario, even though the attackers observe events like other vehicles, they do not send genuine messages about the events they encounter with. Instead, they send fake and false information about an existent or nonexistent event to their neighbours. Attackers modify the event type of a real event in order to mislead their neighbours. They also generate and send fake event messages with event type, event description and event position data in order to gain some advantage on the road. For example, they could decrease the density of the road they have been using by sending fake messages about a nonexistent accident on that road. Vehicles should be aware of that kind of attackers and they must decide whether the received messages from such nodes are trustable or not. Proposing a trust management model against such attacks is the main motivation of this study.

4   The Proposed Method

Trust management models are widely used by researchers in ad hoc networks in order to ensure secure and reliable communication. In such models, each node assigns a trust degree to the messages they receive and/or the nodes that the message is received from. Trust formula is used to calculate such trust degrees by using the available information in the network. However, generally manually generated trust formulas will have limited number of features and, hence cover only a little aspect of network. They will not be able to represent complex properties of VANETs. A trust management model proposed for VANETs should be able to reflect changes in topology and events in the model.

In this study, we investigate the use of genetic programming in order to generate a trust management model automatically in order to efficiently and effectively handle dynamically changing topology and events of VANETs. This trust formula is generated using more features than other studies in the literature. The features are selected to represent complex characteristics of this dynamic environment. The components of the proposed trust management model are described in the following sections.

4.1   Vehicle Type and Weight

Vehicles in VANETs have different roles and objectives on traffic based on their types. They are divided into three groups: police automobiles, public service vehicles and ordinary vehicles. Vehicle type usually indicates trustworthiness of vehicles to some extent. Police automobiles are responsible of controlling the traffic and providing road safety, therefore they are the most trustworthy vehicles in the network. Public service vehicles such as ambulance, bus, engineering vehicle, etc., are usually on duty for ensuring either road safety or efficiency. They are considered as medium level vehicles in the proposed trust model. Ordinary vehicles such as private cars, taxis, etc., are considered as low level vehicles from the trust point of view, since their contribution to road safety is generally lower than others. In order to use this knowledge on trust calculations, a trust feature called vehicle weight $W_V(x)$ is defined as in Eq. 1:

$$W_V(x) = \begin{cases} 1.0, & when\ x\ is\ a\ police\ automobile \\ 0.7, & when\ x\ is\ a\ public\ service\ vehicle \\ 0.5, & when\ x\ is\ an\ ordinary\ vehicle \end{cases} \tag{1}$$

4.2 Event Type and Weight

Events have different impacts on traffic and road safety and require different trustworthiness levels. The most important message type is clearly safety events as described above. Vehicles in VANETs pay attention to messages' importance levels in order to maintain road safety. This information is represented with a trust feature called event weight $W_E(x)$ as defined in Eq. 2:

$$W_E(x) = \begin{cases} 1.0, & when\ x\ is\ a\ safety\ event \\ 0.8, & when\ x\ is\ an\ efficiency\ event \\ 0.5, & when\ x\ is\ an\ infotainment\ event \end{cases} \tag{2}$$

4.3 Trust Evidence

Each term in the trust formula expression is called trust evidence. They represent the characteristics of network including the properties of vehicles and messages. Each vehicle participated into VANET gathers items of evidence about network by using both beacon and event messages. The values of items of trust evidence used in this study are normalized to $[0, 1]$. The trust formula is based on such trust evidence calculated by using messages received from the neighbour nodes. In order to prevent unnecessary computing overhead, the calculation of the trust value takes place only when a vehicle receives an event message. In addition, beacon messages are stored in a sliding window of 5 messages and stale messages are discarded to keep the memory consumption low. Table 3 shows the trust evidence set and Table 4 lists the notations used in the model.

Vehicles calculate the neighbourhood density as the ratio of the number of current neighbours to the encountered maximum number of neighbours by that time. The percentage of newly added neighbours and removed neighbours since the delivery of the last event message is also monitored.

Position and time proximities are important factors to decide whether the trust value of an event message or its sender should be calculated or not. Some messages are not taken into account for the calculation of trust value according to their position and time proximity.

An event could be observed through many messages sent from more than one vehicle. When an event message is received, the receivers wait for a fixed time to receive other messages of the same event. Since these messages are valuable to calculate the trustworthiness of the received messages, there are also some items of trust evidence based on them as shown in Table 3.

4.4 Trust Distribution

The dynamically changing topology of VANETs could cause vehicles to encounter with vehicles that they have not communicated before and had no prior experience about. Therefore, they should prefer to take in consideration the recommendations from their own trustee rather than deciding randomly to trust such newly encountered vehicles or not. Trust distribution plays a vital role to achieve that. Vehicles only forward a message that they have decided to be trustworthy. Before they forward the message, they add their opinions about the message and its sender. This opinion contains both the trust value of the event and the trust value of its sender. Besides these two trust values, the following information about the forwarder node is also added to the event message: its identifier, type, position and velocity. Table 5 shows the forwarded event message format.

When vehicles receive a forwarded event message, they calculate the average trust value of the sender vehicle weighted by the trust values sent by forwarders. The average trust value of the event is also calculated based on the forwarders' opinions about it. All direct and recommended trust evidences are used to compute the combined trust value. These recommended trust evidences are also shown in Table 3.

Table 3. Trust evidence set

Notation	Trust evidence
ND	Neighbourhood density
ANP	Percentage of added neighbours
RNP	Percentage of removed neighbours
EP	Proximity of the receiver to the event
VP	Proximity of the receiver to the source vehicle
TP	Event time proximity
W_V	Weight of the source vehicle
W_E	Weight of the event
ET	Recommendation (trust value) of the event sent by the forwarder
SP	Percentage of the nodes sending the same event
SW	Average weight of the nodes sending the same event
EW	Average weight of the events at the same location
TE	Average weighted forwarder recommendation about event trust
TV	Average weighted trust value of the sender vehicle

Table 4. Notations

Notation	Definition
NN_A	Number of neighbours of vehicle A
MN_A	Maximum number of neighbours of vehicle A
$ND_A = NN_A / MN_A$	Neighbourhood density of vehicle A
AN_A	Added number of neighbours of vehicle A
RN_A	Removed number of neighbours of vehicle A
$ANP_A = AN_A / MN_A$	Percentage of added neighbours of vehicle A
$RNP_A = RN_A / MN_A$	Percentage of removed neighbours of vehicle A
ED_A^X	Distance of vehicle A to the event X
VD_A^B	Distance of vehicle A to the source vehicle B
MD	Maximum allowed distance
$EP_A^X = (MD - ED_A^X) / MD$	Proximity of vehicle A to the event X
$VP_A^B = (MD - VD_A^B) / MD$	Proximity of vehicle A to the source vehicle B
T	Current time
GT_X	Generation time of the event message X
MT	Maximum allowed event time
$TP_X = (MT - (T - GT_X)) / MT$	Proximity of event X to current time
W_V^A	Weight of the source vehicle A
W_E^X	Weight of the event X
VT_A^B	Trust value of vehicle B calculated by vehicle A
ET_A^X	Trust value of event X calculated by vehicle A
SN_A^X	The number of nodes sending the same event X
$SP_A^X = SN_A^X / MN_A$	Percentage of nodes sending the same event X
$SW_A^X = (\sum_{i=1}^{SN_A^X} W_V^i)/SN_A^X$	Average weight of nodes sending the same event
$EW_A^X = (\sum_{i=1}^{SN_A^X} W_E^i)/SN_A^X$	Average weight of events at the same location
$TE_A^X = (\sum_{i=1}^{SN_A^X} VT_A^i * ET_i^X)/SN_A^X$	Average weighted event trust value
$TV_A^X = (\sum_{i=1}^{SN_A^X} VT_A^i * VT_i^B)/SN_A^X$	Average weighted sender trust value

Table 5. Forwarded event message format

Vehicle identifier	Vehicle type	Message time	Vehicle position	Vehicle velocity	Event type	Event description	Event position
Forwarder identifier	Forwarder type	Forwarder position	Forwarder velocity	Forwarder event trust		Forwarder sender trust	

4.5 Trust Update

Vehicles assign trust values not only to event messages but also to the senders of those messages. Messages sent from vehicles that have higher trust value are decided more likely to be trustworthy than messages from untrusted vehicles. At the beginning, the trust value of each vehicle is set to 0.6. Every time a message is received from the sender, its trust value is updated according to the Eq. 3 given below. Interactions with the sender is taken into account for trust update. Let's assume that the vehicle A receives a message sent from the vehicle B. Here, (ET_A^B) represents the trust value of this message and TT refers to the threshold for accepting this message to be forwarded. Where (CT_A^B) shows the current trust value of the sender vehicle B, (NT_A^B) indicates the newly updated trust value of the sender vehicle B calculated by the receiver vehicle A.

$$NT_A^B = \begin{cases} CT_A^B \times \dfrac{ET_A^B}{TT}, & 0 \le ET_A^B < TT \\ CT_A^B + (ET_A^B - CT_A^B) \times (\dfrac{ET_A^B - TT}{1 - TT}), & TT \le CT_A^B \le ET_A^B \le 1 \\ CT_A^B, & TT \le ET_A^B < CT_A^B \le 1 \end{cases} \quad (3)$$

Well-known principle about trust "hard to earn but easy to lose" [6–8] is applied while calculating (NT_A^B). Vehicles must send messages that are more trustworthy than vehicles itself to increase their trust values. Even a message is considered as trustworthy; it will not change the trust value of its sender unless its trust value is higher than vehicle's. Increasing rate of vehicle's trust value is proportional to gap between event trust value and vehicle trust value, and the normalized trust value of the message. In contrast, untrusted messages will decrease the trust value of sender rapidly.

4.6 Evolving Trust Formula by Using Genetic Programming

Genetic programming (GP) [15,16] is a population-based search algorithm inspired by natural evolution. It starts with generating a population of individuals (usually at random) which are candidate solutions for the target problem. Then, each individual is evaluated and assigned with a fitness value that indicates how well this candidate solves or comes close to solving the problem at hand. Until a termination criterion is satisfied, new populations are generated iteratively by using selection, crossover, and mutation operators, as in natural evolution. These genetic operators are used to provide better solutions in the new population.

Each individual, candidate solution for the problem in other words, represents a trust formula, which is generated randomly at first generation. Each individual is represented as a tree in GP. In-order traversal of the tree outputs a candidate trust formula. Terminal nodes are trust evidences in Table 3 and some ephemeral random constants (ERC). Non-terminal nodes consist of mathematical operations listed in Table 6. These operations are implemented to have the result value of [0, 1]. Each individual is assigned a fitness value based on its detection rate of false and fake messages. Higher value of fitness value shows better individuals, so the algorithm tries to increase the fitness value of population using genetic operators. Selection operator probabilistically determines the parent individuals that will be used in the crossover and mutation operators. Better individuals have a higher chance to be selected. Crossover and mutation operators are used on the selected parents to breed new individuals. The crossover operator exchanges different portions of the parents and produces two new child individuals. It aims to create better solutions using good parts of parents. In the mutation operator, some portions of newly generated solutions are changed randomly in order to increase diversity and reach better solutions.

Table 6. Genetic programming operation set

Add $(X + Y) / 2$	Mult $X \times Y$	Square $X \times X$	Cube $X \times X \times X$	Neg $1 - X$
Sub $(X - Y + 1) / 2$	Exp $(e^X - 1) / (e - 1)$			Sqrt $\sqrt{X}$
Sin $(\sin(\pi X - (\pi / 2)) + 1) / 2$		Cos $(\cos(\pi X) + 1) / 2$		

5   Experiments

The experiment consists of two parts: network simulation and the evolution of the trust formula by using genetic programming. The mobility of vehicles on the road, data transfer between vehicles, trust computation of vehicles about other vehicles in neighbourhood and bogus information attack scenario are simulated using the ns-3 network simulator [17]. The ECJ toolkit [18] is used for genetic programming, which automatically generates trust formula as candidate solutions and computes the fitness values of such solutions after running the each evolved formula in the network simulation. By using the evolved trust formula, vehicles determine whether messages are trustworthy or not in a network scenario where both normal and fake messages exist.

5.1   Network Simulation

In the simulation, vehicles are generated and moving according to a real world traffic model taken from a street map in Zurich. This real world traffic model [19] is included in the distribution of ns-3. Each simulation has one of the traffic density settings as low, medium and high and takes 300 s.

Vehicles send two types of messages to others: beacon and event messages. Beacon messages are sent periodically at every second. Event messages are sent when an event occurs in the 100 m range of the node. Beacon messages are stored in a sliding window of 5 messages and processed when an event message is received. Vehicles process these messages and obtain the values of trust evidences listed in Table 3 in order to calculate both the trust value of the sender and the trust value of the message. Trust threshold is set to 0.6 and the event message

Table 7. Network simulation parameters

Name	Value
Simulation area	4.6 km × 3.0 Km street map
Number of vehicles	99 (low), 210 (medium), 370 (high)
Vehicle types	High, medium, normal
Ratio of high level vehicles	5%
Ratio of medium level vehicles	15%
Ratio of attackers	10%
Attack type	Bogus information
Simulation time	300 s
Mobility	Real traffic data model
Beacon interval	1 s
Beacon window's size	5 messages
Beacon messages' size	128 bytes
Event messages' size	256 bytes
Event types	Safety, efficiency, infotainment
Ratio of safety events	10%
Ratio of efficiency events	40%
Event detection range	100 m
Max event distance	500 m
Max event time	1 s
Max delay time	0.2 s
Routing protocol	None
Trust threshold	0.6

is forwarded if the vehicle trusts the message. The forwarding vehicle inserts its own trust value about both the sending vehicle and the message. Table 7 shows the parameters of the network simulation.

5.2   Evolution of Trust Formula

Each individual is run on the network simulation in order to calculate its fitness value. By using the evolved trust formula, vehicles determine whether messages are trustworthy or not in a network scenario where both normal and fake messages exist. A vehicle makes true positive (TP) decision if it decides a malicious event message is untrustworthy. In contrast, if vehicle decides a normal event message is trustworthy, it makes true negative (TN) decision. A vehicle makes false positive (FP) decision if it decides a normal event message is untrustworthy. On the other hand, if vehicle decides a malicious event message is trustworthy, it makes false negative (FN) decision. After obtaining TP, TN, FP and FN

values, precision rate and recall rate are calculated and fitness value of a generated trust formula is determined using F-measure (F) [10], defined as in Eq. 4. It takes value in the interval $[0, 1]$. Table 8 shows the parameters of genetic programming. The parameters not listed here are the default parameters of the ECJ toolkit.

$$F = \frac{2 \times Precision \times Recall}{Precision + Recall} \tag{4}$$

Table 8. Genetic programming parameters

Parameter name	Parameter value
Population size	100 individuals
Maximum generation number	20
Crossover probability	0.7
Mutation probability	0.3
Tournament size	7
Terminal nodes	Trust evidences and ERC
Nonterminal nodes	Add, sub, mult, sin, cos, exp, square, sqrt, cube, neg

5.3   Results and Analysis

The GP algorithm is run five times. In each run, different network settings having different events are applied. The best individual of these runs of GP is evaluated here. The change in the fitness value of the best individuals of different runs is shown in Fig. 1. The best trust formula, which is evolved by using a traffic network under low density (99 vehicles), is given in Table 9. As shown in the formula, 8 of 14 trust evidences, 8 of 10 operations and an ERC have selected in the evolution process.

Table 9. The evolved best trust formula

Mult(Neg(Sqrt(averageWeightedSenderTrustValue)), Add(0.7483884231631781, Mult(Sqrt(Add(Add(neighbourhoodDensity, Sin(receiverToSenderProximity)), Mult(Sqrt(Sqrt(Add(averageEventTypeWeight,Mult(Add(receiverToEventProximity, averageWeightedSenderTrustValue), Add(Neg(Sub(Cos(Add(eventTimeProximity, senderTypeWeight)), Mult(Exp(averageSenderTypeWeight), neighbourhoodDensity))), Exp(averageSenderTypeWeight)))))), Sin(receiverToSenderProximity)))), Sin(receiverToSenderProximity))))

The best trust formula is evaluated on networks with different event positions and varying density patterns from low density of 99 vehicles to high density of 370 vehicles as shown in Figs. 2 and 3 respectively. Even though individuals are

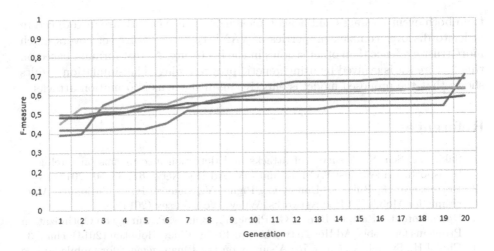

Fig. 1. Change in the fitness value of best individuals over generations

trained in a low density network, the best performances of the formula is achieved on networks under high density. It is an expected result, since a node has a higher number of neighbours on dense networks; they get more recommendations about a node or a message.

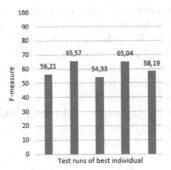

Fig. 2. Performance under low density   **Fig. 3.** Performance under high density

6   Conclusion

This paper presents the first study that explores the use of evolutionary computation techniques to the trust management problem in VANETs. A method based on genetic programming is proposed in order to evaluate data trustworthiness of events in VANETs automatically. A large number of features collected from both event messages and beacon messages in the network are introduced in order to discover complex properties of VANETs. The feature set covers much more trust evidence than other studies in the literature. A trust formula based on this feature set is evolved by using genetic programming. The simulation results shows that the proposed model is effective against bogus information attacks.

References

1. Sakiz, F., Sen, S.: A survey of attacks and detection mechanisms on intelligent transportation systems: VANETs and IoV. Ad Hoc Netw. **61**, 33–50 (2017)
2. Dorronsoro, B., Ruiz, P., Danoy, G., Pigné, Y., Bouvry, P.: Evolutionary Algorithms for Mobile Ad Hoc Networks. Wiley, West Sussex (2014)
3. Dorronsoro, B., Ruiz, P., Danoy, G., Pigné, Y., Bouvry, P.: Survey on Optimization Problems for Mobile Ad Hoc Networks, pp. 49–78. Wiley, Hoboken (2014). chap. 3
4. Cho, J.H., Swami, A., Chen, R.: A survey on trust management for mobile ad hoc networks. IEEE Commun. Surv. Tut. **13**(4), 562–583 (2011)
5. Govindan, K., Mohapatra, P.: Trust computations and trust dynamics in mobile adhoc networks: a survey. IEEE Commun. Surv. Tut. **14**(2), 279–298 (2012)
6. Ma, S., Wolfson, O., Lin, J.: A survey on trust management for intelligent transportation system. In: Proceedings of the 4th ACM SIGSPATIAL International Workshop on Computational Transportation Science, pp. 18–23. ACM (2011)
7. Yu, H., Shen, Z., Miao, C., Leung, C., Niyato, D.: A survey of trust and reputation management systems in wireless communications. Proc. IEEE **98**(10), 1755–1772 (2010)
8. Zhang, J.: A survey on trust management for VANETs. In: IEEE International Conference on Advanced Information Networking and Applications (AINA), 2011, pp. 105–112. IEEE (2011)
9. Wei, Y.-C., Chen, Y.-M.: Efficient self-organized trust management in location privacy enhanced VANETs. In: Lee, D.H., Yung, M. (eds.) WISA 2012. LNCS, vol. 7690, pp. 328–344. Springer, Heidelberg (2012). https://doi.org/10.1007/978-3-642-35416-8_23
10. Van Rijsbergen, C.J.: Information Retrieval. Butterworths. London, UK (1979). http://www.dcs.gla.ac.uk/Keith/Preface.html. Accessed 31 Jan 2019
11. Yao, X., Zhang, X., Ning, H., Li, P.: Using trust model to ensure reliable data acquisition in VANETs. Ad Hoc Netw. **55**, 107–118 (2017)
12. Reina, D.G., Ruiz, P., Ciobanu, R., Toral, S., Dorronsoro, B., Dobre, C.: A survey on the application of evolutionary algorithms for mobile multihop ad hoc network optimization problems. Int. J. Distrib. Sens. Netw. **12**(2), 2082496 (2016)
13. Kusyk, J., Uyar, M.U., Sahin, C.S.: Survey on evolutionary computation methods for cybersecurity of mobile ad hoc networks. Evol. Intel. **10**(3), 95–117 (2018)
14. Tahta, U.E., Sen, S., Can, A.B.: Gentrust: a genetic trust management model for peer-to-peer systems. Appl. Soft Comput. **34**, 693–704 (2015)
15. Koza, J.R.: Genetic Programming: On the Programming of Computers by Means of Natural Selection. MIT Press, Cambridge (1992)

16. Koza, J.R.: Genetic programming as a means for programming computers by natural selection. Stat. Comput. **4**(2), 87–112 (1994)
17. The NS-3 Network Simulator. https://www.nsnam.org/. Accessed 31 Jan 2019
18. Luke, S.: ECJ A Java-based Evolutionary Computation Library (1998). https://cs.gmu.edu/~eclab/projects/ecj/. Accessed 31 Jan 2019
19. Naumov, V., Baumann, R., Gross, T.: An evaluation of inter-vehicle ad hoc networks based on realistic vehicular traces. In: Proceedings of the 7th ACM International Symposium on Mobile Ad Hoc Networking and Computing, pp. 108–119. ACM (2006)

A Matheuristic for Green and Robust 5G Virtual Network Function Placement

Thomas Bauschert[1], Fabio D'Andreagiovanni[2,3]($\boxtimes$), Andreas Kassler[4],
and Chenghao Wang[3]

[1] Chair of Communication Networks, Technische Universität Chemnitz,
09126 Chemnitz, Germany
thomas.bauschert@etit.tu-chemnitz.de
[2] French National Center for Scientific Research (CNRS), Paris, France
[3] Sorbonne Universités, Université de Technologie de Compiègne, CNRS,
Heudiasyc UMR 7253, CS 60319, 60203 Compiègne, France
{d.andreagiovanni,chenghao.wang}@hds.utc.fr
[4] Karlstad University, Universitetsgatan 2, 65188 Karlstad, Sweden
andreas.kassler@kau.se

Abstract. We investigate the problem of optimally placing virtual network functions in 5G-based virtualized infrastructures according to a green paradigm that pursues energy-efficiency. This optimization problem can be modelled as an articulated 0-1 Linear Program based on a flow model. Since the problem can prove hard to be solved by a state-of-the-art optimization software, even for instances of moderate size, we propose a new fast matheuristic for its solution. Preliminary computational tests on a set of realistic instances return encouraging results, showing that our algorithm can find better solutions in considerably less time than a state-of-the-art solver.

Keywords: 5G · Virtual Network Function · Traffic uncertainty ·
Robust Optimization · Matheuristic

1 Introduction

The Fifth Generation of wireless telecommunications systems, widely known as 5G, has attracted a lot of attention in recent times, since it is largely considered a crucial element for a full realization of a digital society and a critical technology to support the deployment of smart cities [1]. 5G is going to offer enhanced service performances unknown to previous wireless technologies, such as data rates of at least 40 Mbps for tens of thousands of users, data rates of 100 Mbps for metropolitan areas, enhanced spectral efficiency and a dramatic reduction

This work has been partially carried out in the framework of the Labex MS2T program. Labex MS2T is supported by the French Government, through the program "Investments for the future", managed by the French National Agency for Research (Reference ANR-11-IDEX-0004-02).

P. Kaufmann and P. A. Castillo (Eds.): EvoApplications 2019, LNCS 11454, pp. 430–438, 2019.
https://doi.org/10.1007/978-3-030-16692-2_29

of latency (see e.g., [1]). In particular, 5G will be strongly based on Network Function Virtualization, according to which network functions run on a set of virtual machines (VMs) that are hosted in cheap commodity hardware servers [2]. This will considerably reduce the cost of network infrastructures, decreasing the need for expensive dedicated hardware. The problem of optimally designing virtual networks, allocating Virtual Network Functions Components (VNFCs) to physical servers and managing the data flows between servers has received great attention in recent times, in particular focusing on adopting a green networking perspective aimed at minimizing the overall power consumption (see e.g., [3–8]). However, while in the available literature purely heuristic solution approaches for virtual network design have been quite widely investigated, the development of hybrid exact-heuristic algorithms exploiting the potentialities of mathematical programming (so-called matheuristic - see [9]) has received very limited attention. By this work, we aim to start to fill this gap by proposing a new matheuristic for the green placement of virtual network function in 5G, while taking into account the uncertainty of function requests (see e.g., [4,6]).

The remainder of this short paper is organized as follows: in Sect. 2, we describe a Binary Linear Programming model for modelling the green and robust placement of VNCFs; in Sect. 3, we present a new mathcuristic to fast solve the placement problem: finally, in Sect. 4, we report preliminary computational results and derive some conclusions.

2   A Binary Linear Program for VNFC Placement

From a modelling point of view, we can essentially describe the topology of the 5G network that we consider through a graph $G(N, L)$, where N is the node set and L is the link set. Each link $\ell \in L$ corresponds to a pair (i, j), where $i, j \in N$ are the nodes it connects. Each link is associated with a bandwidth b_ℓ. The network interconnects a set of servers S and the node to which a server s is connected is denoted by $n(s) \in N$. Each server offers an amount of computational resources (e.g., CPU and RAM): denoting by R the set of resource types, the amount of resources available for each type $r \in R$ at a server $s \in S$ is denoted by a_{sr}. The set of VNFCs is denoted by V and the set of service chains offered in the network is denoted by $\mathcal{C}$. When executed, a VNFC $v \in V$ requires an amount a_{vr} of each resource type $r \in R$. Each chain $C \in \mathcal{C}$ corresponds to a subset of pairs (v_1, v_2) belonging to $V \times V$. The exchange of data between v_1 and v_2 in a pair (v_1, v_2) requires an amount of bandwidth $b_{v_1 v_2}$ in each traversed network link. Concerning power consumption, every node $n \in N$ and link $\ell \in L$ consumes P_n and P_ℓ when used, respectively. Each server $s \in S$ has a consumption that is a linear function in the range $[P_s^{\min}, P_s^{\max}]$.

The optimization problem related to VNFC placement that we consider can be resumed as follows: given a 5G network interconnecting a set of servers, we want to decide how to establish a set of virtual chains in the network, respecting the available resource budget of the servers, while minimizing the overall power consumption. The decisions taken are modelled by the following decision variables:

(1) variables $y_s \in \{0, 1\}$, $\forall s \in S$ representing the activation of a server ($y_s = 1$ if s is turned on and 0 otherwise);
(2) variables $x_{vs} \in \{0, 1\}$, $\forall v \in V, s \in S$ representing the allocation of a VNFC v to server s ($x_{vs} = 1$ if v is allocated to s and 0 otherwise);
(3) variables $z_n \in \{0, 1\}$, $\forall n \in N$ representing the activation of a node n ($z_n = 1$ if n is turned on and 0 otherwise);
(4) variables $w_{ij} \in \{0, 1\}$, $\forall (i, j) \in L$ representing the activation of a link $\ell = (i, j)$ ($w_{ij} = 1$ if $\ell = (i, j)$ is turned on and 0 otherwise);
(5) variables $f_{ij}^{(v_1, v_2)} \in \{0, 1\}$, $\forall (i, j) \in L, (v_1, v_2) \in \bigcup_{C \in \mathcal{C}}$ representing that link (i, j) is used for data exchange between v_1 and v_2 belonging to some $C \in \mathcal{C}$.

These variables are employed in the following Binary Linear Program, denoted by BLP-VP, modelling the VNFC optimal placement problem:

$$\max \sum_{s \in S} \left[P_s^{\min} \cdot y_s + (P_s^{\max} - P_s^{\min}) \cdot \frac{1}{a_{rs}} \cdot \sum_{v \in V} a_{vr} \cdot x_{vs} \right]$$

$$+ \sum_{n \in N} P_n \cdot z_n + \sum_{(i,j) \in L} P_{ij} \cdot w_{ij} \quad \text{(with } r = \text{CPU)} \qquad \text{(BLP-VP)} \qquad (1)$$

$$\sum_{s \in S} x_{vs} = 1 \qquad\qquad v \in V \qquad (2)$$

$$y_s \leq \sum_{v \in V} x_{vs} \qquad\qquad s \in S \qquad (3)$$

$$x_{vs} \leq y_s \qquad\qquad s \in S, v \in V \quad (4)$$

$$\sum_{v \in V} a_{vr} \cdot x_{vs} \leq a_{rs} \cdot y_s \qquad\qquad s \in S, r \in R \quad (5)$$

$$\sum_{(n,i) \in L} b^{v_1,v_2} \cdot f_{ni}^{v_1,v_2} - \sum_{(i,n) \in L} b^{v_1,v_2} \cdot f_{in}^{v_1,v_2} =$$

$$\sum_{s \in S: n(s)=n} b^{v_1,v_2} \cdot (x_{v_1 s} - x_{v_2 s}) \qquad n \in N, (v_1, v_2) \in \bigcup_{C \in \{\mathcal{C}\}} C \quad (6)$$

$$\sum_{(v_1,v_2) \in \bigcup_{C \in \{\mathcal{C}\}} C} b^{v_1,v_2} f_{ij}^{v_1,v_2} \leq B_{ij} w_{ij} \qquad (i, j) \in L \qquad (7)$$

$$w_{ij} \leq z_i \text{ and } w_{ij} \leq z_j \qquad (i, j) \in L \qquad (8)$$

$$f_{ij}^{v_1,v_2} \leq z_i \text{ and } f_{ij}^{v_1,v_2} \leq z_j \qquad (i, j) \in L \qquad (9)$$

$$y_s, x_{vs}, z_n, w_{i,j}, f_{i,j}^{v_1,v_2} \in \{0,1\} \qquad s \in S, v \in V, n \in N,$$

$$(i,j) \in L, (v_1, v_2) \in \bigcup_{C \in \{C\}} C.$$

The previous model is based on the model proposed in [6], to which we refer the reader for a detailed description of all its elements. The objective function (1) pursues the minimization of the total power consumption, expressed as the sum of the power consumed by servers, nodes and links. The constraints (2) impose that each VNFC must be allocated on exactly one server. The constraints (3) and (4) logically link the values of the server activation and VNFC allocation decision variables. The constraints (5) model the resource capacity for each server and resource type, imposing that the overall resource usage of all the VNFCs cannot exceed the capacity of each server. The constraints (6) express the usage of bandwidth on links of the network, under the form of flow conservation constraints with a flow balance in the right hand side that takes into account the variable allocation of VNFCs to servers. The bandwidth capacity of links is modelled by the constraints (7). Finally, the constraints (8) and (9) link the link and node activation decision variables, imposing that links are activated if and only if the corresponding nodes are activated.

Protecting Against Resource Uncertainty. As in [6], we make the resource capacity constraints (5) robust against fluctuations in the resource requests a_{vr}. Indeed, the resource need for virtual chains requests is typically not exactly known in advance. Taking into account such data uncertainty in the model is very important, since by neglecting the possibility of variations in the input data we risk to obtain design solutions that are of bad quality and even infeasible in practice. For a discussion about the effects of the presence of data uncertainty in mathematical optimization and (telecommunications) network design, we refer the reader to the works [10,11]. In order to protect against resource uncertainty, we adopt a Robust Optimization (RO) paradigm (see e.g., [10]). RO, which has been highly appreciated for its high computational efficiency with respect to more traditional paradigms like Stochastic Programming, essentially takes into account data uncertainty by including additional hard constraints in the optimization problem. These constraints have the task of excluding solutions that are vulnerable to input data deviations, maintaining only *robust solutions*. The data deviations that are relevant to the decision maker and against which protection is needed are specified through a so-called *uncertainty set*. The specific RO model that we consider is Γ-Robustness [12], which belongs to the family of *cardinality-constrained* uncertainty sets and, adapting to our case, assumes that each value a_{vr} may vary in a range $[\bar{a}_{vr} - \Delta a_{vr}, \bar{a}_{vr} + \Delta a_{vr}]$ centered on a reference value $\bar{a}_{vr}$ that may deviate up to $\Delta a_{vr} > 0$. Furthermore, the uncertainty model assumes that at most Γ coefficients in every (5) may vary.

Under these modelling assumptions, the robust version of the constraints (5) can be obtained according to the procedure detailed in [6]. Specifically, each constraint (5) must be replaced by the following set of constraints and additional decision variables:

$$\sum_{v \in V} \bar{a}_{vr} \cdot x_{vs} + \left(\Gamma \cdot v_{rs} + \sum_{v \in V} w_{rsv} \right) \leq a_{rs} \cdot y_s$$

$$v_{rs} + w_{rsv} \geq \Delta a_{vr} \cdot x_{vs} \qquad\qquad v \in V$$
$$v_{rs} \geq 0$$
$$w_{rsv} \geq 0 \qquad\qquad v \in V. \qquad (10)$$

The robust model that we solve in what follows is thus (BLP-VP) with (5) replaced with (10). We denote such robust model by (ROB-BLP-VP).

3   A Matheuristic for ROB-BLP-VP

We present here a new matheuristic for optimal VNFC placement that is based on the integration of a Genetic Algorithm (GA) with an *exact* large neighborhood search, namely a search formulated as an optimization problem solved by a state-of-the-art solver such as CPLEX [13]. The solver is also used for completing partial solutions of (ROB-BLP-VP) in an optimal way: for a fixed value configuration of a subset of decision variables, we employ the solver to find a feasible valorization of all the remaining variables while optimizing the objective function. At the basis of this matheuristic there is the consideration that, while a state-of-the-art solver may find difficulties in identifying good quality solutions for ROB-BLP-VP, it is instead able to efficiently identify good solutions for appropriate subproblems of ROB-BLP-VP, derived by fixing the value of a consistent subset of variables.

GAs are widely known metaheuristics that draw inspiration from the evolution of a population (see [14] for an exhaustive introduction to the topic). The individuals of the population represent solutions of the optimization problem and the *chromosome* of an individual corresponds to a valorization of decision variables of a solution. The quality of an individual/solution is assessed through a *fitness function*. A GA begins with the definition of an initial population that then changes through evolutionary mechanisms like crossover and mutation of individuals, until some stopping criterion is met.

3.1   Initialization of the Population

Solution Representation. The first step consists of establishing what the individuals of the population represent. We decided that the chromosome of an individual corresponds with a valorization of the decision variables (y, x) (of ROB-BLP-VP), which represent the server activation and the VNFC allocation. Indeed, such variables are particularly critical for the problem: once their values have been fixed, we obtain a subproblem of (ROB-BLP-VP) that reduces to a kind of robust network flow problem and is easier to be solved by a state-of-the-art optimization solver, returning an optimal solution for (ROB-BLP-VP) using the valorization of (y, x) as basis.

Fitness Function. To assess the quality of an individual, we adopt a fitness function that corresponds to the objective function (1) of (ROB-BLP-VP).

Initial Population. The strategy that we explored to generate the initial group of individuals relies on the following principles: to generate an individual, we randomly activate a number $\sigma < |S|$ of servers and then we randomly assign each VNFC in V to one single activated server, checking that the resource constraints (5) are not violated. In this way, we obtain a valorization $(\bar{y}, \bar{x})$ of the server and allocation variables that we can then complete by solving the remaining sub-problem of (ROB-BLP-VP) through a state-of-the-art solver. By this strategy, we can obtain the optimal solution of (ROB-BLP-VP) for a fixed $(\bar{y}, \bar{x})$. We denote the set of individuals constituting the population at a generic iteration of the algorithm by POP.

3.2   Evolution of the Population

Selection. The individuals chosen for being combined and generating the new individuals are chosen according to a *tournament selection* principle: we first create a number β of (small cardinality) groups of individuals by randomly selecting them from POP. Then the γ individuals in each group associated with the best fitness value are combined through crossover.

Crossover. We form the couples that generate the offsprings according to the following procedure. From the tournament selection, we obtain $\beta \cdot \gamma$ individuals that are randomly paired in couples, each generating one offspring. Assuming that the two parents are associated with chromosomes/partial solutions (y^1, x^1) and (y^2, x^2), the chromosome of the offspring $(y^{\text{off}}, x^{\text{off}})$ is defined according to two rules: (1) if the parents have the same binary value in a position j, then the offspring inherits such value in its position j (i.e., if $(y^1, x^1)_j = (y^2, x^2)_j$ then $(y^{\text{off}}, x^{\text{off}})_j = (y^1, x^1)_j$); (2) if the parents have distinct binary values in a position j, then the offspring inherits a null value (i.e., if $(y^1, x^1)_j \neq (y^2, x^2)_j$ then $(y^{\text{off}}, x^{\text{off}})_j = 0$). Possible violations in the constraints (2) and (5) associated with $(y^{\text{off}}, x^{\text{off}})$ are then repaired. The main rationale at the basis of this procedure is assuming that two solutions having the same valorization of a variable is a good indication that such valorization should be maintained also in the offspring.

3.3   Exact Improvement Search

We attempt at improving the best solution found by the GA through an *exact large neighborhood search*, namely a search that is formulated as a suitable Binary Linear Programming problem solved by a state-of-the-art optimization solver [9]. The search is based on using the effective heuristic RINS (*Relaxation Induced Neighborhood Search* - we refer the reader to [15] for an exhaustive description of it). Specifically, given a partial solution $(\bar{y}, \bar{x})$ of (ROB-BLP-VP) and (y^{LR}, x^{LR}) an optimal solution of a Linear Relaxation (i.e., a solution obtained by removing the integrality requirements on the binary variables), we solve a subproblem of (ROB-BLP-VP) where the value of the j-th component of the vectors (y, x) is fixed according to the following two rules:

$$IF\ (\bar{y}, \bar{x})_j = 0 \wedge (y^{LR}, x^{LR}) \leq \epsilon\ THEN\ (y, x)_j = 0$$

$$IF\ (\bar{y}, \bar{x})_j = 1 \wedge (y^{LR}, x^{LR}) \geq 1 - \epsilon\ THEN\ (y, x)_j = 1$$

The subproblem of (ROB-BLP-VP) subject to such variable fixing is then solved by the state-of-the-art solver, running with a time limit.

4 Preliminary Computational Results

We preliminary assessed the performance of the proposed matheuristic by considering 10 instances that refer to a network made up of 10 nodes to which 50 servers are connected and that are defined for different VNFC features, defined referring to the works [6,8]. To execute the tests, we employed a Windows machine with 2.70 GHz processor and 8 GB of RAM. As optimization solver, we relied on IBM ILOG CPLEX 12.5, which is interfaced through Concert Technology with a C/C++ code. The global time limit imposed to CPLEX to solve (ROB-BLP-VP) is set to 3600 s. The same time limit is set for the matheuristic (denoted here by *MatHeu*), assigning 3000 s to the GA phase and 600 to the improvement phase based on RINS (in which we set $\epsilon = 0.1$). The initial population includes 100 individuals/solutions and, at each iteration, we consider $\beta = 10$ groups from each of which $\gamma = 2$ individuals are chosen.

The results of the computational tests are presented in Table 1, where: *ID* identifies the instance; T^* *(CPLEX)* and T^* *(MatHeu)* are the time (in seconds) that CPLEX and *MatHeu* needs to find the best solution within the time limit, respectively, whereas $\Delta T^*\%$ is the percentage reduction in time that *MatHeu* grants to find a solution that is at least as good as the best solution found by CPLEX. Finally, $\Delta P^*\%$ is the reduction in power consumption that the best solution found by *MatHeu* grants with respect to the best solution found by CPLEX within the time limit.

As highlighted in several works, such as [5,6] even simplified deterministic versions of (ROB-BLP-VP) may prove difficult to solve for state-of-the-art optimization solvers also in the case of instances. We confirm such behaviour in the case of our instances, which highlights the need for fast (heuristic) solution algorithms. On the basis of the results, we can say that *MatHeu*, for all the instances, is able to return a solution that is at least as good as the best solution found by CPLEX within the time limit in 20% less time, on average. Concerning the reduction in consumed power, we can instead notice that *MatHeu* allows to find better quality solution than CPLEX within the time limit, with a reduction in power consumption that can reach 10% and on average is equal to 7.3%.

Table 1. Preliminary computational results

ID	T^* (CPLEX)	T^* (MatHeu)	$\Delta T^*\%$	$\Delta P^*\%$
I1	3322	2580	22.3	5.4
I2	3194	2742	14.1	6.8
I3	3157	2335	26.0	6.2
I4	3552	2905	18.2	10.2
I5	3513	2536	27.8	6.9
I6	3402	2892	14.9	5.8
I7	3475	2642	23.9	8.6
I8	3362	3041	9.5	9.3
I9	3595	2587	28.0	7.6
I10	3488	2769	20.6	5.5

We consider such results remarkable: as future work, they encourage to refine the solution construction mechanism, better exploiting the specific features of (ROB-BLP-VP) to define the rules adopted to generate the initial population and the offspring solutions by crossover. Furthermore, we intend to investigate also the integration of the GA construction phase with other ad-hoc exact large neighborhood search procedures besides RINS.

References

1. Larsson, C.: 5G Networks - Planning, Design and Optimization. Academic Press, Cambridge (2018)
2. Abdelwahab, S., Hamdaoui, B., Guizani, M., Znati, T.: Network function virtualization in 5G. IEEE Commun. Mag. **54**(4), 84–91 (2016)
3. Herrera, J., Botero, J.: Resource allocation in NFV: a comprehensive survey. IEEE Trans. Netw. Serv. Manage. **13**(3), 518–532 (2016)
4. Baumgartner, A., Bauschert, T., D'Andreagiovanni, F., Reddy, V.S.: Towards robust network slice design under correlated demand uncertainties. In: IEEE International Conference on Communications (ICC), pp. 1–7 (2018)
5. Luizelli, M.C., Bays, L.R., Buriol, L.S., Barcellos, M.P., Gaspary, L.P.: Piecing together the NFV provisioning puzzle: efficient placement and chaining of virtual network functions. In: IFIP/IEEE International Symposium on Integrated Network Management (IM), pp. 98–106 (2015)
6. Marotta, A., D'Andreagiovanni, F., Kassler, A., Zola, E.: On the energy cost of robustness for green virtual network function placement in 5G virtualized infrastructures. Comput. Netw. **125**, 64–75 (2017)
7. Mechtri, M., Ghribi, C., Zeghlache, D.: A scalable algorithm for the placement of service function chains. IEEE Trans. Netw. Serv. Manage. **13**(3), 533–546 (2016)
8. Marotta, A., Zola, E., D'Andreagiovanni, F., Kassler, A.: A fast robust approach for green virtual network functions deployment. J. Netw. Comput. Appl. **95**, 42–53 (2017)

9. Blum, C., Puchinger, J., Raidl, G., Roli, A.: Hybrid metaheuristics in combinatorial optimization: a survey. Appl. Soft. Comput. **11**, 4135–4151 (2011)
10. Ben-Tal, A., El Ghaoui, L., Nemirovski, A.: Robust Optimization. Princeton University Press, Princeton (2009)
11. Bauschert, T., Büsing, C., D'Andreagiovanni, F., Koster, A.M.C.A., Kutschka, M., Steglich, U.: Network planning under demand uncertainty with robust optimization. IEEE Commun. Mag. **52**, 178–185 (2014)
12. Bertsimas, D., Sim, M.: The price of robustness. Oper. Res. **52**(1), 35–53 (2004)
13. IBM ILOG CPLEX. http://www-01.ibm.com/software
14. Goldberg, D.: Genetic Algorithms in Search, Optimization & Machine Learning. Addison-Wesley, Reading (1988)
15. Danna, E., Rothberg, E., Le Pape, C.: Exploring relaxation induced neighborhoods to improve MIP solutions. Math. Program. **102**, 71–90 (2005)

Prolong the Network Lifetime of Wireless Underground Sensor Networks by Optimal Relay Node Placement

Nguyen Thi Tam[1,2](✉), Huynh Thi Thanh Binh[1], Tran Huy Hung[1], Dinh Anh Dung[1], and Le Trong Vinh[2]

[1] Hanoi University of Science and Technology, Hanoi, Vietnam
binhht@soict.hust.edu.vn, tranhuyhung1998@gmail.com,
dinhanhdung1996@gmail.com
[2] Vietnam National University, University of Science, Hanoi, Vietnam
{tamnt,vinhlt}@vnu.edu.vn

Abstract. Wireless Underground Sensor Networks (WUSNs) have received attention in the past years because of their popularity and cost-effectiveness when they are used in many practical fields such as military applications, environmental applications, and home applications. In WUSNs, sensors are deployed with limited power, once their power is out of, the sensors are ineffectual. The extension of the network's lifetime is a critical issue in WUSNs, making it a topic of much interest in research. Several approaches have been proposed to keep the sensor nodes active, one of which is deploying relay nodes above ground to transfer data from sensor nodes to the base station. However, this method has faced issues, such as balancing the load of relay nodes and the increased transmission loss between relay nodes and sensor nodes. This paper addresses this concern and proposes two heuristics named Beam Genitor Search and Connection Swap for the relay node placement problem to guarantee load balance among relay nodes and maximize network lifetime. Our experiments show that the proposed methods result in significantly better quality solutions (longer network lifetime) for the problem when compared to the existing methods.

Keywords: Wireless underground sensor setworks · Network lifetime · Relay node placement

1 Introduction

A Wireless Sensor Network (WSN) is a network of multiple sensor nodes (SNs) which have limited energy and processing power as well as transmission range. Sensor nodes can sense different aspects of the environment. In typical applications, the data generated by all sensor nodes is transmitted wirelessly using multi or single hop routing [1]. There has been a lot of researches in wireless sensor networks as a result of the potential for their usage in many different areas,

© Springer Nature Switzerland AG 2019
P. Kaufmann and P. A. Castillo (Eds.): EvoApplications 2019, LNCS 11454, pp. 439–453, 2019.
https://doi.org/10.1007/978-3-030-16692-2_30

such as home automation, security, environmental monitoring, and many more. In the Internet of Things (IoT) era, there are many applications using WSNs, eg. in smart cities and other smart environments, or in efficient agriculture as well as smart farming.

An important issue in WSN is the limited energy, driven in part by the sensor nodes' size. In [2], the authors classify the network lifetime definitions into four categories, namely node-lifetime based Node Lifetime (NL), coverage and connectivity based NL, transmission based NL, and parameterized NL definitions. However, the node-lifetime based NL is the mostly used definition. According to this definition the NL is that the time span from the deployment to the instant when the network considered nonfunctional [3]. The network is non functional when any of the following events occur: when the first sensor nodes are out of energy or when theanumber of sensor nodes is out of energy or when coverage is lost [3, 4].

Yuan et al. [5] considered the deployment of relay nodes (RNs) in three-dimensional wireless sensor networks with the objective of prolonging network lifetime. They considered a *single-tiered* sensor network architecture where each sensor node does not send data to other sensor nodes. These sensors gather and transmit data to base stations or sinks through relay nodes which are deployed above ground. There are three channels considered in the model, namely the underground-underground channel, the underground-above ground channel and the above ground-above ground channel. They also proposed a two-phase method and several heuristics to solve the relay node placement for maximizing the network lifetime problem. In the first phase, a number of required relay nodes are chosen to minimize the transmission loss between all of relay nodes and sensor nodes, while the load balancing constraint among relay nodes is ignored. In the second phase, the chosen relay nodes are load-balanced. Two heuristics were proposed for phase 1 (LURNS) and three heuristics were proposed for phase 2 (LBSNA). However, in Yuan's approach, two phases, which are selected relays phase and connection assignment phase, are solved independently. Since the selected relays from the first phase would chiefly contribute to decide maximum transmission loss in the network. Therefore, solving this phase separately without transmission loss feedback from the second phase might result in poor performance.

In this paper, we propose a single-phase heuristic called Beam Genitor Search, to solve the problem of maximizing the network lifetime and ensuring load balancing among relay nodes. We also propose an improvement phase called Connection Swap to further improve the solutions.

The rest of this paper is organized as follows. Section 2 present the related works. In Sect. 3, we show the problem models. The main contributions of this paper will be presented in Sect. 4. The experimental results and discussion will be given in Sect. 5. Finally, future works and conclusions are provided in Sect. 6.

2   Related Works

Some works focus on solving the maximizing the NL by maximizing the node-lifetime. In [6], the authors considered NL the duration in which α percentage sensor data can be collected by the base station. They used linear programming and formulated the problem of maximizing the α lifetime of wireless sensor networks with solar energy sources. Using relay nodes to offload the radio transmission power consumption by sensor nodes is one way to prolong the NL. The basic principle of the approach is that sensor nodes can reduce its transmission power by transmitting signal to relay nodes which are closer to the base station. In [7], the authors have categorized relay node placement into two main aspects according to network architecture: in flat networks and in hierarchical networks. Based on the network architectures, relay node placement problem can be classified into either *single-tiered* for flat networks or *two-tiered* in hierarchical architecture. In flat networks, a sensor node transmits its data to an RN or to a base station, but does not receive or forward packets from other sensors. In hierarchical architectures, a sensor node can receive or send packets to other sensors.

Lloyd et al. [8] proposed deploying a minimum number of relay nodes in a WSN so that between each pair of sensor nodes, there is a connecting path consisting of relay and sensor nodes. They present a polynomial time 7-approximation algorithm for this problem. In [9,10], the authors considered the relay node placement problem: how to define relay node locations to improve network performance. They formalized the problem of relay node deployment by defining a linear, mixed integer mathematical programming (MIP) model based on a network flow formulation. Then, they used a standard solver to solve the problem.

The usefulness of WSN is not limited to traditional terrestrial applications. While wireless underground sensor networks which are networks consists of many underground sensors are already in use for many of these applications such as monitoring underground mines, water management,... [11]. Most of mentioned works focus on optimizing two-dimensional wireless sensor networks, which assume that the space between the sender and receiver is free from obstacles, or a line-of-sight communication. A major challenge for network lifetime optimization problem (NLOP) in Wireless Underground Sensor Networks (WUSNs) in comparision with those in 2D environment is the wireless communication in multi environments. In [12], the authors addressed the need for a systematic placement of the nodes in a WUSN to minimize transmission cost. They also formulate a nonlinear program to determine the optimal information exaction in a grid-based wireless underground sensor network.

Deploying relay nodes above ground to tranfer data from sensor nodes to the base station is one of the approaches to solve the maximizing the network lifetime problem. However, this approach have to face several challenges such as load balance of relay nodes. This paper address this concern by using combination between Beam Genitor Search and Connection Swap. Beam Search Genitor is the combination of two algorithms: Beam Search and Genitor In Beam Search, partial solutions are ordered according to some heuristics, then only a predetermined number of best solutions are kept as candidates for further searchings [13]. We replaced Beam Search's greedy-based candidate selection strategy with a rank-based selection algorithm, so that the results are more flexible. Among all rank-based selection methods, we chose Genitor because of its constant complexity, which is highly time-effective.

3   Modeling the Problem

We consider a wireless underground sensor network where each node is either a sensor node (SN) or relay node (RN). Sensor nodes collect and transfer data to a base station or sink through relay nodes. We assume that both sensor nodes and relay nodes have limited energy. The goal of our work is to deploy a required number of relay nodes to ensure that the network achieve maximum lifetime.

3.1   Basic Concepts

Definition 1 *(Network Lifetime)*
In a wireless underground sensor network, the network lifetime is the time from initialization to when the first sensor node runs out of energy.

Definition 2 *(Maximizing Network Lifetime Problem - MNLP)*
In wireless sensor networks, a set of relay nodes is placed to maximize the network lifetime. The problem of choosing the relay nodes' positions and their connections to sensor nodes is called the Relay Node Placement for Maximizing Network Lifetime problem.

3.2   Energy Model

In this section, we summarize the energy model in [5]. For each link between a sensor node s_i and the relay node f_j, the energy consumption of the sensor can be calculated as follows:

$$E_{ij} = P_r - G_r - G_t + T_{ij} \tag{1}$$

where E_{ij} is the transmitting energy cost at sensor s_i when connected to relay f_j. P_r, G_r, G_t, and T_{ij} are the receiving energy cost at relay f_j, its transmitting antenna gain, the receiving antenna gain of the sensor s_i, and the transmission loss between sensor s_i and relay f_j, respectively. In [5], the authors assume that the energy consumption of a sensor only depends on T_{ij}.

The transmission loss of the link between sensor s_i and relay f_j is calculated as follows:

$$T_{ij} = T_{AG} + T_{UG} + T_R \tag{2}$$

where T_{UG} is the underground transmission loss, T_{AG} is the aboveground transmission loss and T_R is the refraction loss from soil to air [5,11].

- T_{AG} is the path of loss in free space, which can be calculated as

$$T_{AG} = -147.6 + 20log(d_{ag}) + 20log(f) \tag{3}$$

where d_{ag} is the above-ground transmission distance, and f is the operating frequency in Hertz.
- T_{UG} is the signal loss of the transmission path with length d_{ug}:

$$T_{UG} = 6.4 + 20log(d_{ug}) + 20log(\beta) + 8.69\alpha d \tag{4}$$

The values of α and β in Eq. 4 depend on the dielectric properties of soil medium. These values are calculated using Pepliski's model [14] as follows:

$$\alpha = 2\pi f \sqrt{\frac{\mu_r \mu_0 \epsilon_1 \epsilon_0}{2} \left[\sqrt{1 + (\frac{\epsilon_2}{\epsilon_1})^2} - 1 \right]} \tag{5}$$

$$\beta = 2\pi f \sqrt{\frac{\mu_r \mu_0 \epsilon_1 \epsilon_0}{2} \left[\sqrt{1 + (\frac{\epsilon_2}{\epsilon_1})^2} + 1 \right]} \tag{6}$$

$$\epsilon = \epsilon_1 - j\epsilon_2 \tag{7}$$

$$\epsilon_1 = 1.15 \left[1 + \frac{\rho_b}{\rho_s}(\epsilon_s^{\alpha_1}) + m_v^{\beta_1} \epsilon_{fw1}^{\alpha_1} - m_v \right]^{\frac{1}{\alpha_1}} - 0.68 \tag{8}$$

$$\epsilon_2 = \left[m_v^{\beta_2} \epsilon_{fw2}^{\alpha_1} \right]^{\frac{1}{\alpha_1}} \tag{9}$$

where ϵ, ϵ_1 and ϵ_2 are the relative complex dielectric constant of the soil water mixture, real and imaginary parts of the complex soil's dielectric constant, respectively. ρ_b is the bulk density in grams per cubic centimeter, ρ_s is the specific density of the soil solids, m_v is the water volume fraction, $\alpha_1 = 0.65$ is an empirically determined constant. $\epsilon_s = (1.01 + 0.44\rho_s)^2 - 0.0062$ is the dielectric constant of soil's solid.
β_1 and β_2 are empirically determined constants, which given by

$$\beta_1 = 1.2748 - 0.519S - 0.512C \tag{10}$$

and

$$\beta_2 = 1.3379 - 0.63S - 0.166C \tag{11}$$

where S and C are the sand percentage and clay percentage, respectively. ϵ_{fw1} and ϵ_{fw2} are the real and imaginary parts of the relative dielectric constant of free water given by

$$\epsilon_{fw1} = \epsilon_{w\infty} + \frac{\epsilon_{w0} - \epsilon_{w\infty}}{1 + (2\pi f\tau_w)^2} \tag{12}$$

and

$$\epsilon_{fw2} = \frac{2\pi f\tau_w(\epsilon_{w0} - \epsilon_{w\infty})}{1 + (2\pi f\tau_w)^2} + \frac{\sigma_{eff}(\rho_s - \rho_b)}{2\pi + \epsilon_0 \rho_s m_v} \tag{13}$$

where ϵ_0 is permittivity of free space (F/m), τ_w is the relaxation time for water, ϵ_{w0} is the static dielectric constant for water, and $\epsilon_{w\infty}$ is the high-frequency limit, which is recommended equal 4.9.

$$\sigma_{eff} = 0.0467 + 0.224\rho_b - 0.411S + 0.6614C$$

3.3   Problem Formulation

Relay node placement is the process of selecting y locations out of the total m potential locations given by the model. The relay node placement problem is formulated as finding a location indicator set $P = \{p_1, p_2, ..., p_m\}$, where p_i is a binary indicator that is equal to one if location i is selected and zero otherwise. In this paper, we assume that the network shuts off when a node in n nodes run out of energy. The details of the problem model is given by the authors in [5] as follows:

Input:

- $S = \{s_1, s_2, ..., s_n\}$ is a set of sensor nodes.
- $F = \{f_1, f_2, ..., f_m\}$ is a set of potential locations.
- y is the selected locations $y < m$.
- $T = (T_{ij})_{n \times m}$ is the transmission loss between sensor s_i and relay f_j. These values are calculated using the energy model in Sect. 3.2.

Output:

- P is a set of selected locations.
- $A = (a_{ij})_{n \times m}$ is the connection matrix, where $a_{ij} = 1$ iff s_i is assigned to f_j.

Constraints:

- Each sensor node must be relayed by exactly one relay node.
- Each relay node must connect to $\frac{n}{y}$ sensor nodes.
- Exactly y relay nodes must be selected.

Objective

- Prolong the network lifetime (T_c) by minimizing the largest transmission loss among sensor nodes and relay nodes

$$T_c = max_{i=1}^n (\sum_{j=1}^m T_{ij} * a_{ij}) \rightarrow min \tag{14}$$

The MNL problem has been shown to be NP-hard by the authors in [5]. There are two approaches for solving NP-hard problems: exact algorithms and approximate algorithms. Exact algorithms attempt to find the optimal solutions, but for NP-hard problems they have exponential complexity and not practical for solving large-scale problems. Approximate algorithms do not ensure the best solution, but attempt to come as close as possible to the optimum value in a reasonable amount of time (polynomial complexity). In this section, we propose an approximation algorithm for solving the MNL problem.

4   Proposed Algorithms

The main idea of our first proposed algorithm is to sequentially select a position $f_i \in F$ to set up a relay node, then remove it from consideration. For each chosen f_i, n/y sensors that have the lowest transmission loss values corresponding to f_i (among the set of unassigned sensors) are assigned to f_i. The process is repeated until $|R| = y$. Hence, the final solution would have y relays, each connected to n/y sensors while each sensor connects to a single relay. The solution thus satisfies all the constraints of the Relay Placement problem.

It is apparent that the order of f_i selection dictates the method's performance, as each permutation of F would result in a different solution with varying quality. Going through every permutation is exponentially expensive and naturally infeasible for large-scale problems. Therefore, we apply Beam Search [13], with a pre-determined capacity c to output a feasible approximated solution in polynomial time.

Beam Search maintains a fixed size "beam" of possible directions for the next solution. Instead of visiting every direction as in brute force search, Beam Search clips the beam to its previous size after every expansion, only keeping the best directions. Although it can be efficient in finding a good solution, Beam Search's greedy characteristic might be sub-optimal for our problem. To address this, we use a Genitor random function [15] instead of the standard greedy beam clipping strategy to generate the next beam. We name the modified beam search Beam Genitor Search (BGS) and describe it in detail in Sect. 4.1.

In the final Beam Search iteration, the beam contains a list of at most c feasible solutions. A naive implementation would then select the best solution in the final beam. However, we also propose a fine-tuning method, Connection Swap (CS), which can further improve each of the resulting solutions from BGS. We describe this algorithm in Sect. 4.2.

4.1   Beam Genitor Search

The BGS algorithm takes as input a set of sensors S, a set of possible locations F, the number of sensors n, the number of possible locations m and the number of selected relays y. It outputs a set of selected relays R and a set of connections between relays and sensors eg. $\{\langle f_1, s_2 \rangle, \langle f_2, s_4 \rangle ...\}$.

BGS starts with an empty solution. In the first iteration, we generate m solutions by adding each relay position f_j to the first element of each solution. Each solution is evaluated by their critical loss value, which is the corresponding network's maximum transmission loss. In the following iterations, each solution in the previous iteration would be inserted with another relay position f_j, which has not been included in the solution yet. Algorithm 2 is used to create solutions in the next iteration. It takes a current solution including current selected relay nodes (inp_rns), current connected sensor nodes (inp_sns) and current connections between sensor and relay nodes(inp_cons) and list of sensor nodes S, list of relay positions F and transmission loss matrix T as input. The output is a new solution with a new added relay and list of connections between chosen sensors and relays.

Similar to standard Beam Search, we "clip" the beam by only keeping c solutions. However, instead of selecting c solutions with the best critical losses, we pick the next beam using a Genitor random function [15]. The Genitor function selects better solutions with higher probability, but can also include less quality ones. This provides the beam search with more variance and helps overcome local optima. In each subsequent iteration i, $(m - 1)$ solutions are generated from each solution in the beam, which is then similarly clipped. The process repeats to iteration y, where the beam has c feasible solutions. The details of algorithm are described in Algorithm 1.

Each of the c solution in the beam expands to at most $m - 1$ solutions (1 for each relay that can be added). For each relay considered, we sort through at most n sensors with $O(nlog(n))$ complexity to find the local optimal sensor set and calculate the critical loss value. Then the entire solution set is sorted with complexity $O(c \times (m - 1) \times log(c(m - 1)))$ in the Genitor function. The overall complexity of each iteration is therefore $O(c \times (m - 1) \times n \times log(n))$. As the entire algorithm runs in y iterations, BGS has a complexity of $O(c \times (m - 1) \times n \times log(n) \times y) \le O(c \times m \times n \times log(n) \times y)$.

Algorithm 1. Beam Genitor Search

Input : S, F, T, n, m, y
Output: $R, cons$

1 $capacity \leftarrow 2 \times m$ ▷ Beam capacity;
2 Initialize $beam$ with an empty solution;
3 $i \leftarrow 0$;
4 **while** $i < y$ **do**
5     $next\_beam \leftarrow$ Create succesors for every current solution in $beam$ (algorithm 2);
6   sort($next\_beam$) by successors with critical value;
   /* **Select** c **solutions in** $next\_beam$ */
7     $beam_{i+1} \leftarrow \emptyset$;
8   **while** $|beam_{i+1}| < capacity$ **do**
9        $sc \leftarrow$ a successor chosen from $next\_beam$ by Genitor function;
10       $beam_{i+1} \leftarrow beam_{i+1} \cup sc$
11    $i \leftarrow i + 1$;

/* **Choose the best solution in the last generation of Beam** $beam_Y$ */
12 **for** $sc \in beam_Y$ **do**
13   Improve sc's transmission loss via Connection Swap algorithm 3;

14 $R, cons \leftarrow$ best successor in $beam_Y$;
15 **return** $R, cons$

Algorithm 2. Generate Next Solutions

    Input : inp_rns, inp_sns, inp_cons, S, F, T
    Output: $next\_solution$

1 $next\_solution \leftarrow \emptyset$;
2 **for** $f_i \in F$ **and** $f_i \notin inp\_rns$ **do**
3      $sel\_rns \leftarrow inp\_rns$;
4      $sel\_sns \leftarrow inp\_sns$;
5      $cons \leftarrow inp\_cons$;
6      $sel\_relays \leftarrow sel\_rns \cup f_i$;
7      $ava\_sns \leftarrow \emptyset$ ▷ list of sensors that are feasible to f_i;

      /* Find available sensors for current solution   */
8   **for** $s_j \in S$ **and** $s_j \notin sel\_sns$ **do**
9          $ava\_sns \leftarrow ava\_sns \cup \langle s_j \rangle$;
10     $\text{sort}(ava\_sns)$ by $T_{f_i, s_j} \forall s_j \in sel\_sns$;

      /* Assign N/Y locally optimal sensors to f_i */
11   **for** $s_j \in ava\_sns$ **and** $j < N/Y$ **do**
12        $cons \leftarrow cons \cup (f_j, s_j)$;
13        $sel\_sns \leftarrow sel\_sns \cup \langle s_j \rangle$;
14     $next\_solution \leftarrow next\_solution \cup \langle sel\_rns, sel\_sns, cons \rangle$;

15 **return** $next\_solution$;

4.2 Connection Swap

We propose a heuristic algorithm called the Connection Swap to improve solutions created by BGS. We observe that the greedy scheme for assigning sensors to each relay line 11 in Algorithm 2 might not be optimal for some members of the resulting beam (especially given that the final beam may contain solutions with high critical loss). CS seeks to improve these solutions by rearranging their SN-RN connections.

CS takes input as a list of RNs, a list of SNs and the SN-RN connections. The main idea is to continuously iterate through all SN-RN connections between SNs and RNs and find the critical connection $cons_c$ (the connection with the highest transmission value). We then go through every other connections $cons_i$ and calculate the improvement margin made by swapping $cons_c$ and $cons_i$. If there are swaps with improvement, we apply the best swap and continue to the next iteration. Otherwise, the algorithm stops. Since each iteration either improves the solution or stops the algorithm, CS always converges at a minima.

Connection Swap can be seen as a greedy local search phase, where the local move is the aforementioned swapping procedure. We describe the method in detail in Algorithm 3. Each iteration iterates the connection list twice with a complexity of $O(n)$.

Algorithm 3. Connection Swap

Input : R, *cons*
Output: R, *cons*

1 *improved* $\leftarrow$ *true*;
2 **while** *improved* **do**
3   Find $(f_B, s_B) \in cons$ with maximum $T(f_B, s_B)$;
4      $T_{best} \leftarrow T(f_B, s_B)$;
5   *best_pair* $\leftarrow (f_B, s_B)$;
6   *improved* $\leftarrow$ *false*;
7   **for** $(f_i, s_j) \in cons$ **and** $(f_i, s_j) \neq (f_B, s_B)$ **do**
8   **if** $\max \{T(f_i, s_B), T(f_B, s_j)\} < T_{best}$ **then**
9              $T_{best} \leftarrow \max \{T(f_i, s_B), T(f_B, s_j)\}$;
10   *best_pair* $\leftarrow (f_i, s_j)$;
11   *improved* $\leftarrow$ *true*;
12   Swap (f_B, s_B) and *best_pair*
13 **return** R, *cons*

---

5 Simulation Results

5.1 Set Up

We implemented all 6 combinations of LURNS and LBSNA heuristics in Yuan, Chen, and Yao, along with our proposed algorithm (BGS and CS) in the Java programming language and tested them on a system with an Intel Core i7-2720QM 2.2 Ghz processor and 8GB RAM. The experiments were performed on 30 instances of a medium dataset with an area size of 500×500 and 30 instances of a large dataset with an area size 1000×1000. We generate 6 schemes based on different distributions for each type of datasets. The detail of the schemes will be described in Table 1. Parameters for the algorithms are set as in Table 2.

BGS and its combination with CS are run 30 times on each dataset. We set the Genitor bias to be 10 and $c = 2 \times m$ in our experiments.

Table 1. Data generation schemes

Scheme	Sensor distribution	Relay distribution
rr	Uniform	Uniform
mcr	Many clusters (normal distribution around $n/y \times 2$ random points)	Uniform
fcr	Few clusters ($n/y/2$ clusters)	Uniform
rmc	Uniform	Many clusters ($y \times 2$ clusters)
fmc	Uniform	Few clusters ($y/2$ clusters)
br	Bordered (only placed near the area's borders)	Uniform

Table 2. Parameter values

Parameter	Value
SN burial depth d_{SN}	1 m
RN height h_{RN}	10 m
Radio operating frequency f	300 MHz
Volumetric moisture content of soil m_v	10%
Soil bulk density p_b	1.5 g/cm^3
Solid soil particles density p_s	2.66 g/cm^3
Mass fractions of sand S	50%
Mass fractions of clay C	15%
Attenuation coefficient in air η	3
Soil refractive index n_1	1.55
Air refractive index n_2	1

5.2   Experimental Results

We denote

- *Yuan*: the best results which refer to the solution's maximum transmission loss among all 6 combinations of LURNS and LBSNA heuristic in [5].
- *BGS*, BGS_{best}: the average and best results, respectively, from 30 runs of BGS, each run taking the best solution in the final beam.
- $(BGS+CS)$, $(BGS+CS)_{best}$: the average and best results, respectively, from 30 runs of applying CS to each solution in BGS's final beam, then selecting the best solution.
- *BS*: the result obtained by Beam Search algorithm without Genitor random.

We evaluate performance based on the maximum transmission loss of the solution. Overall, it can be seen that BGS+CS could outperform Yuan's heuristics over most instances. Especially, By applying CS on the solution achieved from BGS and Yuan, the result is significantly improved.

BGS Evaluation. We compare the output between the BGS and the BS in order to evaluate the efficiency of Genitor algorithm [15] in ameliorating the solution quality of BS. Overall, BGS performs better than BS at 8.9% on medium instances and 9.9% on large instances. Figures 1 and 2 indicates that the number of instances which BGS takes over BS is 28 out of 30 large instances, the corresponding number is 29 in medium instances. Especially, Genitor application improve around 15% from BS over *mcr* instances.

CS Evaluation. The CS performance is considered by examining the improvement of CS over BGS and the best results from Yuan's heuristics in [5]. In details, we compare the solution quality between (Yuan + CS) and (BGS + CS).

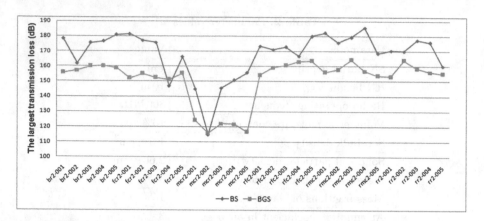

Fig. 1. Comparison between the output results of Beam Search algorithm and the average output results of the Beam Genitor Search and Connection Swap on the medium dataset

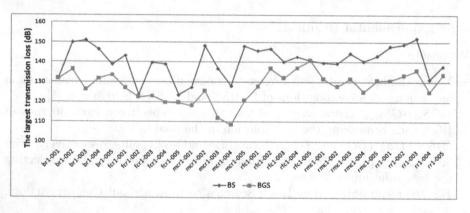

Fig. 2. Comparison between the output results of Beam Search algorithm and the average output results of the Beam Genitor Search and Connection Swap on the large dataset

From Tables 3 and 4, it can clearly be seen that CS improves Yuan's results by 2.6% over medium instances and 3.56% over large instances. The improvement takes place on 30 out of 30 instances. Furthermore, the number is much more significant in improving BGS's solution quality which is 10.8% and 13.6% over all medium and large instances correspondingly. From above examination, the proposed CS algorithm does improve significantly the results of BGS as well as Yuan's heuristics.

BGS + CS Evaluation. In this experiment, we show that the combination of BGS and CS is better than the state-of-the-art by comparing (BGS + CS) with Yuan [5]. In medium dataset, (BGS + CS) outperforms Yuan at 9.66%, this number is 13.5% in large dataset. The experimental results shows that the performance of BGS and CS takes over the existed method.

Table 3. Comparison between Yuan's heuristics, BGS and CS on the medium dataset

Test	Yuan [5]	Yuan + CS	BGS	BGS_{dev}	BGS_{best}	BGS+ CS	BGS+ CS_{best}	BGS+ CS_{dev}	BGS_{best}+ CS
br1-001	136.41	128.74	131.77	0.74	130.77	125.13	**121.16**	1.34	131.05
br1-002	131.07	126.41	136.33	3.89	130.33	111.44	**105.65**	2.22	121.94
br1-003	126.52	125.60	126.22	4.04	118.23	114.14	**109.96**	2.12	121.77
br1-004	127.72	116.05	131.84	4.09	121.38	116.42	**113.11**	1.75	126.26
br1-005	133.42	117.03	133.66	3.99	125.99	119.67	**109.87**	2.98	128.50
fcr1-001	114.24	114.24	126.87	5.56	118.24	107.26	**105.33**	0.71	117.91
fcr1-002	142.65	138.91	122.32	5.94	107.77	107.55	**107.26**	0.74	115.48
fcr1-003	124.66	121.93	123.06	3.37	118.15	109.71	**106.93**	1.70	117.89
fcr1-004	140.30	136.64	119.78	3.13	111.40	107.96	**105.69**	2.01	117.48
fcr1-005	131.20	123.59	119.53	2.10	113.95	103.78	**102.00**	1.44	110.62
mcr1-001	130.11	128.16	118.04	1.57	116.19	116.20	**115.80**	0.24	118.00
mcr1-002	128.45	126.80	125.49	3.83	110.38	111.25	**109.48**	1.90	122.22
mcr1-003	116.68	112.19	111.85	4.16	102.00	97.46	**95.25**	1.15	107.67
mcr1-004	123.68	122.80	108.45	2.97	99.60	99.37	**98.97**	1.01	105.92
mcr1-005	118.93	116.85	120.61	0.86	117.23	111.35	**106.14**	2.94	120.24
rfc1-001	124.12	120.11	127.67	3.83	123.91	113.61	**113.57**	0.09	118.79
rfc1-002	129.73	129.73	136.82	2.35	133.08	127.84	**127.83**	0.04	130.16
rfc1-003	120.42	120.42	131.99	4.66	121.86	117.38	**117.38**	0.00	121.61
rfc1-004	139.12	132.15	137.17	2.89	128.27	128.27	**128.27**	0.00	129.49
rfc1-005	140.81	140.81	140.85	0.23	140.81	140.81	**140.81**	0.00	140.81
rmc1-001	120.12	120.12	131.49	5.35	120.05	113.25	**111.99**	1.27	120.07
rmc1-002	120.43	120.36	127.65	1.86	124.94	116.98	**116.87**	0.25	121.29
rmc1-003	116.00	113.03	131.54	4.77	120.74	114.37	**111.35**	1.55	123.82
rmc1-004	122.14	110.00	124.08	2.46	120.86	117.45	**112.82**	1.02	122.34
rmc1-005	125.60	121.33	130.75	4.49	117.94	111.69	**109.64**	0.97	121.07
rr1-001	124.75	115.99	130.88	4.71	122.82	112.67	**107.30**	2.59	121.81
rr1-002	114.06	114.06	133.30	3.66	127.47	111.20	**105.77**	2.18	121.15
rr1-003	124.31	124.31	135.70	4.49	124.86	112.38	**109.21**	1.64	117.56
rr1-004	123.41	121.17	125.04	4.35	116.12	109.68	**107.67**	1.79	119.32
rr1-005	121.22	121.22	133.50	5.02	125.80	113.61	**109.77**	1.64	124.43
Average	*126.41*	*123.02*	*127.84*	*3.51*	*120.37*	*114.00*	**111.41**	*1.33*	*121.22*

In order to examine the efficiency of the combination of BGS and CS, we combine Yuan's heuristics with the CS and the achieved results are compared with the (BGS + CS) results. The combination between Yuan and CS still does not catch up with the results obtained from (BGS + CS). (BGS + CS) results better than (Yuan + CS) at 8% and 11% in medium dataset and large dataset correspondingly. This indicates that although CS improves both BGS and Yuan results significantly, the CS is more suitable for the proposed BGS algorithm as the overall results achieved from the combination of BGS and CS results much better than the improved Yuan.

Complexity Evaluation. Our proposed algorithms take more time than algorithms in [5] to calculate the solution due to the complexity of BGS, which is $O(c \times m \times n \times log(n) \times y)$, compared to $O((n^3 - y^3) \times m)$. This is an expected tradeoff for higher quality solutions and the increased runtime is still within reasonable limits.

Table 4. Comparison between Yuan's heuristics, BGS and CS on the large dataset

Test	Yuan [5]	Yuan + CS	BGS	BGS_{dev}	BGS_{best}	BGS+ CS	BGS+ CS_{best}	BGS+ CS_{dev}	$BGS_{best}+$ CS
br2-001	146.33	146.10	178.24	5.26	155.60	**128.91**	124.27	1.71	142.01
br2-002	153.44	134.75	161.77	5.44	157.16	128.52	**124.32**	2.10	139.87
br2-003	165.11	138.29	175.61	3.52	159.96	127.32	**123.64**	1.84	139.67
br2-004	155.12	138.90	176.55	5.25	160.10	129.81	**124.62**	2.34	143.31
br2-005	152.98	151.14	180.87	5.28	158.77	131.40	**126.05**	2.27	146.85
fcr2-001	144.99	141.48	181.29	4.46	152.04	128.79	**126.55**	1.69	138.76
fcr2-002	141.91	141.60	177.09	4.58	154.97	130.54	**125.61**	2.03	142.04
fcr2-003	140.42	135.39	175.79	3.23	152.37	128.71	**125.36**	1.36	139.72
fcr2-004	163.75	154.92	146.77	3.64	150.99	132.04	**127.89**	2.09	144.38
fcr2-005	150.47	146.19	166.31	4.94	155.29	132.85	**128.56**	1.50	143.60
mcr2-001	137.40	137.40	144.97	3.52	124.14	110.13	**106.56**	1.92	119.09
mcr2-002	152.36	150.52	114.82	2.44	115.79	108.47	**104.89**	1.04	113.32
mcr2-003	161.14	161.14	145.61	3.62	122.10	109.24	**106.98**	1.22	117.22
mcr2-004	148.53	148.53	150.78	3.38	121.70	111.68	**107.57**	1.98	118.41
mcr2-005	151.59	149.92	155.63	3.10	116.60	108.42	**105.25**	1.46	113.65
rfc2-001	147.59	140.57	173.38	2.67	154.14	136.11	**133.77**	0.96	146.31
rfc2-002	163.56	152.29	171.11	3.71	159.22	144.94	**144.94**	0.00	148.15
rfc2-003	155.36	151.96	172.89	2.85	160.70	142.18	**142.09**	0.37	150.62
rfc2-004	159.48	159.48	166.78	0.93	163.16	159.48	**159.48**	0.00	160.17
rfc2-005	161.32	161.32	179.96	2.69	163.70	155.29	**155.29**	0.00	158.34
rmc2-001	148.45	143.45	182.25	4.72	156.09	132.10	**130.26**	1.00	138.82
rmc2-002	153.64	145.16	175.77	4.61	157.78	141.69	**141.69**	0.00	144.97
rmc2-003	158.71	154.40	179.28	5.70	164.28	136.30	**134.42**	0.66	142.79
rmc2-004	154.25	145.86	185.68	4.22	156.84	136.56	**135.88**	0.88	144.09
rmc2-005	144.08	140.20	168.66	3.28	153.77	132.97	**131.50**	0.88	141.98
rr2-001	142.51	142.51	170.44	2.77	153.13	126.97	**123.86**	1.41	138.25
rr2-002	139.27	132.87	170.10	3.94	164.04	128.70	**125.92**	1.33	140.02
rr2-003	147.94	141.24	177.44	4.88	158.42	132.41	**127.52**	1.74	140.22
rr2-004	134.21	132.96	175.90	4.41	156.25	127.90	**124.57**	1.54	137.62
rr2-005	149.81	140.61	160.21	4.19	154.98	128.29	**125.55**	1.68	138.01
Average	150.86	145.37	168.07	3.91	151.14	130.29	**127.50**	1.30	139.07

6   Conclusions

In this paper, we examined the Relay Placement for Maximizing the Network lifetime problem in WUSNs. *First*, we proposed a novel heuristic method called Beam Genitor Search to solve this problem. *Secondly*, we improved the quality of solution obtained from BGS using an algorithm called Connection Swap. Our experiment results demonstrate that the proposed methods produce better results in all instances of our datasets, at the cost of higher complexity than existing algorithms. This is an expected and acceptable tradeoff and will be studied extensively in the future.

Acknowledgment. This research is funded by Army Research Lab and International Technology Center- Pacific under project "Evolutionary Multitasking for Solving Optimization Problems".

References

1. Akyildiz, I.F., et al.: Wireless sensor networks: a survey. Comput. Netw. **38**(4), 393–422 (2002)
2. Yetgin, H., et al.: A survey of network lifetime maximization techniques in wireless sensor networks. IEEE Commun. Surv. Tutorials **19**(2), 828–854 (2017)
3. Chen, Y., Zhao, Q.: On the lifetime of wireless sensor networks. IEEE Commun. Lett. **9**(11), 976–978 (2005)
4. Dietrich, I., Dressler, F.: On the lifetime of wireless sensor networks. ACM Trans. Sen. Netw. **5**(1), 5:1–5:39 (2009). ISSN: 1550-4859
5. Yuan, B., Chen, H., Yao, X.: Optimal relay placement for lifetime maximization in wireless underground sensor networks. Inf. Sci. **418**, 463–479 (2017)
6. Abu-Baker, A., et al.: Maximizing α-lifetime of wireless sensor networks with solar energy sources. In: Military Communications Conference 2010-MILCOM, pp. 125–129. IEEE (2010)
7. Bari, A.: Relay Nodes in Wireless Sensor Networks: A Survey. University of Windsor (2005)
8. Lloyd, E.L., Xue, G.: Relay node placement in wireless sensor networks. IEEE Trans. Comput. **56**(1), 134–138 (2007)
9. Di Caro, G.A., Feo Flushing, E.: Optimal relay node placement for throughput enhancement in wireless sensor networks. In: Proceedings of the 50th FITCE International Congress (2011)
10. Feo Flushing, E., Di Caro, G.A.: A flow-based optimization model for throughput-oriented relay node placement in wireless sensor networks. In: Proceedings of the 28th Annual ACM Symposium on Applied Computing, pp. 632–639. ACM (2013)
11. Sun, Z.H.I., Akyildiz, I.F., Hancke, G.P.: Dynamic connectivity in wireless underground sensor networks. IEEE Trans. Wirel. Commun. **10**(12), 4334–4344 (2011)
12. Zungeru, A.M., Mangwala, M., Chuma, J.: Optimal node placement in wireless underground sensor networks. Int. J. Appl. Eng. Res. **12**(20), 9290–9297 (2017)
13. Ow, P.S., Morton, T.E.: Filtered beam search in scheduling. Int. J. Prod. Res. **26**(1), 35–62 (1988)
14. Peplinski, N.R., Ulaby, F.T., Dobson, M.C.: Dielectric properties of soils in the 0.3-1.3-GHz range. IEEE Trans. Geosci. Remote Sens. **33**(3), 803–807 (1995)
15. Whitley, L.D., et al.: The GENITOR algorithm and selection pressure: why rank-based allocation of reproductive trials is best. In: ICGA, vol. 89, pp. 116–123. Fairfax, VA (1989)

Neuroevolution and Data Analytics

The Evolution of Self-taught Neural Networks in a Multi-agent Environment

Nam Le[✉], Anthony Brabazon, and Michael O'Neill

Natural Computing Research and Applications Group, University College Dublin,
Dublin, Ireland
namlehai90@gmail.com

Abstract. Evolution and learning are two different forms of adaptation
by which the organism can change their behaviour to cope with problems posed by the environment. The second form of adaptation occurs
when individuals exhibit plasticity in response to environmental conditions that may strengthen their survival. Learning has been shown to
be beneficial to the evolutionary process through the **Baldwin Effect**.
This line of thought has also been employed in evolving adaptive neural networks, in which learning algorithms, such as Backpropagation,
can be used to enhance the adaptivity of the population. Most work
focuses on evolving learning agents in separate environments, this means
each agent experiences its own environment (mostly similar), and has no
interactive effect on others (e.g., the more one gains, the more another
loses). The competition for survival in such settings is not that strong, if
being compared to that of a *multi-agent* (or shared) environment. This
paper investigates an evolving population of *self-taught* neural networks
– networks that can teach themselves – in a shared environment. Experimental results show that learning presents an effect in increasing the
performance of the evolving multi-agent system. Indications for future
work on evolving neural networks are also presented.

Keywords: The Baldwin effect · Neural networks · Neuroevolution ·
Meta-learning · Self-learning

1 Introduction

For many biological organisms, adaptation is necessary for survival and reproduction in an uncertain environment. There are two important kinds of adaptation
that should be distinguished. The first is a change at the genetic level of a population, in which organisms reproduce selectively subject to mechanisms, like
mutation or sexual recombination, which maintain inter-individual variability.
This is usually modeled in terms of biological evolution, which causes changes in
the population from one generation to the next. The second adaptation mechanism, on the other hand, is the phenotypic change at the individual level. This
can be called *lifetime-adaptation* which changes the phenotypic behaviour of the

© Springer Nature Switzerland AG 2019
P. Kaufmann and P. A. Castillo (Eds.): EvoApplications 2019, LNCS 11454, pp. 457–472, 2019.
https://doi.org/10.1007/978-3-030-16692-2_31

organism during its lifetime. Plausibly, lifetime adaptation happens at a quicker pace than the evolutionary process which takes place through the generational timescale, preparing the organism for rapid uncertain environments.

The idea that learning can influence evolution in a Darwinian framework was discussed by psychologists and evolutionary biologists over one hundred years ago ([1,2]), through *A new factor in evolution* called **the Baldwin Effect**. However, it gradually gained more attention since the classic paper by Hinton and Nowlan ([3]) which demonstrated an instance of the Baldwin effect in computer simulation. This line of research motivated the idea of evolving neural networks, or *neuroevolution* (NE), in which one can observe learning and evolution interacting with each other in creating adaptive neural networks [4–6]. Regardless of how learning is implemented, most studies focus on how learning helps evolution to solve an individual problem – this means each agent solves its own problem (and their problems are copies of each other) – there is no mutual interactive effect between learning agents when they learn and compete for survival during their lifetime.

The main aim of this paper is to investigate, through computer simulation, how learning and evolution can provide adaptive advantage in a multi-agent system. We design a neural architecture allowing for lifetime learning, more specifically for the ability to teach oneself. To show the effect of self-taught agents we simulate a simple environment called MiniWorld in which agents have to compete with each others for food in order to survive. In the remainder of this contribution, we initially present some prior research on learning and evolution, including brief literature review on learning and evolving neural networks. We then describe the experiments undertaken. The results from these experiments are analysed and discussed, and finally, conclusions and several interesting future research opportunities are proposed.

2   Learning and Evolution

2.1   The Baldwin Effect

Most organisms need to learn to adapt to their changing environments. Learning is essential for both living organisms, including humans and animals, and artificial learning agents. Organisms, living or digital, need to learn new behaviors to solve tasks that cannot be solved effectively by innate (or genetically encoded) behaviors. Learning shows its powerful advantages when the environment changes [7]. During the lifetime of an individual organism, if its environment changes in a way that its previous knowledge cannot be enough to survive, it has to learn new behaviors or knowledge to adapt to the novel circumstance. Otherwise, it will be unfit, thus having less chance to *survive*.

It is very interesting that the Baldwin Effect was first demonstrated computationally [3] around 20 years before it was first empirically verified in nature [8]. In 1987, the Cognitive Scientist Geoffrey Hinton and his colleague Kevin Nowlan at CMU presented a classic paper [3] to demonstrate an instance of the Baldwin effect in the computer. Hinton and Nowlan (henceforth H&N) used a genetic algorithm [9] to evolve a population in an extreme landscape called *Needle-in-a-haystack*, showing that learning can help evolution to search for the solution when evolution alone fails to do so. An interesting idea that can be extracted from their work is that instead of genetically fixing the genotype, it is wiser to let just a portion of the genotype genetically fixed, and the other be *plastic* that allows for changes through learning. It is these plastic individuals that promote the evolutionary process to search for the optimal solution, although the H&N landscape is static.

The model developed by Hinton and Nowlan, though simple, is interesting, opening up the trend followed by a number of research papers investigating the interaction between learning and evolution. Following the framework of Hinton and Nowlan, there have been a number of other papers studying the Baldwin effect in the NK-fitness landscape, which was developed by Kauffman [10] to model 'tunably rugged' fitness landscapes. Problems within that kind of landscape are shown to fall in NP-completeness category [10]. Several notable studies of the Baldwin effect in the NK-model include work by Mayley [11], and some others [12]. Their results, again, demonstrated that the Baldwin effect does occur and the allowance for lifetime learning, in the form of individual learning, helps evolutionary search overcome the difficulty of a rugged fitness landscape.

2.2   Learning and Evolution in Neural Networks

There have also been several studies investigating the interaction between learning and evolution in Neural Networks. Most use the so-called *Neuroevolution* approach to test if neural network learning facilitates an evolutionary process, often represented by an evolutionary algorithm. Some notable papers on this line of research include [13], and several works by Stefano Nolfi, Domenico Parisi on Evolutionary Robotics [4,6], to name but a few. All of these papers attempted to confirm the existence of the Baldwin effect, by showing how learning interacts with evolution making the system perform better than with evolution alone.

Todd and Miller [14] investigated an imaginary underwater environment in which organisms have to eat food and avoid poison. Each agent was born in one of the two feeding patches, and has to decide whether to consume substances floating by. Those substances can be either food or poison, with colour either red or green. The association between the colour (red or green) and the substance (food or poison) varies between feeding patches. Therefore, an agent has to decide whether to eat or pass a substance based on its sensory experience. Moreover, there is no feedback given to an individual agent that could be used to discriminate between food and poison [14]. Todd and Miller showed that the combination of evolution and learning in the form of *Hebbian Rule* [15] in a simple neural network can do better than both evolution and learning alone in this scenario. One point to be noted here is that this is an imaginary environment, there is no movement in the environment by the agent. The story here can be understood as a population of *disembodied* neural network learning to classify the representation of food and poison.

Another study on this topic includes [13] in which neural networks have to classify food and poison. The neural synaptic connections are both evolved using an evolutionary algorithm and learned by the *Delta Rule* (a simple form of Backpropagation on single layer network). It was also shown that learning enhances the evolving population in terms of food-poison classification. Like the work by Todd and Miller, there is no movement in the environment or the interaction between the neural network and the environment containing food and poison. This is also a disembodied case.

Nolfi and his colleagues simulated a population of simulated embodied *animats* (or artificial organisms) controlled by neural networks situated in a grid-world, with four actions (turn 90° right or left, move one cell forward, or stay still in the current cell). The evolutionary task is to evolve action strategies to collect food effectively, while each agent learns to predict the sensory inputs to neural networks for each time step. Learning was implemented using Backpropagation based on the error between the actual and the predicted sensory inputs. It was shown that learning to predict is able to enhance the evolutionary process and increase the fitness of the population. The same observation was also validated as true when learning performs a XOR function ([6]. In these studies, each animat, or simulated embodied agent, lives in its own copy of the world. Therefore, there is no interaction between these agents while they are foraging during their lifetime.

There have also been several other studies using the idea of learning and evolution in neural networks, including [16–18], showing how the interplay between learning and evolution in evolving neural networks enhances the performance the evolving system. Yet most of the work use *disembodied* and *unsituated* neural networks or there is no interaction between learning agents in the environment – they live in their own copies of the environment, solving their own problems, having no effect on the performance of others. Please refer to [19] for some more recent studies on evolving plastic neural networks (neural networks that learn).

In this study, we investigate the interaction between learning and evolution in a multi-agent system – a system containing multiple situated agents living together and doing their tasks while competing with each other. Each agent is controlled by a neural network but situated (and has a *soft-embodiment*). This means, the way an agent acts and moves in the world affects the subsequent sensory inputs, hence the future behaviour of that agent. We also propose an architecture called self-taught neural network – neural network that can teach itself during the lifetime interaction with the environment. Simulations to investigate between evolution and learning in evolving self-taught neural networks are described in the following section.

3   Simulation Setup

3.1   The Simulated World

We simulate a continuous 2D-world, called **MiniWorld**, with dimension 640×640. It contains 50 food particles randomly uniformly distributed. Each piece of food is represented by a squared image with size 10×10, locating at a random position in range [0, 640] for each dimension. Each food has an energy value. For simplicity, in our simulation we set the energy value of 1 for every food particle. Because the state-space of the world is continuous, not discrete like that of a grid space, and the size of food is much smaller than the dimension of the world, the dispersal of food is sparse enough to complexify the foraging task investigated in this paper. One property of MiniWorld is it has no strict boundary, and we implement the so-called *toroidal* – this means when an agent moves beyond an edge, it appears in the opposite edge. The visualisation of the sample world is shown in Fig. 1.

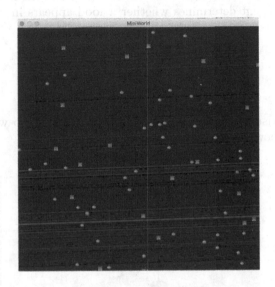

Fig. 1. MiniWorld – The environment of agents and food, 640 × 640.

3.2   Agent

We also simulate an evolving population of 20 agents in MiniWorld. Each agent is represented by an squared image with size 10 × 10, the same size as the food particle. Each agent has a food counter representing the total energy the agent got during its life. The food counter of an agent increases by the energy level of the food that the agent eats (hence by 1 in our simulation). When an agent happens to collide with a food patch, the food is eaten and another food piece randomly spawns in the environment but at a different random location. The collision detection criterion is specified by the distance between the two bodies. This shows how the soft-body affects the movement and the performance of an agent in its environment. By the re-appearance of food, the environment changes as an agent eats a food.

Every agent has a heading (in principle) of movement in the environment. In our simulation, we assume that every agent has a *priori* ability to sense the angle between its current heading and the food if appearing in its visual range. The visual range of each agent is a circle with radius 4. Each agent takes as inputs three sensory information, which can be the binary value 0 or 1, about what it sees from the left, front, and right in its visual range. If there is no food appearing in its visual range, the sensory inputs are all set to 0. If there is food appearing on the left (front, or right), the left (front, or right) sensor is set to 1; otherwise, the sensor is 0.

Let θ (in degree) be the angle between the agent and the food particle in its visual sense. An agent determines whether a food appears in its left, front, or right location in its visual range be the following rule:

$$\begin{cases} 15 < \theta < 45 \implies right \\ \theta \leq 15 \quad or \quad \theta \geq 345 \implies front \\ 315 < \theta < 345 \implies left \end{cases}$$

A examplar visualisation of an agent and its relationship with food in the environment is shown in Fig. 2.

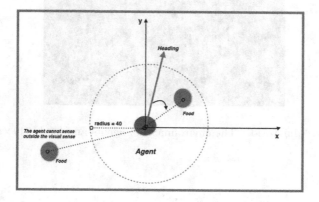

Fig. 2. Agent situated in an environment seeing food

Unlike [4,6] in which each agent has its own copy of the world, we let all agents live in the same MiniWorld. They feed for their own survival during their life. The more an agent eats, the less the chance for others to feed themselves. This creates a stronger competition in the population. When an agent moves for foraging, it changes the environment in which other agents live, changing how others sense the world as well. This forms a more complex dynamics, even in simple scenario we are investigating in this paper.

The default velocity (or speed) for each agent is 1. Every agent has three basic movements: Turn left by 9° and move, move forward by double speed, turn right by 9° and move. For simplicity, these rules are pre-defined by the system designer of MiniWorld. We can imagine the perfect scenario like if an agent sees a food in front, it doubles the speed and move forward to catch the food. If the agent sees the food on the left (right), it would like to turn to the left (right) and move forward to the food particle. The motor action of an agent is guided by its neural network as described below.

3.3   The Neural Network Controller

Each agent is controlled by a fully-connected neural network to determine its movements in the environment. What an agent decides to do changes the world the agent lives in, changing the next sensory information it receives, hence the next behaviour. This forms a sensory-motor dynamics and a neural network acts as a situated cognitive module having the role to guide an agent to behave adaptively, or *Situated Cognition* even in such a simple case like what is presenting in this paper. Each neural network includes 3 layers with 3 input nodes in input layer, 10 nodes in hidden layer, and 3 nodes in output layers.

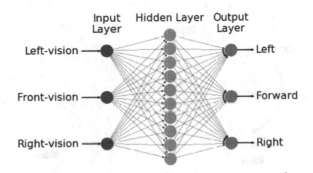

Fig. 3. Neural network controller for each situated agent. Connection weights can be created by evolutionary process, but can also be changed during the lifetime of an agent.

The first layer takes as input what an agent senses from the environment in its visual range (described above). The output layer produces three values as a motor-guidance for how an agent should behave in the world after processing sensory information. The maximum value amongst these three values is chosen as a motor action as whether an agent should turn left, right, or move forward (as described above). All neurons except the inputs use a sigmoidal activation function. All connections (or synaptic strengths) are initialised as Gaussian (0, 1). These weights are first initialised as *innate*, or merely specified by the genotype of an agent, but also have the potential to change during the lifetime of that agent.

The architectural design of neural network controller is visualised in Fig. 3. In fact, the neural architecture as shown in Fig. 3 has no ability to learn, or to teach itself. In the following section, we extend this architecture to allow for self-taught learning agents.

3.4   The Self-taught Neural Architecture

To allow for self-taught ability, the neural controller for each agent now has two modules: one is called **Action Module**, the other is called **Reinforcement Module**. The action module has the same network as previously shown in Fig. 3. This module takes as inputs the sensory information and produces reinforcement outputs in order to guide the motor action of an agent. The reinforcement module has the same set of inputs as the action module, but possesses separate sets of hidden and output neurons. The goal of reinforcement network is to provide *reinforcement signals* to guide the behaviour of each agent. The topology of a neural network in this case is visualised in Fig. 4.

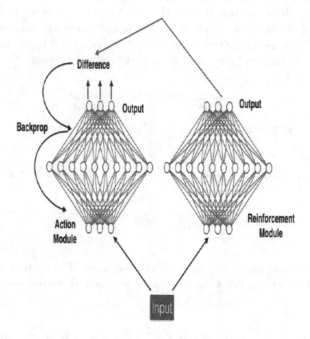

Fig. 4. Self-taught neural network.

The difference between the output of the reinforcement module and the action module is used as the error of the output behaviour of the action module. That error is used to update the weights in action modules through Backpropagation [20]. Through that learning process, the action module approximates its output activation towards the output of the reinforcement module. In fact, the reinforcement and the action modules are not necessary to have the same topology. For convenience, in our simulation we allow the reinforcement module possesses the same neuronal structure as the action module, but has 10 hidden neurons separate from the hidden neurons of the action module, hence the connections. The learning rate is 0.01.

In the following sections, we describe simulations we use to investigate the evolutionary consequence of lifetime learning.

3.5 Simulation 1: Evolution Alone (EVO)

In this simulation, we evolving a population of agents without learning ability. The neural network controller for each agent is the one described in Fig. 3. The *genotype* of each agent is the weight matrix of its neural network, and the evolutionary process takes place as we evolve a population of weights, a common approach in Neuroevolution (NE) [21].

Selection chooses individuals based on the number of food eaten in the foraging task employed as the fitness value. The higher the number of food eaten, the higher the fitness value. For crossover, two individuals are selected as parents, namely $parent_1$ and $parent_2$. The two selected individuals produce one offspring, called *child*. We implement crossover as the more the successful of a parent, the more the chance its weights are copied to the child. The weight matrices of the child can be simply described as the algorithm below.

Algorithm 1. Crossover

```
1:  function CROSSOVER(parent1, parent2)
2:      rate = parent1.fitness/parent2.fitness comment: fitness ratio
3:      child.weights = copy(parent2.weights)
4:      for in ∈ len(child.weights) do
5:          if random() < rate then
6:              child.weights[i] = parent1.weights[i]
7:          end if
8:      end for
9:  end function
```

Once a child has been created, that child will be mutated based on a predefined *mutation rate*. In our work, *mutation rate* is set to 0.05. A random number is generated, if that number is less than *mutation rate*, mutation occurs, and vice versa. If mutation occurs for each weight in the child, that weight is added by a random number from the range [−0.05, 0.05], a slight mutation. After that, the newly born individual is placed in a new population. This process is repeated until the new population is filled 100 new individual agents. No elitism is employed in our evolutionary algorithm.

The population goes through a total of 100 generations, with 10000 time steps per generation. At each time step, an agent does the following activities: Receiving sensory inputs, computing its output, updating its new heading and location. In evolution alone simulation, the agent cannot perform any kind of learning during its lifetime. After that, the population undergoes selection and reproduction processes.

3.6 Simulation 2: Evolution of Self-taught Agents (EVO+Self-taught)

In this simulation, we allow lifetime learning, in addition to the evolutionary algorithm, to update the weights of neural network controllers when agents interact with the environment. We evolve a population of **Self-taught** agents – agents that can teach themselves. The self-taught agent has a self-taught neural network architecture as described previously and shown in Fig. 4. During the lifetime of an agent, the reinforcement modules produce outputs in order to guide the weight-updating process of the action module. Only the weights of action modules can be changed by learning, the weights of reinforcement module are genetically specified in the same evolutionary process as specified above in Evolution alone simulation.

We use the same parameter setting for evolution as in EVO simulation above. At each time step, an agent does the following activities: Receiving sensory inputs, computing its output, updating its new heading and location, and updating the weights in action module by **self-teaching**. After one step, the agent updates its fitness by the number of food eaten. After that, the population undergoes selection and reproduction processes as in Evolution alone.

Remember that we are fitting learning and evolution in a Darwinian framework, not Lamarckian. This means what will be learned during the lifetime of an agent (the weights in action module) is not passed down onto the offspring. Results and analysis are described in the section below.

3.7 Simulation 3: Random Self-taught Agents (Random-Self-taught)

We conduct another simulation in which all agents are self-taught agents – having self-taught networks that can teach themselves during lifetime. What differs from simulation 2 is that at the beginning of every generation, all weights are randomly initialised, rather than updated by an evolutionary algorithm like in simulation 1. The learning agents here are initialised as *blank-slates*, or *tabula rasa*, having no predisposition to learn or some sort of *priori knowledge* about the world. The reason for this simulation is that we are curious whether evolution brings any benefit to learning in MiniWorld. In other words, we would like to see if there is a synergy between evolution and learning, not just how learning can affect evolution.

Experimental results are presented and discussed in the following section.

4 Results and Analysis

4.1 Learning Facilitates Evolution

At first, we compare the eating performance, in terms of the best and the average food eaten, of the population in EVO and EVO+Self-taught simulations. All results are averaged over 30 independent runs.

We look at the dynamics of the eating ability over generations. It can be observed in the left of Fig. 5 that there is a difference in the best eating ability between the two populations, EVO and EVO+Self-taught. We can see that the best eating performance of self-taught population is more stable than that of the evolution alone. Even the distinction does not look obviously significant, but it still has some implication.

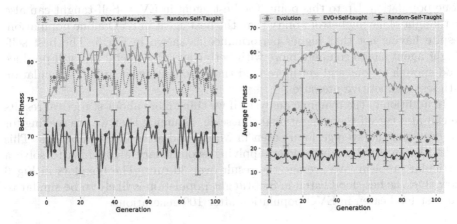

Fig. 5. Comparison of eating ability. (a) Left: Best Fitness (b) Right: Average Fitness

In biology, even a little difference in the dynamic over generations often shows something more interesting than it looks. In this scenario, we know that every agent has to compete with each others for energy during their lifetime, hence for the selection process at the end of each generation. The better movement an agent makes in an environment, the more the chance it approaches the food. The more an agent absorbs, the less the others can get in one time step during the lifetime. From this we can come up with an idea as follows: In an environmental circumstance in which there are quite a few agents without adaptive behaviour, or not having produced adaptive movements, the best agent tends to have more chance to eat because its competitors are *poorer* enough to go against it. In other words, in such a population the competition between individual agents is strong enough but there is little competitive pressure for the best agent because the remaining is inferior than it. On the other hand, if in a population there is not a small portion of agents producing good actions in the environment, the competition between individuals agents is absolutely stronger, even for the best agent.

With this in mind, we are curious whether the population of self-taught agents or that of *innate* agents alone has better performance. We also need to look at the average fitness of the two populations over time. It can be obviously seen in Fig. 5(b) that there is a clear difference between the average eating ability between the two populations of interest. The average agent in EVO+Self-taught eats 20–30 food particles more than the average agent in EVO population alone. This means, on the whole, the EVO+Self-taught multi-agent system eats approximately 400–600 food particles more than the EVO population alone.

Matching the analysis of both the best and the average eating ability, it can be claimed that the combination of evolution and learning results in a higher performance, promoting the whole population, more than evolution alone. The average fitness says an average self-taught agent eats more food and an average agent in EVO population. This also means the ability of self-teaching equips an average agent produces more adaptive actions in MiniWorld. Thus, the competition for food between self-taught agents is higher than that of agents in EVO alone population. Up to this point, the best agent in EVO+Self-taught can also be thought of having better ability than the best agent in EVO alone population, despite having not that higher performance as shown in Fig. 5. The best self-taught agent has to compete more with better average agents, but still produce good enough actions and get more food than the best agent in EVO population with less competitive pressure.

In addition to the dynamic analysis, if we think more about statistical results we can also cut off the generational timescale, assuming that we are seeking for the best and the average food eaten in MiniWorld after 100 generations. This looks like an usual analysis when applying evolutionary algorithms to solve a particular problem. We average the result over 30 runs. The boxplots in Fig. 6 show that the best food eaten in Self-taught population is likely to be similar to the best food eaten in EVO population after 100 generations.

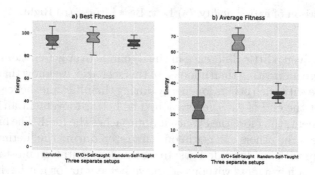

Fig. 6. Boxplot. (a) Left: Best Fitness (b) Right: Average Fitness.

If we look into the average fitness (food eaten) after 100 generations, it can be clearly seen that the self-taught population shows a clear higher performance than the EVO population. This all shows that the combination of learning and evolution increases the adaptivity of the population measured by the number of food eaten in any case.

4.2 Evolution Facilitates Learning

We have seen how learning during lifetime facilitates the evolving population of self-taught agents, having higher performance in a multi-agent environment compared to EVO alone. One curious question here is whether the Baldwin-like Effect has occurred?

This is why we conduct the third simulation in which the neural networks of self-taught agents are all randomly initialised, without the participation of evolution. It can be observed in both the left and the right of Fig. 5 that the population of randomly self-taught agents has lower performance than that of EVO+Self-taught, especially when it comes to the performance of the whole population (average fitness in our scenario). It is also interesting that in our simulation, the **blank-slate** population by learning even cannot outperform the evolving population without learning.

It is plausible here to conclude that learning, as a faster adaptation, can provide more adaptive advantage than the slower evolutionary process when the environment is dynamic like in MiniWorld. However, it is evolution that provides a good base for self-taught agents to learn better adaptive behaviours in future generations rather than learning as *blank-slates* in Random-Self-taught population.

5   Conclusion and Future Work

We have shown how the learning can enhance the performance of the multi-agent system in MiniWorld. The architecture of self-taught neural networks was proposed to illustrate the idea of self-taught agents – learning by oneself without external supervision in the environment. Based on a specific world and parameter settings, the Baldwin-like Effect has been demonstrated. The ability to teach oneself has been shown to provide some sort of adaptive advantage over agents without any learning ability. Simultaneously, evolution has also been shown to facilitate future learning, better than learning as blank-slates without priori knowledge about the world. Computer simulations are simple enough to illustrate the idea of research, yet still have indications for future work.

First, with respect to computational modeling technique the Neuroevolution method used in this paper is a bit highly engineering design – this means the architecture of the neural network (both action module and reinforcement module) is handcrafted by the system designer. Evolution and learning only affect the weights of the fixed topology to find good enough combination of weights. One trivial thought can be varying the structure of the neural networks used in our simulation and see how the performance would be varying. Different methods and algorithms in evolving neural networks (both weights and topology) can be taken into account [19].

Parameter setting of MiniWorld environment can show some more interesting effects than trivially for tuning the performance, and this can be worth future investigations. For example, the number of agents in MiniWorld is 20 and the number of food is kept constant as 50. What if we vary the number of food so that it is much larger than the number of agent? In this scenario, there distribution of food in MiniWorld is denser and, on average, each agent has more chance to get more energy, reducing competition pressure between agents during their lifetime. But what if we set the number of agents more than the number of food in MiniWorld? This would lead to a severe competition between individual agents in the population. Whether the better self-taught learners have more competitive advantage over the others in this scenario is an interesting question to be investigated in the future.

Another way to complexify MiniWorld is to include other substances, for example food and poison in the same environment. The agent then has to solve two tasks: the task of producing adaptive movement to approach substance in the environment and simultaneously the task of discriminating between food and poison to consume. It is plausible to think that having the ability to teach oneself will show more adaptive since there is no external supervision in the environment and relying on evolution alone is not adaptive when the environment is increasingly dynamic.

When it comes to theoretical and biological understanding, learning can also be classified into two types: asocial (or individual) learning and social learning. The former is learning by directly interacting with the environment, e.g. trial-and-error. The latter is learning from others, e.g. imitation learning. The ability to teach oneself can be considered a form of asocial learning. Social learning has been shown to complimentary to asocial learning in some work, including [22–24]. It is often said that the combination of social and asocial learning can result in higher performance than both social and asocial learning alone [22]. Yet in these studies each agent solves its own problem – there is no *real* interactive effect between individual agents. There is no competition for survival between agents during their lifetime. Individual agents compete with each other only through the selection process. The competition in these circumstances is a bit low. One curious question arising here is whether the same observation can be made when asocial and social agents are living in the same environment, sharing the resource of energy and competing for survival during their lifetime. If the competition is too strong, it is plausible to think that learning by oneself seems to be the best strategy when no one is motivated to share *survival information* to others. On the other hand, when the resource is surplus, learning from others can have more advantages. This will be an interesting exploration in the future, which can contribute to the understanding of the evolutionary consequences of learning (asocial and social) in different environments.

Last but not less important, the idea of self-taught neural networks can be powerful when there is no external supervision (or *label* in Machine Learning terminology). The algorithm and technique used in this paper can be a potential technique to solve unsupervised learning or reinforcement learning problems. We are curious whether evolution can provide a better base to learn than learning as blank-slates, to achieve human-level intelligence like what was claimed by DeepMind in games [25].

Acknowledgments. This research is funded by Science Foundation Ireland under Grant No. 13/IA/1850.

References

1. Baldwin, J.M.: A new factor in evolution. Am. Nat. **30**(354), 441–451 (1896)
2. Morgan, C.L.: On modification and variation. Science **4**(99), 733–740 (1896)
3. Hinton, G.E., Nowlan, S.J.: How learning can guide evolution. Complex Syst. **1**, 495–502 (1987)
4. Nolfi, S., Parisi, D., Elman, J.L.: Learning and evolution in neural networks. Adapt. Behav. **3**(1), 5–28 (1994)
5. Stanley, K.O., Bryant, B.D., Miikkulainen, R.: Evolving adaptive neural networks with and without adaptive synapses. In: Proceedings of the 2003 Congress on Evolutionary Computation. IEEE, Piscataway (2003)
6. Parisi, D., Nolfi, S., Cecconi, F.: Learning, behavior, and evolution (1991)
7. Le, N.: How the baldwin effect can guide evolution in dynamic environments. In: 7th International Conference on the Theory and Practice of Natural Computing. IEEE Press, 12–14 December 2018
8. Mery, F., Kawecki, T.J.: The effect of learning on experimental evolution of resource preference in drosophila melanogaster. Evolution **58**(4), 757 (2004)
9. Holland, J.H.: Adaptation in Natural and Artificial Systems, 2nd edn. University of Michigan Press, Ann Arbor (1992). 1975
10. Kauffman, S.A., Weinberger, E.D.: The NK model of rugged fitness landscapes and its application to maturation of the immune response. J. Theoret. Biol. **141**(2), 211–245 (1989)
11. Mayley, G.: Guiding or hiding: explorations into the effects of learning on the rate of evolution. In: Proceedings of the Fourth European Conference on Artificial Life. MIT Press, pp. 135–144 (1997)
12. Bull, L.: On the baldwin effect. Artif. Life **5**(3), 241–246 (1999)
13. Watson, J., Wiles, J.: The rise and fall of learning: a neural network model of the genetic assimilation of acquired traits. In: Proceedings of the 2002 Congress on Evolutionary Computation. IEEE
14. Todd, P.M., Miller, G.F.: Exploring adaptive agency ii: Simulating the evolution of associative learning. In: Proceedings of the First International Conference on Simulation of Adaptive Behavior on From Animals to Animats, pp. 306–315. MIT Press, Cambridge (1990)
15. Hebb, D.: The Organization of Behavior: A Neuropsychological Theory, ser. A Wiley book in clinical psychology. Wiley, New York (1949)
16. Harvey, I.: Is there another new factor in evolution? Evol. Comput. **4**(3), 313–329 (1996)
17. Ackley, D., Littman, M.: Interactions between learning and evolution. In: Langton, C.G., Taylor, C., Farmer, C.D., Rasmussen, S. (Eds.) Artificial Life II, SFI Studies in the Sciences of Complexity. Addison-Wesley, vol. X, pp. 487–509, Reading (1992)
18. Downing, K.L.: The baldwin effect in developing neural networks. In: Proceedings of the 12th Annual Conference on Genetic and Evolutionary Computation, ser. GECCO 2010, pp. 555–562. ACM, New York (2010)
19. Soltoggio, A., Stanley, K.O., Risi, S.: Born to learn: the inspiration, progress, and future of evolved plastic artificial neural networks. Neural Netw. **108**, 48–67 (2018)
20. Rumelhart, D.E., Hinton, G.E., Williams, R.J.: Neurocomputing: foundations of research. In: Anderson, J.A., Rosenfeld, E. (Eds.) ch. Learning Representations by Back-propagating Errors, pp. 696–699. MIT Press, Cambridge (1988)
21. Yao, X.: Evolving artificial neural networks. Proc. IEEE **87**(9), 1423–1447 (1999)

22. Le, N., O'Neill, M., Brabazon, A.: The baldwin effect reconsidered through the prism of social learning. In: 2018 IEEE Congress on Evolutionary Computation (CEC), pp. 1–8, July 2018
23. Rendell, L., Fogarty, L., Hoppitt, W.J., Morgan, T.J., Webster, M.M., Laland, K.N.: Cognitive culture: theoretical and empirical insights into social learning strategies. Trends Cogn. Sci. **15**(2), 68–76 (2011)
24. Le, N., O'Neill, M., Brabazon, A.: Adaptive advantage of learning strategies: a study through dynamic landscape. In: Auger, A., Fonseca, C.M., Lourenço, N., Machado, P., Paquete, L., Whitley, D. (eds.) PPSN 2018. LNCS, vol. 11102, pp. 387–398. Springer, Cham (2018). https://doi.org/10.1007/978-3-319-99259-4_31
25. Mnih, V., et al.: Human-level control through deep reinforcement learning. Nature **518**(7540), 529–533 (2015)

Coevolution of Generative Adversarial Networks

Victor Costa[✉], Nuno Lourenço[✉], and Penousal Machado[✉]

CISUC, Department of Informatics Engineering,
University of Coimbra,
Coimbra, Portugal
{vfc,naml,machado}@dei.uc.pt

Abstract. Generative adversarial networks (GAN) became a hot topic, presenting impressive results in the field of computer vision. However, there are still open problems with the GAN model, such as the training stability and the hand-design of architectures. Neuroevolution is a technique that can be used to provide the automatic design of network architectures even in large search spaces as in deep neural networks. Therefore, this project proposes COEGAN, a model that combines neuroevolution and coevolution in the coordination of the GAN training algorithm. The proposal uses the adversarial characteristic between the generator and discriminator components to design an algorithm using coevolution techniques. Our proposal was evaluated in the MNIST dataset. The results suggest the improvement of the training stability and the automatic discovery of efficient network architectures for GANs. Our model also partially solves the mode collapse problem.

Keywords: Neuroevolution · Coevolution ·
Generative adversarial networks

1 Introduction

Generative adversarial networks (GAN) [1] gained relevance for presenting state-of-the-art results in generative models, mainly in the field of computer vision. A GAN combines two deep neural networks, a discriminator and a generator, in an adversarial training where these networks are confronted in a zero-sum game between them. The generator creates fake samples based on an input distribution. The objective is to deceive the discriminator. On the other hand, the discriminator learns to distinguish between these fake samples and the real input data.

Several works improving the GAN model were recently published, leveraging the quality of the results to impressive levels [2–4]. However, there are still open problems, such as the vanishing gradient and the mode collapse problems, all of them leading to difficulties in the training procedure. Although there are strategies to minimize the effect of those problems, they remain fundamentally unsolved [5,6].

© Springer Nature Switzerland AG 2019
P. Kaufmann and P. A. Castillo (Eds.): EvoApplications 2019, LNCS 11454, pp. 473–487, 2019.
https://doi.org/10.1007/978-3-030-16692-2_32

Another issue, not related only to GANs but also to neural networks in general, is the necessity to define a network architecture previously. In that case, the topology and hyperparameters are usually chosen empirically, thus spending human time in repetitive tasks such as fine-tuning. However, there are approaches that can automatize the design of neural network architectures.

Neuroevolution is the application of evolutionary algorithms to provide the automatic design of neural networks. In neuroevolution, both the network architecture (e.g., topology, hyperparameters and the optimization method) and the parameters (e.g., weights) used in each neuron can be evolved. NeuroEvolution of Augmenting Topologies (NEAT) [7] is a well-known neuroevolution method that evolves the weights and topologies of neural networks. In further experiments, NEAT was also successfully applied in a coevolution context [8]. Moreover, Deep-NEAT [9] was recently proposed to expand NEAT to larger search spaces, such as in deep neural networks.

Therefore, this project proposes a new model, called coevolutionary generative adversarial networks (COEGAN), to combine neuroevolution and coevolution in the coordination of the GAN training algorithm. The evolutionary algorithm is based on DeepNEAT. We extended and adapted this model to work on the context of GANs, making use of the competitive characteristic between the generator and discriminator to apply a coevolution model. Hence, each subpopulation of generators and discriminators evolve following its own evolutionary path. To validate our model, experiments were conducted using MNIST [10] as the input dataset for the discriminator component. The results are the improvement of the training stability and the automatic discovery of efficient network topologies for GANs.

The remainder of this paper is organized as follows: Sect. 2 introduces the concepts of GANs and evolutionary algorithms; Sect. 3 presents our approach used to evolve GANs; Sect. 4 displays the experimental results using this approach; finally, Sect. 5 presents conclusions and proposals for further enhancements.

2   Background and Related Works

This section reviews the concepts employed in this paper and presents works related to the proposed model.

2.1   Evolutionary Algorithms

An evolutionary algorithm (EA) is a method inspired by biological evolution that aims to mimic the same evolutionary mechanism found in nature. In EAs, the population is composed of individuals that represent possible solutions for a given problem, using a high-order abstraction to encode their characteristics [11]. The algorithm works by applying variation operators (e.g., mutation and crossover) to the population in order to search for better solutions.

Neuroevolution. Neuroevolution is the application of evolutionary algorithms in the evolution of neural networks. This approach can be applied to weights, topology and hyperparameters of a neural network. When used to generate a network topology, a substantial benefit is the automation of the architecture design and its parameters [7]. Besides, not only the final architecture is important, but the intermediary models also give their contributions to the final model in form of the transference of their trained weights kept through generations [7]. This automation is even more important with the rise of deep learning, which produces larger models and increases the search space [9,12]. However, large search spaces are also a challenge for neuroevolution. These methods have a high computational complexity that may turn their application unfeasible.

Coevolution. The simultaneous evolution of at least two distinct species is called coevolution [13,14]. There are two types of coevolution algorithms: cooperative and competitive. In cooperative coevolution, individuals of different species cooperate in the search for efficient solutions, and the fitness function of each species are designed to reward this cooperation [15–17]. In competitive coevolution, individuals of different species are competing between them. Consequently, their fitness function directly represents this competition in a way that scores between species are inversely related [8,11,14].

2.2   Generative Adversarial Networks

Generative Adversarial Networks (GAN), proposed in [1], is an adversarial model that became relevant for the performance achieved in generative tasks. A GAN combines two deep neural networks: a discriminator D and a generator G. The generator G receives a noise as input and outputs a fake sample, attempting to capture the data distribution used as input for D. The discriminator D receives the real data and fake samples as input, learning to distinguish between them. These components are trained simultaneously as adversaries, creating strong generative and discriminative components.

The loss function for the discriminator is defined by:

$$J^{(D)}(D, G) = -\mathbb{E}_{x \sim p_{data}}[\log D(x)] - \mathbb{E}_{z \sim p_z}[\log(1 - D(G(z)))]. \qquad (1)$$

The loss function for the generator (non-saturating version proposed in [1]) is defined by:

$$J^{(G)}(G) = -\mathbb{E}_{z \sim p_z}[\log(D(G(z)))]. \qquad (2)$$

In Eq. 1, p_{data} represents the dataset used as input to the discriminator. In Eqs. 1 and 2, z, p_z, G and D represent the noise sample (used as input to the generator), the noise distribution, the generator and the discriminator, respectively.

Besides, several variations of the loss function were proposed to improve the GAN model, such as in WGAN [2] and LSGAN [4]. These variations were studied in [18] in order to access the superiority in respect to the original GAN proposal. The study founds no empirical evidence that these variations are superior to the original GAN mode.

There are two common problems regarding training stability in GANs: vanishing gradient and mode collapse. The vanishing gradient occurs when the discriminator D became perfect and do not commit mistakes anymore. Hence, the loss function is zeroed, the gradient does not flow through the neural network of the generator, and the GAN progress stagnates. In mode collapse, the generator captures only a small portion of the dataset distribution provided as input to the discriminator. This is not desirable once we want to reproduce the whole distribution of the data. Recently, several approaches tried to minimize those problems, but they remain unsolved [5,6].

There are other models extending the original GAN proposal that modify not only the loss function but also aspects of the architecture. The method described by [3] uses a simple strategy to evolve a GAN during the training procedure. The main idea is to grow the model progressively, increasing layers in both discriminator and generator. This mechanism will make the model more complex while the training proceeds, increasing the resolution of images at each phase. However, these layers added progressively are preconfigured, i.e., they are hand-designed and not produced by a stochastic procedure. Thus, the model evolves in a preconfigured way during the training procedure, but the method used in [3] do not use an evolutionary algorithm in this process. Therefore, we can consider this predefined progression as a first step towards the evolution of generative adversarial models.

A very recent model proposes the use of evolutionary algorithms in GANs [19]. Their approach used a simple model for the evolution, using a mutation operator that can change only the loss function of the individuals. Our proposal differs from them by modeling the GAN as a coevolution problem. Besides, in our case, the evolution occurs in the network architecture, and the loss function is the same during the whole method. Nevertheless, a further proposal can incorporate those ideas to evaluate the benefits.

3   Coevolution of Generative Adversarial Networks

We propose a new model called coevolutionary generative adversarial networks (COEGAN). This model combines neuroevolution and coevolution in the coordination of the GAN training algorithm. Our approach is based on DeepNEAT [9], that was extended and adapted to the context of GANs.

In COEGAN, the genome is represented as an array of genes, which are directly mapped into a phenotype consisting of the sequence of layers in a deep neural network. Each gene represents a linear, convolution or transpose convolution layer. Moreover, each gene also has an activation function, chosen from the following set: ReLU, LeakyReLU, ELU, Sigmoid and Tanh. From the specific parameters of each type of gene, convolution and transpose convolution layers only have the number of output channels as a random parameter. The stride and kernel size are fixed as 2 and 3, respectively. The number of input channels is calculated dynamically, based on the previous layer. Similarly, the linear layer only has the number of output features as the random parameter. The number of input features is calculated based on the previous layer. Therefore, only the activation function, output features and output channels are subject to the variation operations.

Figures 1(a) and (b) are examples of a discriminator and a generator genotype, respectively. The discriminator genotype is composed of a convolutional section and followed by a linear section (fully connected layers). As in the original GAN approach, the output of discriminators is the probability of the input sample be a real sample drawn from the dataset. Similarly, the generator genotype is composed of a linear section and followed by a transpose convolutional section. The output of the generator is a fake sample, with the same characteristics (i.e., dimension and channels) of a real sample.

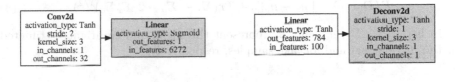

(a) A genotype of a discriminator   (b) A genotype of a generator

Fig. 1. Example of genotypes of a generator and a discriminator.

The overall population is composed of two separated subpopulations: a population of generators, where each G_i represents a generator component in a GAN; and a population of discriminators where each D_j represents a discriminator in a GAN. Furthermore, a speciation mechanism based on the original NEAT proposal is applied to promote innovation in each subpopulation. The speciation mechanism divides the population into species based on a similarity function (used to group similar individuals). Thus, the innovation, represented by the addition of new genes into a genome, causes the creation of new species in order to fit the individuals containing these new genes. Therefore, these new individuals have the chance to survive through generations and reach the performance of older individuals in the population.

The parameters of the layers in the phenotype (e.g., weights and bias) will be trained by the gradient descent method and will not be part of the evolution. The number of parameters to be optimized are too large and evolving them will increase the computational complexity. Therefore, for now, we are interested only in the evolution of the network topology. We plan to develop a hybrid approach that evolves the weight when the gradient descent training stagnates for a number of generations.

3.1 Fitness

For discriminators, the fitness is based on the loss obtained from the regular GAN training method, i.e., the fitness is equivalent to Eq. 1. We have tried to use the same approach for the generator. However, preliminary experiments evidenced that the loss does not represent a good measure for quality in this case. The loss for generators, represented by Eq. 2, is unstable during the GAN training, making it not suitable to be used as fitness in an evolutionary algorithm.

Thus, we selected the Fréchet Inception Distance (FID) [20] as the fitness for generators. FID is the state-of-the-art metric to compare the generative components of GANs and outperforms other metrics, such as the Inception Score [6], with respect to diversity and quality [18]. In FID, an Inception Net [21] (trained on ImageNet [22]) is used to transform images to the feature space (given by a hidden layer of the network). This feature space is interpreted as a continuous multivariate Gaussian. So, the mean and covariance of two Gaussians are estimated using real and fake samples. The Fréchet distance between these Gaussians is given by:

$$FID(x, g) = ||\mu_x - \mu_g||_2^2 + Tr(\Sigma_x + \Sigma_g - 2(\Sigma_x \Sigma_g)^{1/2}).   (3)$$

In Eq. 3, μ_x, Σ_x, μ_g, and Σ_g represent the mean and covariance estimated for the real dataset and for fake samples, respectively.

3.2 Variation Operators

Initially, we have used two types of variation operators to breed new individuals: mutation and crossover. The crossover process uses the transition between convolutional layers and linear (fully connected) layers as the cut point. Figure 2 represents an example of this process. However, preliminary tests evidenced that crossover decreases the performance of the system. We expect to conduct further experiments with other crossover variations to assess the contribution of this operator in our model.

The mutation process is composed of three main operations: add a new layer, remove a layer, and change an existing layer. In the addition operation, a new layer is randomly drawn from the set of possible layers. For discriminators, the available layers are linear and convolution. For generators, the available layers are linear and transpose convolution (also called deconvolution). The remove operation chooses an existing layer and excludes it from the genotype.

The change operation modifies the attributes and the activation function of an existing layer. The activation function is randomly chosen from the set of possibilities. Other specific attributes can be changed depending on the type of the layer. The number of output features and the number of output channels are mutated for the linear and convolution layers, respectively. The mutation of these attributes follows a uniform distribution, with a predefined range limiting the possible values.

It is important to note that if the new gene is compatible with its parent, the parameters (weights and bias) are copied. So, the new individual will carry the training information from the previous generation. This simulates the transfer

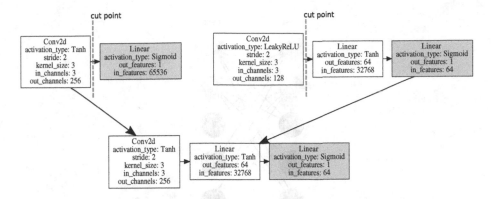

Fig. 2. Example of crossover between discriminators.

learning technique commonly used to train deep neural networks. However, when the attributes of a linear or convolution layer change, the trained parameters are lost. This happens because the setup of weights changes, becoming incompatible with the new layer. Consequently, the weights are not transferred from the parents and the layer will be trained from the beginning.

3.3 Competition Between Generators and Discriminators

In the evaluation step of the evolutionary algorithm, discriminators and generators must be paired to calculate the fitness for each individual in the population. There are several approaches to pair individuals in competitive coevolution, e.g. *all vs. all*, *random* and *all vs. best* [11].

As we want to train the GAN to avoid problems such as the mode collapse and vanishing gradient, we consider the use of the *all vs. best* strategy here. However, we select k individuals rather than only one to promote the diversity in the GAN training. We pair each generator with k best discriminators from the previous generation and, similarly, each discriminator with k best generators. Figure 3 represents an example of this approach with $k = 2$. For the first generation, we assume a random approach, i.e., k random individuals are selected to be paired in the initial evaluation.

The *all vs. all* strategy would also be interesting for our model as it will improve the variability of the environment for both discriminators and generators during the training. However, the trade-off is the time to execute this approach. In *all vs. all*, each discriminator is paired with each generator, resulting in many competitions.

3.4 Selection

For the selection phase, we used a strategy based on NEAT [7]. As in NEAT, we divided the population of generators and discriminators into subpopulations, following a speciation strategy similar to that used in NEAT. Each species contains individuals with similar network structures. For this, we define the

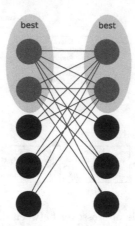

Fig. 3. Representation of the all vs. best competition pattern with $k = 2$.

similarity between individuals based on the parameters of each gene composing the genome. Different of NEAT, we do not use the weights of each layer to differentiate between individuals. Therefore, we calculate the distance δ between two genomes i and j as the number of genes that exist only in i or j. Each species inside the populations of generators and discriminators are clustered based on a threshold δ_t. This threshold is calculated in order to suit the desired number of species. The number of species is a parameter previously chosen.

The selection occurs inside each species. The number of individuals selected inside each species are proportional to the average fitness of the individuals belonging to it. Given the number of individuals to keep in a species, a tournament between k_t individuals is applied to finally select the individuals to breed and compose the next generation.

4   Experiments

In this section, we will evaluate the performance of our method on the MNIST dataset. Normally, the network would be training for several epochs using the whole dataset in the procedure. As this would be an intensive computational task, we will use only a subset of the dataset per generation. This strategy, combined with the transfer of parameters between generations, was sufficient to produce an evolutionary pressure towards efficient solutions and to promote the GAN convergence.

There is no consensus on the metric to represent the quality of samples generated by generative models. However, the Fréchet Inception Distance (FID) was proved to be the best metric when comparing the quality of samples generated by GANs [18]. Therefore, we used the FID score, the same metric used as fitness for generators, to compare our results with the state of the art.

4.1   Experimental Setup

Table 1 describes the parameters used in all experiments reported in this paper.

Table 1. Experimental parameters.

Evolutionary parameters	Value
Number of generations	100
Population size (generators)	20
Population size (discriminators)	20
Crossover rate	0%
Add Layer rate	30%
Remove Layer rate	10%
Change Layer rate	10%
Output features range	[32, 1024]
Output channels range	[16, 128]
k (all vs. best)	3
Tournament k_t	2
FID samples	1000
Genome Limit	4
Species	4
GAN parameters	Value
Batch size	64
Batches per generation	20
Optimizer	RMSProp
Learning rate	0.001

For evolutionary parameters, we chose to execute our experiments for 100 generations. After this number of generations, the fitness stagnates and we expect no improvement of the results. We used 20 individuals for the population of both generators and discriminators. A larger population will probably achieve better results, but the computational cost would be too large. The size of the genome was limited to four layers, also to reduce the computational cost. The number of species used was four, permitting an average of five individuals per species in each subpopulation (generators and discriminators). We empirically defined a probability of 30%, 10% and 10% for the add, remove and change mutations, respectively. As stated before, crossover was not used in the experiments reported in this section.

For the GAN parameters, we choose 64 as batch size, running 20 batches per generation. This amounts to 1280 samples per generation to train discriminators. The optimizer used in the training method was RMSProp [23]. We have also conducted preliminary experiments with Adam [24], but the best results were achieved with RMSProp.

The MNIST dataset was used and we executed each experiment 10 times to achieve the results within a confidence interval of 95%.

4.2 Results

Figure 4 shows the progression of the network through generations. We can see in Fig. 4(a) the average number of layers in the population of generators and discriminators. Because we have limited the genotype to a maximum of four genes, the number of layers rapidly saturates. This is an indication of premature optimization. We can overcome this issue by either increasing the limit or decreasing the growth rate (i.e., reduce the mutation probability). In our tests with crossover activated this problem became even more evident. Figure 4(b) shows the number of genes with the parameters reused in each generation. The linear grow in the amount of reused genes is evidence of the transference mechanism explained in Sect. 3.2. Because we use a strategy similar to transfer learning to keep the trained parameters, this reuse is important to pass the trained weights trough generations.

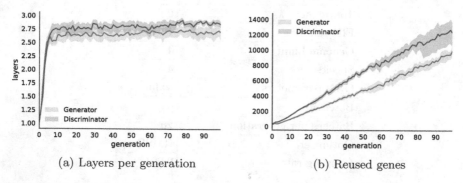

(a) Layers per generation   (b) Reused genes

Fig. 4. Progression of layers and the reuse of parameters with a 95% confidence interval

Figure 5 shows the progression of the fitness for best generators and discriminators. In Fig. 5(a), we can see the fitness for generators reducing through generations with reduced noise. Hence, the chosen fitness, i.e., the FID score, is evidenced as suitable to be used in our evolutionary algorithm. For discriminators (Fig. 5(b)), we can see much more noise, which can harm the selection process in the evolutionary algorithm. This suggests that the choice for the discriminator fitness could be improved.

In the final generation, the mean FID was 49.2, with a standard deviation of 10.5. The high standard deviation clearly shows that we need to increase the number of runs in order to get a more representative value for the FID. Besides that, we can see that this score is much worse than state-of-the-art results.

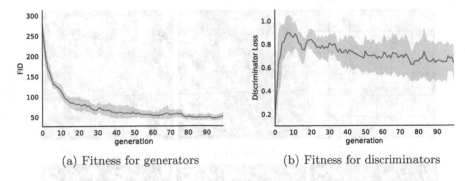

(a) Fitness for generators   (b) Fitness for discriminators

Fig. 5. Fitness for discriminators and generators with a 95% confidence interval

For example, the FID for MNIST was reported in [18] as 6.7, with a standard deviation of 0.3. However, our results showed that the model did not collapse into a single point from the input distribution, which is a common problem in GANs.

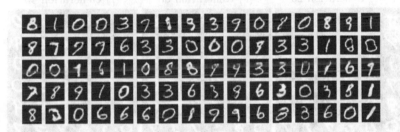

Fig. 6. Samples created by a generator after the evolutionary algorithm.

Figure 6 represents samples generated in one execution, evidencing that our model does not collapse into a single point of the input distribution. We can see this behavior occurring in all executions, which leave us to conclude that our model solves, at least partially, the mode collapse problem. Moreover, all executions reached convergence with equilibrium between the discriminator and the generator, and the vanishing gradient never occurs. This evidences that our proposal brings stability to the training procedure of GANs. Furthermore, our experiments were restricted to a maximum of four layers. A similar number of layers was used by [18]. This evidences that our method does not outperform architectures designed by hand when taking into account only the FID score.

For some executions, the generator captured only a subset of the distribution, which is a form of the mode collapse problem. See in Fig. 6 examples of images created by a generator after the whole evolutionary algorithm. Note that only half of the digits are represented in these samples.

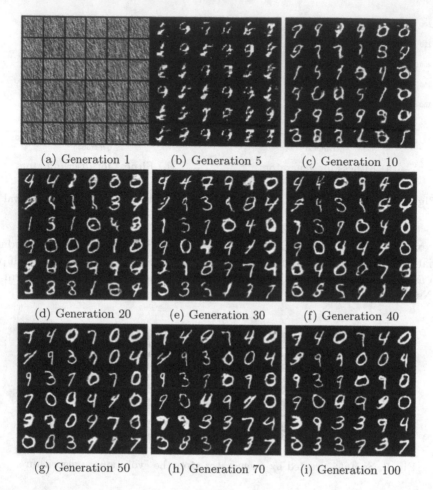

(a) Generation 1   (b) Generation 5   (c) Generation 10

(d) Generation 20   (e) Generation 30   (f) Generation 40

(g) Generation 50   (h) Generation 70   (i) Generation 100

Fig. 7. The progression of samples created by the best generator in generations (a) 1, (b) 5, (c) 10, (d) 20, (e) 30, (f) 40, (g) 50, (h) 70, and (i) 100.

Figure 7 contains generated samples selected to represent the progression of the generator during the evolutionary algorithm. We can see in the first generation only noisy samples, without any structure resembling a digit. After 5 generations (Fig. 7(b)) we can see some structure emerging to form the digits. From the generation 10 onwards we can start to distinguish between the digits, with a progressive improvement of the quality.

Figure 8 presents the network architecture discovered after the last generation. We can see that both components reached the limit of four layers imposed in the experiments. Furthermore, both the generator and the discriminator were composed by a combination of convolutional and linear layers with different activation functions.

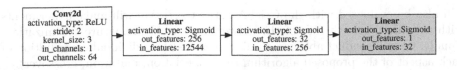

(a) Genotype of a discriminator

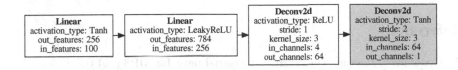

(b) Genotype of a generator

Fig. 8. Best (a) discriminator and (b) generator found after the final generation.

5 Conclusions

Generative adversarial networks (GAN) achieved important results for generative models in the field of computer vision. However, there are stability problems in the training method, such as the vanishing gradient and the mode collapse problems.

We present in this paper a model that combines neuroevolution and coevolution in the coordination of the GAN training algorithm. To design the model, we took inspiration on previous evolutionary algorithms, such as NEAT [8] and DeepNeat [9], and on recent advances in GANs, such as [3].

We made experiments using the MNIST dataset. In all executions, the system reached an equilibrium and convergence, not falling into the vanishing gradient problem. We also evidenced the FID score as a good fitness metric for generators. We used the loss function as fitness for the population of discriminators. However, the results evidenced that a better metric may be necessary to ensure the proper evolution of this population. The results showed that our model partially solves the mode collapse problem. In our experiments, the generated samples always presented some diversity, partially preserving the characteristics of the input distribution. Besides that, our proposal did not outperform the state-of-the-art results, such as presented [18].

Therefore, as future works, we will experiment other metrics to be used as fitness for the discriminator component, such as area under the curve (AUC). We will also assess that the proposed model does not suffer from cyclic issues commonly seen in coevolution models [25, 26]. More experiments should be made with larger populations as well as larger genotype limits. In addition, the model will also be evaluated with the CelebA dataset [27]. We will make experiments with a reduced growth rate of the networks to avoid premature optimization. Finally, we will conduct ablation studies to assess the individual contribution of each aspect of the proposed algorithm (e.g., speciation, mutation, crossover).

Acknowledgments. This article is based upon work from COST Action CA15140: ImAppNIO, supported by COST (European Cooperation in Science and Technology): www.cost.eu.

References

1. Goodfellow, I., et al.: Generative adversarial nets. In: NIPS (2014)
2. Arjovsky, M., Chintala, S., Bottou, L.: Wasserstein generative adversarial networks. In: International Conference on Machine Learning, pp. 214–223 (2017)
3. Karras, T., Aila, T., Laine, S., Lehtinen, J.: Progressive growing of GANs for improved quality, stability, and variation. In: International Conference on Learning Representations (2018)
4. Mao, X., Li, Q., Xie, H., Lau, R.Y., Wang, Z., Smolley, S.P.: Least squares generative adversarial networks. In: 2017 IEEE International Conference on Computer Vision (ICCV), pp. 2813–2821. IEEE (2017)
5. Gulrajani, I., Ahmed, F., Arjovsky, M., Dumoulin, V., Courville, A.C.: Improved training of wasserstein GANS. In: Advances in Neural Information Processing Systems, pp. 5769–5779 (2017)
6. Salimans, T., Goodfellow, I., Zaremba, W., Cheung, V., Radford, A., Chen, X.: Improved techniques for training GANS. In: Advances in Neural Information Processing Systems, pp. 2234–2242 (2016)
7. Stanley, K.O., Miikkulainen, R.: Evolving neural networks through augmenting topologies. Evol. Comput. **10**(2), 99–127 (2002)
8. Stanley, K.O., Miikkulainen, R.: Competitive coevolution through evolutionary complexification. J. Artif. Intell. Res. **21**, 63–100 (2004)
9. Miikkulainen, R., et al.: Evolving deep neural networks. arXiv preprint arXiv:1703.00548 (2017)
10. LeCun, Y.: The MNIST database of handwritten digits (1998). http://yann.lecun.com/exdb/mnist/
11. Sims, K.: Evolving 3D morphology and behavior by competition. Artif. Life **1**(4), 353–372 (1994)
12. Assunção, F., Lourenço, N., Machado, P., Ribeiro, B.: Evolving the topology of large scale deep neural networks. In: Castelli, M., Sekanina, L., Zhang, M., Cagnoni, S., García-Sánchez, P. (eds.) EuroGP 2018. LNCS, vol. 10781, pp. 19–34. Springer, Cham (2018). https://doi.org/10.1007/978-3-319-77553-1_2
13. Hillis, W.D.: Co-evolving parasites improve simulated evolution as an optimization procedure. Physica D **42**(1–3), 228–234 (1990)
14. Rawal, A., Rajagopalan, P., Miikkulainen, R.: Constructing competitive and cooperative agent behavior using coevolution. In: 2010 IEEE Symposium on Computational Intelligence and Games (CIG), pp. 107–114 (2010)
15. García-Pedrajas, N., Hervás-Martínez, C., Muñoz-Pérez, J.: Covnet: a cooperative coevolutionary model for evolving artificial neural networks. IEEE Trans. Neural Netw. **14**(3), 575–596 (2003)

16. García-Pedrajas, N., Hervás-Martínez, C., Ortiz-Boyer, D.: Cooperative coevolution of artificial neural network ensembles for pattern classification. IEEE Trans. Evol. Comput. **9**(3), 271–302 (2005)
17. Gomez, F., Schmidhuber, J., Miikkulainen, R.: Accelerated neural evolution through cooperatively coevolved synapses. J. Mach. Learn. Res. **9**, 937–965 (2008)
18. Lucic, M., Kurach, K., Michalski, M., Gelly, S., Bousquet, O.: Are GANS created equal? a large-scale study. arXiv preprint arXiv:1711.10337 (2017)
19. Wang, C., Xu, C., Yao, X., Tao, D.: Evolutionary generative adversarial networks. arXiv preprint arXiv:1803.00657 (2018)
20. Heusel, M., Ramsauer, H., Unterthiner, T., Nessler, B., Hochreiter, S.: GANS trained by a two time-scale update rule converge to a local nash equilibrium. In: Advances in Neural Information Processing Systems, pp. 6629–6640 (2017)
21. Szegedy, C., Vanhoucke, V., Ioffe, S., Shlens, J., Wojna, Z.: Rethinking the inception architecture for computer vision. In: Proceedings of the IEEE Conference on Computer Vision and Pattern Recognition, pp. 2818–2826 (2016)
22. Russakovsky, O., et al.: Imagenet large scale visual recognition challenge. Int. J. Comput. Vis. **115**(3), 211–252 (2015)
23. Tieleman, T., Hinton, G.: Lecture 6.5-rmsprop: divide the gradient by a running average of its recent magnitude. COURSERA: Neural Netw. Mach. Learn. **4**(2), 26–31 (2012)
24. Kingma, D.P., Ba, J.: Adam: a method for stochastic optimization. In: International Conference on Learning Representations (ICLR) (2015)
25. Ficici, S.G., Pollack, J.B.: A game-theoretic memory mechanism for coevolution. In: Cantú-Paz, E., et al. (eds.) GECCO 2003. LNCS, vol. 2723, pp. 286–297. Springer, Heidelberg (2003). https://doi.org/10.1007/3-540-45105-6 35
26. Monroy, G.A., Stanley, K.O., Miikkulainen, R.: Coevolution of neural networks using a layered pareto archive. In: Proceedings of the 8th Annual Conference on Genetic and Evolutionary Computation, pp. 329–336. ACM (2006)
27. Liu, Z., Luo, P., Wang, X., Tang, X.: Deep learning face attributes in the wild. In: Proceedings of the IEEE International Conference on Computer Vision, pp. 3730–3738 (2015)

Evolving Recurrent Neural Networks for Time Series Data Prediction of Coal Plant Parameters

AbdElRahman ElSaid[1], Steven Benson[2], Shuchita Patwardhan[2],
David Stadem[2], and Travis Desell[1(✉)]

[1] Rochester Institute of Technology, Rochester, NY 14623, USA
{aae8800,tjdvse}@rit.edu
[2] Microbeam Technologies Inc., Grand Forks, ND 58203, USA
{sbenson,shuchita,dstadem}@microbeam.com

Abstract. This paper presents the Evolutionary eXploration of Augmenting LSTM Topologies (EXALT) algorithm and its use in evolving recurrent neural networks (RNNs) for time series data prediction. It introduces a new open data set from a coal-fired power plant, consisting of 10 days of per minute sensor recordings from 12 different burners at the plant. This large scale real world data set involves complex dependencies between sensor parameters and makes for challenging data to predict. EXALT provides interesting new techniques for evolving neural networks, including *epigenetic weight initialization*, where child neural networks re-use parental weights as a starting point to backpropagation, as well as *node-level mutation operations* which can improve evolutionary progress. EXALT has been designed with parallel computation in mind to further improve performance. Preliminary results were gathered predicting the *Main Flame Intensity* data parameter, with EXALT strongly outperforming five traditional neural network architectures on the best, average and worst cases across 10 repeated training runs per test case; and was only slightly behind the best trained Elman recurrent neural networks while being significantly more reliable (*i.e.*, much better average and worst case results). Further, EXALT achieved these results 2 to 10 times faster than the traditional methods, in part due to its scalability, showing strong potential to beat traditional architectures given additional runtime.

Keywords: Neuro-evolution · Recurrent neural networks ·
Time series data prediction

1 Introduction

With the advent of deep learning, the use of neural networks has become widely popular across a variety of domains and problems. However, most of this success currently has been driven by human architected neural networks, which is time

© Springer Nature Switzerland AG 2019
P. Kaufmann and P. A. Castillo (Eds.): EvoApplications 2019, LNCS 11454, pp. 488–503, 2019.
https://doi.org/10.1007/978-3-030-16692-2_33

consuming, error prone and still leaves a major open question: *what is the optimal architecture for a neural network?* Further, optimality may have multiple aspects and changes from problem to problem, as in one domain it may be better to have a smaller yet less accurate neural network due to performance concerns, while in another accuracy may be more important than performance. This can become problematic as many applications of neural networks are evaluated using only a few select architectures from the literature, or may simply just pick an architecture that has shown prior success.

Another issue is that backpropagation is still the de-facto method for training a neural network. While significant performance benefits for certain types of neural networks (*e.g.*, Convolutional Neural Networks) can be gained by utilizing GPUs, other network types, such as recurrent neural networks (RNNs), typically cannot achieve such performance benefits without convolutional components. As backpropagation is an inherently sequential process, the time to train a single large neural network, let alone a variety of architectures, can quickly become prohibitive.

This work introduces a new algorithm, Evolutionary eXploration of Augmenting LSTM Topologies (EXALT), which borrows strategies from both NEAT (NeuroEvolution of Augmenting Topologies [1]) and its sister algorithm, EXACT (Evolutionary eXploration of Augmenting Convolutional Topologies [2,3]) to evolve recurrent neural networks with long short-term memory (LSTM [4]) components. EXALT has been designed with concurrency in mind, and allows for multiple RNNs to be trained in a parallel manner using backpropagation while evolving their structures. EXALT expands on NEAT by having *node-level mutations* which can speed up the evolutionary process, and by utilizing backpropagation instead of an evolutionary strategy to more swiftly train the RNNs. Child RNNs re-use parental weights in an *epigenetic weight initialization* strategy, allowing them to continue training where parents left off, which further improves how quickly the algorithm evolves well performing RNNs.

This work evaluates the performance of EXALT as compared to six traditional neural network architectures (one layer and two layer feed forward neural networks; Jordan and Elman RNNs; and one layer and two layer LSTM RNNs) on a real world dataset collected from a coal-fired power plant. This data set consists of 10 days worth of per minute recordings across 12 sensors; from 12 different burners. The parameters are non-seasonal and potentially correlated, resulting in a highly complex set of data to perform predictions on. This dataset has been made open to encourage validation and reproducibility of these results, and as a valuable research to the time series data prediction research community. Having good predictors for these parameters will allow the development of tools that can be used to forecast and alert plant operators and engineers about poor boiler conditions which may occur as a result of incoming coal and/or current power plant parameters.

Preliminary results predicting the *Main Flame Intensity* parameter of this dataset with the EXALT algorithm are highly promising. K-fold cross validation was done, using each burner file as a test case; and 10 runs of each strategy were repeated for each fold. While the Elman networks were able to be trained to slightly better performance (within 0.0025 mean squared error), on average they were not nearly as reliable. EXALT outperformed all the other network architectures in best, average and worst cases, and while finding more efficient (*i.e.*, smaller) RNNs than the traditional architectures, and was able to do so in significantly less time (between 2 to 10 times faster) operating in parallel across 20 processors. These preliminary results shows the strong potential of this algorithm in evolving RNNs for time series data prediction.

The remainder of this paper is as follows. Section 2 presents related work. Section 3 describes the EXALT algorithm in detail. Section 4 introduces the coal-fired power plant data set, and Sect. 5 provides initialization settings and results for the EXALT algorithm and fixed neural networks. The paper ends with a discussion of conclusions and future work in Sect. 6.

2 Related Work

2.1 Recurrent Neural Networks (RNNs)

RNNs have an advantage over standard feed forward (FF) neural networks (NNs), as they can deal with sequential input data, using their internal memory to process sequences of inputs and use previously stored information to aid in future predictions. This is done by allowing connections between neurons across timesteps, which aids them in predicting more complex data [5]. However, this leads to a more complicated training process as RNNs need to be "unrolled" over each time step of the data and trained using backpropagation through time (BPTT) [6].

In an effort to better train RNNs and capture time dependencies in data, long short-term memory (LSTM) RNNs were first introduced by Hochrieter and Schmidhuber [4]. LSTM neurons provide a solution for the exploding/vanishing gradients problem by utilizing input, forget and output gates in each LSTM cell, which can control and limit the backward flow of gradients in BPTT [7]. LSTM RNNs have been used with strong performance in image recognition [8], audio visual emotion recognition [9], music composition [10] and other areas. Regarding time series prediction, for example, LSTM RNNs have been used for stock market forecasting [11] and forex market forecasting [12]. Also forecasting wind speeds [7,13] for wind energy mills, and even predicting diagnoses for patients based on health records [14].

2.2 Evolutionary Optimization Methods

The EXALT algorithm presented in this work is in part based its sister algorithm, Evolutionary eXploration of Augmenting Convolutional Topologies (EXACT), which has successfully been used to evolve convolutional neural networks (CNNs) for image prediction tasks [2,3]. However, where EXACT evolves feature maps and filters to construct CNNs, EXALT utilizes LSTM and regular neurons along with feed forward and recurrent connections to evolve RNNs. EXALT also utilizes the *epigenetic weight initialization* strategy (see Sect. 3.2 that was shown by EXACT to improve training performance [3].

Other work by Desell and ElSaid [15–17] has utilized an ant colony optimization based approach to select which connections should be utilized in RNNs and LSTM RNNs for the prediction of flight parameters. In particular, this ACO approach was shown to reduce the number of trainable connections in half while providing a significant improvement in predictions of engine vibration [16]. However, this approach works within a fixed RNN architecture and cannot evolve an overall RNN structure.

Several other methods for evolving NN topologies along with weights have been researched and deployed. In [1], NeuroEvolution of Augmenting Topologies (NEAT) has been developed. It is a genetic algorithm that evolves increasingly complex neural network topologies, while at the same time evolving the connection weights. Genes are tracked using historical markings with innovation numbers to perform crossover among different structures and enable efficient recombination. Innovation is protected through speciation and the population initially starts small without hidden layers and gradually grows through generations [18–20]. Experiments have demonstrated that NEAT presents an efficient way for evolving neural networks for weights and topologies. Its power resides in its ability to combine all the four main aspects discussed above and expand to complex solutions. However NEAT still has some limitations when it comes evolving neural networks with weights or LSTM cells for time series prediction tasks as described in [15].

Other more recent work by Rawal and Miikkulainen has utilized tree based encoding [21] and information maximization objectives [22] to evolve RNNs. EXALT differs from this work in a few notable ways, first, the tree-based encoding strategy uses a genetic programming strategy to evolve connections within recurrent neurons, and only utilizes fixed architectures built of layers of evolved node types. On the other hand, the information maximization strategy utilizes NEAT with LSTM neurons instead of regular neurons. EXALT allows the evolution of RNNs with both regular and LSTM neurons, adds new node-level mutation operations and uses backpropagation to train the evolved RNNs (see Sect. 3). Furthermore, it has been developed with large scale concurrency in mind, and utilizes an asynchronous steady-state approach, which has been shown to allow scalability to potentially millions of compute nodes [23].

3 Evolutionary Exploration of Augmenting LSTM Topologies (EXALT)

EXALT has been developed with parallel/concurrent operation in mind. It utilizes a steady state population and generates new RNNs to be evaluated upon request by workers. When a worker completes training a RNN, it is inserted into the population if its fitness (mean squared error on the test data) is better than the worst in the population, and then the worst in the population is removed. This strategy is particularly important as the generated RNNs will have different architectures and will not take the same amount of time to train. By having a master process control the population, workers can complete the training of the generated RNNs at whatever speed they can and the process is naturally load balanced. Further, this allows EXALT to scale to however many processors are available, while having the population size be independent of processor availability, unlike synchronous parallel evolutionary strategies. The EXALT codebase has a multithreaded implementation for multicore CPUs as well as an MPI (the message passing interface [24]) implementation for use on high performance computing resources.

3.1   Mutation and Recombination Operations

RNNs are evolved with edge-level operations, as done in NEAT, as well as with new high level node mutations. Whereas NEAT only requires innovation numbers for new edges, EXALT requires innovation numbers for both new nodes and new edges. The master process keeps track of all node, edge and recurrent edge innovations made, which are required to perform the crossover operation in linear time without a graph matching algorithm. Figure 1 displays a visual walkthrough of all the mutation operations used by EXALT. Nodes and edges selected to be modified are highlighted, and then new elements to the RNN are shown in green. Edge innovation numbers are not shown for clarity. Enabled edges are in black, disabled edges are in grey.

Edge Mutations:

Disable Edge. This operation randomly selects an enabled edge or recurrent edge in a RNN genome and disables it so that it is not used. The edge remains in the genome. As the *disable edge* operation can potentially make an output node unreachable, after all mutation operations have been performed to generate a child RNN genome, if any output node is unreachable that RNN genome is discarded and a new child is generated by another attempt at mutation.

Enable Edge. If there are any disabled edges or recurrent edges in the RNN genome, this operation selects one at random and enables it.

Split Edge. This operation selects an enabled edge at random and disables it. It creates a new node (creating a new node innovation) and two new edges (creating two new edge innovations), and connects the input node of the split edge to the new node, and the new node to the output node of the split edge. The new node is either a regular neuron or LSTM neuron, selected randomly at 50% each.

Add Edge. This operation selects two nodes n_1 and n_2 within the RNN Genome at random, such that $depth_{n_1} < depth_{n_2}$ and such that there is not already an edge between those nodes in this RNN Genome, and then adds an edge from n_1 to n_2. If an edge between n_1 and n_2 exists within the master's innovation list, that edge innovation is used, otherwise this creates a new edge innovation.

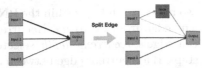

(a) The edge between Input 1 and Output 1 is selected to be split. A new node with innovation number (IN) 1 is created.

(b) Input 3 and Node IN 1 are selected to have an edge between them added.

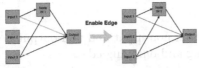

(c) The edge between Input 3 and Output 1 is enabled.

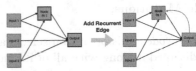

(d) A recurrent edge is added between Output 1 and Node IN 1

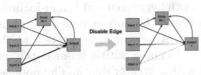

(e) The edge between Input 3 and Output 1 is disabled.

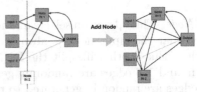

(f) A node with IN 2 is added at a depth between the inputs and Node IN 1. Edges are randomly added to Input 2 and 3, and Node IN 1 and Output 1.

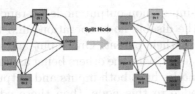

(g) Node IN 1 is split into Nodes IN 3 and 4, which get half the inputs. Both have an output edge to Output 1, because there was only one output from Node IN 1.

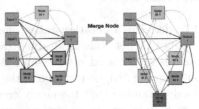

(h) Node IN 2 and 3 are selected to be merged. They are disabled along with their input/output edges. Node IN 5 is created with edges between all their inputs and outputs.

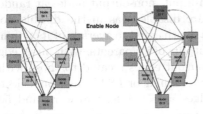

(i) Node IN 1 is selected to be enabled, along with all its input and output edges.

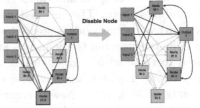

(j) Node IN 5 is selected to be disabled, along with all its input and output edges.

Fig. 1. Edge and node mutation operations. (Color figure online)

Add Recurrent Edge. This operation selects two nodes n_1 and n_2 within the RNN Genome at random and then adds a recurrent edge from n_1 to n_2. Recurrent edges can span multiple time steps, with the edge's *recurrent depth* selected uniformly at random between 1 and 10 time steps. If a recurrent edge between n_1 and n_2 exists within the master's innovation list with the same recurrent depth, that recurrent edge innovation is used, otherwise this creates a new recurrent edge innovation.

Node Mutations:

Disable Node. This operation selects a random non-input and non-output node and disabled it along with all of its incoming and outgoing edges.

Enable Node. This operation selects a random disabled node and enables it along with all of its incoming and outgoing edges.

Add Node. This operation selects a random depth between 0 and 1, non-inclusive. Given that the input node is always depth 0 and the output nodes are always depth 1, this depth will split the RNN in two. A new node is created, at that depth, and 1–5 edges are randomly generated to nodes with a lesser depth, and 1–5 edges are randomly generated to nodes with a greater depth. The node size is set to the average of the maximum input node size and minimum output node size. The new node will be either a regular or LSTM neuron, selected randomly at 50% each. Newly created edges are 50% feed forward and 50% recurrent, selected randomly.

Split Node. This operation takes one non-input, non-output node at random and splits it. This node is disabled (as in the disable node operation) and two new nodes are created at the same depth as their parent. One input and one output edge are assigned to each of the new nodes, with the others being assigned randomly, ensuring that the newly created nodes have both inputs and outputs. If there is only one input or one output edge to this node, then those edges are duplicated for the new nodes. The new nodes will be either a regular or LSTM neuron, selected randomly at 50% each. Newly created edges are 50% feed forward and 50% recurrent, selected randomly.

Merge Node. This operation takes two non-input, non-output nodes at random and combines them. The selected nodes are disabled (as in the disable node operation) and a new node is created with a depth equal to average of its parents. This node is connected to the inputs and outputs of its parents, with input edges created to those with a lower depth, and output edges created to those with a deeper depth. The new node will be either a regular or LSTM neuron, selected randomly at 50% each. Newly created edges are 50% feed forward and 50% recurrent, selected randomly.

Other Operations:

Crossover utilizes two hyperparameters, the *more fit crossover rate* and the *less fit crossover rate*. Two parent RNN genomes are selected, and the child RNN genome is generated from every edge that appears in both parents. Edges that only appear in the more fit parent are added randomly at the *more fit crossover rate*, and edges that only appear in the less fit parent are added randomly at the *less fit crossover rate*. Edges not added by either parent are also carried over into the child RNN genome, however they are set to disabled. Nodes are then added for each input and output of an edge. If the more fit parent has a node with the same innovation number, it is added from the more fit parent.

Clone creates a copy of the parent genome, initialized to the same weights. This allows a particular genome to continue training in cases where further training may be more beneficial than performing a mutation or crossover.

3.2   Epigenetic Weight Initialization

For RNNs generated during population initialization, the weights are initialized uniformly at random between -0.5 and 0.5. Biases and weights for new nodes and edges are initialized randomly with a normal distribution based on the average, μ and variance, σ^2 of the parent's weights. However, RNNs generated through mutation or crossover re-use the weights of their parents, allowing the RNNs to train from where the parents are left off, *i.e.*, *"epigenetic" weight initialization* – these weights are a modification of how the genome is expressed as opposed to a modification of the genome itself.

Additionally, for crossover in the case of where an edge or node exists in both parents, the child weights are generated by recombining the parents weights. Given a random number $-0.5 <= r <= 1.5$, a child's weight w_c is set to $w_c = r(w_{p2} - w_{p1}) + w_{p1}$, where w_{p1} is the weight from the more fit parent, and w_{p2} is the weight from the less fit parent. This allows the child weights to be set along a gradient calculated from the weights of the two parents.

4   Open Data and Reproducibility

The dataset examined in this work is time series data gathered from a coal-fired power plant. The data consists of 10 days of per-minute data readings extracted from 12 of the plant's burners. The data has 12 parameters of time series data:

1. Conditioner Inlet Temp
2. Conditioner Outlet Temp
3. Coal Feeder Rate
4. Primary Air Flow
5. Primary Air Split
6. System Secondary Air Flow Total

7. Secondary Air Flow
8. Secondary Air Split
9. Tertiary Air Split
10. Total Combined Air Flow
11. Supplementary Oil Flow
12. Main Flame Intensity

In order to protect the confidentiality of the power plant which provided the data, along with any sensitive data elements, all identifying data has been scrubbed from the data sets (such as dates, times, locations and facility names). Further, the data has been pre-normalized between 0 and 1 as a further precaution. So while the data cannot be reverse engineered to identify the originating power plant or actual parameter values—it still is an extremely valuable test data set for times series data prediction as it consists of real world data from a highly complex system with interdependent data streams.

In this work, one of the parameters was of key interest for time series data prediction, *Main Flame Intensity*, and was used as the parameter for prediction while gathering the results. In order to further reproducibility of these results and provide this important data set to the time series data prediction research community, it has been made available as part of the EXACT/EXALT GitHub repository, along with instructions on how to use the EXALT code base to recreate these results[1].

5 Results

Two sets of results were gathered predicting *Main Flame Intensity* from the coal plant data set. Six common fixed neural network architectures for time series data prediction were investigated: (1) a one layer feed forward (FF), neural network (NN) (2) a two layer FF NN, (3) an Jordan recurrent neural network (RNN), (4) an Elman RNN, (5) a one layer long short-term memory (LSTM) RNN and (6) a two layer LSTM RNN. K-fold cross validation was performed with 12 folds (*i.e.*, each of the 12 burner data sets was left out to be tested on after training using the other 11 burner data sets). Each NN was trained 10 times for each output data file, resulting in 120 NNs being trained for each NN type. Similarly, EXALT was run 10 times per fold, using each of the 12 burner data sets as testing data, for a total of 120 runs.

Results were gathered using university research computing systems. Compute nodes utilized ranged between 10 core 2.3 GHz Intel®Xeon®CPU E5-2650 v3, 32 core 2.6 GHz AMD Opteron™Processor 6282 SE and 48 core 2.5 GHz AMD Opteron™Processor 6180 SEs, which was unavoidable due to cluster scheduling policies. All compute nodes ran RedHat Enterprise Linux 6.10. This did result in some variation in performance, however discrepancies in timing were overcome by averaging over multiple runs in aggregate. The 720 fixed architecture runs were performed in parallel across 60 compute nodes and took approximately 1,500 compute hours in total. The 120 EXALT runs were performed with each run utilizing 20 processors in parallel, and required 50 compute hours in total.

All neural networks were trained with stochastic backpropagation using the same hyperparameters. Backpropagation was run with a learning rate $\eta = 0.001$, utilizing Nesterov momentum with $\mu = 0.9$ and without dropout, as dropout has been shown in other work to reduce performance when training RNNs for time series prediction [16]. To prevent exploding gradients, gradient clipping

---

[1] https://github.com/travisdesell/exact.

(as described by Pascanu *et al.* [25]) was used when the norm of the gradient was above a threshold of 1.0. To improve performance for vanishing gradients, gradient boosting (the opposite of clipping) was used when the norm of the gradient was below a threshold of 0.05. Initial network weights were randomly initialized uniformly at random between -0.5 and 0.5, however the forget gate bias of the LSTM neurons had 1.0 added to it as this has shown significant improvements to training time by Jozefowicz *et al.* [26]. The fixed NN architectures were trained for 1000 epochs, and EXALT trained 2000 RNNs, with each trained for 10 epochs. As this was in total 20,000 epochs performed in parallel over 20 processors it was seen to be somewhat equivalent to training a single NN for 1000 epochs.

Each EXALT run was done with a population size of 20, and new RNNs were generated via crossover 25% of the time, and by mutation 75% of the time. Mutation operations were performed at the following rates:

1. *clone:* 1/17
2. *add edge:* 1/17
3. *add recurrent edge:* 3/17
4. *enable edge:* 1/17
5. *disable edge:* 3/17
6. *split edge:* 1/17
7. *add node:* 1/17
8. *enable node:* 1/17
9. *disable node:* 3/17
10. *split node:* 1/17
11. *merge node:* 1/17

Mutation rates were chosen in a manner to give mostly equal weighting to each mutation operation. *Add recurrent edge* was given some extra preference as it could be potentially adding recurrent edges with recurrent depths between 1 and 10, which provides a lot of potential options. *Disable edge* and *disable node* were also given extra preference to counteract the RNNs growing quickly, as the other options would put more weight on increasing the RNN size.

Figure 2 shows the minimum, maximum and average progress of the six fixed neural network architectures for each fold, along with the minimum, average and maximum progress of for each EXALT run on each fold. EXALT shows dramatic improvements in reliability and performance over training multiple fixed architecture neural networks. Table 1 presents the aggregate results across each of the folds as well as in total. Two major observations can be made from this, first, the EXALT runs were shown to be much more reliable than training multiple fixed NN architectures, and second, the EXALT runs completed in significantly less time (which was unexpected).

While it was expected that having EXALT evaluate 2000 RNNs each for 10 epochs across 20 nodes in parallel would result in a relatively similar amount of time to training a fixed architectures for 1000 epochs; EXALT runs on average completed more than twice as fast as even the simplest architecture evaluated (a one layer FF NN). Table 2 shows the number of nodes, edges, recurrent edges

Table 1. K-fold cross validation statistics for EXALT and the 6 fixed neural network architectures, presenting the mean squared error and runtime over the 10 repeated trainings. Best results for each fold are shown in bold.

	one layer ff			
	Min	Avg	Max	Time
Fold 0	0.031809	0.044369	0.072142	3658
Fold 1	0.024417	**0.031502**	0.040341	4040
Fold 2	0.020960	0.024908	0.033439	4033
Fold 3	0.033071	0.044107	0.056134	4027
Fold 4	0.030796	0.049311	0.085186	4079
Fold 5	0.033532	0.039205	0.047536	3967
Fold 6	0.010756	0.016743	0.023700	3633
Fold 7	0.030178	0.054017	0.075785	3943
Fold 8	0.019893	0.033458	0.047565	3938
Fold 9	0.016084	0.019077	0.023716	3958
Fold 10	0.023736	0.032435	0.040408	4029
Fold 11	0.041660	0.074404	0.100530	3781
Average	0.026408	0.038628	0.053874	3924

	two layer ff			
	Min	Avg	Max	Time
Fold 0	0.026313	0.042009	0.073753	6670
Fold 1	0.026775	0.033963	0.046181	7542
Fold 2	0.019418	0.028966	0.046257	7480
Fold 3	0.029042	0.051393	0.073627	7615
Fold 4	0.023416	0.037335	0.051478	7639
Fold 5	0.031064	0.039306	0.046585	7616
Fold 6	0.014612	0.016611	0.019820	6345
Fold 7	0.028875	0.045736	0.077376	7222
Fold 8	0.016406	0.031521	0.046914	7547
Fold 9	0.016174	**0.018498**	0.021877	7683
Fold 10	0.025587	0.033352	0.038321	7609
Fold 11	0.036185	0.065018	0.121369	7460
Average	0.024489	0.036976	0.055296	7369

	jordan			
	Min	Avg	Max	Time
Fold 0	0.035064	0.050483	0.097150	3793
Fold 1	0.033920	0.039394	0.043391	3663
Fold 2	0.029067	0.036748	0.046604	3696
Fold 3	0.022927	0.028984	0.034974	3821
Fold 4	0.038322	0.063602	0.098186	3715
Fold 5	0.034472	0.038310	0.043646	3735
Fold 6	0.013130	0.016467	0.020744	3895
Fold 7	0.038538	0.054139	0.090888	3684
Fold 8	0.020665	0.033360	0.043029	3395
Fold 9	0.016776	0.018601	**0.020237**	3439
Fold 10	0.025305	0.028733	**0.032498**	3423
Fold 11	0.055703	0.082065	0.097041	3507
Average	0.030324	0.040907	0.055699	3647

	exalt			
	Min	Avg	Max	Time
Fold 0	0.025360	**0.028749**	**0.030883**	**1675**
Fold 1	0.029976	0.031769	**0.033015**	**1864**
Fold 2	0.021359	**0.023095**	**0.024838**	**2137**
Fold 3	0.018214	**0.019229**	**0.020563**	**1911**
Fold 4	0.020932	**0.023170**	**0.025770**	**1701**
Fold 5	0.030464	0.036091	**0.042542**	**1812**
Fold 6	0.011974	0.012879	**0.013904**	**1763**
Fold 7	**0.016564**	0.019358	**0.020220**	**1847**
Fold 8	0.015867	**0.018151**	**0.020786**	**1885**
Fold 9	0.016922	0.019475	**0.021441**	**1751**
Fold 10	0.020945	0.030016	**0.032662**	**1741**
Fold 11	0.026530	**0.031207**	**0.035073**	**1573**
Average	0.021259	**0.024432**	**0.026808**	**1805**

	elman			
	Min	Avg	Max	Time
Fold 0	0.030173	0.047723	0.073134	6306
Fold 1	**0.014476**	0.035610	0.060415	6225
Fold 2	**0.017132**	0.027319	0.044997	5996
Fold 3	**0.016477**	0.027119	0.033858	5572
Fold 4	**0.017084**	0.029284	0.040682	5848
Fold 5	**0.022649**	**0.031657**	0.045849	5700
Fold 6	**0.008368**	**0.012861**	0.014999	5531
Fold 7	0.018732	0.045840	0.059511	5893
Fold 8	**0.012740**	0.027437	0.043608	6135
Fold 9	**0.013751**	0.018502	0.025968	5957
Fold 10	**0.017572**	**0.028322**	0.038500	6208
Fold 11	**0.024479**	0.053423	0.094839	5717
Average	**0.017803**	0.032092	0.048030	5924

	one layer lstm			
	Min	Avg	Max	Time
Fold 0	**0.017438**	0.052460	0.085376	8221
Fold 1	0.019471	0.037559	0.058379	8552
Fold 2	0.025880	0.039290	0.050615	8837
Fold 3	0.018254	0.025687	0.045868	8118
Fold 4	0.020927	102.586834	512.731000	7887
Fold 5	0.033102	0.043610	0.048107	7943
Fold 6	0.014528	37.755357	188.717000	7764
Fold 7	0.019844	0.034766	0.054353	7860
Fold 8	0.013022	0.098911	0.412826	8039
Fold 9	0.017950	0.035914	0.052342	8069
Fold 10	0.031792	0.035349	0.037908	8038
Fold 11	0.051159	0.076199	0.112534	7998
Average	0.023614	11.735161	58.533859	8110

	two layer lstm			
	Min	Avg	Max	Time
Fold 0	0.057165	0.135696	0.227263	16948
Fold 1	0.020384	0.049007	0.063610	19768
Fold 2	0.026154	0.037273	0.056096	19958
Fold 3	0.020337	0.059560	0.095907	20989
Fold 4	0.038711	0.044041	0.055016	22132
Fold 5	0.024799	0.043924	0.050945	21701
Fold 6	0.014154	0.014464	0.015441	16330
Fold 7	0.026489	0.085277	0.103150	17456
Fold 8	0.022628	0.050065	0.076219	20140
Fold 9	0.026221	0.042297	0.050324	21682
Fold 10	0.028380	0.035213	0.038571	22923
Fold 11	0.052778	0.069439	0.100021	22923
Average	0.029850	0.055521	0.077714	20336

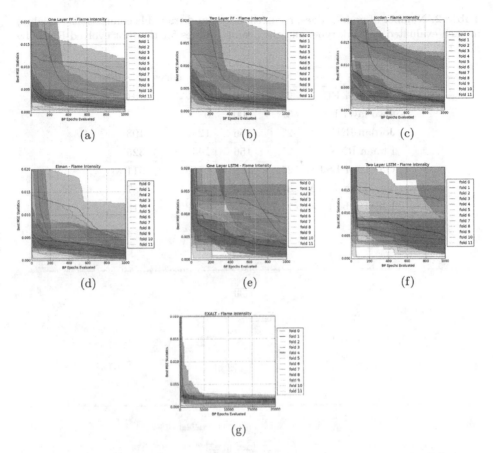

Fig. 2. These plots present the minimum, average, and maximum mean squared errors across each of the 12 folds used by K-fold cross validation by the one layer feed forward NN (a), two layer feed forward NN (b), Jordan RNN (c), Elman RNN (d), one layer LSTM RNN (e), two layer LSTM RNN (f), and by EXALT (g).

and trainable connections (weights) for each neural network type, as well as the average counts of these across the best evolved RNNs by EXALT. Overall, EXALT found well performing RNNs that were much smaller than the fixed network sizes. Figure 3 presents some of the best evolved RNNs. These RNNs dropped out some inputs and were more sparsely connected. Interestingly, they were able to perform very well (better than most of the larger fixed architectures) with only a few hidden nodes and sparse connections.

Table 2. Number of nodes, edges, recurrent edges and trainable connections (weights) in each evaluated network type, and the average values for the best evolved RNNs by EXALT.

	Nodes	Edges	Rec. edges	Weights
One layer FF	25	156	0	181
Two layer FF	37	300	0	337
Jordan RNN	25	156	12	193
Elman RNN	25	156	144	325
One layer LSTM	25	156	0	311
Two layer LSTM	37	300	0	587
EXALT best avg.	14.7	26.2	14.6	81.5

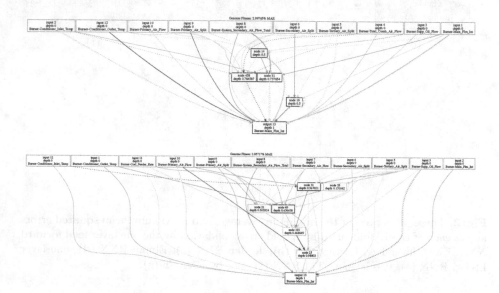

Fig. 3. Two examples of the best RNNs evolved by EXALT. Orange nodes are LSTM neurons, while black nodes are regular neurons. Dotted lines represent recurrent connections, while solid lines represent feed forward connections. Colors of the lines represent the magnitude of the weights (−1.0 is the most blue to 1.0 being the most red). (Color figure online)

So while the Elman network were sometimes able to find the best predictions for some folds, in aggregate these were much more unreliable than utilizing EXALT, which came quite close to these results in the best case. Further, the EXALT runs typically completed in under a third of the time. We expect that running EXALT for a similar length of time will be even closer or outperform these networks.

6 Discussion

Preliminary results for EXALT on this coal fired power plant dataset are very promising. EXALT is quickly able to evolve RNNs that are more efficient (*i.e.*, fewer nodes and trainable connections) than standard RNN architectures, with comparable results. EXALT's best found RNNs outperformed one and two layer feed forward and LSTM neural networks, and had much better average and worst case results than all tested architectures. While Elman networks did find some networks with better results (within a small margin of 0.0025 mean squared error), on average it performed quite a bit worse, and these networks also took over 3 times longer to train—future results providing more time for EXALT to evolve its networks should provide even better results.

This work also introduces a valuable time series dataset gathered from a coal fired power plant, presenting 10 days worth of per minute readings from 12 different burners across 12 different sensors. Having this open large scale real world time series data set of this nature will be very useful for researchers in the field of time series data prediction, and to the authors' knowledge there is not a similar data set available.

There is also potential for significant future work. While the *Main Flame Intensity* parameter was the focus of this work, as having a good predictor for this parameter can help improve plant performance; there are a number of other parameters which can be predicted as well. Further, predictions were only made one time step (*i.e.*, one minute) in the future. Investigating more parameters further in the future along with other data sets will help further demonstrate the effectiveness of the EXALT algorithm.

Using these trained RNNs, the project team aims to develop an advanced tool for coal-fired power plants to actively monitor and manage coal quality and overall boiler conditions that will provide a means to maximize availability and maintain generating capacity while reducing cost. The tool will be used to forecast and alert plant operators and engineers about poor boiler conditions which may occur as a result of incoming coal and/or current power plant parameters.

A more detailed look into how effective the various EXALT mutations are can further improve performance, as well as co-evolution of hyperparameters, which has shown to provide benefits when evolving convolutional neural networks with EXALT's sister algorithm EXACT [2,3]. Additionally, as EXALT converged fairly quickly to a solution in this work, there is potential that methods for increasing speciation may help find better results. One approach would be to utilize mutliple islands evolving in parallel with occasional data transfer, which has been shown by Alba *et al.* to provide significant performance benefits for parallel evolutionary algorithms [27]. Additionally, EXALT was run only utilizing 20 processors and an investigation of its scalability will be interesting.

Overall these preliminary results for EXALT are quite exciting as it provides a parallel algorithm to both train and evolve the structure of RNNs. It can perform parameter selection by dropping out input connections and for the data set tested it generated smaller more accurate RNNs in a shorter amount of time than traditional architectures and backpropagation alone. Further, as it in part utilizes backpropagation, it can be used in conjunction with and stands to benefit from other RNN training methodologies which the machine learning community may develop.

Acknowledgements. This material is based upon work supported by the U.S. Department of Energy, Office of Science, Office of Advanced Combustion Systems under Award Number #FE0031547.

References

1. Stanley, K., Miikkulainen, R.: Evolving neural networks through augmenting topologies. Evol. Comput. **10**(2), 99–127 (2002)
2. Desell, T.: Developing a volunteer computing project to evolve convolutional neural networks and their hyperparameters. In: The 13th IEEE International Conference on eScience (eScience 2017), pp. 19–28, October 2017
3. Desell, T.: Large scale evolution of convolutional neural networks using volunteer computing. CoRR abs/1703.05422 (2017). http://arxiv.org/abs/1703.05422
4. Hochrieter, S., Schmidhuber, J.: Long short term memory. Neural Comput. **9**(8), 1735–1780 (1997)
5. Gers, F.A., Schraudolph, N.N., Schmidhuber, J.: Learning precise timing with LSTM recurrent networks. J. Mach. Learn. Res. **3**(Aug), 115–143 (2002)
6. Werbos, P.J.: Backpropagation through time: what it does and how to do it. Proc. IEEE **78**(10), 1550–1560 (1990)
7. Gers, F.A., Schmidhuber, J., Cummins, F.: Learning to forget: continual prediction with LSTM. Neural Comput. **12**(10), 2451–2471 (2000)
8. Donahue, J., et al.: Long-term recurrent convolutional networks for visual recognition and description. In: Proceedings of the IEEE Conference on Computer Vision and Pattern Recognition, pp. 2625–2634 (2015)
9. Chao, L., Tao, J., Yang, M., Li, Y., Wen, Z.: Audio visual emotion recognition with temporal alignment and perception attention. arXiv preprint arXiv:1603.08321 (2016)
10. Eck, D., Schmidhuber, J.: A first look at music composition using lstm recurrent neural networks. Istituto Dalle Molle Di Studi Sull Intelligenza Artificiale **103**, (2002)
11. Di Persio, L., Honchar, O.: Artificial neural networks approach to the forecast of stock market price movements. Int. J. Econ. Manag. Syst. **1**, (2016)
12. Maknickienė, N., Maknickas, A.: Application of neural network for forecasting of exchange rates and forex trading. In: The 7th international scientific conference Business and Management, pp. 10–11 (2012)
13. Felder, M., Kaifel, A., Graves, A.: Wind power prediction using mixture density recurrent neural networks. In: Poster Presentation gehalten auf der European Wind Energy Conference (2010)
14. Choi, E., Bahadori, M.T., Sun, J.: Doctor AI: Predicting clinical events via recurrent neural networks. arXiv preprint arXiv:1511.05942 (2015)
15. Desell, T., Clachar, S., Higgins, J., Wild, B.: Evolving deep recurrent neural networks using ant colony optimization. In: Ochoa, G., Chicano, F. (eds.) EvoCOP 2015. LNCS, vol. 9026, pp. 86–98. Springer, Cham (2015). https://doi.org/10.1007/978-3-319-16468-7_8

16. ElSaid, A., El Jamiy, F., Higgins, J., Wild, B., Desell, T.: Optimizing long short-term memory recurrent neural networks using ant colony optimization to predict turbine engine vibration. Appl. Soft Comput. **73**, 969–991 (2018)
17. ElSaid, A., Jamiy, F.E., Higgins, J., Wild, B., Desell, T.: Using ant colony optimization to optimize long short-term memory recurrent neural networks. In: Proceedings of the Genetic and Evolutionary Computation Conference, pp. 13–20. ACM (2018)
18. Annunziato, M., Lucchetti, M., Pizzuti, S.: Adaptive systems and evolutionary neural networks: a survey. In: Proceedings of EUNITE02, Albufeira, Portugal (2002)
19. Larochelle, H., Bengio, Y., Louradour, J., Lamblin, P.: Exploring strategies for training deep neural networks. J. Mach. Learn. Res. **10**(Jan), 1–40 (2009)
20. Kandel, E.R., Schwartz, J.H., Jessell, T.M., Siegelbaum, S.A., Hudspeth, A.J.: Principles of Neural Science, vol. 4. McGraw-hill, New York (2000)
21. Rawal, A., Miikkulainen, R.: From nodes to networks: Evolving recurrent neural networks. CoRR abs/1803.04439 (2018). http://arxiv.org/abs/1803.04439
22. Rawal, A., Miikkulainen, R.: Evolving deep LSTM-based memory networks using an information maximization objective. In: Proceedings of the Genetic and Evolutionary Computation Conference 2016, pp. 501–508. ACM (2016)
23. Desell, T.: Asynchronous Global Optimization for Massive Scale Computing. Ph.D. thesis, Rensselaer Polytechnic Institute (2009)
24. Message Passing Interface Forum: MPI: A message-passing interface standard. The International Journal of Supercomputer Applications and High Performance Computing **8**(3/4), 159–416 (Fall/Winter 1994)
25. Pascanu, R., Mikolov, T., Bengio, Y.: On the difficulty of training recurrent neural networks. In: International Conference on Machine Learning, pp. 1310–1318 (2013)
26. Jozefowicz, R., Zaremba, W., Sutskever, I.: An empirical exploration of recurrent network architectures. In: International Conference on Machine Learning, pp. 2342–2350 (2015)
27. Alba, E., Tomassini, M.: Parallelism and evolutionary algorithms. IEEE Trans. Evol. Comput. **6**(5), 443–462 (2002)

Improving NeuroEvolution Efficiency
by Surrogate Model-Based Optimization
with Phenotypic Distance Kernels

Jörg Stork[✉], Martin Zaefferer, and Thomas Bartz-Beielstein

Institute for Data Science, Engineering, and Analytics, TH Köln, Steinmüllerallee 1,
51643 Gummersbach, Germany
{joerg.stork,martin.zaefferer,thomas.bartz-beielstein}@th-koeln.de

Abstract. In NeuroEvolution, the topologies of artificial neural networks are optimized with evolutionary algorithms to solve tasks in data regression, data classification, or reinforcement learning. One downside of NeuroEvolution is the large amount of necessary fitness evaluations, which might render it inefficient for tasks with expensive evaluations, such as real-time learning. For these expensive optimization tasks, surrogate model-based optimization is frequently applied as it features a good evaluation efficiency. While a combination of both procedures appears as a valuable solution, the definition of adequate distance measures for the surrogate modeling process is difficult. In this study, we will extend cartesian genetic programming of artificial neural networks by the use of surrogate model-based optimization. We propose different distance measures and test our algorithm on a replicable benchmark task. The results indicate that we can significantly increase the evaluation efficiency and that a phenotypic distance, which is based on the behavior of the associated neural networks, is most promising.

Keywords: Neuroevolution · Surrogate models · Kernel · Distance · Optimization

1 Introduction

Artificial Neural Networks (ANN) are utilized in many different fields, such as data regression, data classification, or reinforcement learning [1]. In each of these tasks, the network topology is significant for the ANNs performance. Often, only parameters such as the edge weights, number of hidden layers, and number of elements per layer are considered during the optimization of an ANN. In NeuroEvolution (NE), ANNs are generated by Evolutionary Algorithms (EAs). NeuroEvolution allows severe modifications of networks, such as individual connections between neurons and different neuron transfer functions, leading to a large search space of potential topologies. Two example algorithms of this category are NeuroEvolution of Augmenting Topologies (NEAT) [2] and Cartesian Genetic Programming of Artificial Neural Networks (CGPANN) [3,4]. Both NEAT and

© Springer Nature Switzerland AG 2019
P. Kaufmann and P. A. Castillo (Eds.): EvoApplications 2019, LNCS 11454, pp. 504–519, 2019.
https://doi.org/10.1007/978-3-030-16692-2_34

CGPANN may require numerous (>1000) fitness evaluations to find adequate topologies in the large search space. This makes them inefficient for applications with expensive function evaluations such as simulations or real-world experiments, e.g., when an ANN is the controller of a robot that operates in a complex real-time environment. Surrogate Model-based Optimization (SMBO) is frequently used for data efficient optimization of real-world processes, as it has a high evaluation efficiency [5]. Surrogate models are commonly employed in continuous optimization, where Euclidean spaces and distance metrics exist. The more complex discrete search spaces are less often investigated [6,7]. In this work, we want to extend CGPANN to employ Surrogate Model-based NeuroEvolution (SMB-NE) to reduce the load of fitness evaluations. For this task, we defined several distances based on the CGPANN genotypes and a distance that measures, instead of the structure, the difference in behavior of the ANNs. Our approach allows the definition of a phenotypic distance that is indifferent to the size or topology of an ANN. It only requires a representative input set to model the input to output correlation. Our main research questions are:

1. Is SMB-NE able to outperform CGP NeuroEvolution in terms of evaluation efficiency without loss of accuracy?
2. How can we create a representative input set for the phenotypic distance in SMB-NE?

For all experiments, we utilize data-mining classification tasks as a controllable and cheap to evaluate benchmark for the learning efficiency of SMB-NE using few fitness evaluations. This article is structured as follows: Sect. 2 discusses related work. Section 3 illustrates the utilized methods and the SMB-NE algorithm. Section 4 introduces the different distance measures for SMB-NE. In Sect. 5 the experimental setup and the results are shown and further discussed. Finally, in Sect. 6 we conclude the paper and give an outlook to future work.

2   Related Work

This work is an extension of an unpublished study presented at an informal workshop [8]. A recent study by Zaefferer et al. [9] investigated the use of SMBO in Genetic Programming with genotypic and phenotypic distance measures. In contrast to this study, they performed tests comparing an EA with SMBO for a bi-level symbolic regression optimization problem. They concluded, that the SMBO approach is able to outperform the EA in terms of evaluation efficiency. Moreover, the phenotypic distance performed best and was very fast to compute. A first surrogate model for ANN optimization was applied to NEAT by Gaier et al. [10]. They use a surrogate-assisted optimization approach by combining an EA with a surrogate distance-based model, employing a genotypic compatibility distance that is part of NEAT. With this approach, they are also able to increase the evaluation efficiency. To our best knowledge, nobody else used a surrogate model-based approach for the task of NeuroEvolution. Phenotypic distances are also used in other applications, for example, the optimization of job dispatching rules. For this tasks, Hildebrandt and Branke [11] utilize a surrogate-assisted EA, which is able to increase the evolution towards good solutions. The use of surrogates including genotypic distances is more often discussed, e.g., for optimization of fixed neural network topologies in reinforcement learning [12].

3 Data Efficient Optimization for Neuroevolution

3.1 Neuroevolution by Cartesian Genetic Programming

For NeuroEvolution with CGP, we use the C library CGP by A. Turner[1], which also allows the application of NeuroEvolution [13]. The CGP library was modified with function interfaces to the statistical programming language R, distance measures described below, and additional fitness functions. The genotype of a CGPANN individual consists of a fixed number of nodes. Each node has a number of connection genes based on the pre-defined arity with adjacent weight

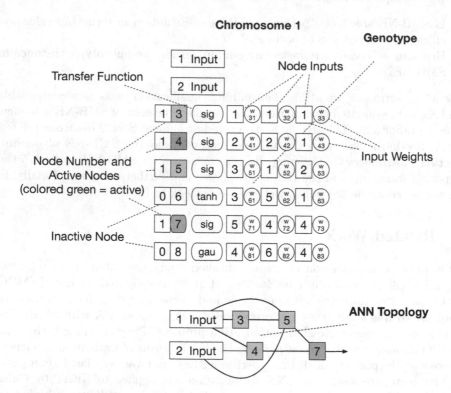

Fig. 1. A CGPANN genotype with two inputs, eight nodes, an arity of three and different transfer functions. Each node has a transfer function, a boolean activity gene and several inputs with adjacent weights. Green nodes are active and part of the encoded ANN. (Color figure online)

---

[1] http://www.cgplibrary.co.uk - accessed:2018-01-12.

genes and a single categorical function gene. Nodes are only connected to preceding nodes. Duplicate connections to nodes are possible and if present, adjacent weights will be added to form a single connection in the resulting ANN. Moreover, each node has a Boolean activity gene, which signals if it is used in the active ANN topology. Figure 1 displays an example for a CGPANN genotype and the encoded ANN with two inputs, eight nodes and an arity of three. In CGP NeuroEvolution, the network topology and weights are optimized using an evolutionary approach, utilizing a (1+4)-Evolution Strategy (ES), i.e., one parent and four new individuals in each generation. In contrast to the standard selection in ES, the rank-based selection process favors the offspring, and thus the new solution, over a parent with the same fitness. Different mutation operators are available, the default is probabilistic random mutation. Inactive genes will not influence the fitness of an ANN.

3.2 Kriging

Our SMBO algorithm is based on a Kriging regression model, which assumes that the observations are samples from a Gaussian process [14]. Kriging is a kernel-based model, i.e., the model uses a kernel, or correlation function, to measure the similarity of two samples. One typical kernel for real-valued samples is the exponential kernel, i.e., $k(x, x') = \exp(-\theta \|x - x'\|_2)$. Here, the kernel parameter θ expresses how quickly the correlation decays to zero, with increasing Euclidean distance $\|x - x'\|_2$ between the samples. The parameter is determined by Maximum Likelihood Estimation (MLE), optimized by using numerical optimization algorithms [14]. It is straightforward to extend kernel-based models to combinatorial search spaces [7,15]. Essentially, the Euclidean distance is replaced by a corresponding distance that applies to the respective search space, with the exponential kernel $k(x, x') = \exp(-\theta d(x, x'))$. The design of appropriate distances $d(x, x')$ for neural networks is in the focus of this paper.

Kriging is frequently employed in SMBO algorithms, because in addition to relatively accurate predictions, it also provides an estimate of the uncertainty of each prediction. The predicted value and the uncertainty estimate are used to compute infill criteria. These infill criteria are supposed to express how desirable it is to evaluate a new candidate solution x. To that end, the uncertainty estimate is integrated to push away from known, well-explored areas, instead preferring solutions that have large uncertainties, yet promising predicted values. One of the most frequently used criteria is Expected Improvement (EI) [16,17]. We employ the Kriging implementation of the R package CEGO [7,18]. It uses distance-based kernels to model data from structured, combinatorial search spaces.

3.3 Surrogate Model-Based NeuroEvolution (SMB-NE)

We extend CGP NeuroEvolution with an SMBO approach, leading to the Surrogate Model-based NeuroEvolution (SMB-NE) strategy outlined below. The strategy is intended to perform a data efficient search by predicting the fitness

---

Algorithm 3.1. Surrogate Model-based Neuroevolution

---

1 **begin**
　　// phase 1: initialization
2　　$t = 1$
3　　**initialize** k neural networks (x_i) at random
4　　**evaluate** their fitness on the objective function
5　　**build** Kriging surrogate model utilizing $s_t(x_i)$ and distance measure $d(x_i, x_j)$;
　　// phase 2: optimization
6　　**while** *not termination-condition* **do**
7　　　**if** $t > 1$ **then**
8　　　　| **rebuild** surrogate model s_t with a set $\mathcal{M}_{\sqcup t} \in \mathcal{D}_t$ of observations
9　　　**end**
10　　　**optimize** EI with a (1+4)-ES to discover improved x_t
11　　　**evaluate** network x_t fitness y_t on the objective function
12　　　**add** evaluated networks to archive $\mathcal{D}_{t+1} = \{\mathcal{D}_t, (x_t, y_t)\}$
13　　　$t = t + 1$
14　　**end**
15 **end**

---

of candidate solutions with the help of a Kriging surrogate model. The algorithm is outlined in 3.1.

The SMB-NE process starts by creating a random initial set of individuals, in our case ANNs, and evaluates them with the objective function. The resulting data is used to learn a Kriging model. For learning the model, we define different genotypic and phenotypic distance measures for ANNs in Sect. 4. With the model, we are able to estimate the EI of an individual. In each following iteration, the model is utilized to suggest new promising ANNs by optimizing the EI criterion with the (1+4)-ES algorithm of CGP. The (1+4)-ES is used to introduce the ability to directly compare CGPANN to SMB-NE, without an additional optimization algorithm with a different operator set, or parametrization influencing the results. In each iteration, the single most promising individual is evaluated and added to the archive $\mathcal{D}_t$. As the archive grows during the optimization process, the computational effort of creating the Kriging model rises with $O(m^3)$, where m is the surrogate model sample size. To keep the computational effort on a feasible level, a subset $\mathcal{M}_t \in \mathcal{D}_t$ of size m is used for the modeling process.

This modeling set $\mathcal{M}_t$ is formed by selecting $\frac{m}{5}$ of the best and $\frac{4*m}{5}$ randomly drawn individuals out of the archive in each iteration. We chose the fractions in the strategy to ensure a balance between exploration and exploitation. If the size of the archive $\mathcal{D}_t$ is smaller than or equal to the size of $\mathcal{M}_t$, all individuals are selected. The influence of the size of the modeling set $\mathcal{M}_t$ is investigated in Sect. 5.2.

4   Proposed Kernels and Distances

In the following, we always use an exponential kernel $k(x, x') = \exp(-\theta d(x, x'))$. Here, x and x' represent ANNs. They consist of the weights x_w, input labels x_i, activity labels x_a, and transfer function labels x_f, i.e., $x = \{x_w, x_i, x_a, x_f\}$. All distances $d(x, x')$ are scaled to $[0, 1]$. In addition, we also considered employing the graph edit distance, but decided against it due to its complexity. Computing the graph edit distance is NP-hard [19].

We illustrate the different distances by providing an example that compares two specific ANNs, with a similar structure as outlined in Fig. 1. For the sake of simplicity and comparability, all weights are set to 1. The two ANNs only differ with respect to two nodes. The transfer function and connections of one active (node 3) and one inactive node (node 8) are changed.

4.1   Genotypic Distance (GD)

The genotypic distance is based solely on the genotype of a CGPANN individual. As the CGPANN genotypes have a fixed structure, the distance of two individuals can be calculated by a row-wise comparison of their nodes. By combining the distances of weights, inputs, activity of nodes, and transfer functions we obtain a distance $d(x, x') = ||x_w - x'_w||_2^2 + H(x_i, x'_i) + H(x_a, x'_a) + H(x_f, x'_f)$, where

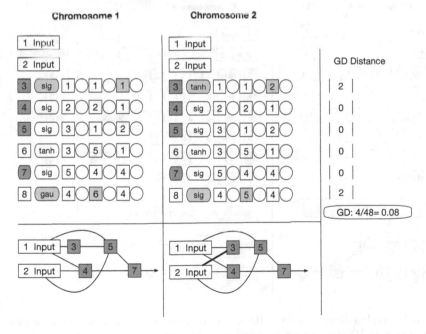

Fig. 2. Example calculation the GD distance for two distances. By introducing only small changes to the genotype, the normalized GD stays rather small.

$H(a, b)$ denotes the Hamming distance, i.e., $H(x_f, x'_f)$ is 0 if the transfer function is identical, else $H(x_f, x'_f)$ is 1. The GD is further normalized by the total number of possible comparisons for the given genotypes. Figure 2 illustrates an example for calculating the GD distance. Although taking the non active nodes into account, the normalized GD is rather small. We further extended the GD by ordering the inputs of each ANN before the calculation to match the weight distances to the correct input if (and only if) two nodes have similar connections, but a different ordering of inputs in the genotype.

4.2 Genotypic ID Distance (GIDD)

The GIDD is intended to solve an important issue of the genotypic distance: different nodes and adjacent functions and weights in one row of a CGPANN genotype are not easily comparable, if their influence on the resulting ANN and also phenotype is considered. The idea behind the GIDD is to only compare those nodes, which are placed in the same position in the ANN and to solve the problem posed by competing conventions, i.e., that a certain ANN topology can be expressed by numerous genotypes. The distance is based on the active topology and creates IDs which are designed to be unique for an equal placement of a node in the ANN. Inactive nodes do not influence the GIDD. Thus, each active node in the ANN is given an ID based on the connections to prior nodes

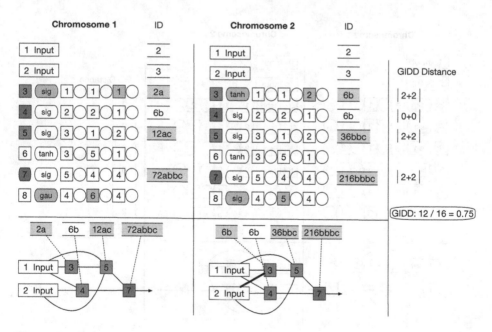

Fig. 3. Example calculation for GIDD. The IDs are based on multiplication of prior node or inputs IDs (set to prime numbers) and the number (a = 1, b = 2, c = 3) for non-duplicate connections. The calculated distance is high because of the different node IDs, which are based on their relative position in the ANN.

or inputs and the number of non-duplicate connections of this node. Then, the distance of nodes can be calculated by a pairwise comparison of all node IDs. If a certain node ID matches for both ANNs, the subgraph prior to this node is analyzed recursively for validation of the ANN up to the position of the matching nodes. If all IDs in the subgraph are identical, we assume that the corresponding nodes have an equal position in the ANN topology. For all nodes that are matched in this way, the Euclidean distance of the weights (x_w) and Hamming distance of the transfer functions (x_f) is computed. A node pair can only be used once for this comparison, as node IDs may be present several times in each individual. If all node IDs of both individuals x and x' are equal, the GIDD is simply the distance $d(x, x') = ||x_w - x'_w||_2^2 + H(x_f, x'_f)$ between all weights and transfer functions. If nodes do not match, for each node not present in the compared ANN, a fixed distance is assigned (in our example 2). Again, the GIDD is normalized by the maximum distance of two individuals. Figure 3 illustrates the calculation in an example. Contrary to the GD, the GIDD reacts strongly to the introduced changes, as they have a large influence on the node relations, which results in different node IDs and puts them to maximum distance.

4.3 Phenotypic Distance (PhD)

The phenotypic distance does not utilize any genotypic information of the ANNs to compute the distance. Instead, it utilizes solely their behavior. In our definition,

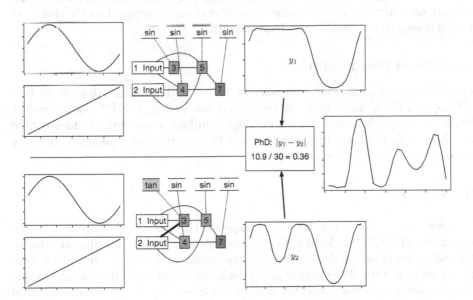

Fig. 4. PhD example with continuous inputs and trigonometric transfer functions. Two input samples, from a sine and linear function of length 30 are fed to two ANNs, which differ in their transfer functions and a single connection. The phenotypic distance is the normalized absolute difference of their output signals.

the phenotype of a neural network is how it reacts to a certain input set, i.e., which outputs are produced. This output is then compared to resolve in a distance measure, which is indifferent to changes to the underlying genotype, including transfer functions, weights or connections, which result in the same behavior. More importantly, it is insensitive to the size of the genotype. The PhD distance utilizes the L1 norm distance to account for large dimensions of the input and output vector and is further normalized by the input set length. While it is indifferent to the network topology in terms of encoding, size, number of connections etc., it is very sensitive to the choice of the input set. Thus, these input sets have to be carefully selected to be representative for the underlying task and/or environment. In Sect. 5.2 we examine the influence of the input set for the task of classification. An example calculation for the distance of small ANNs utilizing PhD is given in Fig. 4. In the example, two continuous input samples, from a sine and linear function of length 30 are fed to two ANNs, where they differ in their transfer functions and a single connection. As this example is intended for understanding the change in the output signals, simple trigonometric functions have been used, which are commonly utilized in Genetic Programming. The small example shows that minor changes in the network topology can result in a large difference in the phenotype. The disadvantage of the PhD is, that it utilizes the ANN outputs (not the function fitness), which have to be computed for each new candidate. Thus, in the SMB-NE model search process, the required ANN evaluations might take a considerable amount of computation time, particular for complex ANNs. This renders the PhD particular interesting if on assumption regarding the objective function is met: that is, that the function calls to evaluate the fitness are, compared to the surrogate model search steps, significantly more expensive.

4.4 Mixed Distance (MD)

Similar to linear combination distance in [9], we utilize a mixed distance of GD, GIDD, and PhD, where each distance receives a weight $\beta_i \in \mathbb{R}^+$ determined by MLE. As the performance of each distance is unknown a-priori, the idea behind the MD is that allows an automatic selection of the most adequate distance measure in each optimization step.

5 Experiments

To assess the ability of SMB-NE to improve the efficiency of optimizing the topologies of ANNs, we decided to perform experiments with classification tasks. Classification is well understood, easily replicable, and does not introduce complex problems with the selection of environments or tasks as in general learning (e.g. reinforcement learning). We limited the experiments to a small budget of function evaluations, which provides a realistic scenario for problems with expensive fitness evaluations, such as real-time learning. The experiments are twofold:

- First, we estimate the ability of SMB-NE to learn an elementary data set comparing the introduced distance measures GD, GIDD, PhD and MD.

- Second, we further research how SMB-NE using the PhD reacts to different inputs sets and surrogate model sizes.

5.1   Comparison of Distance Measures for SMB-NE

Experimental Setup. For the first set of experiments, the well-known IRIS data set is used as an elementary and fast to compute benchmark problem. IRIS has $n = 150$ samples, 4 variables, and 3 evenly distributed ($n = 50$ for each) classes of different flower types (Iris setosa, Iris virginica and Iris versicolor). The focus of this benchmark is the capability of SMB-NE to learn the best network topology to classify the data set with only 250 fitness evaluations. The fitness function is the adjusted classification accuracy: acc $= \sum_{i=1}^{n} a_i$, where $a_i = 1$ if the predicted class is true, otherwise, a_i is the predicted probability for the true class. ANN optimized with Random Search and the inbuilt (1+4)-ES of CGPANN with different mutation rates are being used as baselines. For SMB-NE, all above described distance measures are compared, while for PhD four different input sample sets are tested. As a baseline, we used the complete IRIS data set and additional factorial designs(FD) with small (15) and a large (60) sizes as well as a Latin Hypercube Sample (LHS) with 150 samples. Both the FD and LHS are based on the IRIS variable boundaries. Further, the output of the PhD is adapted by computing the *softmax*, which yields the class probabilities. In this experiment, for the (1+4)-ES of CGPANN and SMB-NE pure probabilistic random mutation is used with a strength of 5%. In SMB-NE, the (1+4)-ES is used in each consecutive iteration alternating between exploitation (local search=L) and exploration (global search=G). Parameters are listed in Table 1. The ANNs in this benchmark were kept rather small with a maximum of 40 nodes and 200 connections. This was intended to keep the search space small, but sufficient for the IRIS data set. Algorithm parameters were not tuned and all experiments were replicated 30 times.

Table 1. Parameter setup, where evaluations denote the initial candidates plus the budget for consecutive evaluations. 40 nodes were used to keep the search space small, but sufficient for the IRIS problem given this small budget. Evaluations of CGP-ANN are due to the underlying (1+4)-ES, where in each iteration 4 new candidates are proposed. SMB-NE proposes only a single one.

Arity	Nodes	Weight range	Function set
5	40	[-1,1]	tanh, softsign, step, sigmoid, gauss
Method	Mutation rate	Evaluations	Surrogate evaluations
Random		250	
CGPANN	5%/15%	$1 + 4 \cdot 63$	
SMB-NE	L:5% G:15%	50+200	L:10+400 G:1000+400

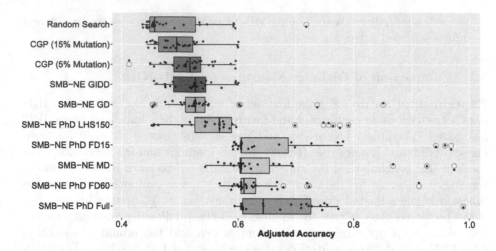

Fig. 5. Results after 250 fitness evaluations with 30 replications, comparing random search, original CGPANN ((1+4)-ES) and different SMB-NE variants. The numbers behind the FD and LHS variant depict the length of the ANN input samples to calculate the PhD. The results were ranked (top down) by median values. Red circles depict outliers. (Color figure online)

Results. Figure 5 visualizes the results. Firstly, the results show that the standard (1+4)-ES of CGPANN performs better than random search, even with the small number of evaluations. A low mutation rate, which depicts a rather local search, seems to be beneficial in this case. SMB-NE utilizing the GD and GIDD distance measures performs only slightly better than the basic CGPANN. With the use of PhD, a significant performance increase is observed, while the PhD with the complete input set performs the best and the LHS the worst. The mixed distance, which also utilizes PhD with complete input sets, cannot benefit from the linear combination, but is able to deliver a close performance to the sole use of PhD. Most runs seem to end up at a local optimum around an accuracy of 66%, which can be explained by the fact, that at this point two out of three classes of the IRIS data set are predicted correctly.

Discussion. An important insight of this experiment is that in comparison to the PhD, the genotypic distances (GD, GIDD) show a poor performance, even for the small genotype size. This fact might be explained by the small correlation between changes to the genotype and the resulting phenotype: small changes to the genes can have a massive impact on the behavior. Moreover, the GD has the problem that the calculated distances do not consider that aligned row-paired nodes in the genotype can have different, not similar placements in the ANN topology. This results in a misleading distance calculation of weights, inputs and transfer function for these nodes. However, even the more complex GIDD, which may be able to avoid this issue, does not seem to provide significantly better results. Further given the fact that the GIDD is computationally very expensive to calculate and has also shown numerical problems for large genotypes (as it utilizes several recursions over the complete ANN), it is further considered as not suitable for the task of SMB-NE. In contrast, the PhD distances show very promising results by directly exploiting the ANN behavior. Importantly, the small factorial input sample, which is very fast to compute, shows a good performance. Given the poor performance of GD and GIDD, the MD is able to automatically select the PhD distance measure and deliver equal performance. For the second set of experiments, we thus focus on exploiting features of SMB-NE utilizing the PhD distance measure.

5.2   Influence of the Input Set and Surrogate Model Size Using SMB-NE with PhD

Experimental Setup. To assess the performance of PhD, we discarded the GD, GIDD, and MD distance and introduced two more complex data sets, the glass and the cancer data set[2]. Both data sets were preprocessed by normalization and subsampling. Glass has 9 variables (material composition), 6 unevenly distributed classes of different glass types and 214 samples, while cancer has 9 variables, 2 classes (cancer yes/no) and 699 samples. Two benchmarks were conducted. The first benchmark investigated the influence of the input sample set used for generating the PhD distance measure. For the benchmark, different input samples were created by design of experiments (DOE) methods. All DOE sets are based on the known variable ranges and are not subsamples of the original data sets. The lengths of the samples are identical for both data sets, as they have the same number of variables. We compared the following:

1. Small and large factorial designs, including main, interaction and quadratic effects, with 55/157 samples each.
2. Small and large Latin Hypercube samples, with 55/110 samples each.
3. The complete datasets as baseline input set, with 214/699 samples each.
4. As algorithm baseline, CGPANN with an increasing number of evaluations (5.5×100, $5.5 \times 1,000$, and $5.5 \times 10,000$).

Table 2. Algorithm parameter setup for the second set of experiments. The size of the genotypes and the number of evaluations was significantly increased.

Arity	Nodes	Weight range	Function set
25	100	[-1,1]	tanh, softsign, step, sigmoid, gauss
Method	Mutation	Evaluations	Surrogate evaluations
CGPANN	Single active	$1 + 4 \cdot 137$ (x10, x100)	
SMB-NE	Single active	50+500	L:10+400 G:1000+400

---

[2] Available in the UCI machine learning repository: https://archive.ics.uci.edu/ml/index.php.

The motivation of this benchmark is as follows: in real-world optimization tasks, often *a priori* information of the the task and/or environment is sparse, as the underlying problem is a black-box problem. Thus, initially no data set is available to serve as an input set and the user has to rely on design of experiment methods to create input data for the PhD. Further, several changes were made to the algorithm setup to account for the more complex classification problems:

- The genotype size was significantly enlarged to 100 nodes with 25 arity, resulting in a maximum of 2500 weights/connections.
- For all compared algorithms, the mutation operator is changed to Goldmans single mutation, which mutates exactly one active node in the genotype (and an arbitrary number of inactive nodes).
- The number of total function evaluations was raised to 550, while fixing the surrogate model set $\mathcal{M}_t$ size to a value of 100. Each sample hereby consists of 80% random and 20% of the most fit individuals from the solution archive $\mathcal{D}_t$.

Table 2 shows the adjusted algorithm parameter setup. In the second, connected benchmark, the influence of the surrogate model size, which is used in each iteration of SMB-NE, to the overall optimization performance is analyzed. The glass data set with the complete input set and cancer with the factorial input set are compared for different model set $\mathcal{M}_t$ sizes.

Results. Figure 6 shows the results of the benchmarks with SMB-NE using PhD for different input sets created by experimental design techniques. For glass, the SMB-NE with the complete input set performed best and on one level with CGPANN with 100 times more real function evaluations. This indicates that SMB-NE is clearly able to improve the evaluation efficiency, if the (best possible) input data set is selected. All DOE input sets performed on an equal level similar to CGPANN with 10 times the evaluations. They are thus also able to significantly increase the evaluation efficiency, while utilizing an input set, which requires (nearly) no a priori information. The results for cancer firstly indicate that the underlying problem seems to include strong local optima, as we can identify certain clusters of different levels of the adjusted accuracy (around 0.8, 0.86, and 0.95). If we only consider only the best cluster, again we can identify that again the complete set performs best, together with CGPANN x100, while the DOE sets are again on one level with CGPANN x10.

Figure 7 shows the results of the benchmark with different surrogate model sizes during sequential optimization runs. In contrast to our expectations, the benchmarks show that the model sample size does not seem to have a significant impact on the performance. Even the small sample sizes perform on the same level.

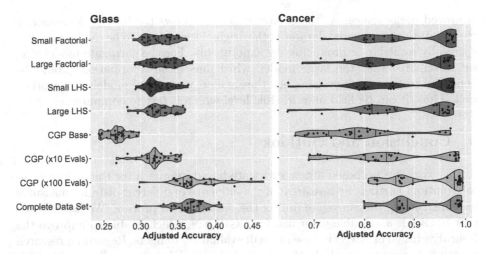

Fig. 6. Violin plot of the benchmarks with SMB-NE using PhD for different input sets created by experimental design techniques. The results are compared to the complete dataset and CGPANN using a different number of total function evaluations

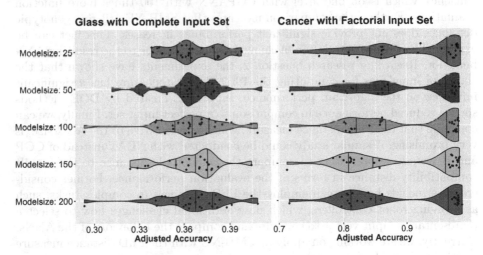

Fig. 7. Results with different surrogate model sizes during sequential optimization run for SMB-NE using PhD. Compared datasets are glass with complete input set for PhD, cancer with factorial input set for PhD.

Discussion. The second set of experiments significantly shows two features of SMB-NE using the PhD distance measure. As expected, the choice of input set has a strong influence on the algorithm performance. The different designs and input sample sizes show a similar performance, which is an unexpected finding, as we would have anticipated that more samples and a larger design would lead to a more precise representational distance for the complete dataset and thus an improved performance. As the complete dataset shows the best performance, a representatively measured dataset of the original task seems to be the best choice, if initially available or producible by experiments. For the surrogate model size, we anticipated that the large model, which has more information, would be beneficial to the search process, but the smaller, more local models also worked well. Thus, it can be held on a feasible level which is fast to compute.

6   Conclusion and Outlook

In this work, we proposed a new surrogate-based approach for the task of NeuroEvolution. Further, we investigated the influence of different distance measures for constructing the surrogate during the optimization process. We have shown that SMBO is a valuable extension for CGPANN, which is able to improve the NeuroEvolution of ANNs in case of small evaluation budgets. Regarding research question 1, we can conclude that SMB-NE with phenotypic distance kernels shows significantly better results than basic CGPANN. Utilizing the PhD distance measure with a perfect input set, we were able to reach an evaluation efficiency which is on one level with CGPANN with 100 times more function evaluations. Further, the comparison indicates that SMB-NE utilizing genotypic distances does not provide significant performance increases. This fact can be explained by the low correlation of changes in the genotype to the resulting ANN behavior. Regarding research question 2, the experiments have shown that the choice of input sets for computing the PhD distance measure has a significant influence on the algorithm performance. Input sets created by DOE methods show a reduced performance in comparison to a perfect input set. Finally, we can conclude that the PhD distance measure, which is insensitive to the ANN size, is very promising. A similar study could be conducted with NEAT instead of CGP and it would be interesting to compare the phenotypic distance to the inbuilt compatibility distance in terms of the evaluation performance. Former considerations include how we can employ the PhD distances in complex tasks, such as evolving robot controllers, which pose additional challenges how to select a representative input vector to compute and compare the behaviors of the ANNs. Currently, we are working on applying SMB-NE with the PhD distance measure to reinforcement learning tasks.

Acknowledgements. This work is supported by the German Federal Ministry of Education and Research in the funding program Forschung an Fachhochschulen under the grant number 13FH007IB6.

References

1. Basheer, I.A., Hajmeer, M.: Artificial neural networks: fundamentals, computing, design, and application. J. Microbiol. Methods **43**(1), 3–31 (2000)
2. Stanley, K.O., Miikkulainen, R.: Evolving neural networks through augmenting topologies. Evol. Comput. **10**(2), 99–127 (2002)

3. Miller, J.F., Thomson, P.: Cartesian genetic programming. In: Poli, R., Banzhaf, W., Langdon, W.B., Miller, J., Nordin, P., Fogarty, T.C. (eds.) EuroGP 2000. LNCS, vol. 1802, pp. 121–132. Springer, Heidelberg (2000). https://doi.org/10.1007/978-3-540-46239-2_9

4. Turner, A.J., Miller, J.F.: Cartesian genetic programming encoded artificial neural networks: a comparison using three benchmarks. In: Proceedings of GECCO 2013, pp. 1005–1012. ACM (2013)

5. Koziel, S., Leifsson, L.: Surrogate-based Modeling and Optimization. Applications in Engineering. Springer, New York (2013). https://doi.org/10.1007/978-1-4614-7551-4

6. Bartz-Beielstein, T., Zaefferer, M.: Model-based methods for continuous and discrete global optimization. Appl. Soft Comput. **55**, 154–167 (2017)

7. Zaefferer, M., Stork, J., Friese, M., Fischbach, A., Naujoks, B., Bartz-Beielstein, T.: Efficient global optimization for combinatorial problems. In: Proceedings of GECCO 2014, pp. 871–878. ACM (2014)

8. Stork, J., Zaefferer, M., Bartz-Beielstein, T.: Distance-based kernels for surrogate model-based neuroevolution. arXiv preprint arXiv:1807.07839 (2018)

9. Zaefferer, M., Stork, J., Flasch, O., Bartz-Beielstein, T.: Linear combination of distance measures for surrogate models in genetic programming. In: Auger, A., Fonseca, C.M., Lourenço, N., Machado, P., Paquete, L., Whitley, D. (eds.) PPSN 2018. LNCS, vol. 11102, pp. 220–231. Springer, Cham (2018). https://doi.org/10.1007/978-3-319-99259-4_18

10. Gaier, A., Asteroth, A., Mouret, J.B.: Data-efficient neuroevolution with kernel-based surrogate models. In: Genetic and Evolutionary Computation Conference (GECCO) (2018)

11. Hildebrandt, T., Branke, J.: On using surrogates with genetic programming. Evol. Comput. **23**(3), 343–367 (2015)

12. Stork, J., Bartz-Beielstein, T., Fischbach, A., Zaefferer, M.: Surrogate assisted learning of neural networks. In: GMA CI-Workshop 2017 (2017)

13. Turner, A.J., Miller, J.F.: Introducing a cross platform open source Cartesian genetic programming library. Genet. Program. Evolvable Mach. **16**(1), 83–91 (2015)

14. Forrester, A., Sobester, A., Keane, A.: Engineering Design via Surrogate Modelling. Wiley, Hoboken (2008)

15. Moraglio, A., Kattan, A.: Geometric generalisation of surrogate model based optimisation to combinatorial spaces. In: Merz, P., Hao, J.-K. (eds.) EvoCOP 2011. LNCS, vol. 6622, pp. 142–154. Springer, Heidelberg (2011). https://doi.org/10.1007/978-3-642-20364-0_13

16. Mockus, J., Tiesis, V., Zilinskas, A.: The application of Bayesian methods for seeking the extremum. In: Towards Global Optimization 2, North-Holland, pp. 117–129 (1978)

17. Jones, D.R., Schonlau, M., Welch, W.J.: Efficient global optimization of expensive black-box functions. J. Global Optim. **13**(4), 455–492 (1998)

18. Zaefferer, M.: Combinatorial Efficient Global Optimization in R - CEGO v2.2.0. https://cran.r-project.org/package=CEGO (2017), https://cran.r-project.org/package=CEGO. Accessed 10 Jan 2018

19. Zeng, Z., Tung, A.K.H., Wang, J., Feng, J., Zhou, L.: Comparing stars: on approximating graph edit distance. Proc. VLDB Endow. **2**(1), 25–36 (2009)

Numerical Optimization: Theory, Benchmarks and Applications

Compact Optimization Algorithms
with Re-Sampled Inheritance

Giovanni Iacca[1]([⊠]) [iD] and Fabio Caraffini[2] [iD]

[1] Department of Information Engineering and Computer Science,
University of Trento, 38123 Povo, Italy
giovanni.iacca@unitn.it

[2] Institute of Artificial Intelligence, School of Computer Science and Informatics,
De Montfort University, Leicester LE1 9BH, UK
fabio.caraffini@dmu.ac.uk

Abstract. Compact optimization algorithms are a class of Estimation of Distribution Algorithms (EDAs) characterized by extremely limited memory requirements (hence they are called "compact"). As all EDAs, compact algorithms build and update a probabilistic model of the distribution of solutions within the search space, as opposed to population-based algorithms that instead make use of an explicit population of solutions. In addition to that, to keep their memory consumption low, compact algorithms purposely employ simple probabilistic models that can be described with a small number of parameters. Despite their simplicity, compact algorithms have shown good performances on a broad range of benchmark functions and real world problems. However, compact algorithms also come with some drawbacks, i.e. they tend to premature convergence and show poorer performance on non-separable problems. To overcome these limitations, here we investigate a possible algorithmic scheme obtained by combining compact algorithms with a non-disruptive restart mechanism taken from the literature, named Re-Sampled Inheritance (RI). The resulting compact algorithms with RI are tested on the CEC 2014 benchmark functions. The numerical results show on the one hand that the use of RI consistently enhances the performances of compact algorithms, still keeping a limited usage of memory. On the other hand, our experiments show that among the tested algorithms, the best performance is obtained by compact Differential Evolution with RI.

Keywords: Compact Optimization · Differential evolution ·
Bacterial foraging optimization · Particle Swarm Optimization ·
Genetic Algorithm

1 Introduction

Compact Optimization [1] is a branch of Computational Intelligence Optimization devoted to the study of optimization algorithms characterized by limited memory requirements. From an algorithmic point of view, compact algorithms belong to the family of the Estimation of Distribution Algorithms (EDAs) [2],

© Springer Nature Switzerland AG 2019
P. Kaufmann and P. A. Castillo (Eds.): EvoApplications 2019, LNCS 11454, pp. 523–534, 2019.
https://doi.org/10.1007/978-3-030-16692-2_35

i.e. algorithms that instead of evolving a population of solutions (as is typically done in population-based optimization algorithms, such as Evolutionary Algorithms and Swarm Intelligence algorithms), build and update a probabilistic model of the distribution of solutions within the search space. Depending on the specific probabilistic model (Gaussian, binomial, etc.), different EDAs can be implemented. In this regard, the specificity of compact algorithms is that they employ a separate distribution for each variable of the problem, and update it as long as the evolutionary process proceeds. Therefore, differently from population-based algorithms where at least n D-dimensional arrays need to be stored in memory (being n the population size and D the problem dimension), compact algorithms need to store only a much more compact "Probability Vector" ($\mathbf{PV}$) that describes the parameters of the probabilistic model. For instance, binary-encoded compact algorithms use as $\mathbf{PV}$ a single D-dimensional array $\mathbf{p} = [p_1, p_2, \ldots, p_D]$. Each $p_i \in [0,1]$, $i = 1, 2, \ldots, D$, represents the probability that the i–th variable has value 1 (i.e., the relative frequency in a corresponding "virtual population" of N_p individuals, with N_p a parameter of the algorithm). Similarly, real-valued compact algorithms based on Gaussian distributions use as $\mathbf{PV}$ two D-dimensional arrays: an array of means $\boldsymbol{\mu} = [\mu_1, \mu_2, \ldots, \mu_n]$ and an array of variances $\boldsymbol{\sigma} = [\sigma_1, \sigma_2, \ldots, \sigma_n]$, describing a (normalized) Gaussian distribution of each variable in the search space.

In the past two decades, the Compact Optimization concept has been declined in a number of compact algorithms, sparkling from the seminal works by Harik et al. [3] and Corno et al. [4], who devised a similar algorithm dubbing it respectively "compact Genetic Algorithm" (cGA) and "Selfish Gene" (SG). The family of compact algorithms was then extended to include improved versions of cGA [5,6], real-valued cGA (rcGA) [7], compact Differential Evolution (cDE) [8] and many of its variants [9–18], compact Particle Swarm Optimization (cPSO) [19], compact Bacterial Optimization (cBFO) [20], and, more recently, compact Teaching-Learning Based Optimization (cTLBO) [21,22], compact Artificial Bee Colony (cABC) [23,24], and compact Flower Pollination Algorithm (cFPA) [25].

Due to their limited usage of memory, compact algorithms are particularly suited for embedded devices, such as Wireless Sensor Networks motes, wearable devices, embedded controllers for robots and industrial plants, etc. Unsurprisingly, the literature abounds with successful examples of compact algorithms applications based on this kind of devices: for instance, cDE has been applied especially in control applications, such as real-time hardware-in-the-loop optimization of a control system for a permanent-magnet tubular linear synchronous motor [8], or real-time trajectory optimization of robotic arms [18,26] and Cartesian robots [17]. In [6], cGA was applied to micro-controller design, while cPSO was used for optimizing a power plant controller in [19]. In [27], cABC was used for topology control in Wireless Sensor Networks. The more recent cTLBO was instead applied to train Artificial Neural Networks in [22].

In this paper, we aim to push forward this research area by tackling the two main drawbacks of compact algorithms, i.e. their tendency to premature convergence (as they do not keep an actual population, they do not maintain explicitly diversity), and a poorer performance on non-separable problems (which is due to the fact that they process each variable separately). To overcome these limitations, we study the effect of a special case of restart named Re-Sampled Inheritance (RI) [28,29], which simply generates a random solution and then recombines it - by using an exponential crossover operator similarly to Differential Evolution - with the best solution detected so far by the compact algorithm. Previous works [29,30] have shown that the RI mechanism is a simple yet effective way to improve the performance of an optimization algorithm, as it allows to escape from local optima while preserving some of the information from the current best, thus guaranteeing a kind of inheritance and avoiding an excessively disruptive restart (compared to other restart mechanisms). Our hypothesis is then that on the one hand the re-sampling should allow compact algorithms to escape from local optima, on the other hand the inheritance mechanism -since it processes blocks of variables- should enable an overall performance improvement especially (but not only) on non-separable problems. Moreover, as the RI mechanism only needs to sample a random solution and recombine it with the current best, it does not require any additional memory with respect to a compact algorithm, therefore allowing to keep a low memory consumption.

To assess the effect of RI on different compact algorithms, we apply it separately to four (real-valued) compact algorithms taken from the literature, namely cDE (specifically, its 'light" version [11]), rcGA, cPSO, and cBFO, and perform extensive tests on the CEC 2014 benchmark [31] in 10, 50 and 100 dimensions.

The rest of this paper is organized as follows: Sect. 2 presents the background concepts on Compact Optimization. Section 3 describes the general algorithmic scheme which combines compact algorithms with Re-Sampled Inheritance. The numerical results are then presented and discussed in Sect. 4. Finally, Sect. 5 concludes this work.

2   Background

In the rest of this paper, we focus on real-valued compact algorithms as they have been shown to perform better than their binary-encoded counterparts [7]. In case of real values, the general structure of a compact algorithm is quite straightforward and can be described as follows. First of all, for each i-th variable a Gaussian Probability Distribution Function (PDF) is considered, truncated within the interval $[-1, 1]$, with mean μ_i and standard deviation σ_i taken from the Probability Vector $\mathbf{PV} = [\boldsymbol{\mu}, \boldsymbol{\sigma}]$. The height of the PDF is normalized in order to keep its area equal to 1.

At the beginning of the optimization process, for each design variable i, $\mu_i = 0$ (unless Re-Sampled Inheritance is used, see next section) and $\sigma_i = \lambda$, where λ is a large positive constant (e.g. $\lambda = 10$), such that it simulates a uniform distribution (thus exploring the search space). Subsequently, a starting individual, **elite**, is generated by sampling each i-th variable from the corresponding PDF. For more details about the sampling mechanism, see [7].

Then, the iterative process starts. At each step, depending on the specific compact algorithm, a candidate solution $\mathbf{x}$ is generated by sampling one or more individuals from the current $\mathbf{PV}$. E.g., in rcGA [7], $\mathbf{x}$ is obtained by generating a single individual and recombining it with **elite** with binomial crossover. In cDE [8], **x** is obtained by generating a mutated individual (for instance sampling three individuals from **PV** and applying the rand/1 DE mutation), and then recombining it with the current **elite** by using either binomial or exponential crossover. Other compact algorithm paradigms, such as cPSO and cBFO, use the same mechanism for sampling new individuals, but apply different algorithmic operators inspired from the corresponding biological metaphor to generate a new candidate solution **x**. In all cases, the fitnesses of **elite** and **x** are compared and, according to the chosen elitism scheme (persistent or non-persistent, see [1]), **elite** is replaced by **x**. Furthermore, the fitness comparison is used to update the **PV**, i.e. it changes its μ and σ values by "moving" the Gaussian PDF towards the better solution and "shrinking" the PDF around it. Details for this update mechanism are given in [7]. This iterative process is executed until a certain stop condition is met. The pseudo-code of a general real-valued compact algorithm is given in Algorithm 1.

Algorithm 1. General structure of a compact algorithm

---

1 initialize $\mathbf{PV} = [\boldsymbol{\mu}, \boldsymbol{\sigma}]$;
2 generate **elite** by means of **PV**;
3 **while** *stop condition is not met* **do**
4   generate candidate solution **x** (according to the specific operators);
5   compare fitness of **x** and **elite**;
6   **if** *elite replacement condition is true* **then**
7 | elite = x;
8   **end if**
9   update **PV**;
10 **end while**
11 Return: **elite**

---

3   Compact Optimization Algorithms with Re-Sampled Inheritance

The general scheme of a compact algorithm with Re-Sampled Inheritance is shown in Algorithm 2. The only difference w.r.t. the original compact algorithm shown in Algorithm 1 is the RI component, which enables the restart mechanism with inheritance, and is activated at the end of each execution of the compact algorithm (that is continued for a given % of the total budget).

The Re-Sampled Inheritance (see [28,29] for more details) first randomly generates a solution **x** from a uniform distribution within the given search space. Then, it recombines **x** with the current best solution $\mathbf{x_{best}}$ by applying the exponential crossover used in Differential Evolution. More specifically, a random initial index is selected in $[0, D)$, and the corresponding variable is copied from $\mathbf{x_{best}}$ into **x**. Then, as long as a (uniform) random number $rand(0, 1)$ is less than or equal to Cr, the design variables from $\mathbf{x_{best}}$ are copied into the corresponding positions of **x**, starting from the initial index. Cr, the crossover rate, is a parameter affecting the number of variables inherited from $\mathbf{x_{best}}$, and is set as in [11], i.e. $Cr = 1/\sqrt[D]{2}$, where $D\alpha$ is the expected number of variables that are copied from $\mathbf{x_{best}}$. As soon as $rand(0,1) > Cr$, the copy process is interrupted. The copy is handled as in a cyclic buffer, i.e. when the D-th variable is reached during the copy process the next to be copied is the first one. When the copy stops, the fitness of $\mathbf{x}$ is compared with that of $\mathbf{x_{best}}$. If the newly generated solution $\mathbf{x}$ outperforms the current best solution, the latter gets updated (i.e. $\mathbf{x_{best}} = \mathbf{x}$). The compact algorithm is then restarted after setting its mean value μ (that is used in the Probability Vector $\mathbf{PV}$) equal to the new restarted point, i.e. $\mu = \mathbf{x}$. This way, the new initial distribution is centered in a new point which, despite being randomly sampled, still contains some inheritance from the current best solution. At the end of each compact optimization routine, see Algorithm 1, an elite solution is returned and compared for replacement against the current best solution, as shown in Algorithm 2.

Algorithm 2. Compact algorithm with Re-Sampled Inheritance

---

1 generate a random solution $\mathbf{x}$ in the search space and set $\mathbf{x_{best}} = \mathbf{x}$;
2 **while** *stop condition is not met* **do**
3 $\quad$ // Compact algorithm
4 $\quad$ set $\mu = \mathbf{x}$ and run compact algorithm as in Algorithm 1 (for a % of the budget);
5 $\quad$ **if elite** *is better than* $\mathbf{x_{best}}$ **then**
6 $\quad\quad$ $\mathbf{x_{best}} = \mathbf{elite}$;
7 $\quad$ **end if**
8 $\quad$ // Re-Sampled Inheritance
9 $\quad$ generate a random solution $\mathbf{x}$ (from a uniform distribution);
10 $\quad$ generate $i = round(D \cdot rand(0,1))$;
11 $\quad$ $\mathbf{x}[i] = \mathbf{x_{best}}[i]$;
12 $\quad$ $k = 1$;
13 $\quad$ **while** $rand(0,1) \leq Cr$ **and** $k < D$ **do**
14 $\quad\quad$ $i = i + 1$;
15 $\quad\quad$ **if** $i == D$ **then**
16 $\quad\quad\quad$ $i = 1$;
17 $\quad\quad$ **end if**
18 $\quad\quad$ $\mathbf{x}[i] = \mathbf{x_{best}}[i]$;
19 $\quad\quad$ $k = k + 1$;
20 $\quad$ **end while**
21 $\quad$ **if** $\mathbf{x}$ *is better than* $\mathbf{x_{best}}$ **then**
22 $\quad\quad$ $\mathbf{x_{best}} = \mathbf{x}$;
23 $\quad$ **end if**
24 **end while**

The rationale behind the RI mechanism is then to restart the algorithm from a partially random solution, i.e. a solution that is randomly generated but still inherits part of the variables from the current best. This way the restart is not entirely disruptive, but preserves at least a block of (an expected number of) $D\alpha$ variables. This partial inheritance allows the algorithm to keep some information from one restart and the next one, but also to escape from local optima.

From this point of view, the RI mechanism shares some resemblance with the Iterated Local Search (ILS) methods [32], that try to apply a small perturbation to the best-so-far solution during restart (in fact, as small as possible to not disrupt it much, but as large as needed to allow the local search to converge to a different local optimum). However, ILS has been especially designed for (and applied to) combinatorial optimization, while here we focus on continuous optimization.

4   Numerical Results

In the following we present the numerical results obtained on the CEC 2014 benchmark [31]. This benchmark is composed of 30 functions, with different properties in terms of separability, ill-conditioning, and landscape structure. In particular, it is worth noting that except for f_8 (Shifted Rastrigin's Function) and f_{10} (Shifted Schwefel's Function), all CEC 2014 benchmark functions are non-separable. Therefore this benchmark is particularly suited for testing the performance of optimization algorithms on non-separable problems.

We considered the following four real-valued compact algorithms, with the parametrization proposed in their original papers:

- cDE "light" [11] with exponential crossover and parameters: $N_p = 300$, $F = 0.5$, and $\alpha_m = 0.25$;
- rcGA [7], with persistent elitism and parameters: $N_p = 300$;
- cPSO [19], with parameters: $N_p = 300$, $\phi_1 = 0.2$, $\phi_2 = 0.07$, $\phi_3 = 3.74$, $\gamma_1 = 1$, and $\gamma_2 = 1$;
- cBFO [20], with parameters: $N_p = 300$, $C_i = 0.1$, and $N_s = 4$.

As for the corresponding versions with RI (dubbed, respectively as RIcDE, RIrcGA, RIcPSO and RIcBFO), the same parametrization was kept for the compact optimazion process while the RI component was parametrized with $\alpha = 0.05$ (such that only 5% of the variables are inherited, on average, from the current best). A number of fitness function calls equal to 25% of the total computational budget was assigned to execute the compact algorithm after each restart. It should be noted that these are the only two parameters of the RI mechanism and they were empirically set after having observed their effect in preliminary experiments.

Furthermore, in order to assess the effect of the RI mechanism w.r.t. a simple random restart without any form of inheritance, we included in our experimental setup also four variants of the same compact algorithms where the restart was applied, with the same period of the RI variants (25% of the total computational budget), by simply applying a uniform re-sampling of a new solution $\mathbf{x}$ within the search space, and restarting the compact algorithm by setting $\boldsymbol{\mu} = \mathbf{x}$. We dub these compact algorithms with random restart, respectively, as RecDE, RercGA, RecPSO and RecBFO.

Finally, to provide a baseline for all the compact algorithms with/without restart tested in this paper, we evaluated the performance of a simple Random Walk (RW) algorithm where at each step a new solution is generated by applying a uniform re-sampling within the search space. From our numerical results (see Table 5) it can be seen that its performance is -as expected- considerably worse than any of the compact algorithms considered in the experiments, thus highlighting that the "compact" logic is more than a mere random sampling and performs significantly better w.r.t. pure uniform random searches.

To assess the scalability of all the algorithms, we performed experiments in 10, 50 and 100 dimensions. Thus, the total experimental setup consists of 13 algorithms (4 compact algorithms, 4 RI variants, 4 variants with random restart, and RW) and $30 \times 3 = 90$ optimization problems (i.e. 30 functions each tested in three different dimensionalities). On each benchmark function, each algorithm was executed for 30 independent runs, to collect statistics on the fitness values obtained in each run at the end of the allotted computational budget. Each run was executed for a total budget of $5000 \times D$ function evaluations, being D the problem dimension.

In the following, for the sake of brevity we will show only a compact representation of the main experimental results[1]. For that, we will use the sequentially rejective Holm-Bonferroni procedure [33, 34]. This procedure consists of the following: considering N_{TP} test problems (in our case, 90) and N_A optimization algorithms, the performance obtained by each algorithm on each problem is computed. This is measured as average of the best fitness values obtained by the algorithm on that problem over multiple (in our case, 30) independent runs, at the end of the computational budget (in our case, $5000 \times D$ function evaluations). Then, for each problem a score R_i is assigned to each algorithm, being N_A the score of the algorithm displaying the best performance (i.e., assuming minimization, the minimum average of the fitness values) on that problem, $N_A - 1$ the score of the second best, and so on. The algorithm displaying the worst performance scores 1. These scores are then averaged, for each algorithm, over the whole set of N_{TP} test problems. The algorithms are sorted on the basis of these average scores. Indicating with R_0 the rank of an algorithm taken as reference, and with R_j for $j = 1, \ldots, N_A - 1$ the rank of the remaining algorithms, the values z_j are calculated as:

$$z_j = \frac{R_j - R_0}{\sqrt{\frac{N_A(N_A+1)}{6N_{TP}}}}. \tag{1}$$

By means of the z_j values, the corresponding cumulative normal distribution values p_j are derived. These p_j values are then compared to the corresponding δ/j where δ is the confidence interval, set to 0.05: if $p_j > \delta/j$, the null-hypothesis (that the algorithm taken as reference has the same performance as the j-th algorithm) is accepted, otherwise is rejected as well as all the subsequent tests.

---

[1] Detailed numerical results are available at: http://www.cse.dmu.ac.uk/~fcaraf00/NumericalResults/RICompactOptResults.pdf.

Table 1. Holm-Bonferroni procedure (reference: RIcDE, Rank = 2.63e+00)

j	Optimizer	Rank	z_j	p_j	δ/j	Hypothesis
1	RecDE	2.33e+00	−2.85e+00	2.21e−03	5.00e−02	Rejected
2	cDE	1.03e+00	−1.52e+01	2.44e−52	2.50e−02	Rejected

Table 2. Holm-Bonferroni procedure (reference: RIrcGA, Rank = 2.53e+00)

j	Optimizer	Rank	z_j	p_j	δ/j	Hypothesis
1	RercGA	2.47e+00	−6.32e−01	2.64e−01	5.00e−02	Accepted
2	rcGA	1.00e+00	−1.45e+01	3.07e−48	2.50e−02	Rejected

Let us first consider, for each compact algorithm, how the corresponding algorithms with RI and random restart perform w.r.t. the original compact algorithm without restart. Tables 1, 2, 3 and 4 show, respectively, the results of the Holm-Bonferroni procedure (in this case with $N_A = 3$) on cDE, rcGA, cPSO and cBFO based algorithms. The tables display the ranks, z_j values, p_j values, and corresponding δ/j obtained by this procedure. In each case we considered as reference algorithm the corresponding algorithm with RI, whose rank is shown in parenthesis in each table caption. Moreover, we indicate in each table whether the null-hypothesis (that the algorithm taken as reference has the same performance as each other algorithm in the corresponding table row) is accepted or not.

From these Holm-Bonferroni procedures, we can observe that, except for the case of cPSO (where, quite surprisingly, cPSO shows the same performance as the corresponding algorithms with RI and random restart) in all other cases the algorithms with RI score a better rank than their corresponding compact algorithms. It is also interesting to note that, while in the case of cDE RIcDE performs better also than RecDE, on the other compact algorithms it results that the RI variant is statistically equivalent to the variant with random restart (note that the null-hypothesis is accepted in those cases). This equivalence between RI and random restart in the case of rcGA, cPSO and cBFO might be due to parametrization used for RI (number of restarts and number of variables inherited from the current best), as well as the different algorithmic logics used by these algorithms compared to cDE. In general though, these observations demonstrate that the use of restarts, and, especially in the case of cDE, the use of RI is beneficial in terms of optimization performance.

Table 3. Holm-Bonferroni procedure (reference: RIcPSO, Rank = 1.99e+00)

j	Optimizer	Rank	z_j	p_j	δ/j	Hypothesis
1	RecPSO	2.01e+00	2.11e−01	5.83e−01	5.00e−02	Accepted
2	cPSO	2.00e+00	1.05e−01	5.42e−01	2.50e−02	Accepted

Table 4. Holm-Bonferroni procedure (reference: RIcBFO, Rank = 1.84e+00)

j	Optimizer	Rank	z_j	p_j	δ/j	Hypothesis
1	RecBFO	1.88e+00	3.16e−01	6.24e−01	5.00e−02	Accepted
2	cBFO	1.54e+00	−2.85e+00	2.21e−03	2.50e−02	Rejected

Finally, we provide an overall comparison of all the 12 compact optimization algorithms, in addition to the Random Walk algorithm. The resulting Holm-Bonferroni procedure is reported in Table 5, where RIcDE is considered as reference algorithm (as it shows the highest rank) and $N_A = 13$. In this case, except for RecDE, all the hypotheses are sequentially rejected, meaning that when all the algorithms are considered together, RIcDE is statistically equivalent to RecDE (although it shows a slightly higher rank), but it shows a statistically better performance (on average, on the entire set of tested problems) than all other algorithms under study. As expected, the Random Walk algorithm performs worse than all other papers. Moreover, the rank shows that each compact algorithm with RI (or random restart) performs better (on average) than the corresponding compact algorithm, confirming the fact that the RI component is beneficial to all the compact algorithms considered in our experimentation.

Table 5. Holm-Bonferroni procedure (reference: RIcDE, Rank = 1.09e+01)

j	Optimizer	Rank	z_j	p_j	δ/j	Hypothesis
1	RecDE	1.06e+01	−6.41e−01	2.61e−01	5.00e−02	Accepted
2	RecBFO	8.56e+00	−4.38e+00	5.86e−06	2.50e−02	Rejected
3	RIcBFO	8.52e+00	−4.44e+00	4.40e−06	1.67e−02	Rejected
4	RIrcGA	8.38e+00	−4.71e+00	1.22e−06	1.25e−02	Rejected
5	RercGA	8.30e+00	−4.86e+00	5.93e−07	1.00e−02	Rejected
6	cBFO	8.07e+00	−5.29e+00	6.04e−08	8.33e−03	Rejected
7	cDE	8.03e+00	−5.35e+00	4.30e−08	7.14e−03	Rejected
8	rcGA	5.68e+00	−9.74e+00	1.05e−22	6.25e−03	Rejected
9	RecPSO	3.97e+00	−1.29e+01	1.73e−38	5.56e−03	Rejected
10	RIcPSO	3.97e+00	−1.29e+01	1.73e−38	5.56e−03	Rejected
11	cPSO	3.89e+00	−1.31e+01	2.61e−39	5.00e−03	Rejected
12	RW	1.43e+00	−1.76e+01	6.80e−70	4.55e−03	Rejected

5 Conclusions

In this paper we have presented an algorithmic scheme for solving continuous optimization problems on devices characterized by limited memory. The proposed scheme is based on a combination of a compact algorithm with a restart mechanism based on Re-Sampled Inheritance (RI). We tested this scheme on four different compact algorithms presented in the literature (namely: cDE, rcGA, cPSO, and cBFO) and performed numerical experiments on a broad range of benchmark functions in several dimensionalities. Our experiments show that the use of RI consistently enhances the performances of compact algorithms, still keeping a limited usage of memory. In addition to that, we noted that among the tested algorithms the best performance was obtained by cDE with Re-Sampled Inheritance.

In future works, we will further investigate the effect of the parametrization on the proposed compact algorithms with Re-Sampled Inheritance, focusing in particular on the influence of the number of restarts, as well as the number of variables inherited from the best individual at each restart. We will also investigate alternative inheritance mechanisms, for instance based on binomial crossover or exponential crossover on shuffled variables.

Acknowledgments. This project has received funding from the European Union's Horizon 2020 research and innovation programme under grant agreement No. 665347.

References

1. Neri, F., Iacca, G., Mininno, E.: Compact optimization. In: Zelinka, I., Snášel, V., Abraham, A. (eds.) Handbook of Optimization. Intelligent Systems Reference Library, vol. 38, pp. 337–364. Springer, Heidelberg (2013). https://doi.org/10.1007/978-3-642-30504-7_14
2. Larrañaga, P., Lozano, J.A.: Estimation of Distribution Algorithms: A New Tool for Evolutionary Computation. Kluwer Academic Publishers, Boston (2001)
3. Harik, G.R., Lobo, F.G., Goldberg, D.E.: The compact genetic algorithm. IEEE Trans. Evol. Comput. **3**(4), 287–297 (1999)
4. Corno, F., Reorda, M.S., Squillero, G.: The selfish gene algorithm: a new evolutionary optimization strategy. In: ACM Symposium on Applied Computing, pp. 349–355 (1998)
5. Ahn, C.W., Ramakrishna, R.S.: Elitism-based compact genetic algorithms. IEEE Trans. Evol. Comput. **7**(4), 367–385 (2003)
6. Gallagher, J.C., Vigraham, S., Kramer, G.: A family of compact genetic algorithms for intrinsic evolvable hardware. IEEE Trans. Evol. Comput. **8**(2), 111–126 (2004)
7. Mininno, E., Cupertino, F., Naso, D.: Real-valued compact genetic algorithms for embedded microcontroller optimization. IEEE Trans. Evol. Comput. **12**(2), 203–219 (2008)
8. Mininno, E., Neri, F., Cupertino, F., Naso, D.: Compact differential evolution. IEEE Trans. Evol. Comput. **15**(1), 32–54 (2011)
9. Iacca, G., Mallipeddi, R., Mininno, E., Neri, F., Suganthan, P.N.: Global supervision for compact differential evolution. In: IEEE Symposium on Differential Evolution, pp. 1–8 (2011)
10. Iacca, G., Mallipeddi, R., Mininno, E., Neri, F., Suganthan, P.N.: Super-fit and population size reduction in compact differential evolution. In: IEEE Workshop on Memetic Computing, pp. 1–8 (2011)
11. Iacca, G., Caraffini, F., Neri, F.: Compact differential evolution light: high performance despite limited memory requirement and modest computational overhead. J. Comput. Sci. Technol. **27**(5), 1056–1076 (2012)

12. Iacca, G., Mininno, E., Neri, F.: Composed compact differential evolution. Evol. Intel. **4**(1), 17–29 (2011)
13. Iacca, G., Neri, F., Mininno, E.: Opposition-based learning in compact differential evolution. In: Di Chio, C., et al. (eds.) EvoApplications 2011, Part I. LNCS, vol. 6624, pp. 264–273. Springer, Heidelberg (2011). https://doi.org/10.1007/978-3-642-20525-5_27
14. Iacca, G., Neri, F., Mininno, E.: Noise analysis compact differential evolution. Int. J. Syst. Sci. **43**(7), 1248–1267 (2012)
15. Jewajinda, Y.: Covariance matrix compact differential evolution for embedded intelligence. In: IEEE Region 10 Symposium, pp. 349–354 (2016)
16. Mallipeddi, R., Iacca, G., Suganthan, P.N., Neri, F., Mininno, E.: Ensemble strategies in compact differential evolution. In: IEEE Congress on Evolutionary Computation, pp. 1972–1977 (2011)
17. Neri, F.: Memetic compact differential evolution for cartesian robot control. IEEE Comput. Intell. Mag. **5**(2), 54–65 (2010)
18. Neri, F., Iacca, G., Mininno, E.: Disturbed exploitation compact differential evolution for limited memory optimization problems. Inf. Sci. **181**(12), 2469–2487 (2011)
19. Neri, F., Mininno, E., Iacca, G.: Compact particle swarm optimization. Inf. Sci. **239**, 96–121 (2013)
20. Iacca, G., Neri, F., Mininno, E.: Compact bacterial foraging optimization. In: Rutkowski, L., Korytkowski, M., Scherer, R., Tadeusiewicz, R., Zadeh, L.A., Zurada, J.M. (eds.) EC/SIDE -2012. LNCS, vol. 7269, pp. 84–92. Springer, Heidelberg (2012). https://doi.org/10.1007/978-3-642-29353-5_10
21. Yang, Z., Li, K., Guo, Y.: A new compact teaching-learning-based optimization method. In: Huang, D.-S., Jo, K.-H., Wang, L. (eds.) ICIC 2014. LNCS (LNAI), vol. 8589, pp. 717–726. Springer, Cham (2014). https://doi.org/10.1007/978-3-319-09339-0_72
22. Yang, Z., Li, K., Guo, Y., Ma, H., Zheng, M.: Compact real-valued teaching-learning based optimization with the applications to neural network training. Knowl.-Based Syst. **159**, 51–62 (2018)
23. Banitalebi, A., Aziz, M.I.A., Bahar, A., Aziz, Z.A.: Enhanced compact artificial bee colony. Inf. Sci. **298**, 491–511 (2015)
24. Dao, T.-K., Chu, S.-C., Nguyen, T.-T., Shieh, C.-S., Horng, M.-F.: Compact artificial bee colony. In: Ali, M., Pan, J.-S., Chen, S.-M., Horng, M.-F. (eds.) IEA/AIE 2014. LNCS (LNAI), vol. 8481, pp. 96–105. Springer, Cham (2014). https://doi.org/10.1007/978-3-319-07455-9_11
25. Dao, T.K., Pan, T.S., Nguyen, T.T., Chu, S.C., Pan, J.S.: A compact flower pollination algorithm optimization. In: International Conference on Computing Measurement Control and Sensor Network, pp. 76–79 (2016)
26. Iacca, G., Caraffini, F., Neri, F., Mininno, E.: Robot base disturbance optimization with compact differential evolution light. In: Di Chio, C., et al. (eds.) EvoApplications 2012. LNCS, vol. 7248, pp. 285–294. Springer, Heidelberg (2012). https://doi.org/10.1007/978-3-642-29178-4_29
27. Dao, T.K., Pan, T.S., Nguyen, T.T., Chu, S.C.: A compact artificial bee colony optimization for topology control scheme in wireless sensor networks. J. Inf. Hiding Multimed. Signal Process. **6**(2), 297–310 (2015)
28. Caraffini, F., Iacca, G., Neri, F., Picinali, L., Mininno, E.: A CMA-ES super-fit scheme for the re-sampled inheritance search. In: IEEE Congress on Evolutionary Computation, pp. 1123–1130 (2013)

29. Caraffini, F., Neri, F., Passow, B.N., Iacca, G.: Re-sampled inheritance search: high performance despite the simplicity. Soft Comput. **17**(12), 2235–2256 (2013)
30. Caraffini, F., Iacca, G., Yaman, A.: Improving (1+1) covariance matrix adaptation evolution strategy: a simple yet efficient approach. In: International Global Optimization Workshop (2018)
31. Liang, J., Qu, B., Suganthan, P.: Problem definitions and evaluation criteria for the CEC 2014 special session and competition on single objective real-parameter numerical optimization. Zhengzhou University, Zhengzhou China and Technical Report, Nanyang Technological University, Singapore, Computational Intelligence Laboratory (2013)
32. Lourenço, H.R., Martin, O.C., Stützle, T.: Iterated local search: framework and applications. In: Gendreau, M., Potvin, J.Y. (eds.) Handbook of Metaheuristics. International Series in Operations Research & Management Science, vol. 272, pp. 363–397. Springer, Cham (2010). https://doi.org/10.1007/978-3-319-91086-4_5
33. Garcia, S., Fernandez, A., Luengo, J., Herrera, F.: A study of statistical techniques and performance measures for genetics-based machine learning: accuracy and interpretability. Soft Comput. **13**(10), 959–977 (2008)
34. Holm, S.: A simple sequentially rejective multiple test procedure. Scand. J. Stat. **6**(2), 65–70 (1979)

Particle Swarm Optimization: Understanding Order-2 Stability Guarantees

Christopher W. Cleghorn$^{(\boxtimes)}$ (iD)

Department of Computer Science, University of Pretoria, Pretoria, South Africa
ccleghorn@cs.up.ac.za

Abstract. This paper's primary aim is to provide clarity on which guarantees about particle stability can actually be made. The particle swarm optimization algorithm has undergone a considerable amount of theoretical analysis. However, with this abundance of theory has come some terminological inconstancies, and as a result it is easy for a practitioner to be misguided by overloaded terminology. Specifically, the criteria for both order-1 and order-2 stability are well studied, but the exact definition of order-2 stability is not consistent amongst researchers. A consequence of this inconsistency in terminology is that the existing theory may in fact misguide practitioners instead of assisting them. In this paper it is theoretically and empirically demonstrated which practical guarantees can in fact be made about particle stability. Specifically, it is shown that the definition of order-2 stability which accurately reflects PSO behavior is that of convergence in second order moment to a constant, and not to zero.

Keywords: Particle swarm optimization · Stability analysis · Stability criteria

1 Introduction

Particle swarm optimization (PSO), originally developed by Kennedy and Eberhart [1], has become a widely used optimization technique [2]. Given PSO's success, a substantial amount of theoretical work has been performed on the stochastic search algorithm to try and predict and understand its underlying behavior [3–9].

One aspect of PSO that has attracted a great deal of theoretical research is that of particle convergence. Given the complexity of PSO's internal dynamics, much of the early research was performed under a number of modeling assumptions. The original works on particle convergence [3,4,10,11] considered a deterministic model where the stochastic terms in PSO's update equations were replaced with constants. In a deterministic context the concept of particle convergence in $\mathbb{R}^n$ is well defined. Newer research has focused on modeling

© Springer Nature Switzerland AG 2019
P. Kaufmann and P. A. Castillo (Eds.): EvoApplications 2019, LNCS 11454, pp. 535–549, 2019.
https://doi.org/10.1007/978-3-030-16692-2_36

PSO's dynamics with the stochasticity still present. In a stochastic context simply stating a sequence is convergent is ambiguous. This ambiguity arises from the fact that there exist numerous types of stochastic convergence, ranging from convergence in a nth order moment, convergence in probability, to almost sure convergence, to name a few.

In current theoretical PSO research, two different definitions for order-2 stability of particle positions have arisen. The first definition is in line with the concept of convergence of a second order moment to a constant, whereas the second definition makes the stronger requirement that the second order moment of the sequence must not only converge, but rather that is must converge to zero. This overloading is potentially problematic as existing stability research provides criteria on the control parameters of PSO that will ensure order-2 stability. However, in practice, which type of order-2 stability can be actually expected may be unclear.

The concept of particle stability is vital for effective use of PSO, as it was shown by Cleghorn and Engelbrecht [12] that parameter configurations that resulted in particle instability almost always caused PSO to perform worse than random search. Given the relationship between particle stability and performance it is important to fully understand the practical implications of particle stability on the swarm. The concept of stability is also vital to PSO variants. Many popular PSO variants have also undergone stability analysis [6,13–17].

In this paper, guided by existing theory, it is empirically demonstrated which practical guarantees can in fact be made about particle stability. Specifically, which definition of order-2 stability accurately reflects these guarantees is demonstrated.

A description of PSO is given in Sect. 2. Section 3 briefly summarizes the existing theoretical results on PSO stability, and shows why convergence to a zero second order moment cannot be guaranteed by the theory. The experimental setup is presented in Sect. 4, followed by the experimental results and a discussion thereof in Sect. 5. Section 6 presents a summary of the findings of this paper.

2   Particle Swarm Optimization

Particle swarm optimization (PSO) was originally inspired by the complex movement of birds in a flock. The variant of PSO this section focuses on uses the inertia coefficient, as proposed by Shi and Eberhart [18], which is referred to as PSO in this paper.

The PSO algorithm is defined as follows: Let $f : \mathbb{R}^d \to \mathbb{R}$ be the objective function that the PSO algorithm aims to find an optimum for, where d is the dimensionality of the objective function. For the sake of simplicity, a minimization problem is assumed from this point onwards. Specifically, an optimum $o \in \mathbb{R}^d$ is defined such that, for all $x \in \mathbb{R}^d$, $f(o) \leq f(x)$. In this paper the analysis focus is on objective functions where the optima exist. Let $\Omega(t)$ be a set of N particles in $\mathbb{R}^d$ at a discrete time step t. Then $\Omega(t)$ is said to be the particle swarm at time t. The position $\boldsymbol{x}_i$ of particle i is updated using

$$\boldsymbol{x}_i(t+1) = \boldsymbol{x}_i(t) + \boldsymbol{v}_i(t+1), \tag{1}$$

where the velocity update, $\boldsymbol{v}_i(t+1)$, is defined as

$$\boldsymbol{v}_i(t+1) = w\boldsymbol{v}_i(t) + c_1\boldsymbol{r}_1(t) \otimes (\boldsymbol{y}_i(t) - \boldsymbol{x}_i(t)) + c_2\boldsymbol{r}_2(t) \otimes (\hat{\boldsymbol{y}}_i(t) - \boldsymbol{x}_i(t)), \tag{2}$$

where $r_{1,k}(t), r_{2,k}(t) \sim U(0,1)$ for all t and $1 \leq k \leq d$. The operator $\otimes$ is used to indicate component-wise multiplication of two vectors. The position $\boldsymbol{y}_i(t)$ represents the "best" position that particle i has visited, where "best" means the location where the particle had obtained the lowest objective function evaluation. The position $\hat{\boldsymbol{y}}_i(t)$ represents the "best" position that the particles in the neighborhood of the i-th particle have visited. The coefficients c_1, c_2, and w are the cognitive, social, and inertia weights, respectively. A full algorithm description is presented in Algorithm 1.

A primary feature of the PSO algorithm is social interaction, specifically the way in which knowledge about the search space is shared amongst the particles in the swarm. In general, the social topology of a swarm can be viewed as a graph, where nodes represent particles, and the edges are the allowable direct communication routes. The fixed topologies star, ring, and Von Neumann, are frequently used in PSO [19]. A number of dynamic topologies have also been proposed. The interested reader is referred to the work of [20] for an in-depth discussion on dynamic topologies.

3   Theoretical Results on Particle Stability

This section briefly discusses the state of PSO particle stability analysis. Specific focus is placed on the research pertaining to the second order moment of particles.

The PSO algorithm has undergone a considerable amount of theoretical research. In particular the study of long term particle dynamics has attracted a lot of attention. Despite PSO's simplicity as an optimizer, accurately modeling the algorithm in a tractable fashion has been non trivial, with much of the early research requiring multiple simplifications in the modeling of PSO. The two primary aspects of PSO that have been subject to this simplification are the removal of stochasticity (deterministic assumption) [3,4,7,21] and the fixing of the personal and neighborhood best positions (stagnation assumption) [22–25]. A detailed discussion on when and where each modeling assumption was used in PSO literature can be found in [9].

Recent works on particle stability have catered for the stochasticity of PSO and focused on the weakening of the stagnation assumption [8,9,26]. Currently, the weakest assumption used in the study of particle stability is the non-stagnate distribution assumption [9]. Under the non-stagnate distribution assumption, personal and neighborhood best positions are modeled as sequences of random variables that are convergent in first and second order moments to a constant (specifically in expectation and variance).

Algorithm 1. PSO algorithm

Create and initialize a swarm, $\Omega(0)$, of N particles uniformly within a predefined hypercube of dimension d.
Let f be the objective function.
Let y_i represent the personal best position of particle i, initialized to $x_i(0)$.
Let $\hat{y}_i$ represent the neighborhood best position of particle i, initialized to $x_i(0)$.
Initialize $v_i(0)$ to $\mathbf{0}$.
Let $t = 0$
repeat
    for all particles $i = 1, \cdots, N$ **do**
        if $f(x_i) < f(y_i)$ **then**
            $y_i = x_i$
        end if
        for all particles $\hat{i}$ with particle i in their neighborhood **do**
            if $f(y_i) < f(\hat{y}_i)$ **then**
                $\hat{y}_i = y_i$
            end if
        end for
    end for
    $t = t + 1$
    for all particles $i = 1, \cdots, N$ **do**
        update the velocity of particle i using equation (2)
        update the position of particle i using equation (1)
    end for
until stopping condition is met

Research on PSO stability has focused on deriving criteria for PSO's control parameters that will ensure order-1 and order-2 stability of particle positions; where order-1 and order-2 stability of a stochastic sequence are defined as follows:

Definition 1. *Order*-1 *stability*
The sequence (s_t) *in* $\mathbb{R}^d$ *is order-1 stable if there exists an* $s_E \in \mathbb{R}^d$ *such that*

$$\lim_{t \to \infty} E[s_t] = s_E \tag{3}$$

where $E[s_t]$ *is the expectation of* s_t.

Definition 2. *Order*-2 *stability*
The sequence (s_t) *in* $\mathbb{R}^d$ *is order-2 stable if there exists a* $s_V \in \mathbb{R}^d$ *such that*

$$\lim_{t \to \infty} V[s_t] = s_V \tag{4}$$

where $V[s_t]$ *is the variance of* s_t.

Specifically, under varying degrees of model simplification, the authors Poli [24], Liu [8], Bonyadi and Michalewicz [26], and Cleghorn and Engelbrecht [9], have derived the following necessary and sufficient criteria for order-1 and order-2 stability of PSO's particles with update equations (1) and (2):

$$0 < c_1 + c_2 < \frac{24 \left(1 - w^2\right)}{7 - 5w} \quad \text{and} \quad |w| < 1. \tag{5}$$

where $c_1 = c_2$.

However, in the works of Liu [8] and Bonyadi and Michalewicz [26] order-2 stability is defined in a more restrictive manner. Specifically, the limit of the second order moment, s_V, is required to be zero. At first glance the result of Liu [8] and Bonyadi and Michalewicz [26] appears stronger. However, under closer inspection it becomes clear that the only way to ensure convergence of variance to zero is to impose additional requirements of the personal and neighborhood best positions. Specifically, Poli [27] showed that the component-wise limit point for particle variance is defined as

$$V[x_{ij}] = \frac{c(w + 1)}{4c(5w - 7) - 48w^2 + 48}(\hat{y}_{ij} - y_{ij})^2, \tag{6}$$

where $c = c_1 = c_2$. It is clear from Eq. (6) that, when the criteria of equation (5) are satisfied, the only way for the variance of particle positions to become zero is if $\hat{y}_{ij} = y_{ij}$ for each $1 \leq j \leq d$. This implies that for every particle i in the swarm to be have a zero variance the personal and neighborhood best positions must become the same position during the run. A similar argument for PSOs with arbitrary distributions can be found in [9]. In practice the distance between the personal and neighborhood best positions cannot be guaranteed to approach zero, which will be demonstrated in this paper.

For the sake of clarity in the field, it is proposed that the restrictive version of order-2 stability where s_V must be zero, should be referred to as order-2* stability, with the aim of providing clearer guidance for PSO practitioners. For the rest of this paper this new terminology will be utilized.

It is worth mentioning that convergence to a non-zero variance can be seen in a positive light. If the variance of particle positions collapses to zero, the swarm has ceased searching, as any newly generated particle positions can only be sampled from a fixed point. Whereas, if the positional variance is non-zero, it is still possible for a swarm to be usefully searching, as newly visited particle position are sampled from an area of non-zero measure. Said informally, a non-zero positional variance means that the swarm never gives up.

In order to show that the criteria of equation (5), for order-2 stability, cannot ensure order-2* stability the following proposition is proved.

Proposition 1. *There exists no control coefficients, w, c_1, c_2 satisfying equation (5) that ensure order-2* stability for any objective function f.*

Proof: All that is required is the construction of an f that demonstrates that for any w, c_1, c_2 satisfying equation (5) the particle variance as defined in Eq. (6) cannot become 0 for all particles. From this point on the proof focuses on a particle which does not have its personal best positions equal to the global best position yet. Consider the 1-dimensional objective function $f(x) = x^4 - x^2$, which has two global minimum at $x = \frac{-1}{\sqrt{2}}$ and $x = \frac{1}{\sqrt{2}}$ and is illustrated in Fig. 1. Let $\hat{y}_i = \frac{-1}{\sqrt{2}}$, $y_i = \frac{1}{\sqrt{2}}$, and Γ_i be the distribution, parameterize by w, c_1, c_2, $v_i(t-1)$, $x_i(t-1)$, $\hat{y}_i$, and y_i, that describes the possible next positions of particle i, namely $x_i(t)$. For the new positions $x_i(t) \sim \Gamma_i$ to be accepted as a new global best position $f(x_i(t))$ would need to be strictly less than $f(\frac{-1}{\sqrt{2}}) = -\frac{1}{4}$, which is not possible since $x = \frac{-1}{\sqrt{2}}$ is a global minimum. As a results $\hat{y}_{ij}$ cannot change. Following the same line of argument it is clear that y_{ij} can also not change, as such $(\hat{y}_{ij} - y_{ij})^2$ will remain constant and equal to $(\frac{-1}{\sqrt{2}} - \frac{1}{\sqrt{2}})^2 = 2$. Since $\frac{c(w+1)}{4c(5w-7)-48w^2+48}$ cannot be zero if the criteria of equation (5) are satisfied [27], it follows that order-2* stability cannot occur. □

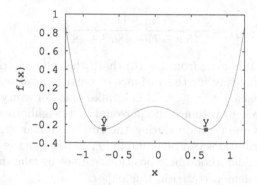

Fig. 1. $f(x) = x^4 - x^2$, with $\hat{y} = \frac{-1}{\sqrt{2}}$ and $y = \frac{1}{\sqrt{2}}$

It is conceivable that while order-2* stability cannot be is theoretical guaranteed for all objective functions, it may occur for non-artificially designed ones like that used in Proposition 1. To demonstrate that order-2* stability cannot be guaranteed in practice either it is empirically shown in Sects. 4 and 5 that order-2* stability cannot be guaranteed by selecting control parameter that satisfy equation (5) for an array of well known objective function.

4 Experimental Setup

This section summarizes the experimental procedure used in this paper.

The experiment utilized a population size of 20 and the gbest topology. The optimization is allowed to run for 5000 iterations, so as to provide a fair indication of long term particle behavior. Particle positions were initialized within $[-100, 100]^d$ and velocities were initialized to $\mathbf{0}$ [28]. Particle personal and neighborhood best positions were only updated with feasible solutions. The analysis is done in 5, 10, 20, and 30 dimensions.

The experiment is conducted over the parameter region corresponding to the criteria of order-1 and order-2 stability as defined in Eq. (5). A total of 504 stable parameter configurations were tested with spacing of 0.1 between each considered $c_1 + c_2 > 0$ and each considered $w > -1$. The experiment is conducted using the 11 base objective functions from the CEC 2014 problem set [29]. The functions are as follows: Ackley (F1), Bent Cigar (F2), Discus (F3), Expanded Griewank plus Rosenbrock (F4), Griewank (F5), HappyCat (F6), HGBat (F7), High Conditioned Elliptic (F8), Katsuura (F9), Rastrigin (F10), and Rosenbrock (F11).

For each considered parameter configuration the following measurement is made:

$$\Gamma(t+1) = \frac{1}{N} \sum_{i=1}^{N} \|\hat{\boldsymbol{y}}_i(t+1) - \boldsymbol{y}_i(t)\|_\infty, \tag{7}$$

where $\|\mathbf{x}\|_\infty = \max_{1 \le i \le N} |x_i|$ is the infinity norm. This norm is selected as it makes direct comparison of results across different dimensionality easy. The choice of the infinity norm is done without loss of generality, since the convergence of a sequence under any arbitrary norm in $\mathbb{R}^n$ implies convergence under all other norms, as all norms are equivalent in a finite dimensional Banach space [30]. If Γ does not go to zero, it implies that each particle cannot be order-2* stable, since Eq. (6) could not be zero for all particles. Furthermore, it should be noted that if the criteria of equation (5) for order-1 and order-2 stability are satisfied, the term $\frac{c(w+1)}{4c(5w-7)-48w^2+48}$ is never zero. It is known that for all parameter configurations tested, particles will present as order-1 and order-2 since the criteria of equation (5) has been empirically verified without the presence of simplifying assumptions by Cleghorn and Engelbrecht [31].

The results reported in Sect. 5 are derived from 50 independent runs for each objective function, each tested parameter configuration, and each considered dimensionality.

5   Experimental Results and Discussion

This section presents the results of the experiments described in Sect. 4.

It is known that each tested parameter configuration is both order-1 and order-2 stable as they satisfy equation (5). Given this knowledge all particles in a swarm are classified as order-2* stable only if $\Gamma(5000) = \Gamma \le \epsilon$, where $\epsilon = 0.001$. The allowance for an ϵ of 0.001 is relatively relaxed. However, if a configuration does not even ensure that $\Gamma \le 0.01$ then it is very clearly not order-2* stable.

For each tested dimension a table is presented to summarize the recored Γ measurements across the 11 base objective functions of the CEC 2014 problem set [29]. Specifically, the number of parameter configurations that resulted in $\Gamma \le \epsilon$ (order-2* stable) and that resulted in $\Gamma > \epsilon$ (not order-2* stable) are presented. Additionally, the average and maximum value for Γ are presented from the 504 parameter configurations. The reported values for Γ are comparable

Table 1. Γ measurements in 5-dimensions

Function	Number of $\Gamma \leq \epsilon$	Number of $\Gamma > \epsilon$	Average Γ	Max Γ
F1	0	504	11.4421	39.3982
F2	504	0	0.0000	0.0007
F3	504	0	0.0000	0.0001
F4	213	291	0.0352	0.3289
F5	71	433	7.6169	34.1461
F6	150	354	0.1320	1.1043
F7	155	349	0.1161	1.1066
F8	504	0	0.0000	0.0000
F9	306	198	11.6192	61.9315
F10	248	256	0.8043	3.7139
F11	297	207	1.4899	32.1101

across different dimensionalities, given the use of the infinity norm. Furthermore, the resulting Γ can be easily related to the initialized component-wise size of the search space. Namely, the largest possible value for $\|\hat{\boldsymbol{y}}_i(t+1) - \boldsymbol{y}_i(t)\|_\infty$ is 200. The results for 5, 10, 20, and 30 dimensions are reported in Tables 1, 2, 3 and 4 respectively.

It is immediately apparent from Table 1 through Table 4 to that there are numerous parameter configurations that are both order-1 and order-2 stable but are not order-2* stable. It is also clear from Table 1 that there is also not a subset of the region defined in Eq. (5) that can guarantee order-2* stability. Specifically, in the case of the Ackley (F1) objective function there are 0 parameter configu-

Table 2. Γ Measurements in 10-dimensions

Function	Number of $\Gamma \leq \epsilon$	Number of $\Gamma > \epsilon$	Average Γ	Max Γ
F1	0	504	25.1646	130.9834
F2	504	0	0.0000	0.0035
F3	504	0	0.0000	0.0033
F4	451	53	0.0089	0.3157
F5	249	255	2.9112	16.2970
F6	364	140	0.0926	1.1898
F7	376	128	0.0844	1.3799
F8	504	0	0.0000	0.0012
F9	290	214	10.0291	56.4448
F10	275	229	0.6743	3.6561
F11	457	47	0.4510	11.8847

Table 3. Γ measurements in 20-dimensions

Function	Number of $\Gamma \le \epsilon$	Number of $\Gamma > \epsilon$	Average Γ	Max Γ
F1	1	503	28.0691	154.5342
F2	504	0	0.0000	0.0012
F3	503	1	0.0001	0.0242
F4	474	30	0.0084	0.4301
F5	488	16	0.0044	0.2557
F6	430	74	0.1214	2.3300
F7	440	64	0.0648	1.2652
F8	504	0	0.0000	0.0050
F9	289	215	7.2412	35.7945
F10	326	178	0.5773	3.4318
F11	470	34	0.1958	7.0119

Table 4. Γ measurements in 30-dimensions

Function	Number of $\Gamma \le \epsilon$	Number of $\Gamma > \epsilon$	Average Γ	Max Γ
F1	0	504	30.0851	181.9311
F2	502	2	0.0001	0.0328
F3	501	3	0.0068	3.3226
F4	477	27	0.0068	0.3774
F5	504	0	0.0000	0.0035
F6	452	52	0.1015	2.5492
F7	459	45	0.0593	4.5566
F8	503	1	0.0001	0.0250
F9	276	228	6.0534	27.9352
F10	339	165	0.5417	3.4578
F11	475	29	0.1412	5.5532

rations that are order-2* stable in 5, 10, or 30 dimensions. In the 20 dimensional case only 1 configuration is classified as order-2* as illustrated in Table 3.

In 5, 10, and 20 dimensions all parameter configurations are order-2* stable for the Bent Cigar (F2) and High Conditioned Elliptic (F8) objective functions as illustrated in Tables 1, 2 and 3 respectively. However, in the 30 dimensional case, only for the Griewank (F5) function were all parameter configurations order-2* stable as illustrated in Table 4. Interestingly, for some objective functions, such as Griewank (F5), an increase in dimensionality makes order-2* stability more obtainable, where as the opposite was true for some objective functions, such as the Katsuura (F9) objective function.

It should also be noted that in the case where order-2* stability does not occur the actual particle variance may be relatively large, given the relationship between particle variance and Eq. (6). The most extreme case occurs with the Ackley (F1) objective function which had the largest maximum and average Γ measurements across all dimensions. Specifically, in 10 dimensions the average Γ value was 25.1646, which is 12.58% of the maximum component-wise distance, indicating that on average, particle behavior is far from order-2* stable. In 30 dimensions the average Γ value increased to 30.0851, a substantial 15% of the maximum component-wise distance as illustrated in Table 4. A similar, though less extreme situation occurs with the Katsuura (F9) objective function which has a relatively large average Γ value across 5, 10, 20, and 30 dimensions.

A snapshot of the parameter configurations for which order-2* stability occurred is provided for Ackley (F1), Griewank (F5), HappyCat (F6), High Conditioned Elliptic (F8), Katsuura (F9), and Rosenbrock (F11) in Figs. 2(a)–(f) for the 10 dimensional case and in Figs. 3(a)–(f) for the 30 dimensional case.

In the 10-dimensional case the Ackley (F1) and High Conditioned Elliptic (F8) objective functions present the most extreme behavior, as illustrated in Figs. 2(a) and (d) respectively. Under the Ackley (F1) objective function the particles are never order-2* stable whereas under the High Conditioned Elliptic (F8) objective function the particles are order-2* stable for all configurations satisfying equation (5). In all other snapshots, of the 10 dimensional case, there are some parameter configurations that do present as order-2* stable and some that do not, despite being both order-1 and order-2 stable. It is interesting to note that there is not a clear pattern present in Figs. 2(a)–(f) with regards to when order-2* stability will occur. Specifically, in case of the Katsuura (F9) and Griewank (F5) objective functions the region of order-2* stable parameter configurations excludes a curved region excluding many parameter configurations with low cognitive and social weight values as illustrated in Figs. 2(e) and (b). Whereas in the case of the HappyCat (F6) and Rosenbrock (F11) objective functions the region of order-2* stable parameter configurations excludes a number of configurations with large positive inertia values, while configuration with low cognitive and social coefficient weights are included as illustrated in Figs. 2(c) and (f). One of the most striking difference between the 10 and the 30 dimensional cases is that of Griewank (F5). In the 10 dimensions there are numerous configuration which do not exhibit order-2* stability, while in the 30 dimensions all configurations exhibit order-2* stability as illustrated in Figs. 2(b) and 3(b) respectively.

It should be noted that providing PSO with considerably more iterations than 5000 does not lead to order-2* stability presenting itself. A fact that is theoretically evident from Proposition 1. However, for a more practical illustration, consider PSO optimizing the Ackley (F1) objective function, with coefficients $w = 0.729$, $c_1 = 1.494$, and $c_2 = 1.494$, that satisfy the stability criteria of equation (5). The Γ measurement is plotted across 10^6 iteration, averaged over 50 runs, in Fig. 4. The $\Gamma(t)$ measurement clearly decreases as the iterations, t, increase. However, the rate of decrease is rapidly slowing, with $\Gamma(t)$ still over

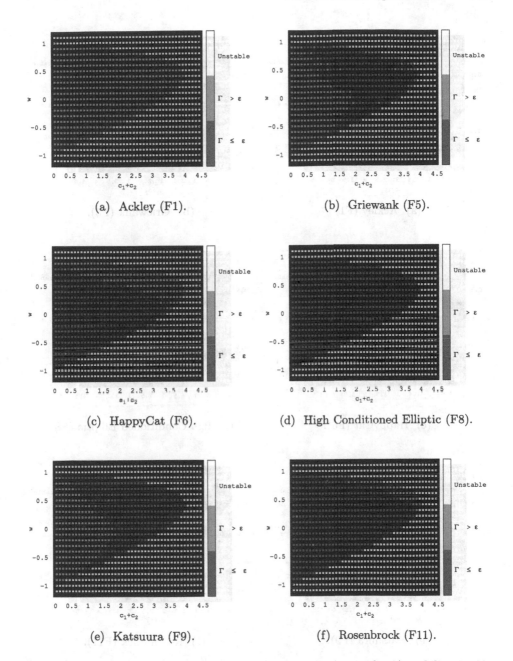

(a) Ackley (F1).

(b) Griewank (F5).

(c) HappyCat (F6).

(d) High Conditioned Elliptic (F8).

(e) Katsuura (F9).

(f) Rosenbrock (F11).

Fig. 2. Summary of parameter configurations that lead to order-2* stability in 10-dimensions. If a parameter configuration has $\Gamma \leq \epsilon$ it is said to be order-2* stable. Parameter configurations that violate the criteria for order-1 and order-2 stability as defined in Eq. (5) are marked as unstable. When $\Gamma > \epsilon$ and Eq. (5) is satisfied the parameter configuration is said to be order-1 and order-2 stable but not order-2* stable

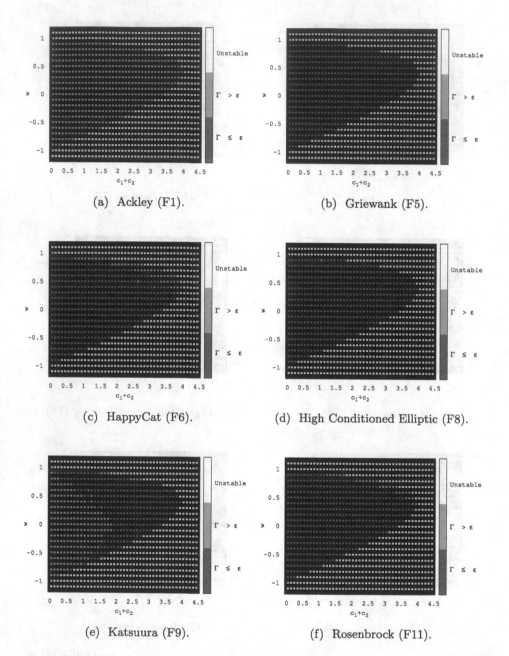

(a) Ackley (F1).

(b) Griewank (F5).

(c) HappyCat (F6).

(d) High Conditioned Elliptic (F8).

(e) Katsuura (F9).

(f) Rosenbrock (F11).

Fig. 3. Summary of parameter configurations that lead to order-2* stability in 30-dimensions. If a parameter configuration has $\Gamma \leq \epsilon$ it is said to be order-2* stable. Parameter configurations that violate the criteria for order-1 and order-2 stability as defined in Eq. (5) are marked as unstable. When $\Gamma > \epsilon$ and Eq. (5) is satisfied the parameter configuration is said to be order-1 and order-2 stable but not order-2* stable

8.7 after 10^6 iteration in the 10 dimensional case. While this is not proof that Γ cannot eventually become 0 on the Ackley (F1) objective function, there is certainly no guarantee of it given Proposition 1 for an arbitrary objective function. From a practical perspective it is clear that one should not expect order-2* stability to occur even if order-2 stability criteria are met.

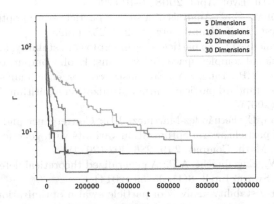

Fig. 4. Long term average $\Gamma(t)$ of a swarm on the Ackley (F1) objective function in 5, 10, 20, and 30 dimensions under the coefficients $w = 0.729$, $c_1 - 1.494$, and $c_2 = 1.494$.

What is clear from Tables 1, 2, 3 and 4 and Figs. 2 and 3 is that whether or not a swarm's particles will be order-2* stable is problem dependent. Specifically, in practice utilizing the criteria of equation (5) can only guarantee order-1 and order-2 stability and not the more restrictive order-2* stability which required a zero positions variance.

6 Conclusion

In this paper it is theoretically and empirically demonstrated which practical guarantees can in fact be made about particle stability in the particle swarm optimization (PSO) algorithm. Specifically, it is shown that the definition of order-2 stability which accurately reflects PSO behavior is that of convergence in second order moment to a constant, and not to zero. It was empirically shown that convergence in second order moment to 0 (order-2* stability), cannot be guaranteed by using the criteria for order-1 and order-2 stability. Furthermore, the occurrence of order-2* stability is shown to be problem dependent, and not directly controllable by coefficient selection, excluding of course the trivial case of zero valued social and cognitive weights.

This paper provided clarification on the meaning of order-2 stability and the guarantees on particle stability that can be made. This clarification will allow PSO practitioners to be better guided by the wealth of existing PSO theory.

References

1. Kennedy, J., Eberhart, R.: Particle swarm optimization. In: Proceedings of the IEEE International Joint Conference on Neural Networks, pp. 1942–1948. IEEE Press, Piscataway, NJ (1995)
2. Poli, R.: Analysis of the publications on the applications of particle swarm optimisation. J. Artif. Evol. Appl. **2008**, 1–10 (2008)
3. Ozcan, E., Mohan, C.K.: Analysis of a simple particle swarm optimization system. Intell. Eng. Syst. Artif. Neural Netw. **8**, 253–258 (1998)
4. Clerc, M., Kennedy, J.: The particle swarm-explosion, stability, and convergence in a multidimensional complex space. IEEE Trans. Evol. Comput. **6**(1), 58–73 (2002)
5. Jiang, M., Luo, Y.P., Yang, S.Y.: Stochastic convergence analysis and parameter selection of the standard particle swarm optimization algorithm. Inf. Process. Lett. **102**(1), 8–16 (2007)
6. García-Gonzalo, E., Fernández-Martinez, J.L.: Convergence and stochastic stability analysis of particle swarm optimization variants with generic parameter distributions. Appl. Math. Comput. **249**, 286–302 (2014)
7. Cleghorn, C.W., Engelbrecht, A.P.: A generalized theoretical deterministic particle swarm model. Swarm Intell. **8**(1), 35–59 (2014)
8. Liu, Q.: Order-2 stability analysis of particle swarm optimization. Evol. Comput. 1–30 (2014). https://doi.org/10.1162/EVCO_a_00129
9. Cleghorn, C.W., Engelbrecht, A.P.: Particle swarm stability: a theoretical extension using the non-stagnate distribution assumption. Swarm Intell. **12**(1), 1–22 (2018)
10. Trelea, I.C.: The particle swarm optimization algorithm: convergence analysis and parameter selection. Inf. Process. Lett. **85**(6), 317–325 (2003)
11. Van den Bergh, F., Engelbrecht, A.P.: A study of particle swarm optimization particle trajectories. Inf. Sci. **176**(8), 937–971 (2006)
12. Cleghorn, C.W., Engelbrecht, A.P.: Particle swarm optimizer: the impact of unstable particles on performance. In: Proceedings of the IEEE Symposium Series on Swarm Intelligence, pp. 1–7. IEEE Press, Piscataway, NJ (2016)
13. Cleghorn, C.W., Engelbrecht, A.P.: Fully informed particle swarm optimizer: convergence analysis. In: Proceedings of the IEEE Congress on Evolutionary Computation, pp. 164–170. IEEE Press, Piscataway, NJ (2015)
14. Cleghorn, C.W., Engelbrecht, A.P.: Unified particle swarm optimizer: convergence analysis. In: Proceedings of the IEEE Congress on Evolutionary Computation, pp. 448–454. IEEE Press, Piscataway, NJ (2016)
15. Cleghorn, C.W., Engelbrecht, A.P.: Fitness-distance-ratio particle swarm optimization: stability analysis. In: Proceedings of the Genetic and Evolutionary Computation Conference, pp. 12–18. ACM Press, New York, NY (2017)
16. Cleghorn, C.W., Engelbrecht, A.P.: Particle swarm variants: standardized convergence analysis. Swarm Intell. J. **9**(2–3), 177–203 (2015)
17. Bonyadi, M.R., Michalewicz, Z.: Analysis of stability, local convergence, and transformation sensitivity of a variant of particle swarm optimization algorithm. IEEE Trans. Evol. Comput. **20**(3), 370–385 (2016)
18. Shi, Y., Eberhart, R.C.: A modified particle swarm optimizer. In: Proceedings of the IEEE Congress on Evolutionary Computation, pp. 69–73. IEEE Press, Piscataway, NJ (1998)
19. Engelbrecht, A.P.: Particle swarm optimization: Global best or local best. In: Proceedings of the 1st BRICS Countries Congress on Computational Intelligence, pp. 124–135. IEEE Press, Piscataway, NJ (2013)

20. Bonyadi, M.R., Michalewicz, Z.: Particle swarm optimization for single objective continuous space problems: a review. Evol. Comput. **25**(1), 1–54 (2016)
21. Ozcan, E., Mohan, C.K.: Particle swarm optimization: surfing the waves. In: Proceedings of the IEEE Congress on Evolutionary Computation, vol. 3, pp. 1939–1944. IEEE Press, Piscataway, NJ (July 1999)
22. Kadirkamanathan, V., Selvarajah, K., Fleming, P.J.: Stability analysis of the particle dynamics in particle swarm optimizer. IEEE Trans. Evol. Comput. **10**(3), 245–255 (2006)
23. Gazi, V.: Stochastic stability analysis of the particle dynamics in the PSO algorithm. In: Proceedings of the IEEE International Symposium on Intelligent Control, pp. 708–713. IEEE Press, Piscataway (2012)
24. Poli, R., Broomhead, D.: Exact analysis of the sampling distribution for the canonical particle swarm optimiser and its convergence during stagnation. In: Proceedings of the Genetic and Evolutionary Computation Conference, pp. 134–141. ACM Press, New York, NY (2007)
25. Blackwell, T.: A study of collapse in bare bones particle swarm optimization. IEEE Trans. Evol. Comput. **16**(3), 354–372 (2012)
26. Bonyadi, M.R., Michalewicz, Z.: Stability analysis of the particle swarm optimization without stagnation assumption. IEEE Trans. Evol. Comput. **PP**(1), 1–7 (2015)
27. Poli, R.: Mean and variance of the sampling distribution of particle swarm optimizers during stagnation. IEEE Trans. Evol. Comput. **13**(4), 712–721 (2009)
28. Engelbrecht, A.P.: Roaming behavior of unconstrained particles. In: Proceedings of the 1st BRICS Countries Congress on Computational Intelligence, pp. 104–111. IEEE Press, Piscataway, NJ (2013)
29. Liang, J.J.K., Qu, B.Y., Suganthan, P.N.: Problem definitions and evaluation criteria for the CEC 2014 special session and competition on single objective real-parameter numerical optimization. Technical report 201311, Computational Intelligence Laboratory, Zhengzhou University and Nanyang Technological University (2013)
30. Kreyszig, E.: Introductory Functional Analysis with Applications. Wiley, Wiltshire (1978)
31. Cleghorn, C.W., Engelbrecht, A.P.: Particle swarm convergence: an empirical investigation. In: Proceedings of the Congress on Evolutionary Computation, pp. 2524–2530. IEEE Press, Piscataway, NJ (2014)

Fundamental Flowers: Evolutionary Discovery of Coresets for Classification

Pietro Barbiero[1] and Alberto Tonda[2,3]

[1] Politecnico di Torino, Torino, Italy
`pietro.barbiero@studenti.polito.it`
[2] Université Paris-Saclay, Saclay, France
[3] UMR 782 INRA, Thiverval-Grignon, France
`alberto.tonda@inra.fr`

Abstract. In an optimization problem, a *coreset* can be defined as a subset of the input points, such that a good approximation to the optimization problem can be obtained by solving it directly on the coreset, instead of using the whole original input. In machine learning, coresets are exploited for applications ranging from speeding up training time, to helping humans understand the fundamental properties of a class, by considering only a few meaningful samples. The problem of discovering coresets, starting from a dataset and an application, can be defined as identifying the minimal amount of samples that do not significantly lower performance with respect to the performance on the whole dataset. Specialized literature offers several approaches to finding coresets, but such algorithms often disregard the application, or explicitly ask the user for the desired number of points. Starting from the consideration that finding coresets is an intuitively multi-objective problem, as minimizing the number of points goes against maintaining the original performance, in this paper we propose a multi-objective evolutionary approach to identifying coresets for classification. The proposed approach is tested on classical machine learning classification benchmarks, using 6 state-of-the-art classifiers, comparing against 7 algorithms for coreset discovery. Results show that not only the proposed approach is able to find coresets representing different compromises between compactness and performance, but that different coresets are identified for different classifiers, reinforcing the assumption that coresets might be closely linked to the specific application.

Keywords: Machine learning · Coresets · Evolutionary computation · Explain AI · Multi-objective optimization

1 Introduction

The concept of *coreset*, originally from computational geometry, has been redefined in the field of machine learning (ML) as the minimal number of input samples from which a ML algorithm can obtain a good approximation of the

© Springer Nature Switzerland AG 2019
P. Kaufmann and P. A. Castillo (Eds.): EvoApplications 2019, LNCS 11454, pp. 550–564, 2019.
https://doi.org/10.1007/978-3-030-16692-2_37

behavior it would have on the whole original dataset. In other words, a coreset represents a fundamental subset of an available training set, that is sufficient for a given ML algorithm to obtain a good performance, or even the same performance it would have if trained on the whole training set [1]. This definition is intentionally generic, as it encompasses different tasks, ranging from classification, to regression, to clustering, that entail different measures of performance. The practical applications of coresets range from speeding up training time, to obtaining a better understanding of the data itself, making it possible for human experts to analyze only a few data points, instead of dealing with prohibitive amounts of samples.

Discovering coresets for a specific task is an open research line, and specialized ML literature reveals a considerable number of approaches, among which the most popular are Bayesian Logistic Regression (BLR [2]), Greedy Iterative Geodesic Ascent (GIGA [3]), Frank-Wolfe (FW [4]), Forward Stagewise (FSW [5]), Least-angle regression (LAR [6,7]), Matching Pursuit (MP [8]), and Orthogonal Matching Pursuit (OMP [9]). Interestingly, very often such algorithms require the user to specify the desired number N of points in the coreset; or assume, for simplicity, that a good coreset is independent from the task and/or the ML algorithm selected for that task. However, the problem of finding a coreset can be intuitively framed as multi-objective, as the quality of the results is probably dependent on the number of points included in the coreset; and minimizing the number of selected points goes against maximizing performance. Moreover, it also seems likely that different ML algorithms would actually need coresets of different size and shape to operate at the best of their possibilities.

Starting from these two intuitions, we present an evolutionary approach to coreset discovery for classification tasks. Starting from a given training set, a state-of-the-art multi-objective evolutionary algorithm, NSGA-II [10], is set to find the coresets representing the best trade-offs between amount of points (to be minimized) and classifier error (to be minimized), for a specific classification algorithm. The resulting Pareto front will then represent several coresets, each one a different optimal compromise between the objectives, and a human expert will then be able to not only select the coreset more suited for their needs, but also obtain extra information on the ML algorithm's behavior from observing its decrease in performance as the number of coreset points decreases.

Experimental results on two iconic classification benchmarks show that the proposed approach is able to best several state-of-the-art coreset discovery algorithms in literature, obtaining results that also allow the classifier to generalize better on an unseen test set, belonging to the same benchmark. Moreover, a meta-analysis of the results on different classifiers shows that indeed, while some of the points selected are common to different algorithms, the choice of points in the coreset is heavily dependent on the classifier selected for the classification task, with similarities between classifiers belonging to the same families.

2   Background

In this section, we briefly recall the basics of machine learning, coreset discovery, and multi-objective evolutionary algorithms, that are necessary to introduce the scope of our work.

2.1   Machine Learning and Classification

Machine learning techniques are defined as algorithms able to *improve their performance on a given task over time through experience* [11]. In other words, such algorithms create models that are trained on user-provided (training) data, and are then able to provide predictions on unseen (test) data. The essence of ML is to frame a learning task as an optimization task, to then find a near-optimal solution for the optimization problem, relying upon the training data. Popular machine learning techniques range from decision trees [12], to logistic regression [13], to artificial neural networks [14].

A classic task for ML is *classification*, where a single instance of measurements of several *features*, called *sample*, is to be associated to one (or more) pre-defined *classes*, representing different groups. Usually ML techniques position hyperplanes (also called *decision boundaries*) in the feature space, and later use them to decide how to classify a given sample. Such decision boundaries are placed to maximize technique-specific metrics, whose value depends on the efficacy of the boundary with respect the (labeled) training data.

2.2   Coresets

Coresets were originally studied in the context of computational geometry, and defined as a small set of points that approximates the shape of a larger point set. In ML the definition of coreset is extended to intend a subset of the input samples, such that a good approximation to the original input can be obtained by solving the optimization problem directly on the coreset, rather than on the whole original set of input samples [1].

Finding coresets for ML problems is an active line of research, with applications ranging from speeding up training of algorithms on large datasets [15] to gaining a better understanding of the algorithm's behavior. Unsurprisingly, a considerable number of approaches to coreset discovery can be found in the specialized literature. In the following, a few of the main algorithms in the field, that will be used as a reference during the experiments, are briefly summarized: Bayesian Logistic Regression (BLR [2]), Greedy Iterative Geodesic Ascent (GIGA [3]), Frank-Wolfe (FW [4]), Forward Stagewise (FSW [5]), Least-angle regression (LAR [6,7]), Matching Pursuit (MP [8]) and Orthogonal Matching Pursuit (OMP [9]).

BLR is based on the idea that finding the optimal coreset is too expensive. In order to overcome this issue, the authors use a k-clustering algorithm to obtain a compact representation of the data set. In particular, they claim that samples that are bunched together could be represented by a smaller set of points, while samples that are far from other data have a larger effect on inferences. Therefore, the BLR coreset is composed of few samples coming from tight clusters plus the outliers.

The original FW algorithm applies in the context of maximizing a concave function within a feasible region by means of a local linear approximation. In Sect. 4, we refer to the Bayesian implementation of the FW algorithm designed for core set discovery. This technique, described in [16], aims to find a linear combination of approximated likelihoods (which depends on the core set samples) that is similar to the full likelihood as much as possible.

GIGA is a greedy algorithm that further improves FW. In [3], the authors show that computing the residual error between the full and the approximated likelihoods by using a geodesic alignment guarantees a lower upper bound to the error at the same computational cost.

FSW [5], LAR [6,7], MP [8] and OMP [9] were all originally devised as greedy algorithms for dimensionality reduction. The simplest is FSW which projects high-dimensional data in a lower dimensional space by selecting one at a time the feature whose inclusion in the model gives the most statistically significant improvement. MP instead includes features having the highest inner product with a target signal, while its improved version OMP at each step carries an orthogonal projection out. Similarly, LAR increases the weight of each feature in the direction equiangular to each one's correlations with the target signal. All these procedures could be applied to the transpose problem of feature selection, that is approximation of core sets.

Very often these algorithms start from the assumption that the coreset for a given dataset will be independent from the ML pipeline used. This premise might not always be correct, as the optimization problem underlying, for example, a classification task, might vary considerably depending on the algorithm used. It is also important to notice that the problem of finding the coreset, given a specific dataset and an application, can be naturally expressed as multi-objective: on the one hand, the user wishes to identify a set of core points as small as possible; but on the other hand, the performance of the algorithm trained on the coreset should not differ from its starting performance, when trained on the original dataset. For this reason, multi-objective evolutionary algorithms could be well-suited to this task.

2.3   Multi-objective Evolutionary Algorithms

When dealing with problems that feature contrasting objectives, there is no single optimal solution to be found. Each candidate solution, in fact, represents a different compromise between conflicting aims. Nevertheless, it is still possible to find *optimal trade-offs*, for which it is not possible to improve an objective without degrading the others. The set of such optimal compromises is called Pareto front. Multi-objective evolutionary algorithms (MOEA) have been particularly successful in tackling problems with contradictory objectives and obtaining good approximations of the Pareto front. One of the most known MOEAs is the Non-Sorting Genetic Algorithm II (NSGA-II) [10], that exploits a crowding mechanism to evenly spread candidate solutions on the Pareto front, a procedure that is considerably efficient for problems with 2–3 objectives.

3   Proposed Approach

We propose to exploit evolutionary computation to explore the space of all subsets of samples in a given training dataset in a classification task, searching for those subset that do not reduce classification accuracy. Such samples would then represent a coreset, a collection of points that is necessary and sufficient to correctly define the decision boundaries for a target classifier, given a target dataset.

A candidate solution in our problem is a set of samples to be kept from the training set. The genome of an individual is thus defined as a binary array of size equal to the training set, with a '1' in a given position meaning that the sample corresponding to that index will become part of the coreset, and a '0' meaning that the sample will be removed.

Intuitively, it is easier to maintain a classifier's accuracy while keeping a large number of samples from its training set, rather than just a small quantity. For this reason, it is more appropriate to frame the optimization problem as multi-objective, with the conflicting aims of *i. minimizing the number of samples in the coreset*, and *ii. maximizing classifier's accuracy on the whole dataset*. The first objective is measured by simply counting the number of '1's, i.e. the number of samples in the coreset, in each candidate solution. For the second objective, the target classifier is trained on the reduced training set, and its accuracy is then tested on the whole set of training points, including all samples removed from the training set by the candidate solution. A scheme of the individual evaluation is presented in Fig. 1.

4   Experimental Evaluation

In order to empirically validate the proposed approach, we evaluate the methodology with 6 algorithms, chosen to be representative of both hyperplane-based and ensemble, tree-based classifiers: `BaggingClassifier` [17], `Gradient BoostingClassifier` [18], `LogisticRegression` [13], `RandomForest Classifier` [19], `RidgeClassifier` [20], `SVC` (Support Vector Machines) [21]. All classifiers are implemented in the `scikit-learn`[1] [22] Python module and use default parameters. A fixed seed has been set for all those that exploit pseudo-random elements in their training process, such as `RandomForestClassifier`, as for our objective it is important that the classifier will follow the same training steps, albeit with a reduced training set. A similar result could have been obtained by repeating multiple times the training process of classifiers containing random elements. In this first batch of experiments, that constitute our proof of concept, we selected the former option to reduce computational effort.

---

[1] scikit-learn: Machine Learning in Python, http://scikit-learn.org/stable/.

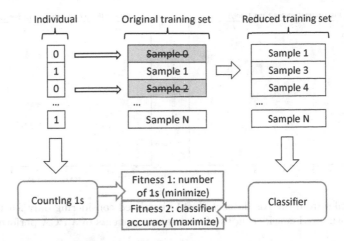

Fig. 1. Scheme of the proposed fitness evaluation. A given dataset is randomly split between training and test set. Candidate solutions, encoded as binary strings, are used to reduce the training set, by removing samples whose index corresponds to a '0' in the binary string. The target classifier is then trained on the reduced training set, while its accuracy is measured on all the original training samples.

The implementation of NSGA-II, used during the evolutionary process, is taken from the `inspyred`[2] [23] Python module. NSGA-II uses a binary tournament selection, $\mu = 100$, $\lambda = 100$, a one-point crossover, a bit-flip mutation, and a stop condition set at 100 generations. Each individual can remove between 1 and 90 samples from the training dataset, so valid candidate genomes can contain between 1 and 90 '0's.

The results obtained by the proposed approach are then compared against the 7 coreset discovery algorithms BLR [2], GIGA [3], FW [4], MP [9], OMP [9], LAR [6, 7], and FSW [5], described in more detail in SubSect. 2.2. Some of the algorithms require the user to specify the number N of desired points in the coreset: in order to provide a fair comparison, N is set to the size of the highest-accuracy coreset found by the proposed approach in the corresponding experiment.

All experiments are performed on the well-known Iris dataset [24], comprising 150 samples from 3 different classes. The samples are randomly split between a 99-sample training set and a 51-sample test set. As shown in Table 1, all considered classifiers perform reasonably well on the dataset, as Iris is a benchmark that presents no particular difficulty for most algorithms. In a first batch of experiments (Iris-2), we consider only two features of the dataset, in order to offer the reader a more intuitive visual assessment of the results. In a second set of tests (Iris-4), we consider all four features, and rely upon *Principal Component Analysis* (PCA) to support the visualization and discussion. The datasets are shown in Fig. 2. The code used in this work is freely available in the BitBucket repository https://bitbucket.org/evomlteam/evolutionary-core-sets.

---

[2] inspyred: Bio-inspired Algorithms in Python, https://pythonhosted.org/inspyred/.

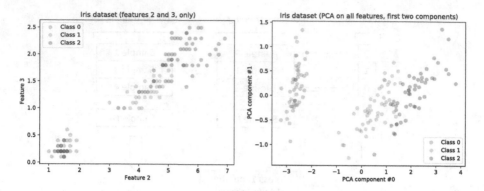

Fig. 2. Samples from the Iris dataset **(left)** visualized considering only features 2 and 3, **(right)** visualized considering the first two components of a PCA performed on all features.

Table 1. Initial performance (classification accuracy) of the considered classifiers on the Iris dataset (using two features and all, respectively).

Classifier	Iris (2 features)			Iris (all features)		
	Overall	Training	Test	Overall	Training	Test
GradientBoostingClassifier	0.9800	0.9899	0.9608	0.9867	1.0	0.9608
BaggingClassifier	0.9733	0.9899	0.9412	0.9867	1.0	0.9608
LogisticRegression	0.8533	0.8081	0.9412	0.9733	0.9596	1.0
RandomForestClassifier	0.9733	0.9899	0.9412	0.9867	1.0	0.9608
RidgeClassifier	0.8000	0.7677	0.8627	0.8667	0.8586	0.8824
SVC	0.9733	0.9697	0.9804	0.9733	0.9697	0.9804

4.1 Iris 2-Feature Validation

The first column of Figs. 3 and 4 shows the Pareto front obtained by the proposed approach for all the classifiers on Iris-2. It is noticeable how the trade-offs found are different in both size and accuracy, depending on the considered algorithm. The second and third columns of the figures portray the frequency of appearance of samples in the training set inside the candidate solutions on the Pareto fronts of the corresponding experiment. It is easy to observe that several points appear among all candidate coresets found in the same experiment.

Table 2. Test performance of the considered classifiers and coreset algorithms on the Iris-2 dataset.

	Training	EvoCore	BLR	GIGA	FW	MP	OMP	FSW	LAR
BaggingClassifier	0.9798	1.0000	0.8824	0.8824	0.8039	0.9412	0.9412	0.9412	0.9412
GradientBoostingClassifier	0.9596	0.9804	0.8627	0.8431	0.8824	0.9608	0.9608	0.9608	0.9608
LogisticRegression	0.9697	0.9804	0.6471	0.7059	0.6667	0.9804	0.9804	0.9804	0.9804
RandomForestClassifier	0.9798	0.9804	0.9020	0.8824	0.8039	0.9412	0.9412	0.9412	0.9412
RidgeClassifier	0.9798	0.9608	0.6667	0.6667	0.6667	0.9020	0.9020	0.9020	0.9020
SVC	0.9697	0.9804	0.6667	0.8824	0.8627	0.9412	0.9412	0.9412	0.9412
Average	**0.9731**	**0.9804**	**0.7712**	**0.8105**	**0.7810**	**0.9444**	**0.9444**	**0.9444**	**0.9444**

Table 3. Test performance of the considered classifiers and coreset algorithms on the Iris-4 dataset.

	Training	EvoCore	BLR	GIGA	FW	MP	OMP	FSW	LAR
BaggingClassifier	0.9596	0.9804	0.7059	0.9412	0.8824	0.9608	0.4118	0.8431	0.4118
GradientBoostingClassifier	0.9697	0.9804	0.7647	0.7255	0.8824	0.8824	0.6471	0.8824	0.6471
LogisticRegression	0.9798	1.0000	0.7843	0.8824	0.8039	0.8431	0.8039	0.7059	0.8039
RandomForestClassifier	0.9394	0.9608	0.7255	0.8627	0.9020	0.9804	0.7647	0.7843	0.7647
RidgeClassifier	0.9798	1.0000	0.7255	0.9020	0.8824	0.8824	0.7451	0.7451	0.7451
SVC	0.9697	0.9608	0.7843	0.9608	0.9608	0.9608	0.6275	0.6471	0.6275
Average	**0.9663**	**0.9804**	**0.7484**	**0.8791**	**0.8856**	**0.9183**	**0.6667**	**0.7680**	**0.6667**

Table 2 presents the accuracy of the most accurate coreset found by the proposed approach, against coresets discovered by state-of-the-art algorithms in ML literature. Not only the proposed approach outperforms the accuracy of all algorithms on the training set, but it also enables the classifiers to create decision boundaries that generalize better, obtaining improved results on the unseen test set as well.

4.2   Iris 4-Feature Validation

The first column of Figs. 5 and 6 shows the Pareto front obtained by the proposed approach for all the classifiers on Iris-4. Again, the trade-offs found appear different in both size and accuracy, depending on the considered algorithm. As we observed for Iris-2, the second and third columns of the figures portraying the frequency of appearance of samples in the training set inside the candidate solutions on the Pareto fronts show several points appearing among all candidate coresets found in the same experiment.

Table 3 presents the accuracy of the most accurate coreset found by the proposed approach, against coresets discovered by state-of-the-art algorithms in ML literature. Again, the proposed approach outperforms the accuracy of all algorithms on both the training set and the unseen test set.

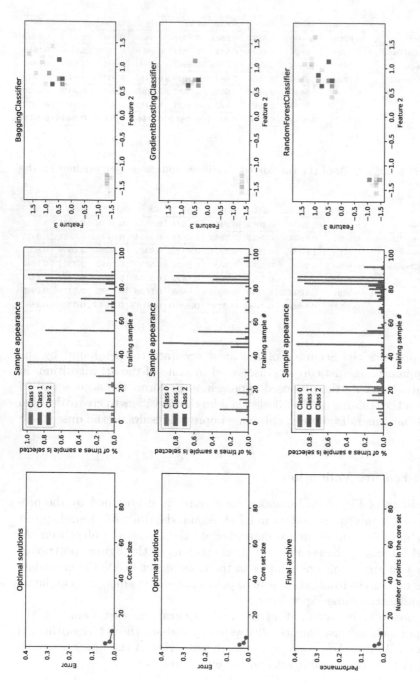

Fig. 3. For each experiment on dataset Iris-2, Pareto fronts showing optimal solutions (**left**), bar plots (**center**) and scatter plots (**right**) displaying the frequency of appearance of samples inside coresets. Bar height in bar plots and color saturation in scatter plots are directly proportional to the frequency of appearance of a specific sample. The first row refers to BaggingClassifier, the second one to GradientBoostingClassifier, and the third one to RandomForestClassifier.

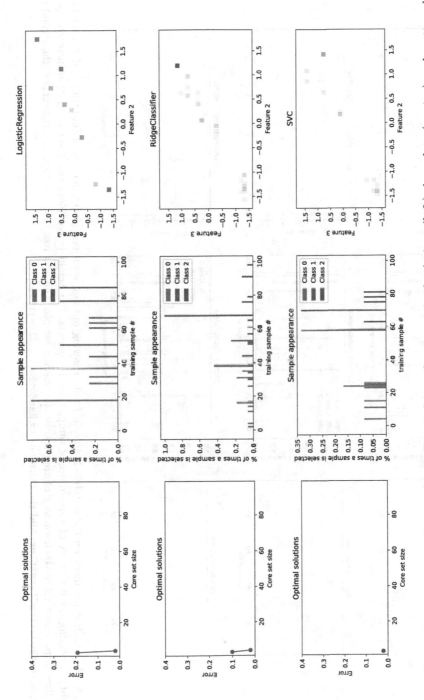

Fig. 4. For each experiment on dataset Iris-2, Pareto fronts showing optimal solutions (**left**), bar plots (**center**) and scatter plots (**right**) displaying the frequency of appearance of samples inside coresets. Bar height in bar plots and color saturation in scatter plots are directly proportional to the frequency of appearance of a specific sample. The first row refers to LogisticRegression, the second one to RidgeClassifier, and the third one to SVC.

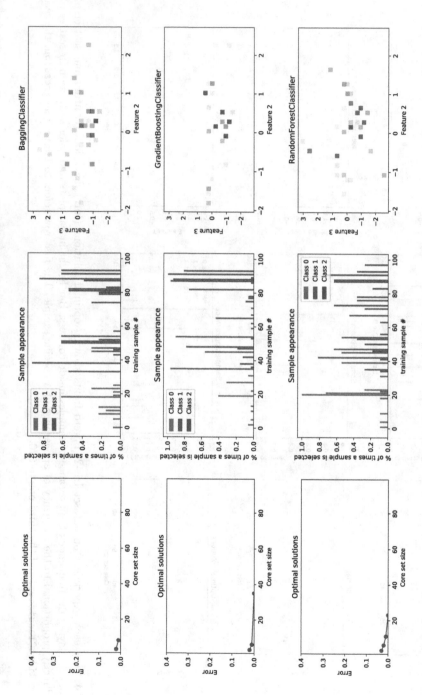

Fig. 5. For each experiment on dataset Iris-4, Pareto fronts showing optimal solutions (**left**), bar plots (**center**) and scatter plots (**right**) displaying the frequency of appearance of samples inside coresets. Bar height in bar plots and color saturation in scatter plots are directly proportional to the frequency of appearance of a specific sample. The first row refers to BaggingClassifier, the second one to GradientBoostingClassifier, and the third one to RandomForestClassifier.

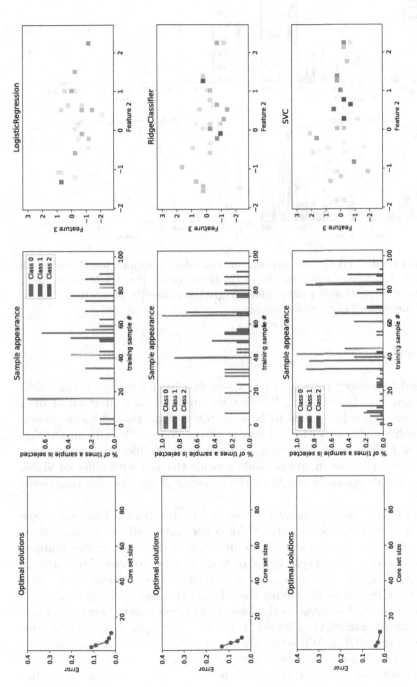

Fig. 6. For each experiment on dataset Iris-4, Pareto fronts showing optimal solutions (**left**), bar plots (**center**) and scatter plots (**right**) displaying the frequency of appearance of samples inside coresets. Bar height in bar plots and color saturation in scatter plots are directly proportional to the frequency of appearance of a specific sample. The first row refers to LogisticRegression, the second one to RidgeClassifier, and the third one to SVC.

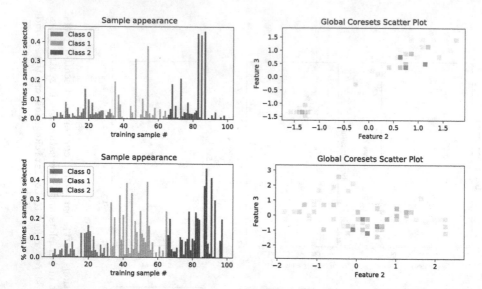

Fig. 7. Global results, taking into account all candidate solutions found during all experiments on Iris-2 (**top row**) and Iris-4 (**bottom row**). Bar height in bar plots (**left**) and color saturation in scatter plots (**right**) are directly proportional to the frequency of appearance of a specific sample.

5   Discussion

The proposed framework proved to be extremely effective in discovering high-quality coresets for classification. Framing the problem as a function of both application and algorithm seems to be a better method than a more general approach, as highlighted by the differences in the coresets found using different classifiers. Furthermore, obtaining a Pareto front of different compromises, instead of a single solution, might not only provide the user with different viable alternatives, but also provide insight on the behavior of the classifier considered for the task.

Nevertheless, this methodology presents a few drawbacks. The main issue of the approach is computational time. All other coreset discovery algorithms in literature are able to return solutions in a few seconds on regular laptops, while the proposed method typically takes from minutes to tens of minutes. In fact, as the MOEA trains a classifier for each individual evaluation, everything depends on the classifier itself. Among the selected classifiers, a huge variance in training time is noticable between the fastest (`RandomForestClassifier`) and the slowest (`GradientBoostingClassifier`), with corresponding repercussions on the speed of the MOEA. While this problem can be mitigated by parallelizing evaluations, we argue that even the current performance is not a great obstacle, as usually coresets are computed once, and then used multiple times. Given that not only the quality of the coresets discovered by the MOEA is higher, but that also several trade-offs are delivered in place of a single solution, we contend that the proposed approach is a preferable alternative to other solutions in literature.

From an overall analysis of the results on the two benchmarks Iris-2 and Iris-4, it is noticeable that, while points frequently appearing in candidate solutions on the Pareto front are generally different between classifiers, algorithms based on decision trees (`RandomForestClassifier`, `GradientBoosting Classifier`, `BaggingClassifier`) and algorithms based on linear hyperplanes (`LogisticRegression`, `RidgeClassifier`, `SVC`) tend to use similar points, see the second and third column of Figs. 3, 4, 5, and 6. The two families seem to prefer points on the boundaries between classes (decision trees) or close to the class centroid (linear hyperplanes), respectively. Furthermore, taking into account all coresets found during all the experiments, Fig. 7 shows how, despite differences between algorithms, a few samples are consistently selected among most of the candidate solutions found. Such samples might have a relevance going beyond the scope of the single technique, being instead excellent representatives of the class they belong to. Not surprisingly, the most common samples for each class are in part close to the class centroid, and in part close to the class boundaries, fully defining the shape of the training samples of the class.

6   Conclusions and Future Works

In this work, a novel methodology for coreset discovery in classification tasks is presented. The proposed approach relies on multi-objective evolutionary optimization to find the best compromises between size of the coreset (to be minimized) and classification accuracy (to be maximized). The approach is shown to outperform state-of-the-art coreset discovery algorithms in literature on classical classification benchmarks, and despite its considerably longer computational time, we argue that it is a preferable alternative as the quality of the coresets found is higher, and the algorithm is also more informative for the user.

Future works will analyze the behavior of evolutionary core set discovery for high-dimensional datasets, where visual expert validation is impractical, and explore the possibilities of the proposed approach for regression tasks.

References

1. Bachem, O., Lucic, M., Krause, A.: Practical coreset constructions for machine learning. arXiv preprint arXiv:1703.06476 (2017)
2. Huggins, J.H., Campbell, T., Broderick, T.: Coresets for scalable bayesian logistic regression. In: 30th Annual Conference on Neural Information Processing Systems (NIPS) (2016). https://arxiv.org/pdf/1605.06423.pdf
3. Campbell, T., Broderick, T.: Bayesian coreset construction via greedy iterative geodesic ascent. In: International Conference on Machine Learning (ICML) (2018). https://arxiv.org/pdf/1802.01737.pdf
4. Clarkson, K.L.: Coresets, sparse greedy approximation, and the Frank-Wolfe algorithm. ACM Trans. Algorithms 6(4), 63 (2010). http://citeseerx.ist.psu.edu/viewdoc/download?doi=10.1.1.145.9299&rep=rep1&type=pdf

5. Efroymson, M.A.: Multiple regression analysis. In: Ralston, A., Wilf, H.S. (eds.) Mathematical Methods for Digital Computers. Wiley, New York (1960)

6. Efron, B., Hastie, T., Johnstone, I., Tibshirani, R.: Least angle regression. Ann. Stat. **32**(2), 407–451 (2004). https://arxiv.org/pdf/math/0406456.pdf

7. Boutsidis, C., Drineas, P., Magdon-Ismail, M.: Near-optimal coresets for least-squares regression. Technical report (2013). https://arxiv.org/pdf/1202.3505.pdf

8. Mallat, S., Zhang, Z.: Matching pursuits with time-frequency dictionaries. IEEE Trans. Signal Process. **42**(12), 3397–3415 (1993)

9. Pati, Y., Rezaiifar, R., Krishnaprasad, P.: Orthogonal matching pursuit: recursive function approximation with applications to wavelet decomposition. In: Proceedings of 27th Asilomar Conference on Signals, Systems and Computers, pp. 40–44 (1993). http://ieeexplore.ieee.org/document/342465/

10. Deb, K., Pratap, A., Agarwal, S., Meyarivan, T.: A fast and elitist multiobjective genetic algorithm: NSGA-ii. IEEE Trans. Evol. Comput. **6**(2), 182–197 (2002)

11. Samuel, A.L.: Some studies in machine learning using the game of checkers. IBM J. Res. Dev. **3**(3), 210–229 (1959)

12. Breiman, L., Friedman, J., Stone, C.J., Olshen, R.A.: Classification and Regression Trees. CRC Press, Boca Raton (1984)

13. Cox, D.R.: The regression analysis of binary sequences. J. Roy. Stat. Soc.: Ser. B (Methodol.) **20**(2), 215–242 (1958)

14. Goodfellow, I., Bengio, Y., Courville, A.: Deep Learning. MIT press, Massachusetts (2016)

15. Tsang, I.W., Kwok, J.T., Cheung, P.M.: Core vector machines: fast SVM training on very large data sets. J. Mach. Learn. Res. **6**, 363–392 (2005)

16. Campbell, T., Broderick, T.: Automated Scalable Bayesian Inference via Hilbert Coresets (2017). http://arxiv.org/abs/1710.05053

17. Breiman, L.: Pasting small votes for classification in large databases and on-line. Mach. Learn. **36**(1–2), 85–103 (1999)

18. Friedman, J.H.: Greedy function approximation: a gradient boosting machine. Ann. Stat. **29**(5), 1189–1232 (2001)

19. Breiman, L.: Random forests. Mach. Learn. **45**(1), 5–32 (2001)

20. Tikhonov, A.N.: On the stability of inverse problems. Dokl. Akad. Nauk SSSR. **39**, 195–198 (1943)

21. Hearst, M.A., Dumais, S.T., Osman, E., Platt, J., Scholkopf, B.: Support vector machines. IEEE Intell. Syst. Appl. **13**(4), 18–28 (1998)

22. Pedregosa, F., et al.: Scikit-learn: machine learning in Python. J. Mach. Learn. Res. **12**, 2825–2830 (2011)

23. Garrett, A.: inspyred (version 1.0.1) inspired intelligence (2012). https://github.com/aarongarrett/inspyred

24. Fisher, R.A.: The use of multiple measurements in taxonomic problems. Ann. Eugenics **7**(2), 179–188 (1936)

Robotics

Influence of Local Selection and Robot Swarm Density on the Distributed Evolution of GRNs

Iñaki Fernández Pérez and Stéphane Sanchez[✉]

University of Toulouse - IRIT - CNRS - UMR5505, Toulouse, France
fernandezperez.inaki@gmail.com, stephane.sanchez@irit.fr

Abstract. Distributed Embodied Evolution (dEE) is a powerful approach to learn behaviors in robot swarms by exploiting their intrinsic parallelism: each robot runs an evolutionary algorithm, and locally shares its learning experience with other nearby robots. Given the distributed nature of this approach, dEE entails different evolutionary dynamics when compared to standard centralized Evolutionary Robotics. In this paper, we investigate the distributed evolution of Gene Regulatory Networks (GRNs) as controller representation to learn swarm robot behavior, which have been extensively used for the evolution of single-robot behavior with remarkable success. Concretely, we use dEE to evolve fixed-topology GRN swarm robot controllers for an item collection task; this constitutes the first work to evolve GRNs in distributed swarm robot settings. To improve our understanding of such distributed GRN evolution, we analyze the fitness and the behavioral diversity of the swarm over generations when using 5 levels of increasing local selection pressure and 4 different swarm sizes, from 25 to 200 robots. Our experiments reveal that there exist different regimes, depending on the swarm size, in the relationship between local selection pressure, and both behavioral diversity and overall swarm performance, providing several insights on distributed evolution. We further use a metric to quantify selection pressure in evolutionary systems, which is based on the correlation between number of offspring and fitness of the behaviors. This reveals a complex relationship on the overall selection pressure between the ability or ease to spread genomes (or environmental pressure), and the fitness of the behavior (or task-oriented (local) pressure), opening new research questions. We conclude the paper by discussing the need for developing specialized statistical tools to facilitate the analysis of the large and diverse amount of data relevant to distributed Embodied Evolution.

Keywords: Distributed Embodied Evolution · Swarm robotics ·
Gene Regulatory Networks · Evolutionary Robotics ·
Behavioral diversity

© Springer Nature Switzerland AG 2019
P. Kaufmann and P. A. Castillo (Eds.): EvoApplications 2019, LNCS 11454, pp. 567–582, 2019.
https://doi.org/10.1007/978-3-030-16692-2_38

1   Introduction

Evolutionary approaches have been widely used to optimize robot controllers, in the shape of Artificial Neural Networks (ANNs) and Gene Regulatory Networks (GRNs), among other representations. On the other hand, distributed Embodied Evolution (dEE) refers to a family of evolutionary algorithms to learn behaviors in a swarm of robots. These methods exploit the parallelism in the swarm by having each robot run a separate instance of an evolutionary algorithm with an internal local population and exchanging the results of such learning when meeting other robots (generally genomes and fitness values). dEE results in different evolutionary dynamics as compared to standard centralized evolution, because the internal populations on which selection is applied are built over time through local communication, and depend thus on genome spread, which is a result of a complex interaction between a number of factors, *e.g.* communication range and frequency, swarm density, or obstacle density.

GRNs have widely shown their usefulness in artificial evolution as controller representation to evolve robot behavior, in addition to other applications. As such, and as part of a research to gain insights on the benefits and disadvantages in dEE of different controller representations (*e.g.* ANNs or GRNs), we perform a set of experiments where a swarm of simulated robots adapts to an item collection task by evolving GRNs with distributed evolution.

In this paper, we investigate the impact on the quality (*i.e.* performance, fitness), and on the behavioral diversity (*i.e.* how different the actual behaviors performed by the robots in the swarm are at a given generation), of the swarm size (number of robots) and task-driven selection strength (of the local selection operator), when evolving GRNs with a standard distributed Embodied Evolutionary algorithm. Concretely, we use dEE to evolve fixed-topology GRN swarm robot controllers for an item collection task; this constitutes the first work to evolve GRNs in distributed swarm robot settings. We measure the quality of the behaviors, as well as the behavioral diversity of the swarm over generations when using 5 levels of increasing selection strength for a local selection tournament operator, parametrized by $\theta_{sp} \in \{0, .25, .5.75, 1\}$ (from random selection to local elitism), and 4 different swarm sizes, 25, 50, 100, and 200 robots.

Our experiments reveal complex interactions between genome spread (as favored by higher robot swarms), and local, task-driven selection pressure, with respect to both behavioral diversity and swarm fitness, which provides several insights on distributed evolution. For example, behavioral diversity and performance tends to be more unstable and have a larger variance in smaller swarms than in bigger ones. To further analyze our results, we use a recent metric to quantify selection pressure in evolutionary systems, which is based on the correlation between number of offspring and fitness of the behaviors, and accounts for the decentralized settings of dEE.

This follow-up analysis reveals the existence of a complex relationship on the overall selection pressure between the ability or ease to spread genomes (or environmental pressure), and the fitness of the behavior (or task-oriented (local) pressure), which opens new research questions.

Since robot density, as one of the possible factors promoting faster genome spread, increases environmental selection pressure, this interacts in a non-linear manner with task-driven local selection pressure, thus impacting the process of distributed adaptation, which is a testable hypothesis for further research. We conclude the paper by making a claim for the need for developing specialized statistical tools to facilitate the analysis of the large and diverse amount of data relevant to distributed Embodied Evolution.

2   Related Work

Distributed Embodied Evolution [1] is an algorithmic family for learning swarm robot control. In these approaches, each robot runs on-board a separate instance of an Evolutionary Algorithm (EA), including a local population for each one of them. This population is progressively built by exploiting the intrinsic parallelism of the robot swarm: as a robot is evaluating its current controller's fitness, it locally broadcasts both the genome corresponding to the controller, and the current fitness estimate. This sharing of learning experience could be a sort of social learning, where learning experience is shared among individuals when meeting, which is known as *mating* operator, specific of dEE. Selection and variation are locally applied by each robot based on its corresponding local population. While these algorithms are relatively new approaches to learn swarm behavior, several insights have been gained regarding the different evolutionary dynamics that this approach entails. For example, in [2,3], the authors show that, in some settings, the higher the intensity of selection pressure applied by robots on local population, the higher the performance reached by the swarm, which is opposed to the centralized case. The authors further show that this is due to better diversity maintenance in dEE. In [4], the authors investigate the evolution of altruism in dEE for swarms of robots without any objective or fitness, *i.e.* only with environmental pressure pushing toward behaviors that spread more genomes. In [5], using dEE, the authors successfully evolve behaviors for a complex item collection task that requires coordination between multiple robots, which is challenging for a distributed algorithm. Since fitness must be computed at the level of individual robots, in collaborative tasks, it may be impossible to assign adequate rewards to robots successfully engaging in the collaboration, an instance of the spatial credit-assignment problem [6].

Gene Regulatory Networks (GRN) are biological structures that control the internal behavior of living cells. They regulate gene expression by enhancing and inhibiting the transcription of certain parts of the DNA. For this purpose, the cells use protein sensors dispatched on their membranes; these provide crucial information to guide the cells through their cycle. Many modern computational models of these networks exist and the capability of artificial GRNs controllers has been displayed in different studies. Joachimczak and Wrobel used GRNs in a robotic foraging problem [7], where a two-wheeled robot must collect randomly placed food particles and avoid poisonous ones. The authors in [8] showed the capabilities of a GRN on a single pole problem. The authors in [9,10] evolved

GRN embodied controllers to perform phototaxis with two wheeled robots. Cussat-Blanc and Harrington [11] demonstrated competitive performance with SARSA on a suite of common reinforcement learning benchmarks, including mountain car, maze navigation, and acrobat. In [12] a GRN controller is evolved to competitively drive a virtual racing car in TORCS game. These former works mainly focused on using GRN to control a unique robot or agent. Some works use GRNS as controller in multi-robots or multi-agents setup. In [13] multi-cellular creatures are developed from single cells with individual GRN controllers. [14] evolved GRNs as decentralized controllers to achieve a simple clustering task with underwater robots. In [15], GRNs regulate high-level behaviors of a team two-wheeled robots to produce efficient and robust interception strategies.

This paper is the follow-up of our previous study [16] on using embodied evolved GRN controllers in a swarm of robots performing an item collection task.

3 Methodology

Our goal in this work is to analyze the results of the of GRNs as robot control formalism to be evolved using a distributed Embodied Evolutionary approach. Here, we describe the computational properties of the specific type GRN used in this study, as well as its corresponding encoding for evolution and variation (mutation and crossover) evolutionary operators. Subsequently, we describe the distributed EE algorithm (mEDEA [17] with local selection pressure [3]), run by each robot in our experiments, while focusing on mating and (local) selection operators, and the differences with more classical centralized EAs. Finally, we describe the metrics used in our experiments to answer our research questions, in terms of overall performance, behavioral diversity, and selection pressure trends.

3.1 GRN and Encoding

The artificial GRN model used in this work is based on Banzhaf's model [18]. It has been designed for computational purposes only and not to simulate biological networks. A detailed description of this model with involved dynamics equations and a complete study of its variants, can be found in [19].

The artificial GRN is composed of multiple proteins, which interact via evolved properties: developing a GRN genotype into its corresponding phenotype yields a function based on the combination of elementary computations, which maps a set of *protein* inputs (sensors) to a set of *protein* outputs (motors). These properties, called tags, are: the protein *identifier*, the *enhancer identifier* used to calculate the enhancing matching factor between two proteins, the *inhibitor identifier* used to calculate the inhibiting matching factor between two proteins. The identifiers are real values between 0.0 and 1.0. The proteins also have a non-evolved *type*, either *input*, *output*, or *regulator*, that indicates their role in the network.

Each protein has a concentration, representing the use of this protein and proving state to the network. For *input* proteins, the concentration is given by the environment and is unaffected by other proteins. *output* protein concentrations are used to determine actions in the environment; these proteins do not affect others in the network. The bulk of the computation is performed by *regulatory* proteins, an internal protein whose concentration is influenced by other *input* and *regulatory* proteins. With this structure, the dynamics of the GRN are computed by using the protein tags in order to provide output concentrations.

One computational step of the GRN changes protein i concentration as in the following differential equation, with the affinity metric a (that estimates matching between proteins), the influence function f (that determines the influence of one protein onto another), and the normalization function n:

$$\frac{dc_i}{dt} = n(\frac{\delta}{N} \sum_{j}^{N} c_j(f(a^+(i,j)) - f(a^-(i,j)))) \tag{1}$$

where N is the number of proteins in the network and δ is one of the two constants that determine the speed of reaction of the GRN.

In [19], the agent in the Ship escape experiment is similar to our robots. We will use the same optimal set of a, f and n functions.

a is the original affinity metric as in [18]:

$$a^+(i,j) = \beta(1.0 - |enh_j - id_i| - u_{max}^+)$$
$$a^-(i,j) = \beta(1.0 - |inh_j - id_i| - u_{max}^-) \tag{2}$$

where id_x is the identifier, enh_x is the enhancer identifier and inh_x is the inhibitor identifier of protein x, u_{max}^+ is maximum enhancing and u_{max}^- is maximum inhibiting affinities between all protein pairs, β is one of the two constants that determine the speed of reaction of the GRN. Constants β and δ determine the speed of reaction of the regulatory network. The higher these values, the more sudden the transitions in the GRN. The lower they are, the smoother the transitions. For this paper, they are evolved as part of the GRN chromosome and are both kept within the range $[0.5, 2.0]$.

As in [20], f is a hyperbolic tangent function:

$$f(a(i,j)) = tanh(a(i,j)) + 1 \tag{3}$$

n, the normalization of protein concentrations at each step, is such that the output and regulatory protein concentrations sum to 1. This makes the output layer of the GRN function similarly to the softmax layer of modern ANNs.

In this paper, GRNs are evolved with a distributed evolutionary algorithm. The crossover operator relies on an alignment (based on the one in NeuroEvolution of Augmenting Topologies, NEAT [21]) between two selected parent GRNs, as proposed in Gene Regulatory Network Evolution Through Augmenting Topologies (GRNEAT [22]) and modified in [19]. The alignment crossover compares individual genes before selection for a new individual with the distance metric $D_{prot}(i,j)$:

$$D_{prot}(i,j) = a|id_i - id_j| + b|enh_i - enh_j| + c|inh_i - inh_j| \qquad (4)$$

where $a = 0.75$, $b = 0.125$, and $c = 0.125$.

Proteins were aligned in each GRN during comparison first based on *type* and secondly based on minimum $D_{prot}(A, B)$. Aligned proteins are randomly selected from either parent with probability $p_{cross} = 0.5$. The only mutation operation used during this work is an equiprobable modification of the identifier id_x, of the enhancer identifier enh_x or of the inhibitor identifier inh_x of a randomly chosen protein x.

The algorithm used in our experiments corresponds to mEDEA with task-driven selection pressure [2,17]. Each robot in the swarm runs an independent instance of the algorithm. At every moment, a robot carries an active genome corresponding to its current neurocontroller, which is randomly initialized at the beginning of each experiment. A robot executes its controller for some time T_e, while estimating its fitness and continuously broadcasting the active genome and its current fitness estimate to other nearby robots (and vice versa). Once T_e timesteps are elapsed, the robot stops and selects a parent genome using a given selection operator. The selected genome is mutated and replaces the active genome (no crossover is used), the local population l is emptied, and a new generation starts.

$g_a \leftarrow random()$;
while *true* **do**
    $l \leftarrow \emptyset, f \leftarrow 0$;
    for $t \leftarrow 1$ *to* T_e **do**
        $exec(g_a)$;
        $f \leftarrow evaluate()$;
        $broadcast(g_a, f)$;
        $l \leftarrow l \bigcup listen()$;
    end
    $l \leftarrow l \bigcup \{(g_a, f)\}$;
    $selected \leftarrow select(l)$;
    $g_a \leftarrow mutate(selected)$;
end

Algorithm 1. Vanilla distributed Embodied Evolutionary algorithm run by every robot in the swarm. It corresponds to mEDEA with task-driven selection pressure.

We use parametrized tournament selection operator, that, given a parameter $\theta_{sp} \in [0, 1]$ and a local population, selects the genome with the best fitness in a random θ_{sp} fraction of the population. The parameter θsp influences selection pressure by determining the actual tournament size, and the higher the tournament size, the stronger the selection pressure. If $\theta_{sp} = 0$, the fitness is disregarded and selection is random, while if $\theta_{sp} = 1$, the best genome in the population is selected (maximal selection pressure). Each experiment consists in running this algorithm for an item collection task, with a given θ_{sp}, and swarm size $|S|$.

At each generation, in addition to measuring the swarm's average fitness, we measure behavioral diversity using the metric of dispersion among a set of behaviors proposed in [3]. It is based on a distance metric that measures how different are the motor outputs computed by two neurocontrollers when confronted with the same set of inputs, averaged between each pair of behaviors in the swarm. We use this diversity metric to evaluate at each generation how diverse are the behaviors at the level of the swarm.

Finally, we use the P_τ metric for quantifying the intensity of selection pressure in evolutionary systems, recently proposed in [23]. It uses a Kendall-τ-b correlation test with correction for ties, to measure the correlation between number of surviving offspring and a quantifiable trait or quantity. Here, we use the fitness as such trait, and thus, P_τ reflects to which point high-fitness behaviors produce more offspring. Our goal is to study the interplay between environmental and task-driven pressure, and their impact on overall P_τ.

4   Experiments

Here, we present the experimental settings to analyze the impact of swarm size and local selection pressure, on fitness quality and behavioral diversity in the evolving swarm.

Table 1. Experimental parameters. This includes varying parameters: 5 different levels of local selection pressure, T_e, and 4 swarm sizes $|S| \in \{25, 50, 100, 200\}$, are evaluated.

Parameter	Value(s)
# Robots	25, 50, 100, 200
θ_{sp}	0, .25, .5, .75, 1
# Items	100
Envir. radius	$\approx 1000\,px$
Sensor range	$40\,px$
# runs	30
Generations	250
T_e	500 $steps$

In each experiment, a swarm of $|S|$ robotic agents is deployed in a circular environment containing 100 food items. Each robot runs a copy of the algorithm presented in the previous section, to evolve GRNs, and uses a given θ_{sp} as parameter for the tournament selection operator.

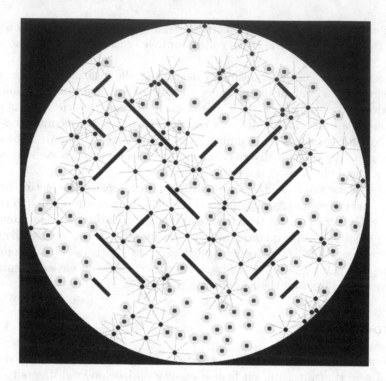

Fig. 1. Simulated environment: enclosed circular arena containing obstacles (black bars), food items (blue circles), and a swarm of robots (black circles with thin hairlines representing sensor orientation, 100 robots in this snapshot). (Color figure online)

The parameter settings of our experiments are summarized in Table 1. Figure 1 shows the simulated environment used in our experiments. The environment contains robots (black dots with hairlines representing sensors) food items (blue dots), which reappear at a random location when robots collect them to keep the number of items constant; and obstacles (black bars). The experiments are all performed using RoboRobo simulator [24], a fast and simple 2D simulator, well adapted to research in distributed evolution of swarm robot behavior.

To parametrize both variation operators, and to decide among several GRN architecture variants, related work on the evolution of GRNs for similar agent models was studied, and preliminary experiments were performed. This led to a GRN with no bias input protein, and with 4 output nodes using differential coding to control both wheels of a robot, $i.e.\ w_l = \frac{out_l^+ - (out_l^-)}{out_l^+ + (out_l^-)}$. The GRNs had no internal structure, that is, GRNs were not initialized with regulatory proteins, were they added during evolution. Additionally, uniform noise of an amplitude of 10% of the maximum speed was added to the motors. This constitutes one of the few works dealing with noise during the distributed evolution of swarm robot behavior. 30 runs of each of the $5 \times 4 = 20$ experiments were run. The next section presents the main highlights of the results, then discussed in Sect. 6.

5   Results and Analysis

Figures 2 and 3 present the swarm average fitness (average number of collected items per robot), for a selected set of experiments, namely with minimal and maximal ($\theta_{sp} \in \{0,1\}$, the higher the parameter, the more opaque the figure), and either 25 and 200 robots, or all the swarm sizes, 25, 50, 100 (with median full lines and interquartile shaded areas over the 30 independent runs), and 200 (respectively, in yellow, blue, green, and red, with median full lines over the 30 runs only). Other results are not shown here, to improve readability. The analysis of the trend of swarm fitness over time reveals dampened noise impact on larger swarms, as evidenced by the smaller variance and higher stability of red graphs. Smaller swarms, especially $|\mathbf{S}| = 25$, reach lower fitness values with all the intensities of local selection pressure. On the other hand, for all the different swarm sizes and $\theta_{sp} = 0.0$, in which local selection pressure is random, thus disregarding fitness, there is improvement in the item collection task. This provides further evidence supporting the work by Bredèche and collaborators on distributed swarm robot behavior evolution with only environment-driven selection pressure for survival: the pressure to adapt and survive by itself pushes evolution toward navigating behaviors that maximize mating opportunities, thus collecting items as a byproduct of fast navigation.

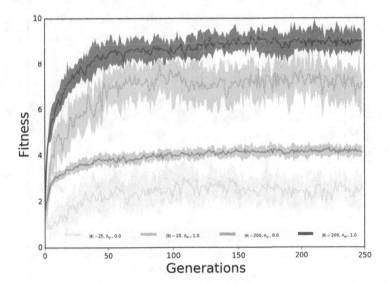

Fig. 2. Swarm fitness (*i.e.* average number of collected items per robot) over generations for $|\mathbf{S}| \in \{25, 200\}$, and $\theta_{sp} \in \{0.0, 1.0\}$.

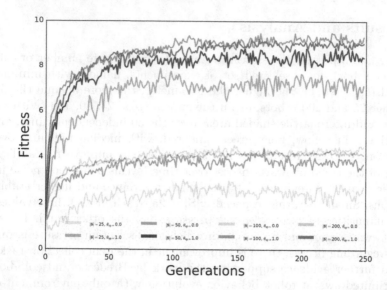

Fig. 3. Swarm fitness (*i.e.* average number of collected items per robot) over generations for $|\mathbf{S}| \in \{25, 50, 100, 200\}$, and $\theta_{\mathrm{sp}} \in \{0.0, 1.0\}$. Only median values are shown for readability. (Color figure online)

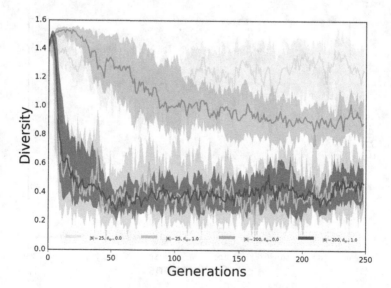

Fig. 4. Global diversity (*i.e.* average pairwise behavioral distance between the behavioral descriptors of the active controllers in the swarm) over generations for $|\mathbf{S}| \in \{25, 50, 100, 200\}$, and $\theta_{\mathrm{sp}} \in \{0.0, 1.0\}$.

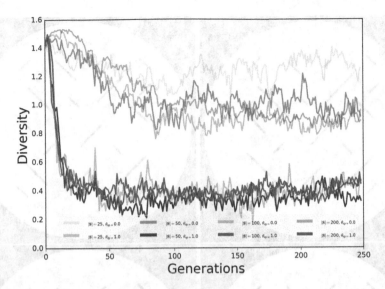

Fig. 5. Global diversity (*i.e.* average pairwise behavioral distance between the behavioral descriptors of the active controllers in the swarm) over generations for $|S| \in \{25, 50, 100, 200\}$, and $\theta_{sp} \in \{0.0, 1.0\}$. Only median values are shown for readability. (Color figure online)

Figures 4 and 5 present the behavioral diversity of the swarm over time, as described in Sect. 3, for the same selected experiments as above. The trend of the diversity for $\theta_{sp} = 1.0$ is similar for both $|S| = 25$ and $|S| - 200$: rapid convergence to a similar stable value of behavioral diversity in the swarm. However, in the case of $|S| = 25$, the rapid decrease of diversity is more unstable, having a larger variance than for $|S| = 200$. As for the case of $\theta_{sp} = 0.0$, loss of diversity is slower than for stronger selection pressure, for all swarm sizes. For larger swarm sizes (*cf.* also Fig. 3), the decrease in diversity is slower when compared to $|S| = 25$, being also less prone to statistical size effects of instability.

Figure 6 shows the trajectories of the robots in the swarm at the beginning of evolution, at generation 50, 100, and 200, of a run with $|S| = 100$ and $\theta_{sp} = 1.0$. The figures show a progressively larger covering of the entire environment, with trajectories becoming smoother, arguably more efficient, over time, especially at generation 200, but this requires further investigation.

Figures 7 and 8 show, for the same set of selected variants, the selection pressure coefficient P_τ, and provide a value aggregating the impact of local task-driven selection pressure, and environmental pressure for genome propagation. For example, Fig. 8 shows that, for $\theta_{sp} = 1.0$, there is an inverse, nonlinear relationship between swarm size and induced selection pressure (*i.e.* P_τ). Though more investigation is required to ascertain the cause of this phenomenon,

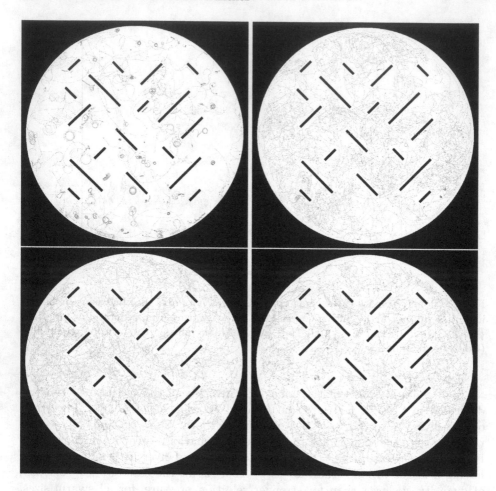

Fig. 6. Example of trajectories during one run of evolution with $|\mathbf{S}| = 100$ and $\theta_{\mathrm{sp}} = 1.0$ at generations $G = 0$, $G = 50$, $G = 100$, and $G = 200$, from left to right.

we hypothesize that this is due to the fact that: large swarms, *e.g.* 200, induce more genome propagation, thus increasing environmental selection pressure. As such, local populations contain more genomes that compete with each other, thus converging more rapidly to a situation when fitness is homogeneous and does not provide an evolutionary advantage in terms of offspring. As another example of complex dynamics in terms of selection pressure, Fig. 7 shows a much larger variance and instability for the selection pressure with 25 robots than with 200. This is due to statistical instability in lower sample sizes.

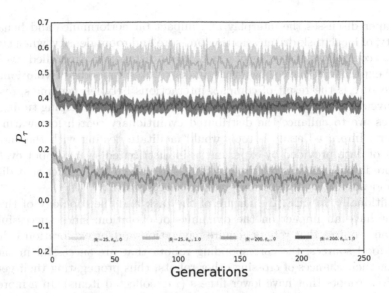

Fig. 7. Selection pressure P_τ (*cf.* [23]) over generations for $|\mathbf{S}| \in \{25, 200\}$, and $\theta_{\mathrm{sp}} \in \{0.0, 1.0\}$.

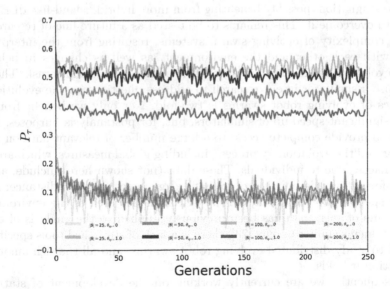

Fig. 8. Selection pressure P_τ (*cf.* [23]) over generations for $|\mathbf{S}| \in \{25, 50, 100, 200\}$, and $\theta_{\mathrm{sp}} \in \{0.0, 1.0\}$. Only median values are shown for readability. (Color figure online)

6 Conclusion

This paper discusses the interplay and impact on performance and behavioral diversity, of local, task-driven selection pressure and robot density, when evolving GRNs as controllers in swarms of robots using distributed Embodied Evolution.

Our experiments have shown that, in distributed Embodied Evolution, selection pressure is the result of complex interactions between genome spread and task-driven selection pressure. This informs future research, helping tuning selection pressure to enhance the distributed evolutionary search for swarm robot behaviors. Improved analysis tools would facilitate dealing with the enormous amount of data provided by experiments in distributed swarm robot evolution, and more investigation is required to fully understand the impact that different types and intensities of selection pressure have on dEE.

Additionally, in dEE, the nature of the task and the topology of the environment have an impact on the dynamics of evolution: since successful item collection requires that robots acquire navigation and/or exploration behaviors to find food sources, the corresponding robots that are successful in the task augment their chances of crossing other robots, thus propagating their genomes more than robots that have lower fitness (*i.e.* collected items). In a more cluttered environment, or when performing a task that requires dispersion (*e.g.* maximal surveillance coverage), genome exchange becomes scarcer. Consequently, we hypothesize that such environments can lead to an extreme loss of diversity, as supported by [3,16], making purely distributed evolutionary algorithms more prone to stagnation, possibly benefiting from more hybrid, island-like dEE algorithms to overcome it. This remains to be tested as a future line of research.

The complexity of evolving swarm systems, resulting from the interplay of agents with each other and the environment, as well as changes in behavior during evolution, makes the analysis of their behavior an arduous task. The reason to use a simulator in our research to address questions on the evolutionary dynamics of evolving robot swarms is two-fold. Not only we benefit from the typical significant speed-up of simulators, but, for postanalysis purposes, simulators also provide complete access to a large number of relevant data on robot behavior and the evolutionary process, including global measures, which are otherwise inaccessible to individuals. These data (not shown here) include, among other information, the number of exchanged genomes, covered distance, robot average speed, robot collisions, density of robots per area of the environment. While some of such data has been previously applied to the analysis of evolutionary systems similar to ours [23], there is a lack of statistical tools specifically tailored to study distributed evolving robot swarms, and all the rich amount of information it holds.

Consequently, we are currently working on the development of statistical analysis and visualization tools for the aforementioned data, in order to further investigate the interplay of environmental and local, task-driven selection pressures when evolving behaviors in swarms of robots using distributed Embodied Evolution.

References

1. Watson, R.A., Ficici, S.G., Pollack, J.B.: Embodied evolution: distributing an evolutionary algorithm in a population of robots. Robot. Auton. Syst. **39**(1), 1–18 (2002)
2. Fernández Pérez, I., Boumaza, A., Charpillet, F.: Comparison of selection methods in on-line distributed evolutionary robotics. In: Proceedings of the International Conference on the Synthesis and Simulation of Living Systems (Alife 2014), pp. 282–289. MIT Press, New York, July 2014. http://mitpress.mit.edu/books/artificial-life-14
3. Fernández Pérez, I., Boumaza, A., Charpillet, F.: Maintaining diversity in robot swarms with distributed embodied evolution. In: ANTS Conference, the Eleventh International Conference on Swarm Intelligence, October 2018
4. Bredèche, N., Montanier, J.M., Liu, W., Winfield, A.: Environment-driven distributed evolutionary adaptation in a population of autonomous robotic agents. Math. Comput. Model. Dyn. Syst. **18**(1), 101–129 (2012). http://hal.inria.fr/inria-00531450
5. Fernández Pérez, I., Boumaza, A., Charpillet, F.: Learning collaborative foraging in a swarm of robots using embodied evolution. In: ECAL 2017–14th European Conference on Artificial Life. Inria, Lyon, France, September 2017. https://hal.archives-ouvertes.fr/hal-01534242. nominated to the best paper award at ECAL2017 (4 nominees over 100+ papers)
6. Weiss, G.: Multiagent Systems: A Modern Approach to Distributed Artificial Intelligence. MIT press, Cambridge (1999)
7. Joachimczak, M., Wrobel, B.: Evolving gene regulatory networks for real time control of foraging behaviours. artificial life XII. In: Proceedings of the 12th International Conference on the Synthesis and Simulation of Living Systems, pp. 348–355 (2010)
8. Nicolau, M., Schoenauer, M., Banzhaf, W.: Evolving genes to balance a pole. In: Esparcia-Alcázar, A.I., Ekárt, A., Silva, S., Dignum, S., Uyar, A.Ş. (eds.) EuroGP 2010. LNCS, vol. 6021, pp. 196–207. Springer, Heidelberg (2010). https://doi.org/10.1007/978-3-642-12148-7_17
9. Quick, T., Nehaniv, C.L., Dautenhahn, K., Roberts, G.: Evolving embodied genetic regulatory network-driven control systems. In: Banzhaf, W., Ziegler, J., Christaller, T., Dittrich, P., Kim, J.T. (eds.) ECAL 2003. LNCS (LNAI), vol. 2801, pp. 266–277. Springer, Heidelberg (2003). https://doi.org/10.1007/978-3-540-39432-7_29
10. Roli, A., Manfroni, M., Pinciroli, C., Birattari, M.: On the design of boolean network robots. EvoApplications 2011. LNCS, vol. 6624, pp. 43–52. Springer, Heidelberg (2011). https://doi.org/10.1007/978-3-642-20525-5_5
11. Cussat-Blanc, S., Harrington, K.: Genetically-regulated neuromodulation facilitates multi-task reinforcement learning. In: Proceedings of the 2015 on Genetic and Evolutionary Computation Conference - GECCO 2015, No. 1, pp. 551–558. ACM Press, New York (2015)
12. Sanchez, S., Cussat-Blanc, S.: Gene regulated car driving: using a gene regulatory network to drive a virtual car. Genet. Program. Evolvable Mach. **15**(4), 477–511 (2014). https://doi.org/10.1007/s10710-014-9228-y
13. Cussat-Blanc, S., Luga, H., Duthen, Y.: From single cell to simple creature morphology and metabolism. In: Artificial Life XI, pp. 134–141 (2008)
14. Taylor, T.: A genetic regulatory network-inspired real-time controller for a group of underwater robots. In: Proceedings of the Eighth Conference on Intelligent Autonomous Systems (IAS-8), pp. 403–412. IOS Press (2004)

15. Delecluse, M., Sanchez, S., Cussat-Blanc, S., Schneider, N., Welcomme, J.B.: High-level behavior regulation for multi-robot systems. In: Proceedings of the Companion Publication of the 2014 Annual Conference on Genetic and Evolutionary Computation GECCO Comp 2014, pp. 29–30. ACM, New York (2014). http://doi.acm.org/10.1145/2598394.2598454
16. Fernández Pérez, I., Sanchez, S.: Influence of mating mechanisms in distributed evolution for collective robotics. In: EVOSLACE Workshop at ALIFE2018: Workshop on the Emergence and Evolution of Social Learning, Communication, Language and Culture in Natural and Artificial Agent, Tokyo, Japan (2018)
17. Bredeche, N., Montanier, J.-M.: Environment-driven embodied evolution in a population of autonomous agents. In: Schaefer, R., Cotta, C., Kołodziej, J., Rudolph, G. (eds.) PPSN 2010. LNCS, vol. 6239, pp. 290–299. Springer, Heidelberg (2010). https://doi.org/10.1007/978-3-642-15871-1_30. http://hal.inria.fr/inria-00506771
18. Banzhaf, W.: Artificial regulatory networks and genetic programming. In: Riolo, R., Worzel, B. (eds.) Genetic Programming Theory and Practice, vol. 6, pp. 43–61. Springer, Boston (2003). https://doi.org/10.1007/978-1-4419-8983-3_4
19. Disset, J., Wilson, D.G., Cussat-Blanc, S., Sanchez, S., Luga, H., Duthen, Y.: A comparison of genetic regulatory network dynamics and encoding. In: Proceedings of the Genetic and Evolutionary Computation Conference GECCO 2017, pp. 91–98. ACM, New York (2017). http://doi.acm.org/10.1145/3071178.3071322
20. Knabe, J., Nehaniv, C., Schilstra, M., Quick, T.: Evolving biological clocks using genetic regulatory networks. In: Artificial Life X : Proceedings of the Tenth International Conference on the Simulation and Synthesis of Living Systems (Alife 10), pp. 15–21 (2006)
21. Stanley, K.O., Miikkulainen, R.: Evolving neural networks through augmenting topologies. Evol. Comput. **10**(2), 99–127 (2002)
22. Cussat-Blanc, S., Harrington, K., Pollack, J.: Gene regulatory network evolution through augmenting topologies. IEEE Trans. Evol. Comput. **19**(6), 823–837 (2015)
23. Haasdijk, E., Heinerman, J.: Quantifying selection pressure. Evol. Comput., 1–23 (2017)
24. Bredèche, N., Montanier, J.M., Weel, B., Haasdijk, E.: Roborobo! a fast robot simulator for swarm and collective robotics. CoRR abs/1304.2888 (2013)

Body Symmetry in Morphologically Evolving Modular Robots

T. van de Velde[1]([✉]), C. Rossi[2], and A. E. Eiben[3]

[1] University of Amsterdam, Amsterdam, The Netherlands
timon11444@hotmail.com
[2] Centre for Automation and Robotics UPM-CSIC, Madrid, Spain
[3] Vrije Universiteit Amsterdam, Amsterdam, The Netherlands

Abstract. Almost all animals natural evolution has produced on Earth have a symmetrical body. In this paper we investigate the evolution of body symmetry in an artificial system where robots evolve. To this end, we define several measures to quantify symmetry in modular robots and see how these relate to fitness that corresponds to a locomotion task. We find that, although there is only a weak correlation between symmetry and fitness over the course of a single evolutionary run, there is a positive correlation between the level of symmetry and maximum fitness when a set of runs is taken into account.

Keywords: Evolutionary robotics · Modular robots · Symmetry

1 Introduction

A particular field within Evolutionary Computing [1,2] is Evolutionary Robotics [3] that is "useful both for investigating the design space of robotic applications and for testing scientific hypotheses of biological mechanisms and processes" [4]. In general, a robot consists of a body with sensors and actuators that constitute the physical makeup of the robot (morphology, hardware), and a brain, which contains the operating logic of the robot (controller, software). The vast majority of Evolutionary Robotics studies is concerned with evolving robot controllers in fixed bodies. The evolution of robot morphologies has only been addressed in a handful of publications [5–9].

A recent development is fueled by the vision of the Evolution of Things [2,10] aiming at evolving robots in the real world, enabled by recent advances in technology (in particular, 3D-printing and rapid prototyping). However, as of today, a system where robots can reproduce and evolve has not been engineered yet. A generic system architecture for evolving robots in real-time and real-space has been established in [11], but, currently, mostly software implementations of this framework [12,13] have been explored. Exceptions are the proof-of-concept installation that demonstrated a physical 'robot baby' parented by two robots, relying on several simplifications and handwork in the reproduction step [14] and the setup proposed in [9], where robots are physically constructed and evaluated.

© Springer Nature Switzerland AG 2019
P. Kaufmann and P. A. Castillo (Eds.): EvoApplications 2019, LNCS 11454, pp. 583–598, 2019.
https://doi.org/10.1007/978-3-030-16692-2_39

In general, there are many parallels between evolutionary algorithms and natural evolution [2]. In nature, most living creatures exhibit forms of symmetry along the longitudinal axis, known as bilateral symmetry. Even though there do not appear to be any physical laws that state that this should be so, theories that explain bilateral symmetry have been proposed (see, e.g., [15]).

In this paper we examine the relationship between symmetry and fitness in a morphologically evolving robot population. The robots are based on the RoboGen system [8]. Their bodies are composed of predefined modules, their brains are artificial neural networks, and their fitness is determined by their ability to locomote. Our main hypothesis is that symmetrical robots perform better, and therefore evolution promotes symmetry.

To this end, we present three measures to quantify the level of symmetry in robots. Furthermore, we distinguish online evolution and offline evolution (see [13]) and run experiments with all six setups using the Revolve simulator [13]. These experiments are to provide an answer to the following research question: is there any relation between symmetry and fitness?

2   Robot Symmetry

To our best knowledge, not much work has been done on symmetry in evolutionary robotics. Certainly, it has been noted that modular robots may end up showing symmetry after having been evolved for locomotion in a straight line [7]. Other research specifically uses symmetry to evolve robot morphologies. For instance, by mirroring limbs along a spine [16] or by employing a mutation operator which can symmetrically replicate branches of robots [17]. Symmetry has also been used as a human evaluation criterion of evolved robots, where the ability of generative encoding to create symmetrical soft robots is seen as a positive quality [18]. Additionally, the importance of symmetry in locomotion has been shown to influence robot controller evolution through HyperNEAT [19]. Despite all this, not much work can be found that specifically measures symmetry in robots. There is, however, a lot of work that has been done in computer vision to detect symmetry [20,21]. Symmetry detection has even been used in robotics for tasks such as real-time object tracking [22]. Zabrodsky et al. [23,24] define a robust way to quantify symmetry based on how much an object would have to change to achieve true symmetry. The most relevant work to this study is that of Miras et al. [25] that defines several morphological descriptors for RoboGen-like robots in the Revolve system. This set of descriptors contains one that reflects a basic notion of mirror-symmetry. In this paper, we will use this measure as the *base symmetry*.

The bodies of the robots we use here consist of a number of basic components arranged to form larger structures. They are defined by trees consisting of nodes and edges. Because trees are acyclical, so are the robots defined by them. The robots furthermore have some other requirements; they must be planar (i.e. flat), and need to possess at least one motor unit. In addition, the maximum number of parts a robot may have is limited. The nodes of the tree represent components, their parameters (if applicable) and their orientation, while edges specify which components connect to each other. Table 1 shows an overview of the components used; for more details we refer to [25].

Table 1. Overview of used components.

Component	Connectivity	Parameters	Actuators	Sensors	Colour
Core	4	-	-	6	Yellow
Fixed brick	4	-	-	-	Blue
Parametric bar	2	3	-	-	Brown
Passive hinge	2	-	-	-	Purple
Active hinge	2	-	1	-	Orange

The controllers of the robots are fully connected recurrent neural networks with a single hidden layer. The input nodes are sensor inputs, and the output nodes are signals for the actuators. Hidden nodes and output nodes can have one of three different activation functions; linear, sigmoid, or sinusoidal. The sinusoidal activation function will ignore any input to the node and produce a sinusoid defined by three parameters for period, phase, and gain.

Our definitions of symmetry are based on a two-dimensional representation introduced in [25]. In this representation, each robot part appears as a coloured square, its colour indicating the part type (see Table 1). As each component can have at most four connection slots, we can draw squares adjacent to each other where components connect. Figure 1 shows an example of a robot and its 2D representation.

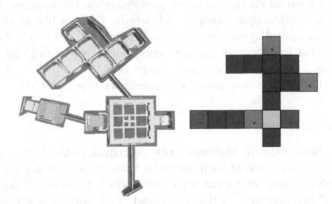

Fig. 1. Two representations of a robot. Left: the robot as rendered by the simulator. Right: its two dimensional approximation used to define symmetry measures.

2.1 Symmetry Measure of Miras *et al.*

Miras *et al.* [25] define a symmetry measure which we used as a base value to which novel symmetry measures are compared. Based on the 2D image of a robot, the core component is selected as the center of the robot. From there, a vertical line is drawn. This line will be the spine of the robot. Components on the spine do not factor into the final symmetry score. The rest of the algorithm is as follows:

```
total_components = 0
for each component:
    if component_a is on spine:
        continue
    total_components = total_components + 1
    coordinates_a = coordinates of component_a
    coordinates_b = coordinates_a mirrored in spine
    if another component exists on coordinates_b:
        matches = matches + 1
symmetry = matches / total_components
```

Because in the robots there is not a preferred longitudinal axis, this algorithm is performed twice. Once for a horizontal spine, and once for a vertical spine. The highest symmetry score of the two determines the final symmetry value for the robot. Note that the final calculation of symmetry is the ratio between matching components and total components. As such, the symmetry value will be 0 if the robot is not symmetrical at all, and 1 if it is completely symmetrical.

Figure 2 shows an example of the symmetry calculation in a robot. Computing its symmetry score, we would first draw a vertical line through its core component, the yellow square. Then, for all components that are not on the vertical axis, we would check if another part exists on the opposite side of the spine. With the vertical spine, that would only be the two blocks just adjacent the core. It's horizontal symmetry score would thus be $2/9 = 0.222$. If we now draw a horizontal line through the core component, we would end up with 8 matching components and a symmetry score of $8/14 = 0.571$. Choosing the larger value, this robot is now assigned a symmetry score of 0.571.

2.2 Novel Symmetry Measures

Several limitations exist to the previously described method. For one, snake-like robots that have most of their parts on their spine will not be accurately assessed, as all components on the spine are ignored. Secondly, it does not take into account if two components that are found on the opposite side of the symmetry line are the same component, or different ones. And finally, symmetry might develop along other lines than just the vertical and horizontal ones that pass through the core component. As such several possible solutions have been implemented.

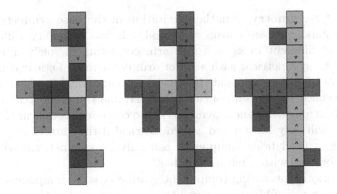

Fig. 2. Example of symmetry calculation. The first image is the robot in question, the middle image is for calculating horizontal symmetry, and the right image is for calculating the vertical symmetry. Grey components are not counted, red components have no symmetrical match, and green components do have a match. This robot has a symmetry value of 0.571 on a scale from 0 to 1. (Color figure online)

Symmetry Lines. No particular reason exists that should cause the core component to be at the center of the robot, other than that it is used as the root of the tree defining the robot. Should the core component be located in one of the robot's extremities, only drawing mirror lines from this point appears to limit the accuracy of the symmetry approximation. An example of this effect can be seen in Fig. 3. As before, the grey components represent the symmetry mirror line and core component in yellow. The third image shows that the robot is more symmetrical along some other mirror line that does not cross the core component. To account for this effect, the novel symmetry measures iterates over every possible symmetry line, to find the line that provides the highest symmetry value.

Fig. 3. Three images of the same robot. The first is just the robot. The second is the robot with highest symmetry as found by drawing a mirror line through the core component; symmetry value 0.36. The third image is the highest possible symmetry that can be found with a freely movable mirror line; symmetry value 0.8.

Component Symmetry. Another limitation of the base symmetry measure is that all components are considered equal, whereas in reality, different components fulfill different tasks. A robot arm consisting of solely hinges should perhaps not be symmetrical with a robot arm consisting of static components.

Two ways to take component differences into account have been developed. One will be called *strong* component symmetry, and the other *weak* component symmetry. In strong component symmetry, two components that are locationally symmetrical will only be counted as symmetrical if they are of the same type. For instance, a fixed brick component can only be symmetrical with another fixed brick, but not with a parametric bar.

Another way of measuring component symmetry is by component similarity. There are two types of hinge components; one is passive and the other is active. An argument can be made that they serve a similar purpose; to make movement possible in the robot. On the other hand, the fixed brick and the parametric bar serve only to connect other components, although they do so in very different ways. Nevertheless, it is possible to create a weak component symmetry by using these similarities. On locational matches, Table 2 is used to determine the symmetry value of the match. An example of this calculation can be seen in Fig. 4.

Table 2. Look-up table of similarity scores used for weak component symmetry.

	Parametric bar	Brick	Active hinge	Passive hinge
Parametric bar	1	0.5	0	0
Brick	0.5	1	0	0
Active hinge	0	0	1	0.5
Passive hinge	0	0	0.5	1

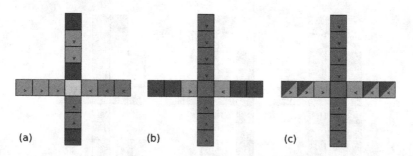

(a)   (b)   (c)

Fig. 4. (a) Image of the robot. (b) Image of symmetry calculation with strong component symmetry. Matching components are green, non-matching components are red. Symmetry value 0.33. (c) Image of symmetry calculation with weak component symmetry. Matching components are green, similar components are red and green. Symmetry value 0.66. (Color figure online)

Spinal Symmetry. Up until now, we have ignored the components that are located directly on the mirror line. In many cases this line could be considered the spine of the robot. There is one relatively common case in which ignoring the symmetry of the spine itself is not desirable. In particular in snake-like robots, where most of the components are aligned along a single axis, and represent a large percentage of the robot's total number of components. Intuitively, we would say that the robot of Fig. 5 is reasonably symmetrical, but ignoring the spine, we end up with a symmetry value of 0.

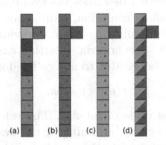

Fig. 5. (a) Snake-like robot. (b) Symmetry calculation without counting the spine. There is only one red component, thus a symmetry of 0. (c) Symmetry calculation while also counting the spine results in a score of 0.91. (d) Symmetry calculation with discounted spine. The red and green components all count as half components for a symmetry score of 0.83. (Color figure online)

To correct this, we want to make sure the spine also influences the symmetry. However, assigning full symmetry points would skew the results as well. Other components in the robot have to satisfy both locational symmetry and component symmetry. The parts on the mirror line would always be symmetrical with themselves. To further adjust this bias, components on the spine will only count as half a component. Keeping mind that our formula for symmetry was:

`symmetry = matches / total_components`

In Fig. 5, doing full spinal symmetry would then result in 10 parts being symmetrical, with a final symmetry score of $10/11 = 0.91$. Discounted spinal symmetry would mean that those 10 symmetrical components now only count as half components; both as `matches` and `total_components`. The resulting symmetry score then follows from $5/6 = 0.83$. Reducing the value of `total_components` is necessary to assure symmetry can still reach a value of 1 if the robot is completely symmetrical. Also note that the spine is nothing more than the location of the mirror line. This line is still shifted throughout the robot to find the highest fitness value.

Final Measures. To summarise, we now have two symmetry measure variations which we can compare to our base symmetry. The two novel measures implement all techniques described above, but differ in *weak* and *strong* component symmetry. Consequently they shall be referred to as such.

3 Experimental Setup

There are two distinct evolutionary approaches available in Revolve; offline and online. The following subsection describes the evolutionary process as specified in Hupkes *et al.* [13]. In offline evolution, individuals are evaluated separately and only once, whereas with online evolution, individuals will be continuously evaluated until they die by some criteria. In either case, the fitness function is based on how well the robots are able to move. The formula is as follows:

$$f(\rho) = v + 5s. \tag{1}$$

Here, $f(\rho)$ is the fitness value of robot ρ. The velocity of the robot, or the total distance travelled since the beginning of the evaluation is denoted as v. Thus, if a robot moves back and forth repeatedly, the value v increases. The variable s is the speed of the robot, and is obtained by measuring the distance between the robot's location at the start of evaluation and the robot's location at the end of evaluation.

This fitness value is then used for selecting parents through tournament selection. This means that out of the entire population, several random individuals are selected, and out of those, the individual with the highest fitness is selected to be a parent. This process is repeated to select the other parent, with the only requirement that both are distinct robots. Through crossover, a new robot is generated and born into the simulator. The simulation ends after a set amount of births. The difference between offline and online solution has some consequences to population management.

During offline evolution, each individual is inserted into the simulator separately. They are then evaluated according to the fitness function for some period of time. Afterwards, their fitness is recorded, the robot removed and a new one added. After an entire generation has been evaluated, an equal amount of offspring is produced. Out of these two generations, the fittest individuals are kept, and the rest removed. This way, the size of the living generation remains constant. Specific settings are listed in Table 3.

Table 3. Settings for offline evolution.

Offline evolution	
Generation size	15 individuals
Evaluation time	12 s
Parent selection	4-tournament
Survivor selection	Best 15
Stopping criteria	3000 births

With online evolution, the entire initial population is inserted into the world at once. They are randomly placed within a circular area called the *birth clinic*. The individuals are then continually evaluated within a sliding time-window. They will have matured after having been evaluated through an entire time-window, but the fitness value will keep updating throughout the robot's lifespan. The simulator is set up so that new robots are born at regular intervals. New robots are simply inserted into the world, along with the already existing population. Robots can only be selected as a parent if they are mature.

The simulator will, at certain fixed time intervals, remove robots that do not have a fitness value over a certain threshold. To ensure the success of the simulation, robots will not be killed if that would mean that the population size drops below a set minimum. On the other end of the spectrum, if a maximum number of living individuals is exceeded, a percentage of the population with the lowest fitness will be culled. Specific settings are listed in Table 4.

Table 4. Settings for online evolution.

Online evolution	
Initial population	15 individuals
Evaluation window	12 s
Parent selection	4-tournament
Birth interval	15 s
Kill interval	30 s
Survivor selection	>70% average fitness
Minimum population size	8 individuals
Maximum population size	30 individuals
Population culling	Weakest 70%
Stopping criteria	3000 births

The process of robot generation works through several sequential steps. First of all, the total number of components is decided by randomly drawing from a normal distribution defined by $\mu_{\text{initial parts}}$ and $\sigma_{\text{initial parts}}$. This number is required to be between R_{minimum} and R_{maximum}. Starting from the core component, a random attachment slot is chosen to attach another part to. This part, in turn, is also randomly selected, as is its attachment slot that will be attached to the robot. If the selected part has parameters to be set, as would be the case with the parametric bar joint, these are also initialised at random. This process continues until the total number of components has been reached, or until there are no further attachment slots available. As previously stated, the neural network of the robot will have its input and output neurons determined by the presence of sensors and actuators in the robot's body. The number of hidden neurons will be a randomly chosen value between 0 and a variable h_{max}. For every hidden and output node, a random activation with randomised parameters is chosen.

Finally, all weights in the fully connected recurrent neural network are initialised to 0. They are then with a random chance $p_{\text{connect neurons}}$ set to a value between $[0, 1]$. Parameters for robot generation are listed in Table 5.

Full offspring generation consists of a specific sequence. Every step happens only with a probability. For instance, it is possible to set different possibilities for subtree removal and subtree duplication. If the result of a step produces a robot that violates set restrictions as described in Sect. 2 (that is, planarity, number of motor units, total number of parts), crossover is aborted and no offspring is created. The following crossover steps exist, each step continuing with the result of the previous step:

- **One point crossover**: A node is selected on parent a, and replaced with a random node from parent b.
- **Subtree removal**: A random subtree is selected and removed.
- **Subtree duplication**: A random subtree is duplicated.
- **Subtree swap**: Two random subtrees are swapped.
- **Hidden neuron and neural connection removal**: Each hidden neuron or neural connection has a probability to be removed.
- **Parameter mutation**: Each parameter, for either body parts or neurons is updated with a randomly drawn value. They are changed according to $(1 - \epsilon)\pi + \epsilon\pi*$, where π is the current parameter value, $\pi*$ a new randomly drawn value, and ϵ the mutation rate.
- **Hidden neuron and neural connection addition**: With a probability, new hidden neurons and neural connections are added.
- **Part addition**: A new part is added to the robot body.

Table 5. Parameters and restriction values for robot generation.

Parameter	Value
$\mu_{\text{initial parts}}$	12
$\sigma_{\text{initial parts}}$	5
R_{minimum}	3
R_{maximum}	30
h_{max}	10
$p_{\text{connect neurons}}$	0.1

Table 6. All probabilities and parameters used during crossover.

Parameter	Value
$p_{\text{delete subtree}}$	0.05
$p_{\text{duplicate subtree}}$	0.1
$p_{\text{swap subtree}}$	0.05
$p_{\text{remove brain connection}}$	0.05
$p_{\text{delete hidden neuron}}$	0.05
$\epsilon_{\text{body mutation}}$	0.05
$\epsilon_{\text{brain mutation}}$	0.1

Table 6 lists the probabilities of the crossover actions. The probabilities of neuron addition neural connection addition and part addition are tuned in such a way that the number of parts, number of neurons, and number of neuron connections stay the same on average and do not, for instance, tend to zero [13].

The setup of this experiment is to determine whether there is any relationship between a robot's fitness and its symmetry. By running many simulations and collecting data on the born individuals' fitnesses and morphologies (out of which we extract their symmetry values), we can compute linear correlations between the two. Data was collected both by running online evolution and offline evolution.

Combining these data, we have several possible measurements we can do. For each separate simulation run, we can try to see if the symmetry of robots increases as their fitness increases. This can be done simply by verifying whether a linear correlation between the two exists. The same linear correlation can be calculated not only on separate runs, but also on the combined datasets of all runs.

The population size used during experiments is quite small; around 15 individuals for online evolution, and a fixed 15 individuals for offline evolution. As such, there exists a possibility that the global morphology of the entire population during a simulation run converges, meaning that all individuals end up with a similar morphology. In this case it is interesting to make comparisons between the runs. One way to do this, is to create another dataset by taking the average symmetry and highest found fitness of each run, and computing the linear correlation between those two values.

Finally, all of these experiments have 3 different measures for symmetry available. There are 2 novel symmetry measures and the base symmetry. For further use of symmetry measures, it would be useful to confirm whether there are any differences. One measure might correlate better with fitness than others, and would therefore be more likely to be useful when symmetry is used to modify the fitness function during the simulation.

4   Experimental Results

4.1   Fitness and Symmetry Correlations

Thirty runs were performed in both offline and online evolution. By computing fitness and the three symmetry measures for each individual we can examine the correlation between them. Table 7 shows the averaged correlations of all the separate runs. From this we can see that the correlation between symmetry and fitness is low. Most of the values are around 0, indicating a random relationship between fitness and symmetry. It appears that symmetry does not generally increase as a run progresses, even though fitness does.

Table 8 shows some slightly more positive numbers. These values were computed by first concatenating the fitness and symmetry values of each run and then finding the correlations, rather than correlating each run separately and then averaging. The difference between Tables 7 and 8 can be explained by the fact that different runs achieve different maximum fitness values. By averaging correlations of the separate runs, the difference in fitness between runs is removed. By not doing so, runs with higher fitness values essentially have a higher impact while calculating correlations.

Finally, the difference between online and offline evolution can be explained by the way new robots are added to the simulation. Because in offline evolution only fitter individuals are added to the database, there is more statistical weight for individuals with higher fitness. Or, in other words, the average fitness of the recorded individuals in offline evolution is higher than that of online evolution.

Table 7. Correlation values averaged over all separate runs.

Symmetry	Correlations	
	Offline	Online
Base	−0.184	−0.162
Weak	−0.059	−0.052
Strong	−0.037	−0.040

Table 8. Correlation values as calculated over the entire dataset.

Symmetry	Correlations	
	Offline	Online
Base	0.035	−0.071
Weak	0.152	0.018
Strong	0.191	0.030

4.2 Average Symmetry and Maximum Fitness Correlations

In the previous analyses, we have not found a strong correlation between fitness and symmetry. However, visual inspection of the generated robots suggests that a general morphology is settled on during each run. It is therefore possible that some morphologies have a higher potential for fitness. We can verify this by correlating the maximum obtained fitness values and average symmetries of each run. The resulting graphs can be seen in Fig. 6. The corresponding correlation numbers can be found in Table 9.

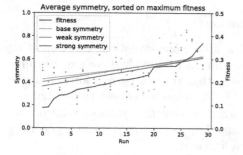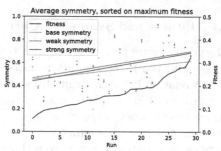

Fig. 6. Relationship between the maximum fitness and the average symmetry in 30 independent runs ordered by the maximum fitness at termination. Symmetry values are represented by scatter points and a linear fit. Left: offline evolution. Right: online evolution.

Table 9. Correlation values as calculated between the maximum fitness values of runs and their average symmetries.

Symmetry	Correlations			
	Offline		Online	
	r	p-value	r	p-value
Base	0.300	0.110	0.235	0.210
Weak	0.447	0.013	0.380	0.038
Strong	0.493	0.006	0.393	0.032

We can now see stronger correlations between the average symmetry of a run and its maximum obtained fitness. Strong symmetry correlates a little better than weak symmetry, and both of them better than base symmetry. It should be noted that strong symmetry is always more conservative than weak symmetry. That is, for any robot the strong symmetry score is generally lower than the weak symmetry score. This difference is greatest when symmetry is low. At higher symmetry values, the measures get closer together.

As in Table 8, there is some difference between online and offline evolution, with lower values for online evolution. Aside from the bookkeeping differences between the two approaches as mentioned before, we can now also see that the maximum achieved fitness of each run in online evolution is lower than in offline evolution. A difference likely caused by the different population management approaches, as noted in [13].

5   Discussion

Throughout all the offline simulations, population size was maintained at the default setting of 15 individuals. This is not a particularly big number and may have had some effects on the results, such as the body morphology quickly converging. Effective population sizes were even smaller in online evolution. As opposed to offline evolution, where every generation spawns a new generation and the 15 best individuals of both generations are kept, online evolution just periodically kills weaker individuals. The practical consequence of this is that oftentimes there will only be a mature population of close to 8 individuals. Premature convergence could also be averted by reducing evolutionary pressure.

The different ways of population management and record keeping also have a profound effect on the average fitness scores. In offline evolution, only the best 15 robots out of 15 children + 15 parents have their fitness recorded in the database every generation. Whereas in online evolution, all new individuals are kept, even if their fitness scores are a lot lower than those of the previous generation. This is why average fitness scores are a lot lower in online evolution than in offline evolution. The effective population size and population management may also explain why the maximum fitness values of online evolution simulations are lower than those of offline evolution.

5.1   Symmetry Limitations

The first limiting factor for the symmetry measures is the conversion of the robots' tree representation to the 2D graph representation. As mentioned in Sect. 2, not all components have the same size, some are positioned at an angle, and some are not static. All of these factors are ignored in the 2D representation, leading to some fundamental differences between the two representations. It is difficult to estimate how much this affects the results without doing an overhaul on this conversion.

There are also parts of the symmetry evaluation that are based on personal design choices. For instance, symmetry detection works by shifting the mirror lines to find the highest symmetry value available in the robot. This means, that the core component, which contains all the sensors, may not be located at the center of the robot. If symmetry is to be evaluated and used during runtime as part of a fitness function, it may be more beneficial to actually keep the core component centered. One case in which this may be true is if visual sensors are situated in the core component.

Another design choice has been to only count components that are located on the mirror line as half components. This was done to value the robots' extremities more than their spine, and to compensate for the fact that the specific components always match with themselves. Whether this is has been the right choice may need to be examined further.

Finally, mirror line placement is currently constrained to only be along the centers of the 2D components. Allowing the lines to be placed on the borders between components would allow for certain robot configurations to be rated as more symmetrical than they would be now. For instance a robot consisting of two components might be more symmetrical if the mirror line was to be placed in between its components, rather than through their center.

6   Conclusions and Further Work

Over the course of single runs, there do not seem to be any strong correlations between symmetry and fitness. The average correlation values tend to zero, indicating a random relationship between the two. However, when we take into account that each run quickly settles on specific morphologies that only change slowly over time, we can find more interesting figures. Correlating the maximum attained fitness of a run with the entire population's average symmetry values, we observed a positive correlation with fitness. There was a positive correlation up to 0.49 for offline evolution, and 0.39 for online evolution. Out of the three measures strong symmetry seems to correlate best with fitness, though the difference with weak symmetry is minor.

One way to examine the cause of the disparity between single run correlations and maximum fitness correlations would be to seed the simulator with hand-crafted symmetric robots. If these runs attain high fitness values, it may indicate that either the genome is unsuited for developing symmetry by itself or that the simulator parameters may need to be changed.

Another limitation of our current setup is that newborn robots do not learn after birth; their neural networks only change randomly during crossover. We chose for this setup to see an isolated, clean effect of evolution. However, in general, any neural connection that worked well in a parent robot, may not work after crossover has changed large portions of the body layout. This will cause the robot to perform poorly, even if the given body may be better suited for locomotion than the bodies of the parents. Future investigations shall therefore include lifetime learning capabilities to the robots at least during their infancy, e.g., attending a 'Robot School' [26]. In this way, the problem of having good but poorly controlled morphologies should be mitigated. In addition, this might also allow body morphology to change more throughout the simulation runs.

Finally, a promising line of investigation is to experiment with exploiting the symmetry and fitness relationship to produce better robots faster. This could be done through a multi-objective approach based on an objective that measures the locomotion abilities and one that measuring symmetry.

References

1. Eiben, A., Smith, J.: Introduction to Evolutionary Computing, 2nd edn. Springer, Heidelberg (2015). https://doi.org/10.1007/978-3-662-05094-1
2. Eiben, A., Smith, J.: From evolutionary computation to the evolution of things. Nature 521(7553), 476–482 (2015)
3. Bongard, J.C.: Evolutionary robotics. Commun. ACM 56(8), 74–83 (2013). http://doi.acm.org/10.1145/2493883
4. Floreano, D., Husbands, P., Nolfi, S.: Evolutionary robotics. In: Siciliano, B., Khatib, O. (eds.) Springer Handbook of Robotics, vol. Part G. 61, pp. 1423–1451. Springer, Heidelberg (2008). https://doi.org/10.1007/978-3-540-30301-5_62
5. Sims, K.: Evolving 3D morphology and behavior by competition. Artif. Life 1(4), 353–372 (1994). https://doi.org/10.1162/artl.1994.1.353
6. Sims, K.: Evolving virtual creatures. In: Proceedings of the 21st Annual Conference on Computer Graphics and Interactive Techniques SIGGRAPH 1994, pp. 15–22. ACM, New York (1994). https://doi.org/10.1145/192161.192167
7. Lipson, H., Pollack, J.B.: Automatic design and manufacture of robotic lifeforms. Nature 406, 974–978 (2000)
8. Auerbach, J., et al.: Robogen: robot generation through artificial evolution. In: Artificial Life 14: Proceedings of the Fourteenth International Conference on the Synthesis and Simulation of Living Systems, pp. 136–137 (2014)
9. Brodbeck, L., Hauser, S., Iida, F.: Morphological evolution of physical robots through model-free phenotype development. PLoS One 10(6), e0128444 (2015)
10. Eiben, A.E., Kernbach, S., Haasdijk, E.: Embodied artificial evolution - artificial evolutionary systems in the 21st century. Evol. Intell. 5(4), 261–272 (2012). http://www.few.vu.nl/~ehaasdi/papers/EAE-manifesto.pdf
11. Eiben, A.E., et al.: The triangle of life: Evolving robots in real-time and real-space (2013)
12. Eiben, A.E.: EvoSphere: the world of robot evolution. In: Dediu, A.-H., Magdalena, L., Martín-Vide, C. (eds.) TPNC 2015. LNCS, vol. 9477, pp. 3–19. Springer, Cham (2015). https://doi.org/10.1007/978-3-319-26841-5_1

13. Hupkes, E., Jelisavcic, M., Eiben, A.E.: Revolve: a versatile simulator for online robot evolution. In: Sim, K., Kaufmann, P. (eds.) EvoApplications 2018. LNCS, vol. 10784, pp. 687–702. Springer, Cham (2018). https://doi.org/10.1007/978-3-319-77538-8_46

14. Jelisavcic, M., et al.: Real-world evolution of robot morphologies: a proof of concept. In: Artificial Life (2017)

15. Werner, E.: The origin, evolution and development of bilateral symmetry in multicellular organisms. Quantitative Biology ArXiv:1207.3289 (2012)

16. Marbach, D., Ijspeert, A.J.: Online optimization of modular robot locomotion. In: IEEE International Conference Mechatronics and Automation 2005, vol. 1, pp. 248–253, July 2005

17. Faiña, A., Bellas, F., López Peña, F., Duro, R.: Edhmor: evolutionary designer of heterogeneous modular robots. Eng. Appl. Artif. Intell. **26**, 2408–2423 (2013)

18. Cheney, N., MacCurdy, R., Clune, J., Lipson, H.: Unshackling evolution: Evolving soft robots with multiple materials and a powerful generative encoding. In: Proceedings of the 15th Annual Conference on Genetic and Evolutionary Computation GECCO 2013, pp. 167–174. ACM, New York (2013). https://doi.org/10.1145/2463372.2463404

19. Clune, J., Beckmann, B.E., Ofria, C., Pennock, R.T.: Evolving coordinated quadruped gaits with the hyperneat generative encoding. In: 2009 IEEE Congress on Evolutionary Computation, pp. 2764–2771, May 2009

20. Sun, C., Sherrah, J.: 3D symmetry detection using the extended gaussian image. IEEE Trans. Pattern Anal. Mach. Intell. **19**(2), 164–168 (1997)

21. Mitra, N.J., Guibas, L.J., Pauly, M.: Partial and approximate symmetry detection for 3D geometry. ACM Trans. Graph. **25**(3), 560–568, July 2006. https://doi.org/10.1145/1141911.1141924

22. Li, W.H., Zhang, A.M., Kleeman, L.: Bilateral symmetry detection for real-time robotics applications. Int. J. Robot. Res. **27**(7), 785–814 (2008). https://doi.org/10.1177/0278364908092131

23. Zabrodsky, H., Peleg, S., Avnir, D.: A measure of symmetry based on shape similarity. In: Proceedings 1992 IEEE Computer Society Conference on Computer Vision and Pattern Recognition, pp. 703–706, June 1992

24. Zabrodsky, H., Peleg, S., Avnir, D.: Symmetry as a continuous feature. IEEE Trans. Pattern Anal. Mach. Intell. **17**(12), 1154–1166 (1995)

25. Miras, K., Haasdijk, E., Glette, K., Eiben, A.E.: Search space analysis of evolvable robot morphologies. In: Sim, K., Kaufmann, P. (eds.) EvoApplications 2018. LNCS, vol. 10784, pp. 703–718. Springer, Cham (2018). https://doi.org/10.1007/978-3-319-77538-8_47

26. Rossi, C., Eiben, A.: Simultaneous versus incremental learning of multiple skills by modular robots. Evol. Intell. **7**(2), 119–131 (2014)

Trophallaxis, Low-Power Vision Sensors and Multi-objective Heuristics for 3D Scene Reconstruction Using Swarm Robotics

Maria Carrillo[1], Javier Sánchez-Cubillo[2], Eneko Osaba[2],
Miren Nekane Bilbao[1], and Javier Del Ser[1,2(✉)]

[1] University of the Basque Country (UPV/EHU), Bilbao, Spain
[2] TECNALIA Research & Innovation, Derio, Spain
javier.delser@tecnalia.com

Abstract. A profitable strand of literature has lately capitalized on the exploitation of the collaborative capabilities of robotic swarms for efficiently undertaking diverse tasks without any human intervention, ranging from the blind exploration of devastated areas after massive disasters to mechanical repairs of industrial machinery in hostile environments, among others. However, most contributions reported to date deal only with robotic missions driven by a single task-related metric to be optimized by the robotic swarm, even though other objectives such as energy consumption may conflict with the imposed goal. In this paper four multi-objective heuristic solvers, namely NSGA-II, NSGA-III, MOEA/D and SMPSO, are used to command and route a set of robots towards efficiently reconstructing a scene using simple camera sensors and stereo vision in two phases: explore the area and then achieve validated map points. The need for resorting to multi-objective heuristics stems from the consideration of energy efficiency as a second target of the mission plan. In this regard, by incorporating energy trophallaxis within the swarm, overall autonomy is increased. An environment is arranged in V-REP to shed light on the performance over a realistically emulated physical environment. SMPSO shows better exploration capabilities during the first phase of the mission. However, in the second phase the performance of SMPSO degrades in contrast to NSGA-II and NSGA-III. Moreover, the entire robotic swarm is able to return to the original departure position in all the simulations. The obtained results stimulate further research lines aimed at considering decentralized heuristics for the considered problem.

Keywords: Swarm robotics · Scene reconstruction · Stereo vision ·
Energy trophallaxis · Multi-objective heuristics

© Springer Nature Switzerland AG 2019
P. Kaufmann and P. A. Castillo (Eds.): EvoApplications 2019, LNCS 11454, pp. 599–615, 2019.
https://doi.org/10.1007/978-3-030-16692-2_40

1 Introduction

In the last decade the proliferation of Swarm Robotics is rising sharply in different application scenarios, energetically propelled by the progressive technological maturity gained by Unmanned Aerial Vehicles (UAV), autonomous cars and micro-electronic/micro-mechanic devices, among others [1–3]. Although the spectrum of scenarios faced by swarm robotics is very diverse, a significant fraction is focused on disaster/hostile environments, where the robotic swarm must operate with human intervention kept at its minimum due to the harsh, characteristics and uncertainty held in such situations [4,5]. As a result, the person commanding the robotic mission is assigned to status monitoring and captured data assessment, hence delegating operational decisions in the swarm itself or, alternatively, in communications equipment (e.g. gateway).

In most cases related to the scenario exposed above, the mission is mostly driven by a measure of efficiency that numerically quantifies the progress of the robotic swarm in completing the tasks in which the mission is defined [6]. Examples of this measure are many and diverse in nature, ranging from the explored area to the degree of completion of every task. From a mathematical perspective, an optimal mission plan could be formulated as a dynamic optimization problem embracing such a measure as its single optimization objective. However, such a formulation neglects other operational aspects that may remove on the mission efficiency. Energy consumption is arguably the most intuitive factor in this regard, especially for unattended scenarios as the one in hand. To minimize its relevance for a given mission, attention is being paid in the community working on Swarm Robotics to developments at different levels aimed at fostering energy efficiency, including lightweight materials, battery technologies and/or low-power sensing devices for exploration. These *local* improvements for energy efficiency can be complemented by *trophallaxis*, a concept borrowed from synergistic relationships observed in Nature [7,8] by which robots are endowed with energy donation capabilities that make the swarm more energetically autonomous.

This work addresses an exploration problem where the goal is to guide a robotic swarm so that they efficiently reconstruct a three-dimensional mesh of points that represent the scene over which it is deployed. The scenario incorporates several modeling assumptions for the sake of a better energy efficiency of the entire swarm, which can be further considered as the main novel ingredients of this work with respect to the state of the art. First, robots within the swarm are equipped with low-power RGB cameras instead of other sensors with higher energy consumption rates (e.g. LIDAR). The swarm approach adds redundancy and higher failure resilience to the whole, while offering a scalable solution to affordable costs (each camera is less expensive and less power consuming than a laser-based system). As a result, robots sense collaboratively their surroundings by taking pictures, delivering and fusing them in the control center by means of stereo vision processing. Secondly, energetic trophallaxis is considered by allowing certain robots to recharge the batteries of the rest of the swarm, for which their route will be also considered as a variable to be dynamically optimized during the mission. Beyond the formulated problem itself, a third aspect that brings novelty to this study refers to the use of multi-objective heuristic algorithms for its computationally efficient solving. In particular we will benchmark and quantitatively verify the performance of four different multi-objective methods widely known by the community – namely, NSGA-II [9], SMPSO [10], MOEA/D [11], and NSGA-III [12]. We will show that the considered heuristics excel at balancing between the *quality* of the exploration and its energy efficiency by dynamically tracing the routes to be followed by each member of the swarm. We therefore deem the approximation of the Pareto front between both objectives attained by such heuristic solvers to be a valuable operational asset for the person controlling the mission, as it is an indicator assessing the risk of battery depletion taken when the swarm is forced to explore as much as possible with their remaining battery level.

The rest of this paper is structured as follows. Section 2 reviews background literature related to this work, whereas Sect. 3 poses the addressed problem. Next, Sects. 4 and 5 introduce the considered solvers and the simulated scenario. Results are discussed in Sect. 6, and Sect. 7 concludes the paper.

2   Related Work

As sketched in Sect. 1, this work elaborates on three different branches related to Swarm Robotics, namely: (1) energy trophallaxis; (2) 3D reconstruction of a scene through cost-efficient vision sensors and stereo vision image processing; and (3) multi-objective optimization. These three technologies collide together in our modeled scenario, with implications imprinted in the formulation of the underlying optimization problem. To the best of our knowledge, this is the first time these three concepts are used together in a Swarm Robotics context.

Remarkable scientific effort has been made in recent years regarding energy efficiency in Robotic Swarms. An early example can be found in [13], where the importance of having a robust swarm is discussed, defining robustness as the ability of the robotic swarm to conduct a complex mission avoiding, at the same time, the total drainage of their batteries. Authors of this paper also proposed a practical solution for enhancing the robustness of a swarm, which relied in the use of power stations. In connection with this initial work, trophallaxis within Swarm Robotics was first approached in [14,15]. This idea gravitates on individual energy donations between the robots of the swarm. This collective energy distribution strategy has been widely employed thereafter. An illustrative use case was reported in [16] for a dust cleaning swarm of robots. Another instance putting trophallaxis to practice was contributed in [17], in which this altruistic behavior is materialized in a specific robot architecture named CISSBot. More recently, in [18] an exploratory swarm of robots is deployed, using trophallaxis as one of the key concepts for their efficient scouting.

Another efficient procedure to energy consumption is the one recently proposed in [19], where an energy-aware movement strategy for aerial microrobots was optimized using Particle Swarm Optimization (PSO) heuristics. The designed approach considers energy levels of the aerial robots for their efficient movement. Despite no charging mechanism is implemented in this solution, the designed method renders a considerable reduction of the total energy consumption, making the robotic swarm more reliable and robust. Furthermore, as authors highlighted in [20], an approach often used for dynamic energy charging lies on the deployment of removable chargers or power banks. This concept has emerged a promising performance, despite having its own disadvantages, such as the weight increase of the robot equipment. As a result of these negative aspects, several studies have been recently undertaken aimed at circumventing them.

Back in connection to the developed system, a three-dimensional point representation of the explored scene is built by inferring the depth of objects from stereo image pairs (epipolar geometry). Indeed, vision has become one of the cheapest, most challenging and promising means for robots to perceive the environment [21]. A single LIDAR system is known to be capable of a comprehensive 3D-scene reconstruction, however power and reconstruction time get compromised through this solution. Recent studies have evinced how an autonomous navigation system can be implemented using only a stereo camera and a low-cost GPS receiver [22]. Therefore, instead of using expensive and power-consuming sensors, software-based methods implementing visual odometry, pose estimation, obstacle detection, local path planning and a waypoint follower are used. We partly embrace low-cost RGB sensors as a cost-efficient approach for a robotic swarm to reconstruct a scene. We use the principle of measurement distances with two cameras, namely binocular stereo or epipolar geometry [23]: if two or more images are taken from different camera positions, three-dimensional structures of the scene can be extracted on the basis of triangulation principles. Before triangulation can be applied, cameras must be calibrated and images rectified, for which intrinsic parameters of the camera are required to be known a priori. If camera parameters are instead unknown but constant over time, we assume the motion of the robot holding the camera to be unrestricted, thus allowing for the application of the self-calibration method proposed by [24]. This enables performing stereo correspondence by finding homologous points in the two images, for which many solvers have been proposed [25]. For our scenario in Swarm Robotics, we review the method in [23] for multiple-view stereo vision, based on (i) the positions of the viewpoints are known; (ii) directions from each viewpoint to the target are known; or (iii) both previous conditions hold.

A method that performs 3D object reconstruction based on multiple viewpoints of the same scene is explained in [26]. The method uses a space-carving solver equipped with a lighting compensating dissimilarity measure, then refined by an artificial Ant Colony Optimization (ACO) solver. Instead of reconstructing the objects' shape under the assumption that the existence of the object is known, our approach overrides this assumption by initially focusing on maximizing the area that the robotic swarm can explore and reconstruct until a given threshold of the explored is attained. Upon meeting this thresholding value, the objective related to the quality of the exploration is changed to the maximization of the number of points coherently conformed by stereo vision. Both exploration objectives are assessed jointly with a measure of the remaining energy margin, which is also to be maximized for improving the robustness of the swarm and for lowering the risk of battery depletion. We therefore disregard object details and only require to approximate the visual hull of the imaged object.

3 Problem Statement

In accordance with the schematic diagram shown in Fig. 1, we consider a swarm $\mathcal{N}$ composed by $|\mathcal{N}| = N$ robots with positions $\{\mathbf{r}_n^{\Delta,t}\}_{n=1}^N \doteq \{(x_n^{\Delta,t}, y_n^{\Delta,t}, z_n^{\Delta,t})\}_{n=1}^N$ and orientations $\{\theta_n^{\Delta,t}\}_{n=1}^N, \forall \theta \in [-\pi, \pi]$ (rad) (with t denoting the time stamp). Without loss of generality, the swarm is assumed to be spread over a quadrangular prism volume $V^{\square} \doteq [X_{min}, X_{max}] \times [Y_{min}, Y_{max}] \times [Z_{min}, Z_{max}]$. Each robot $n \in \{1, \ldots, N\}$ is equipped with two vision sensors $S_n^{\Delta} = \{s_n^{\Delta,left}, s_n^{\Delta,right}\}$, allowing for the coverage of a pyramidal volume $V_n^{\Delta,t} \subset V^{\square}$ rooted on the location of the robot at time stamp t. Considering that $V_n^{\Delta,t}$ has a quadrangular pyramid form, with its symmetry axis aligned with the optical axis of any vision sensor of base $X_s \times Z_s$ and height D_f (namely, the focal distance of the sensors), then:

$$V_n^{\Delta,t} = \left\{ (x_s^t, y_s^t, z_s^t) \in V^{\square} : \right.$$

$$\left. y_s^t \leq D_f; \frac{-Z_s}{2D_f} y_s^t \leq z_s^t \leq \frac{Z_s}{2D_f} y_s^t; \frac{-X_s}{2D_f} y_s^t \leq x_s^t \leq \frac{X_s}{2D_f} y_s^t \right\}, \quad (1)$$

where (x_s^t, y_s^t, z_s^t) denote the coordinates associated to the left sensor $s_n^{\Delta,left}$ of robot $n \in \mathcal{N}$, with the origin fixed in its apex, and with the Y-axis associated to the sensor's focal axis (z_s^t is parallel to the absolute coordinate system Z axis).

For each pair of images taken by sensors S_n^{Δ} of any robot $n \in \mathcal{N}$ at time t, a set of points $\mathcal{P}_n^t$ of $|\mathcal{P}_n^t| = P_n^t$ are obtained through stereo vision. These points are characterized by their coordinates $\{\mathbf{r}_p^t\}_{p=1}^{P_n^t} \doteq \{(x_p^t, y_p^t, z_p^t) \in V^{\square}\}_{p=1}^{P_n^t}$; coordinates of the left sensors that detected the corresponding points at time t, expressed as $\{\mathbf{r}_{s,p}^t\}_{p=1}^{P_n^t} \doteq \{(x_{s,p}^t, y_{s,p}^t, z_{s,p}^t) \in V^{\square}\}_{p=1}^{P_n^t}$; orientations of those sensors $\{\theta_{s,p}^t\}_{p=1}^{P_n^t}$ (which are the same for the right and left sensors of the corresponding robot); and pixel colors of the points in RGB code $\{\mathbf{c}_p^t\}_{p=1}^{P_n^t} \doteq \{(R_p^t, G_p^t, B_p^t)\}_{p=1}^{P_n^t}$.

Every point set $\mathcal{P}_n^t$ can be sorted into different subsets, attending to their coordinates. Let $\mathcal{G}$ be a regular mesh built by rectangular prisms of sides $\mathcal{E}_x \times \mathcal{E}_x \times \mathcal{E}_z$, where $\mathcal{E}_\rho = (\rho_{max} - \rho_{min})/\alpha_\rho$ ($\rho \in \{x, y, z\}$), with $\alpha_x, \alpha_y, \alpha_z \in \mathbb{N}$ denoting the number of divisions along every axis, so that $G = |\mathcal{G}| = \alpha_x \cdot \alpha_y \cdot \alpha_z$. The size of each prism in $\mathcal{G}$ represents the spatial granularity (level of detail) at which the scene is reconstructed. Since stereo vision with two cameras is not totally precise and prone to false positives, each prism in the grid must be validated to avoid false positives. In this paper, the validation criteria relies on two different conditions. Specifically, for the validation of a prism, at least another point p' in g should be detected (1) from a position located at a distance of at least r_{ref} from the location from which other points was located; and/or (2) with another orientation with a minimum difference of θ_{ref}. This detection should happen at the same time instance t' (detected by another robot) or at another previous time instance t (detected by the same or another robot). In mathematical terms, if one of the following conditions hold:

$$\left\{ \begin{array}{l} |\mathbf{r}_{s,p'}^{t'} - \mathbf{r}_{s,p}^{t}| \leq r_{ref} \\ |\theta_{s,p'}^{t'} - \theta_{s,p}^{t}| \leq \theta_{ref} \end{array} \right\} \forall p \in [1, P_n^{t'}), \forall t \in [0, t'] \tag{2}$$

then η_g will equal 0 (and 1 otherwise), being η_g a binary function indicating whether the prism g has been validated.

In another vein, an additional set of $M \leq N$ robots $\mathcal{M}$ with battery recharging capabilities is deployed over the same scene, jointly with the rest of the robotic swarm $\mathcal{N}$, with coordinates $\{\mathbf{r}_m^{\odot,t}\}_{m=1}^{M} \doteq \{(x_m^{\odot,t}, y_m^{\odot,t}, z_m^{\odot,t})\}_{m=1}^{M}$. Two different conditions should be fulfilled for robot $m \in \mathcal{M}$ to recharge the battery of robot $n \in \mathcal{N}$. The first condition dictates that their distance $d_{m,n}^t$ must fall below a certain threshold D_{max} (area in ■ in Fig. 1), i.e.:

$$d_{m,n} = \sqrt{\left(x_m^{\odot,t} - x_n^{\Delta,t}\right)^2 + \left(y_m^{\odot,t} - y_n^{\Delta,t}\right)^2} \leq D_{max}, \tag{3}$$

whereas the second rule determines that the above distance must be maintained for a minimum of T_{min} seconds, modeling the power plug coupling/uncoupling maneuver to align connectors. If these two conditions are met, energy is transferred from the recharging robot $m \in \mathcal{M}$, to $n \in \mathcal{N}$, so that a fraction $\beta^{\Delta T}$ of the maximum battery capacity B_{max} of the robot n is transferred from m during the time interval ΔT.

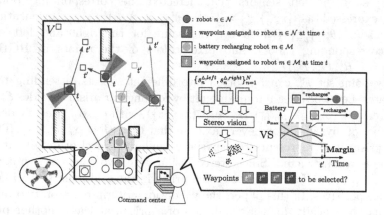

Fig. 1. Schematic diagram of the robotic swarm scenario tackled in this paper, for $N = 6$ exploring and $M = 1$ recharging robots.

Furthermore, the sensors and the movement of the robot itself involve a battery consumption. Each pair of sensors consume a constant rate of δ_{on} units of power and δ_{img} units for each pair of images captured. The movement of the robot implies also a consumption, fixed in γ units per arm of distance. Therefore, in a time gap ΔT measured from time t, the remaining amount of battery $B_n^{\Delta,t+\Delta T}$ in robot n can be mathematically expressed as:

$$B_n^{\Delta,t+\Delta T} =$$
$$\min\left\{B_n^{\Delta,t} + I_D I_T \beta^{\Delta T} B_{max} - \gamma V_n^{\Delta} \Delta T - \delta_{on} \Delta T - \delta_{img} \xi_{img}^{s,\Delta T}, B_{max}\right\}, \quad (4)$$

where V_n^{Δ} denotes the speed of the robot (in units of distance per unit of time), and $\xi_{img}^{s,\Delta T}$ corresponds to the number of image pairs taken by sensor s associated to robot n during the time interval ΔT. Finally, I_D and I_T are auxiliary binary functions such that $I_D = 1$ if $d_{m,n}^{t'} < D_{max} \ \forall t' \in [t, t + \Delta T]$, and $I_T = 1$ if $\Delta T \geq T_{min}$ (0 otherwise in both cases). In the above expression B_{max} stands for the nominal maximum battery load (in units of energy) of the robot model, which is considered equal throughout the entire swarm $\mathcal{N}$.

Bearing all these definitions in mind, the main objective of the optimization problem is to find an optimal set of waypoints over time for the exploring robots $\mathbf{W}^{\Delta,t,\looparrowright} \doteq \{\mathbf{w}_n^{\Delta,t,\looparrowright}\}_{n=1}^{N} = \{(x_n^{\Delta,t,\looparrowright}, y_n^{\Delta,t,\looparrowright})\}_{n=1}^{N}$ and for the battery recharging robots $\mathbf{W}^{\odot,t,\looparrowright} \doteq \{\mathbf{w}_m^{\odot,t,\looparrowright}\}_{m=1}^{M} = \{(x_m^{\odot,t,\looparrowright}, y_m^{\odot,t,\looparrowright})\}_{m=1}^{M}$. However, we place an explicit focus on discovering a set of Pareto-optimal different routes balancing quantitative measures of exploration and the *risk of no return* of the swarm. Intuitively, the larger the area explored by the robotic swarm is, the higher the chances of any robot not to have enough battery to return to its initial point $\{(x^{\Delta,0}, y^{\Delta,0})\}$ will be. This risk is crucial in practice for the situations in which this work is framed (i.e. disaster events requiring no human intervention).

Mathematically the quality of the exploration task is defined on the basis of an initial metric related to the explored area, which is replaced by a second metric once a sufficiently large area has been explored. The first is calculated by projecting the analyzed volume defined in (1) onto the XY plane. Thus, the total area $A^T(t)$ examined by the robotic swarm at time t' is given by:

$$A^T(t') = \frac{1}{A_{max}} \bigcup_{t=1}^{t'} \bigcup_{s=1}^{N} A_s^{\Delta,t}, \quad (5)$$

where it should be clarified that some objects can be spread over the scenario, reason for which the total explored area might not reach the total area value A_{max}. Accordingly, the established threshold to attend to the second task of this objective is to reach a stationary zone, where the total explored area is not practically incremented from iteration to iteration. When this situation is achieved, the determination of waypoints $\mathbf{W}^{\Delta,t,\looparrowright}$ and $\mathbf{W}^{\odot,t,\looparrowright}$ depends on non-validated points (namely, those points in $\mathcal{G}$ with $\eta_g = 1$). For this task, it is assumed that robots' vision covers 360°. This way, its vision volume $V_s^{\bowtie}$ is the revolution of volume in Expression (1) around the absolute Z axis. By letting

$\overline{P_s^{\bowtie,t}}$ denote the average non-validated points for a pair of sensors s located at a certain position at an instance t, and $\mathcal{G}_s^{\bowtie,t}$ the subset of prisms contained by the vision volume at instant t (with cardinality $|\mathcal{G}_s^{\bowtie,t}| = G_s^{\bowtie,t}$), then:

$$\overline{P_s^{\bowtie,t}}(\mathcal{G}_s^{\bowtie,t}, \mathbf{w}_n^{\Delta,t,\looparrowright}) = \frac{\sum_{g=1}^{G_s^{\bowtie,t}} \eta_g}{G_s^{\bowtie,t}}, \tag{6}$$

that, extended to all explorer robots within the swarm, renders the second measure of exploration quality as:

$$\overline{P^{\bowtie}}(t) = \sum_{n,s=1}^{N} \overline{P_s^{\bowtie,t}}(\mathcal{G}_s^{\bowtie,t}, \mathbf{w}_n^{\Delta,t,\looparrowright}). \tag{7}$$

The *risk of no return* can be mathematically modeled by computing the battery margin $B_n^{\blacktriangle,t}$ expected to be left for each robot assuming the movement to the assigned waypoint and the safe return to $\{(x^{\Delta,0}, y^{\Delta,0})\}$. Thus, considering that route optimization is performed at time t, the value of the battery margin $B_n^{\blacktriangle,t}$ for the robot $n \in \mathcal{N}$ when commanded to go to waypoint $\mathbf{w}_n^{\Delta,t,\looparrowright} = (x_n^{\Delta,t,\looparrowright}, y_n^{\Delta,t,\looparrowright})$ can be computed as:

$$B_n^{\blacktriangle,t}(\mathbf{r}_n^{\Delta,t}, \mathbf{w}_n^{\Delta,t,\looparrowright}, \{\mathbf{r}_m^{\Delta,t}\}_{m=1}^M, \{\mathbf{w}_m^{\Delta,t}\}_{m=1}^M) = B_n^{\Delta,t} - B_n^{\Delta,t+\Delta T_{\mathbf{r},\mathbf{w}}+\Delta T_{\mathbf{w},\mathbf{r}_0}}, \tag{8}$$

where $\Delta T_{\mathbf{r},\mathbf{w}}$ and $\Delta T_{\mathbf{w},\mathbf{r}_0}$ are the times needed by the robot $n \in \mathcal{N}$ to travel from its current situation $\mathbf{r}_n^{\Delta,t}$ to the established waypoint $\mathbf{w}_n^{\Delta,t,\looparrowright}$, and return to its initial position $\{(x^{\Delta,0}, y^{\Delta,0})\}$. This estimation is made by assuming that the robot performs the trip without colliding with any object placed in the scenario nor with any other robot along its trajectory. Noteworthy is also to mention that the battery expenditure as per (4) takes into account the power consumed by the robot dynamics and sensors' operations, and also the time periods along the path during which the relative position between battery recharging robots and robot $n \in \mathcal{N}$ fulfill conditions I_D and I_T required to recharge the battery of robot n on the move. The total duration of such recharging periods can be computed as $\sum_{(t_s,t_e)\in\mathcal{T}_n^{\Delta,t}}(t_e - t_s)$ over the set of periods $\mathcal{T}_n^{\Delta,t}$, defined as:

$$\mathcal{T}_n^{\Delta,t} \doteq (t_s, t_e) \in [t, t + \Delta T_{\mathbf{r},\mathbf{w}} + \Delta T_{\mathbf{w},\mathbf{r}_0}] \text{ such that:} \tag{9}$$

$$(1)\ t_e > t_s; \tag{10}$$

$$(2)\ \exists m \in \mathcal{M} : d_{mn}^{t'} \leq D_{max} \forall t' \in [t_s, t_e]; \tag{11}$$

$$(3)\ t_e - t_s \geq T_{min}, \tag{12}$$

with $[t_s', t_e'] \cap [t_s'', t_e''] = \emptyset \ \forall (t_s', t_e'), (t_s'', t_e'') \in \mathcal{T}_n^{\Delta,t}$. Therefore, the swarm-wide battery margin $B^T(t)$ at time t so as to keep the aforementioned risk to its minimum is given by:

$$B^T(t) = \min_{n \in \mathcal{N}} \{\max \{0, B_n^{\blacktriangle,t}(\mathbf{r}_n^{\Delta,t}, \mathbf{w}_n^{\Delta,t,\looparrowright}, \{\mathbf{r}_m^{\Delta,t}\}_{m=1}^M, \{\mathbf{w}_m^{\Delta,t}\}_{m=1}^M)\}\}, \tag{13}$$

from which the problem addressed in this work can be formally posed as:

$$\underset{\mathbf{W}^{\odot,t,\looparrowright},\,\mathbf{W}^{\Delta,t,\looparrowright}}{\text{maximize}} \; \left\{ [A^T(t), \overline{P^{\bowtie}}(t)], B^T(t) \right\}, \tag{14}$$

namely, the simultaneous maximization of two conflicting objectives: the quality of the 3D scene mapping performed by the robotic swarm through stereo vision; and the minimum expected battery margin over the robots should it be commanded to return to the initial deployment point after reaching the enforced waypoint. It should be noted that each of the orientations $\theta_n^{\Delta,t}$ of robots $n \in \mathcal{N}$ are selected in a greedy fashion such that the average number of non-validated points for the corresponding selected waypoint is maximized.

4   Considered Multi-objective Solvers

The optimization problem stated above can addressed in a computationally efficient manner by using evolutionary multi-objective heuristics. Specifically four different algorithms have been selected to assess their performance: NSGA-II [9], SMPSO [10], MOEA/D [11] and NSGA-III [12]. The choice of this heterogeneous group of methods gravitates on two reasons. On the one hand, all have been utilized profusely in the literature to efficiently solve real-valued multi-objective problems similar to the one tackled in this paper. On the other hand, both Pareto-dominance based and decomposition-based evolutionary solvers can be found among this algorithmic benchmark. It is also worth mentioning that two of these approaches, namely NSGA-II and SMPSO, have widely demonstrated an outstanding effectiveness when dealing with bi-objective optimization problems. Furthermore, MOEA/D and NSGA-III are also appropriate for many-objective problems; therefore their inclusion in the benchmark is not only for comparative purposes, but also a test point towards more elaborated problem formulations with an increased number of objectives[1]. Each heuristic solver within the benchmark is now detailed:

- NSGA-II (*Non-dominated Sorting Genetic solver II*) is a generational solver which employs a Pareto ranking mechanism to guide an evolutionary search process towards a Pareto front with progressively enhanced diversity and optimality. It also defines a crowding distance density estimator, which is crucial to measure the diversity of the produced solutions.
- SMPSO (*Speed-constrained Multi-objective Particle Swarm Optimization*) is a Swarm Intelligence solver whose main feature is the use of a velocity constraint to prevent particles (solutions) to surpass the limits of the solution space. Furthermore, SMPSO stores an external archive of bounded size where non-dominated solutions replaced during the run are saved. This archive is also used to select the leader.

---

[1] Future research paths in this regard will be sketched in Sect. 7.

- MOEA/D (*Multi-Objective Evolutionary solver Based on Decomposition*) is a steady-state evolutionary solver which relies on an aggregative approach. The cornerstone of this approach is to decompose the problem into a group of single-objective subproblems. Then, these subproblems are faced simultaneously by taking into account information of a set of neighbors.
- NSGA-III (*Non-dominated Sorting Genetic solver III*) is a reference-point-based many-objective evolutionary technique, which follows the same structure than NSGA-II, but incorporating specific novel aspects: NSGA-III emphasizes on non-dominated population of individuals, yet still close to a set of supplied reference points.

In order to attain the main goal of this study is to conduct an experimental comparison between the selected four multi-objective methods, several multi-objective quality indicators can be measured over time. Specifically the so-called hypervolume indicator achieved by every solver will be computed over time for a total of 20 independent experiments run for every solver over a reference scenario. Statistics drawn from these indicators can be then utilized to compare among such algorithms and eventually find performance gaps, whose statistical significance can be verified by resorting to non-parametric hypothesis tests. The next section will provide further details on the methodology adopted for carrying out the experimental part of this work.

5   Simulation Setup

In order to check and compare the performance of the considered approaches, a simulation framework has been built with V-REP and Python 3.5 at its core. V-REP [27] is a renowned software platform that allows realistically designing and performing experimental studies with robotic devices. Given the computational resources demanded by this platform, we have kept the dimensions of the scenario reduced to $N = 6$ exploring robots, and a single battery-recharging node ($M = 1$). This robotic swarm is deployed over a $10 \times 10\,\text{m}^2$ square area, with a height of 2.7 m. Furthermore, the maximum distance to recharge batteries has been established to $D_{max} = 1$ m, while the minimum time that a robot must stay coupled to the battery-recharging node has been set to $T_{min} = 3$ s.

Each robot within the swarm is an *hexapod* (i.e. a robot with six mechanical legs arranged in an hexagonal chasis) with a fixed diameter size of 0.5 m. This specific robot model was selected due to its better adjusted dynamics to harsh and disaster environments. Robot speeds are set to $V_n^\triangle = 3.5$ cm/s $\forall n \in \{1,\ldots,6\}$ and $V_1^\odot = 2.3$ cm/s. Regarding the amount of battery recharged per second, a value of 1% with respect to the nominal maximum capacity B_{max} has been considered for mapping nodes, while the recharging robot is endowed with a total battery capacity of $10 \cdot B_{max}$. Moreover, battery consumption rates have been fixed to $\gamma = 1.5\%$ per meter, $\delta_{on} = 0.1\%$ per second, and $\delta_{img} = 0.05\%$ per captured image pair, all with respect to B_{max}. The decision making criterion considered to select a solution among the estimated Pareto fronts over time is to choose the solution (waypoints) whose associated battery margin is closest to 20% of B_{max}. If there is no solution

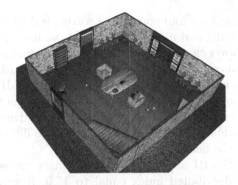

Fig. 2. Snapshot of the scene considered in this work.

with a margin greater than this threshold, robots are forced to return to the departure position (i.e. the location from which the swarm originally departed). The reference scene, which is depicted in Fig. 2, emulates an indoor environment with several physical objects that should be mapped by the robotic swarm by taking pictures, delivering and feeding them to a stereo vision processing pipeline, which is performed at the control center.

On the other hand, as explained above each robot is equipped with two photographic cameras placed in the frontal side of the node, and horizontally separated by $b = 0.4$ m. It is worth noting that cameras belonging to different robots have the same characteristics and are configured equivalently. Focal angle and distance have been set to $\psi_0 = 50°$ and $D_f = 2$ m, respectively. Thus, the width of the captured image in the XY plane is $X_s = 2 \cdot D_f \cdot \tan(\psi_0/2) = 1.865$ m, while its height in the Z axis is $Z_s = 1.865$ m. Finally, in order to validate a point that a pair of cameras has detected, the minimum distance between these two cameras and the minimum angle between their directions have been set to $r_{ref} = 0.05$ m and $\theta_{ref} = 5°$, respectively.

Table 1. Parameter setting of the solvers considered in the experimental benchmark.

solver	Parameter	Value
All	Population/swarm size	100 solutions (92 NSGA-III)
	Evaluations	1000
	Independent runs	10
	Mutation	Polynomial mutation
	| Probability	0.1
	| Distribution index η_m	20.0
SMPSO	Archive size	100
	Mutation perturbation	0.5
	Density estimator	Crowding distance
NSGA-II NSGA-III	Selection	Binary Tournament Selection
	Crossover	Simulated binary crossover
	| Probability	1.0
	| Distribution index η_m	15
MOEA/D	Differential evolution scheme	rand/1/bin
	| CR	1.0
	| F	0.5
	Neighborhood size	10
	Neighborhood selection probability	0.8
	Max. number of replaced solutions	1

Parameter settings of the four considered solvers is depicted in Table 1. Each technique has been configured using conventional parameter settings utilized in the literature, but always ensuring that the subsequent performance comparison is fair in terms of computational complexity. In this regard, NSGA-III uses a population size of 92 individuals, whereas in the case of SMPSO the size of the swarm has been set in 100, with the capacity of its external archive established in 92. This archive utilizes the crowding distance as its density estimator. Furthermore, the population of MOEA/D counts with 100 different individuals, all of them chosen from the final population.

NSGA-II and NSGA-III apply a simulated binary crossover, with a probability of 1.0 and a distributed index equal to 15.0. Regarding the mutation procedure, all methods use a polynomial mutation function with a probability of 0.1 and a distributed index equal to 20.0. Additionally, MOEA/D follows a *rand/1/bin* differential evolution scheme, with $F = 0.5$ and $CR = 1.0$. For this same solver, the neighborhood selection probability, maximum number of replaced solutions and neighborhood size have been fixed to 0.8, 1 and 10, respectively. Finally, methods compute a maximum of 5000 function evaluations per execution, ensuring a fair comparison between them.

6   Results and Discussion

Outcomes rendered by each heuristic solver over the reference scene in Fig. 2 can be inspected in Figs. 3, 4 and 5. To begin with, Fig. 3 (uppermost plot) illustrates a boxplot of the hypervolume values scored by every solver in the benchmark as a function of time. Specifically, every multi-objective algorithm is run again once all nodes have reached their waypoint commanded by the control center

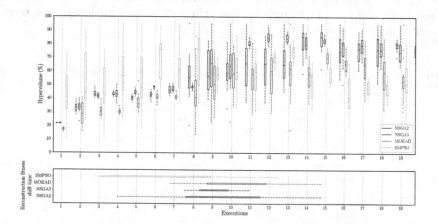

Fig. 3. Boxplot of the hypervolume scores obtained by each algorithm in the benchmark over their first 20 executions. In the plot below, boxplot showing when the exploration quality objective shifts from the explored area as per Expression (5) to the total number of validated points given in Expression (7).

after the previous execution. Therefore, values 1, 2 and forth in the X axis of this figure denote the run index of every algorithm, but might correspond to different instants in time.

From a general perspective the hypervolume attained by the fronts estimated by every solver increases over the duration of the mission, which is a reliable indicator that the robotic swarm progressively reaches better exploration levels. A deeper analysis reveals that SMPSO emerges as the algorithm featuring better exploration capabilities in early stages of the mission, scoring notably higher (yet less statistically stable) hypervolume values over this phase. On the contrary, MOEAD clearly lags behind, with remarkably lower hypervolume scores. However, a change is noted when the quantification of the mission accomplishment passes from area to the number of validated points, which occurs in a variable run index for every algorithm as depicted in the bottom plot of Fig. 3: in this second phase the performance of SMPSO degrades severely in contrast to NSGA-II and NSGA-III, which thereafter render significantly better hypervolume scores than the rest of algorithms in the benchmark. This superiority was buttressed by a Wilcoxon test performed off-line between every pair of result sets, for every run index and with a 95% confidence interval: NSGA-II and NSGA-III were proven to be superior to MOEAD and SMPSO for more than 95% of the runs, but at the same time no clear winner was found when comparing the scores attained by NSGA-II and NSGA-III.

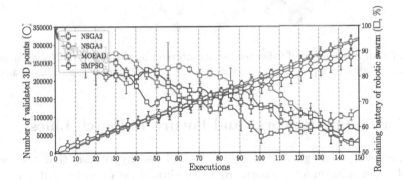

Fig. 4. Dual plot showing the average total number of validated points versus the average remaining battery through 150 executions of the solvers with error bars. Curves with ○ correspond to the Y axis on the left (validated points), whereas values of the curves with □ must be read from the Y axis on the right (battery margin). (Color figure online)

We follow the discussion by inspecting Fig. 4, which depicts the evolution of the number of validated points achieved by the robotic swarm during the duration of the mission for each solver. This plot also includes the overall remaining battery level (in %) for every algorithm. First it is insightful to note that the battery level does not decrease monotonically during the time spanned by mission as an effect of the energy donation enabled by robotic trophallaxis. This is particularly relevant in the case of NSGA-III, which helps the algorithm attain higher values of the average remaining battery levels all over the mission. Differences between algorithms regarding the number of validated points are significantly lower, with MOEAD and NSGA-II still scoring best (without statistical significance, though). Interestingly, all average remaining battery levels overpass the 50% over the depicted mission time span, meaning that the entire robotic swarm is able to return to the original departure position.

Finally, Fig. 5 exemplifies the output shown to the controller of the mission after one run of the NSGA-III solver has been completed. The obtained cloud of validated points permits to infer the contents of the explored scene by only processing images captured by low-cost sensors and by keeping the risk of no return to its minimum during the mission.

Fig. 5. Exemplifying grid of validated points inferred by the NSGA-III solver in one of the runs (a miniature plot of the simulated scene is provided for ease of assessment).

7   Concluding Remarks and Future Research Lines

This manuscript has gravitated on the postulated hypothesis that the person controlling a mission with swarm robotics should be provided not only with information about the degree of completion of the mission itself, but should also handle information about other related aspects (e.g. risk of battery depletion or *no return*) that are of utmost relevance in operations over harsh environments. In this regard, energy efficiency has been considered in this work by including several novel modeling assumptions with respect to the state of the art: (1) the adoption of energy trophallaxis within the robotic swarm; (2) the use of low-cost, low-consumption vision sensors and the processing of the captured information through stereo vision processing pipelines; and (3) the consideration of the battery margin (a quantitative measure closely linked to the risk of no return) as another mission goal to be jointly optimized with the quality of the exploration undertaken by the robotic swarm. In order to efficiently infer the mission plan (waypoints) that optimally balances the Pareto trade-off between both objectives, several multi-objective heuristics have been utilized to guide a swarm of robots over a simulated indoor scene. The obtained results are insightful in regards to the need for jointly assessing both objectives dynamically.

Several additional research lines have been planned for the near future in connection to this work. The first one is the inclusion of additional bioinspired multi-objective heuristic solvers to the performed experimentation, using also additional multi-objective indicators for their comparison. Another research objective is the conduction of a more extensive experimentation, with further real-world scenarios and a greater number of simulations. Finally, the most challenging research direction to be followed gravitates on distributing the intelligence among the robots in order to realize a true robotic swarm, namely, a swarm of robots that communicate to each other and exchange information, deciding on an optimal set of waypoints without requiring a centralized command center as the one assumed in this work. To this end distributed optimization approaches based on message passing and local hypothesis testing will be considered such as Stochastic Diffusion Search [28]. By distributing the intelligence and by taming latest advances in multi-view 3D scene reconstruction, robots will be able to communicate and disseminate their captured imagery in a decentralized manner, yielding what we will refer to as *swarmoscopic vision* in future contributions. Additional interesting further research could be related with the hybridization of our methods other fields, such as the coverage path planning [29,30].

Acknowledgements. This work was supported by the Basque Government through the EMAITEK program.

References

1. Beni, G.: From swarm intelligence to swarm robotics. In: Şahin, E., Spears, W.M. (eds.) SR 2004. LNCS, vol. 3342, pp. 1–9. Springer, Heidelberg (2005). https://doi.org/10.1007/978-3-540-30552-1_1
2. Tan, Y., Zheng, Z.: Research advance in swarm robotics. Defence Technol. **9**(1), 18–39 (2013)
3. Ben-Ari, M., Mondada, F.: Swarm robotics. In: Elements of Robotics, pp. 251–265. Springer, Cham (2018). https://doi.org/10.1007/978-3-319-62533-1_15
4. Wong, C., Yang, E., Yan, X.T., Gu, D.: Autonomous robots for harsh environments: a holistic overview of current solutions and ongoing challenges. Syst. Sci. Control Eng. **6**(1), 213–219 (2018)
5. Wong, C., Yang, E., Yan, X.T., Gu, D.: An overview of robotics and autonomous systems for harsh environments. In: International Conference on Automation and Computing, pp. 1–6 (2017)
6. Barca, J.C., Sekercioglu, Y.A.: Swarm robotics reviewed. Robotica **31**(3), 345–359 (2013)
7. Korst, P., Velthuis, H.: The nature of trophallaxis in honeybees. Insectes Soc. **29**(2), 209–221 (1982)
8. Hamilton, C., Lejeune, B.T., Rosengaus, R.B.: Trophallaxis and prophylaxis: social immunity in the carpenter ant camponotus pennsylvanicus. Biol. Lett. **7**(1), 89–92 (2011)

9. Deb, K., Pratap, A., Agarwal, S., Meyarivan, T.: A fast and elitist multiobjective genetic algorithm: NSGA-II. IEEE Trans. Evol. Comput. **6**(2), 182–197 (2002)
10. Nebro, A.J., Durillo, J.J., García-Nieto, J., Coello Coello, C., Luna, F., Alba, E.: SMPSO: a new PSO-based metaheuristic for multi-objective optimization. In: IEEE Symposium on Computational Intelligence in Multicriteria Decision-Making, pp. 66–73 (2009)
11. Zhang, Q., Li, H.: MOEA/D: a multiobjective evolutionary algorithm based on decomposition. IEEE Trans. Evol. Comput. **8**(11), 712–731 (2008)
12. Jain, H., Deb, K.: An evolutionary many-objective optimization algorithm using reference-point based nondominated sorting approach, Part II: handling constraints and extending to an adaptive approach. IEEE Trans. Evol. Comput. **18**(4), 602–622 (2014)
13. Haek, M., Ismail, A.R., Basalib, A.O.A., Makarim, N.: Exploring energy charging problem in swarm robotic systems using foraging simulation. Jurnal Teknologi **76**(1), 239–244 (2015)
14. Schmickl, T., Crailsheim, K.: Trophallaxis among swarm-robots: a biologically inspired strategy for swarm robotics. In: IEEE/RAS-EMBS International Conference on Biomedical Robotics and Biomechatronics, pp. 377–382 (2006)
15. Schmickl, T., Crailsheim, K.: Trophallaxis within a robotic swarm: bio-inspired communication among robots in a swarm. Auton. Robots **25**(1–2), 171–188 (2008)
16. Melhuish, C., Kubo, M.: Collective energy distribution: maintaining the energy balance in distributed autonomous robots using trophallaxis. Distrib. Auton. Robot. Syst. **6**, 275–284 (2007)
17. Schiøler, H., Ngo, T.D.: Trophallaxis in robotic swarms-beyond energy autonomy. In: International Conference on Control, Automation, Robotics and Vision, pp. 1526–1533 (2008)
18. Carrillo, M., et al.: A bio-inspired approach for collaborative exploration with mobile battery recharging in swarm robotics. In: Korošec, P., Melab, N., Talbi, E.-G. (eds.) BIOMA 2018. LNCS, vol. 10835, pp. 75–87. Springer, Cham (2018). https://doi.org/10.1007/978-3-319-91641-5_7
19. Mostaghim, S., Steup, C., Witt, F.: Energy aware particle swarm optimization as search mechanism for aerial micro-robots. In: IEEE Symposium Series on Computational Intelligence, pp. 1–7 (2016)
20. Ismail, A.R., Desia, R., Zuhri, M.F.R.: The initial investigation of the design and energy sharing algorithm using two-ways communication mechanism for swarm robotic systems. In: Phon-Amnuaisuk, S., Au, T.W. (eds.) Computational Intelligence in Information Systems. AISC, vol. 331, pp. 61–71. Springer, Cham (2015). https://doi.org/10.1007/978-3-319-13153-5_7
21. Bonin-Font, F., Ortiz, A., Oliver, G.: Visual navigation for mobile robots: a survey. J. Intell. Rob. Syst. **53**(3), 263 (2008)
22. Hong, S., Li, M., Liao, M., van Beek, P.: Real-time mobile robot navigation based on stereo vision and low-cost GPS. Electron. Imaging **2017**, 10–15 (2017)
23. Sugihara, K.: Three principles in stereo vision. Adv. Robot. **1**(4), 391–400 (1986)
24. Pollefeys, M., Koch, R., Gool, L.V.: Self-calibration and metric reconstruction inspite of varying and unknown intrinsic camera parameters. Int. J. Comput. Vis. **32**(1), 7–25 (1999)
25. Mattoccia, S., De-Maeztu, L.: A fast segmentation-driven algorithm for accurate stereo correspondence. In: International Conference on 3D Imaging, pp. 1–6 (2011)
26. Chrysostomou, D., Gasteratos, A., Nalpantidis, L., Sirakoulis, G.C.: Multi-view 3D scene reconstruction using ant colony optimization techniques. Meas. Sci. Technol. **23**(11), 114002 (2012)

27. Rohmer, E., Singh, S.P., Freese, M.: V-REP: a versatile and scalable robot simulation framework. In: International Conference on Intelligent Robots and Systems (IROS), pp. 1321–1326. IEEE (2013)

28. De Meyer, K., Slawomir, N.J., Mark, B.: Stochastic diffusion search: partial function evaluation in swarm intelligence dynamic optimisation. In: Swarm Intelligence Dynamic Optimisation, pp. 185–207. Springer, Heidelberg (2006)

29. Zhu, D., Tian, C., Sun, B., Luo, C.: Complete coverage path planning of autonomous underwater vehicle based on GBNN algorithm. J. Intell. Robot. Syst. 1–13 (2018). https://link.springer.com/article/10.1007/s10846-018-0787-7

30. Horvátha, E., Pozna, C., Precup, R.E.: Robot coverage path planning based on iterative structured orientation. Acta Polytechnica Hungarica 15(2), 231–249 (2018)

Evolving Robots on Easy Mode: Towards a Variable Complexity Controller for Quadrupeds

Tønnes F. Nygaard[✉], Charles P. Martin, Jim Torresen, and Kyrre Glette

University of Oslo, Oslo, Norway
tonnesfn@ifi.uio.no

Abstract. The complexity of a legged robot's environment or task can inform how specialised its gait must be to ensure success. Evolving specialised robotic gaits demands many evaluations—acceptable for computer simulations, but not for physical robots. For some tasks, a more general gait, with lower optimization costs, could be satisfactory. In this paper, we introduce a new type of gait controller where complexity can be set by a single parameter, using a dynamic genotype-phenotype mapping. Low controller complexity leads to conservative gaits, while higher complexity allows more sophistication and high performance for demanding tasks, at the cost of optimization effort. We investigate the new controller on a virtual robot in simulations and do preliminary testing on a real-world robot. We show that having variable complexity allows us to adapt to different optimization budgets. With a high evaluation budget in simulation, a complex controller performs best. Moreover, real-world evolution with a limited evaluation budget indicates that a lower gait complexity is preferable for a relatively simple environment.

Keywords: Evolutionary robotics · Real-world evolution ·
Legged robots

1 Introduction

Robots are used in more and more demanding and changing environments. Being able to adapt to new situations, unexpected events, or even damage to the robot itself can be crucial in many applications. Robots that are able to learn and adapt their walking will be able to operate in a much wider range of environments.

Selecting a suitable gait controller for a robot learning to walk can be very challenging, especially when targeting hardware platforms. A controller is often chosen early in the design process of a robot, and is used in a wide range of different evaluation budgets and environments. Simple controllers produce gaits

---

This work is partially supported by The Research Council of Norway under grant agreement 240862.

P. Kaufmann and P. A. Castillo (Eds.): EvoApplications 2019, LNCS 11454, pp. 616–632, 2019.
https://doi.org/10.1007/978-3-030-16692-2_41

with a limited diversity. More complex gait controllers are able to produce a wider range of gaits, with higher variance in performance and behaviors.

Controllers that are too complex might exhibit bootstrap problems, where the initial random population does not contain a suitable gradient towards better solutions [1]. Random solutions might also exhibit a high probability of the robot falling, making it more challenging to evolve in hardware. Another important factor is the larger and more complex search space, which might require more evaluations to converge than practically possible without simulations [2].

A controller can be made simpler by embedding more prior knowledge, for instance by reducing the allowable parameter ranges of the controller. When the size of the search space is reduced, fewer evaluations are needed, and with more conservative parameter ranges, falling can be greatly reduced. Reducing the gait complexity too much, however, leaves the system with a very narrow and specialized controller that might not be able to produce gaits with the varied behaviors needed to adapt to new environments or tasks, and limitations set by human engineers might discard many near-optimal areas of the search space.

Being able to find the right complexity balance when designing a controller can be very challenging. Any choice made early in the design process might not suit future use, and picking a single controller complexity for all different uses might end up being a costly compromise reducing performance significantly. We have experienced this challenge in our own work where experiments are performed with a four-legged mammal-inspired robot with self-modifying morphology in both simulation and hardware [2]. Balancing the need for a low complexity controller when evolving morphology and control in few evaluations in hardware without falling, and evolution in complex and dynamic environments requiring exotic ways of walking in simulations, has proven impossible with our earlier controller design [3].

In this paper, we introduce a new controller where the complexity can be set by a single parameter that addresses this limitation. We use a dynamic genotype-phenotype mapping, illustrated in Fig. 1, where higher complexity controllers map the genotypic space to a larger controller space than lower complexity controllers. This allows a more flexible gait either when an evaluation budget allows for longer evolutionary runs, or when the added flexibility is needed for coping with difficult environments. Less flexible gaits can be used when there is a stricter evaluation budget, for instance in real-world experiments. We have investigated the controller in simulation with our four-legged mammal-inspired robot, and found that different gait complexities are optimal under different evaluation budgets. We also verified this through initial tests on the physical robot in the real world. This suggests that our new controller concept will be useful for coping with the competing demands of freedom versus ease-of-learning, especially important when evolving on both virtual and real-world robots.

The contribution of this paper is as follows: We introduce the concept of a variable complexity gait controller, and show how this can be implemented for a quadruped robot. We then demonstrate its value through experiments in

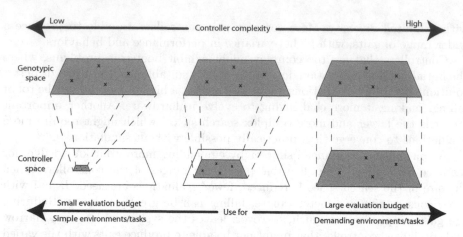

Fig. 1. This diagram shows the concept of a variable complexity controller. The genotypic space is always the same size, but the mapping to controller space is changed by the controller complexity parameter, giving safer and more conservative gaits at lower controller complexities.

simulation, and verify the results with preliminary testing on a physical robot in the real world.

2 Background

Evolutionary robotics uses techniques from evolutionary computation to optimize the brain or body of a robot. It can be used directly to improve the performance of a robot, or to study biological processes and mechanisms. When optimizing the brain of a robot, high-level tasks like foraging, goal homing or herding can be evolved, or lower level functions like sensory perception or new walking gaits. Optimizing the body of a robot allows adaptation to different tasks or environments, and research has shown that the complexity of evolved bodies mirror the complexity of the environments they were evolved in [4].

Several different types of optimization algorithms from evolutionary computation are used to optimize robot control. The most common is the Genetic Algorithm (GA) [5], which uses genetic operators like mutation and recombination to optimize gait parameters. It is often done using multiple objectives, in many cases achieving a range of solutions with different trade-offs in conflicting objectives, including speed and stability [6], or even speed, stability, and efficiency [7]. Evolutionary Strategies (ES) feature self-adaptation, by adding the mutation step size to the individuals. This has been shown to speed up the search, and in some cases outperform traditional EA approaches, when evolving quadrupedal robot gaits [8]. Genetic Programming (GP) represents individuals as tree structures rather than vectors, and has been shown to outperform simple GA algorithms when used to evolve quadruped gaits [9]. Quality-Diversity algorithms aim to build up an archive of solutions that exhibit different behaviors or characteristics that all perform as well as possible [10]. This set of diverse individuals then serves as a pool of solutions that can be searched through to find solutions to new problems, like a robot adapting to a broken leg [11].

Optimizing how a robot walks can be very difficult, and one of the biggest challenges is the bootstrap problem [1]. It can be very hard to start optimizing a robot gait if none of the random individuals tested initially provides a gradient towards good solutions. This is mostly a problem when optimizing in hardware, with much harder time constraints and potential physical damage to the robot. It can, however, also affect simulations, where initial individuals without any ability to solve a task can completely remove the selective pressure from the fitness functions needed for evolution to succeed.

There is a wide range of gait controller types used in evolutionary robotics, depending on what is being optimized. They are often divided into two categories, based on whether they work in the joint space, or Cartesian space [12]. A gait can either be represented as a few discrete poses with trajectories generated automatically between them, or as a continuous function that specifies the position or joint angles at all times. Some gait controllers use simple parameterized functions that control the joint space of the robot [11,13]. Other gait controllers used in evolutionary experiments consist of a parameterized spline that defines each legs trajectory in Cartesian space. Evolution optimizes either the position of the spline points directly [8], or some higher level descriptors [6,14]. Other controllers are based on central pattern generators of different architectures and models [15]. Some produce neural networks using techniques such as Compositional Pattern Producing Networks (CPPN), which has an inherent symmetry and coordination built-in. This can lead to gaits far surpassing the performance of hand-designed gaits based on parameterized functions [16].

The field of neuro-evolution often evolves the structure of the neural networks making up the gait controller, in addition to the connection weights. This goes against the general trend in other fields, where the complexity of gait controllers is most often kept static. Togelius defines four different categories [17]. *Monolithic evolution* uses a single-layered controller with a single fitness function. *Incremental evolution* in neuro-evolution has several fitness functions, but still one controller layer. *Modularised evolution* has more controller layers, but a single fitness function. *Layered evolution* uses both several controller layers, and several fitness functions. When evolving the complexity of a network, it has been shown that new nodes should be added with zero-weights [18], allowing evolution to gradually explore the added complexity.

3 Implementation

3.1 Robot

The experiments in this paper were performed on a simulated version of "DyRET", our four legged mammal-inspired robot with mechanical self-reconfiguration [3]. The robot platform is a fully certified open source hardware project, with source and details available online[1]. We use the Robot Operating System (ROS) framework for initialization and communication, and the simulated version runs on the Gazebo physics simulator. The robot and its simulated counterpart can be seen in Fig. 2.

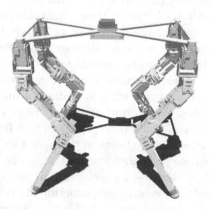

Fig. 2. The physical robot to the left, and the simulated robot to the right.

The robot uses Dynamixel MX-64 servos from Robotis in the hip joints, and Dynamixel MX-106 servos for the two lower joints. Its legs consist of two custom linear actuators each that allow reconfiguration of the leg lengths during operation. More mechanical details can be found in our previous work [3], and is not included here due to space constraints and the fact that we are mainly using a simulated version for our experiments.

3.2   Control

In our earlier experiments, we used a fairly standard parameterized spline-based gait controller working in Cartesian space. We have used the controller for evolving both control and morphology on the physical robot, with a complex search space with many degrees of freedom. This required us to have a low complexity controller, but that meant it was not flexible enough to give us more complex gaits when we had higher evaluation budgets, such as when using simulations. Our goal was for the new controller to be adaptable to fit whatever needs we currently have or might have in the future, with a controller complexity that could be changed with a single parameter.

---

[1] https://github.com/dyret-robot/dyret_documentation.

The Gait Controller. Since this gait is used on a physical mammal-inspired robot, the property of being learnable without excessive falling is important, and a much bigger challenge than for spider-inspired robots. We believe that a controller operating in joint space would not allow robust enough gaits at low controller complexity for our robot, so we chose to implement it in Cartesian space. There are many ways a gait can result in a fall, but ensuring that all legs on the ground are moving in the same direction with the same speed severely limits the chance of falling. Complementing this with a wide leg stance gives a good base to build a parameterizable gait controller on. Ensuring that only one leg is in the air at a time, and that the robot is always using the proper leg lift order, further helps the robot to remain stable.

Leg Trajectory. The control system uses standard inverse kinematics to get the individual joint angles from the calculated positions. The leg trajectory is parameterized using an interpolating looping cubic Hermite spline, which intersects five control points. A simple example trajectory can be seen in Fig. 3. The start and end point of the spline are on the ground, while the other three points define how the leg moves forward through the air. The leg moves in a straight line on the ground, parallel to the body of the robot, so only two parameters decide their positions. The three points in the air are all three dimensional, with sideways movement being mirrored between left and right legs. This gives a total of 11 parameters that define the spline shape.

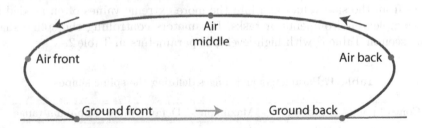

Fig. 3. A simple example leg trajectory, seen from the side. The tip of the leg follows this path when the robot walks. The front of the robot is to the left.

The two control points along the ground are sorted so that they always move the leg backward, while the order of the three control points in the air is chosen with an order resulting in the shortest possible spline. This ensures that no looping or self-intersection can happen, and allows all gait parameters to be set without constraints. A parameter for lift duration specifies the time the leg uses to lift back to the front, given in percentage of the gait period, while the frequency parameter gives the number of gait periods per second.

Balancing Wag. In addition to positions generated for individual legs, a balancing wag is added to all legs. Due to the leg lift order, this can not be a simple circular motion, but needs different frequencies for the two axes. The movement allows the robot to lean away from the leg it is currently lifting, and gives better stability. Equation 1 shows how the wag is defined, with t defining the current time, and T the gait period. 0.43 is a factor to offset the movement between the two wag axes to align them with the gait. It has a phase offset (W_ϕ) that allows for tuning to dynamic effects of the robot, while amplitude can be set separately for the two directions (A_x/A_y).

$$W_x = \frac{A_x}{2} * tanh(3 * sin(\frac{2\pi * (t + (W_\phi * T))}{T}))$$
$$W_y = \frac{A_y}{2} * tanh(3 * sin(\frac{2\pi * (t + (W_\phi + 0.43) * \frac{T}{2})}{\frac{T}{2}})) \tag{1}$$

Complexity Scaling. The complexity of the controller can be modified by a single parameter, from 0 to 100%. There are many ways to provide a scaling of the complexity of the controller, but we chose to implement this using a dynamic genotype-phenotype mapping that varies the range of gait parameters linearly with the controller complexity. All controller parameters have a center value, that together with the minimum range gives the allowable range at controller complexity 0%. These have been chosen so they represent a very conservative and safe controller that should work well in most conditions, based on traditional robotics techniques and earlier experience with the robot. Using a more complex controller by allowing a large range of values, however, allows the controller to deviate from the safe values and into the more extreme values often needed for more complex environments or tasks. Parameters controlling the spline shape can be seen in Table 1, with high-level gait parameters in Table 2.

Table 1. Parameters and ranges defining the spline shape

Control point	Minimum	Maximum	Default value	Minimum range
Ground front	−150	150	50	50
Ground back	−150	150	−100	50
Air 1	[−25, −150, 10]	[25, 150, 80]	[0, 75, 30]	[0, 50, 10]
Air 2	[−25, −150, 10]	[25, 150, 80]	[0, 0, 50]	[0, 0, 10]
Air 3	[−25, −150, 10]	[25, 150, 80]	[0, −75, 50]	[0, 50, 10]

Examples of splines with different gait complexities can be seen in Fig. 4. For complexities of 0, the splines are fairly conservative, but even though the parameter ranges are low, they do show some variation in their basic shapes. The higher complexity gaits have spline shapes that are much more unconventional, though sorting the control points to minimize spline length does remove self-intersections to keep all trajectories feasible. Please note that the plot shows the commanded position to the robot, and that the actual leg trajectory can be very

Table 2. Parameters and ranges of gait parameters.

Parameter	Minimum	Maximum	Default value	Minimum range
Wag phase	$-\pi/2$	$\pi/2$	0	0.2
Wag amplitudes	0	50	0	5
Lift duration	0.05	0.20	0.175	0.05
Frequency	0.25	1.5	–	–

different than commanded, due to the mechanical and control properties of the actuators, and the dynamics of the system. Very complex shapes that appear unintuitive for human engineers might end up giving much smoother and higher performing gaits in the real world than expected.

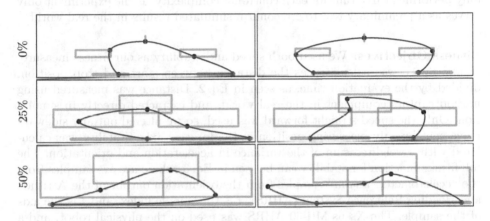

Fig. 4. Examples of leg trajectory splines generated at different gait complexities. These are seen from the side of the robot, with the front of the robot to the left of the plot. The red boxes show the range of possible control point positions. (Color figure online)

3.3 Evolutionary Setup

Here we describe the setup we used for evolving the controllers, as well as how we evaluated them. We evolved controllers for both stable and fast forward walking on flat ground.

Evolutionary Algorithm and Operators. We used the NSGA-II evolutionary algorithm, running on the Sferes2 evolutionary framework. We chose this algorithm since we are optimizing both speed and stability, but would not like to choose the specific trade-off between the two objectives before optimization. NSGA-II features a mechanism to increase the crowding distance in the Pareto front, which gives a wide range of trade-offs to pick from.

Gaussian mutation was used with a mutation probability of 100% and a sigma of 1/6. No recombination operators were used.

Early experimentation showed a big difference in the number of evaluations before convergence for different controller complexities, which suggested the need for different population sizes. We tested a range of different population sizes at the minimum and maximum complexity, as well as a few points in between, and found that a population of eight at zero complexity, and 64 at full complexity worked best. Population sizes for all intermediary complexities were set linearly, and rounded to the nearest power of two. Tests showed that runs at all gait complexities converge to a satisfactory degree after 8192 evaluations.

We performed 25 runs for each controller complexity in simulations to gain a good estimate of the performance. Each simulated run took about 11 h, and we used about 10,000 CPU core hours on the simulation for the experiments featured in the paper. Experiments in the real world take a lot longer, so we only performed three runs for each controller complexity, as the experiment only serves as a preliminary test to see confirm simulated results in the real world.

Fitness Objectives. We used both speed and stability as our fitness measurements. Speed was calculated as the distance between start and stop position, divided by the evaluation time, as seen in Eq. 2. Distance was measured using motion capture equipment in the real world, and extracted directly in simulation. Only the speed straight forward was used, so we filtered out any sideways movement by only measuring position in the forward axis. Stability was calculated with a weighted sum of the variance in acceleration and orientation. The full fitness function for stability can be seen in Eq. 3, where acc are samples from the accelerometer, ang are samples from the orientation output of the Attitude and Heading Reference System (AHRS), i is the sample index, and j is the axis of the sample. The Xsens Mti-30 AHRS was used on the physical robot, and a virtual version of the same was used in simulation.

$$F_{speed} = \frac{\|P_{end} - P_{start}\|}{time_{end} - time_{start}} \tag{2}$$

$$G(A_j) = \sqrt{\frac{1}{n} \sum_{i=1}^{n} (A_{j,i}^2 - \overline{A_j}^2)}$$

$$F_{stability} = -\left(\alpha * \sum^{axes} G(Acc_{axis}) + \sum^{axes} G(Ang_{axis}) \right) \tag{3}$$

Evaluation. We ran all our simulations on the Gazebo physics simulator. Each gait was evaluated in simulation by walking forwards 1 m, with a timeout of 10 s. The position and pose of the robot were reset between all evaluations.

Evaluating and comparing the performance of different optimization runs can be challenging when doing multi-objective optimization. This is especially true when using an algorithm like NSGA-II, that has a mechanism for stretching out the Pareto front, making it hard to compare the two objectives separately. Therefore, we instead looked at the hypervolume [19] when comparing populations. The hypervolume measures the volume (or area, in the case of two objectives) of the dominated part of the objective space. The lower bound of stability was set to -1 for the hypervolume calculation, while speed was capped to 0 m/min.

4   Experiments and Results

We present the results of experiments in simulation and on a real-world robot. These experiments are simplified and performed with as many variables removed as possible. The robot's task is to walk straight forward, and the environment is a flat surface with medium friction, both in simulation and the real world.

4.1   Finding the Maximum Needed Complexity

First, we wanted to investigate whether there is a maximum controller complexity needed for the environment and task we are using. Since neither is very challenging, we do not expect the need for very complex controllers. We ran full evolutionary runs at a range of gait complexities.

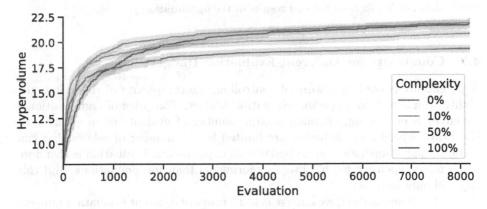

Fig. 5. Hypervolume from evolutionary runs with selected gait complexities. The solid lines show the means, with 95% confidence interval in the shaded areas.

Figure 5 shows how the hypervolume progresses over evaluations. This shows that the lower complexity controllers converge quicker, but are not able to achieve the same performance as the higher complexity controllers. The 50% and 100% complexity controllers end up with the same performance, though the 100% complexity controller takes considerably longer to converge.

The details of the last evaluations of the runs are better illustrated with the boxplots, seen in Fig. 6. These show the distribution of the hypervolumes achieved at the end of all the optimization runs. The hypervolume improves for gait complexities from 0% to 40%, but there is no improvement between 40% and 50%. 100% complexity has a wider spread than the others, which might be beneficial in some applications, but the median performance is no better than the 40–50% complexity.

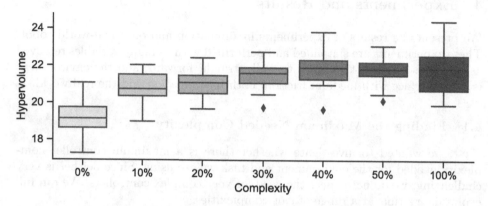

Fig. 6. Hypervolume for runs with gait controller complexities ranging from 0% to 100%, showing details from the end results of the optimization process.

4.2   Complexity for Different Evaluation Budgets

A potentially rewarding feature of controlling the complexity of the gait is the ability to adapt it to a specific evaluation budget. The price of computational resources is decreasing, enabling a large number of evaluations in simulation. Hardware experiments, however, are limited by the number of robots that can be built, maintained, and supervised during experiments. Evaluation is therefore much more expensive for hardware experiments than for simulations, and this gap will only increase.

For this investigation, we have selected a range of different evaluation budgets to test. We have previously used 64 and 128 evaluations in our hardware experiments [2,14], and 512, 2048 and 8192 evaluations gives a range more appropriate for simulation experiments.

Figure 7 shows how the controller complexity affects achieved hypervolume for the different budgets. For the shortest two simulation cases, with 64 and 128 evaluations, hypervolume is highest at 10% complexity. Budgets 512 and 2048 achieve the best performance around 30%, while the long simulation case performs best at 40%–100%.

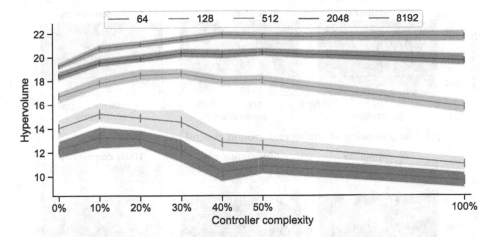

Fig. 7. This figure shows how different controller complexities affects achievable hypervolume for different evaluation budgets. The vertical lines show the standard deviation, while the shaded areas show the 95% confidence intervals.

4.3 Analyzing Resulting Populations

Figure 8 shows which parameters are tested at various parts of the search. Some parameters, like the y position of the back ground control point, end up close to their conservative estimate, and do not exploit their additional freedom from the higher complexity in our simple experiments, as seen in Fig. 8a. Other parameters, like the y position of the front ground control point, do use more of their available range, although it is still close to its original estimate. In Fig. 8c, the search with 50% controller complexity seems to maximize the x position of the third air control point in the spline, while with the whole area available in the 100% complexity controller, it ends up minimizing it.

4.4 Initial Hardware Testing

We also did evolutionary runs using this new controller on the physical robot in the real world with 64 evaluations per run, using eight generations of eight individuals. We decided to test a controller complexity of 0%, as well as 50%, which is the highest complexity we were confident in using on the physical robot without excessive risk of physical damage to the system. We also tested 20%, which gives us another data point between these, and was among the two top performing complexities in simulation with this evaluation budget. The results can be seen in Fig. 9, where we can see the same general trends as in the simulator. Controller complexities 0% and 20% both did well, and we are not able to separate the two with the limited number of evaluations we were able to do in hardware. 50% controller complexity, however, does considerably worse than the other two, just as we saw in simulation.

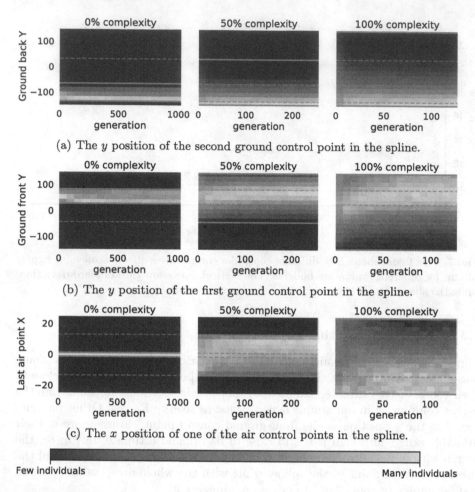

(a) The y position of the second ground control point in the spline.

(b) The y position of the first ground control point in the spline.

(c) The x position of one of the air control points in the spline.

Few individuals   Many individuals

Fig. 8. Values of a select few parameters throughout the optimization run. The solid red lines show the range of the parameters, and the dashed red lines trace the range from the other complexities to ease comparison. (Color figure online)

5   Discussion

The performance differences in Fig. 7 suggest that choosing the right controller complexity for an evaluation budget can be very important, especially when that budget is small. Lower complexity controllers fall less, so if optimization is done in hardware, this could also be taken into account when deciding on the complexity. We did a simple grid-search for our experiments since we were only investigating the controller, but more advanced search algorithms could be performed to further optimise the choice of complexity.

We used different population sizes when evolving with different complexities in our experiments. Our controller was designed to be evolved with evaluation

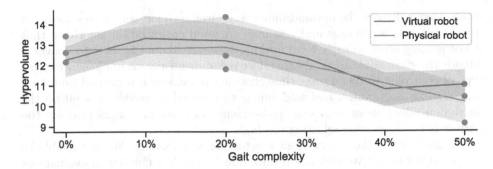

Fig. 9. This figure shows the performance on the real robot, compared to the simulated version seen in Fig. 7. The added green dots are the resulting hypervolume from each of the runs in hardware. (Color figure online)

budgets as small as 32 or 64 evaluations when doing real world experiments, and with budgets larger than 8192 when evolving in simulation. Limiting the population size to the smallest budget would give a very unrealistic measurement of performance for the larger budgets, and thus we chose suitable population sizes for the different complexities through simple trial and error. This does obfuscate the results to a degree, but we feel this gives the most fair comparison. The evolutionary operators would likely also be slightly different, but they were kept the same as they affect the search to a much smaller degree.

The parameter for the x position of the third air control point, seen in Fig. 8c, seems to be maximized at 50% complexity, but be minimized at 100% complexity. This is most likely due to interactions between different parameters. At half complexity, the optimal value might be towards the top of the parameter range. At full complexity, however, new ranges for the other parameters are opened up, allowing better performance for lower parts of the range.

The choice of centers and minimum ranges for each gait parameter greatly affect the performance of lower complexity gait controllers. The choice should be based on conservative values that are assumed to work sufficiently in all environments, not on optimal values for a single environment. Evolution is often used to adapt to changes in environments or tasks. If the centers and ranges were chosen after optimal solutions were found, they would most likely not perform well when things change, and one might as well just select the top performing individuals from simulation directly. In our case, we selected these values before doing the optimization, and several are far from optimal. This can be seen in Fig. 6, where the performance of the low complexity controller is much worse than for the higher complexity ones. This is by design, as safe and conservative parameters that work for all environments rarely do very well in any of them.

The choice of maximum ranges also affect the outcome, but not to as high a degree as the center and minimum range. Limiting the ranges too much means the controller will never be able to achieve the potential increase in performance from that specific controller feature. Having ranges that are too large, with values that will never be optimal under any circumstance, serves to slow down the search, and waste time and resources. A good optimization algorithm that is not getting stuck in early local optima, however, should be able to converge outside these infeasible areas. We therefore recommend anyone implementing this type of controller to spend some time choosing parameter centers and minimum ranges to be conservative and safe, but not be afraid to overshoot a bit on the maximum allowable range, as the consequence of choosing ranges that are too narrow is far worse than selecting too high.

Figure 9 shows the results from the testing in hardware. We are unable to say anything definitive with the results due to the low number of evaluations and the relatively high degree of noise, but it does support what we found in simulation. Not only did the 50% controller complexity perform worse, like predicted in simulation, but we also experienced qualitatively more extreme gaits, and actually had to pause the evolutionary runs at several times to repair the robot after damage. We also experienced several falls with the 50% controller complexity, but no falls or damage at the two lower complexities, supporting our original assumption that the gait values were conservative and safe.

We consider this type of controller to be very useful for researchers doing gait optimization in the real world on physical robots, as the reality gap can often times make it impractical or impossible to directly use individuals from simulation in the real world. Simulations can be used to find approximate upper bounds of the needed complexity as we saw in Fig. 6, but even more useful is being able to tune the complexity to the limited evaluation budget used in hardware, as seen in Fig. 7. We also expect that more demanding or dynamic environments and tasks might be able to exploit higher complexities better than what we experienced in our experiments, which only included forward walking in straight lines on even terrain.

6   Conclusion and Future Work

In this paper, we introduced our new gait controller with variable complexity. We tested the controller in simulation, and found that different gait complexities are optimal for different evaluation budgets. We also did preliminary tests on a physical robot in the real world that supported our findings. Being able to change the controller complexity allows a researcher to use less complex controllers when optimizing gait on a physical robot, and increase the complexity when needed for demanding environments, or when doing longer optimization in simulations.

One natural extension of our work is to use our variable complexity controller in incremental evolution. Since this controller offers a continuous complexity parameter, the difficulty can be gradually increased for each generation. Since an increase in difficulty follows a known set of rules, all individuals can keep their phenotypic values between generations, even when parameter ranges are expanded. This allows evolution to gradually explore the added complexity, in the same way that has been shown to be optimal for neuro-evolution [18]. The controller complexity can also be changed during the evolutionary process as part of evolutionary strategies, or be controlled during robot operation as part of lifelong learning.

We have only tested this controller in a single environment in simulation where complexities over 50% were not needed. It would be interesting to test it in more challenging and dynamic environments to see if controllers with higher complexities are able to use the increased parameter ranges to actually increase performance. Doing a more thorough investigation into the parameters selected might yield ranges or values that act limiting on the fully complex controller, and would allow even more flexible gaits. Analyzing the individual leg trajectories evolved would also be interesting, and could shed light on the matter from a different perspective. Investigating how evolutionary meta-parameters interact with the complexity would be interesting, including population size and evolutionary operators. Adding sensing and allowing the robot to choose which complexity is needed for its current environment is also worth exploring.

References

1. Mouret, J., Doncieux, S.: Overcoming the bootstrap problem in evolutionary robotics using behavioral diversity. In: IEEE Congress on Evolutionary Computation, pp. 1161–1168, May 2009
2. Nygaard, T.F., Martin, C.P., Samuelsen, E., Torresen, J., Glette, K.: Real-world evolution adapts robot morphology and control to hardware limitations. In: Proceedings of the Genetic and Evolutionary Computation Conference. ACM (2018)
3. Nygaard, T.F., Martin, C.P., Torresen, J., Glette, K.: Self-modifying morphology experiments with DyRET: dynamic robot for embodied testing. In: IEEE International Conference on Robotics and Automation (ICRA), May 2019
4. Auerbach, J.E., Bongard, J.C.: Environmental influence on the evolution of morphological complexity in machines. PLoS Comput. Biol. **10**(1), e1003399 (2014)
5. Gong, D., Yan, J., Zuo, G.: A review of gait optimization based on evolutionary computation. Appl. Comput. Intell. Soft Comput. **2010**, 12 (2010)
6. Golubovic, D., Hu, H.: GA-based gait generation of Sony quadruped robots. In: Proceedings of the 3rd IASTED International Conference on Artificial Intelligence and Applications (AIA) (2003)
7. Moore, J.M., McKinley, P.K.: A comparison of multiobjective algorithms in evolving quadrupedal gaits. In: Tuci, E., Giagkos, A., Wilson, M., Hallam, J. (eds.) SAB 2016. LNCS (LNAI), vol. 9825, pp. 157–169. Springer, Cham (2016). https://doi.org/10.1007/978-3-319-43488-9_15
8. Hebbel, M., Nistico, W., Fisseler, D.: Learning in a high dimensional space: fast omnidirectional quadrupedal locomotion. In: Lakemeyer, G., Sklar, E., Sorrenti, D.G., Takahashi, T. (eds.) RoboCup 2006. LNCS (LNAI), vol. 4434, pp. 314–321. Springer, Heidelberg (2007). https://doi.org/10.1007/978-3-540-74024-7_28
9. Seo, K., Hyun, S., Goodman, E.D.: Genetic programming-based automatic gait generation in joint space for a quadruped robot. Adv. Robot. **24**(15), 2199–2214 (2010)
10. Pugh, J.K., Soros, L.B., Szerlip, P.A., Stanley, K.O.: Confronting the challenge of quality diversity. In: Proceedings of the 2015 Annual Conference on Genetic and Evolutionary Computation, pp. 967–974. ACM (2015)

11. Cully, A., Clune, J., Tarapore, D., Mouret, J.B.: Robots that can adapt like animals. Nature **521**(7553), 503 (2015)
12. de Santos, P.G., Garcia, E., Estremera, J.: Quadrupedal Locomotion: An Introduction to the Control of Four-Legged Robots. Springer, London (2007). https://doi.org/10.1007/1-84628-307-8
13. Nygaard, T.F., Samuelsen, E., Glette, K.: Overcoming initial convergence in multi-objective evolution of robot control and morphology using a two-phase approach. In: Squillero, G., Sim, K. (eds.) EvoApplications 2017. LNCS, vol. 10199, pp. 825–836. Springer, Cham (2017). https://doi.org/10.1007/978-3-319-55849-3_53
14. Nygaard, T.F., Torresen, J., Glette, K.: Multi-objective evolution of fast and stable gaits on a physical quadruped robotic platform. In: IEEE Symposium Series on Computational Intelligence (SSCI), December 2016
15. Ijspeert, A.J.: Central pattern generators for locomotion control in animals and robots: a review. Neural Netw. **21**(4), 642–653 (2008)
16. Yosinski, J., Clune, J., Hidalgo, D., Nguyen, S., Zagal, J.C., Lipson, H.: Evolving robot gaits in hardware: the hyperNEAT generative encoding vs. parameter optimization. In: ECAL, pp. 890–897 (2011)
17. Togelius, J.: Evolution of a subsumption architecture neurocontroller. J. Intell. Fuzzy Syst. **15**(1), 15–20 (2004)
18. Tomko, N., Harvey, I.: Do not disturb: recommendations for incremental evolution. In: Proceedings of ALIFE XII, the 12th International Conference on the Synthesis and Simulation of Living Systems (2010)
19. Knowles, J.D., Corne, D.W., Fleischer, M.: Bounded archiving using the Lebesgue measure. In: The 2003 Congress on Evolutionary Computation, vol. 4, December 2003

Introducing Weighted Intermediate Recombination in On-Line Collective Robotics, the $(\mu/\mu_{\mathrm{W}}, 1)$-On-line EEA

Amine Boumaza[✉]

Université de Lorraine, CNRS, Inria, LORIA, 54000 Nancy, France
amine.boumaza@loria.fr

Abstract. Weighted intermediate recombination has been proven very useful in evolution strategies. We propose here to use it in the case of on-line embodied evolutionary algorithms. With this recombination scheme, solutions at the local populations are recombined using a weighted average that favors fitter solutions to produce a new solution. We describe the newly proposed algorithm which we dubbed $(\mu/\mu_{\mathrm{W}}, 1)$-On-line EEA, and assess it performance on two swarm robotics benchmarks while comparing the results to other existing algorithms. The experiments show that the recombination scheme is very beneficial on these problems.

Keywords: On-line embodied EA · Swarm robotics ·
Weighted intermediate recombination

1 Introduction

Embodied evolutionary robotics (EER) [1], aims to learn collective behaviors for a swarms of agents, where evolution is distributed on the agents that adapt on-line to the task [2]. Each agent runs an EA onboard and exchange genetic material with other agent when they meet. Selection and variation are performed locally on the agents and successful genes, whose offspring survive throughout many generations, are those that adapt to the task and also maximize mating opportunities while minimizing the risk for their vehicles.

In this paper we propose the $(\mu/\mu_{\mathrm{W}}, 1)$-On-line EEA which adapts the well known weighted recombination scheme from evolution strategies (ES) to the on-line EER setting. It has been shown that recombination allows to speed up progress of ES and improves robustness against selection errors [3]. These properties are the result of the *genetic repair principal* (GR) which reduces the effect of the harmful components of mutations as a result of the averaging process. Furthermore, [4] extends these conclusions in the case of noisy fitness environments and argues that combined with higher mutation steps, recombination can reduce the signal to noise ratio in the evaluation.

Our main motivation is to investigate if these properties can also benefit EER algorithms. We study the impact of this recombination scheme and show that

© Springer Nature Switzerland AG 2019
P. Kaufmann and P. A. Castillo (Eds.): EvoApplications 2019, LNCS 11454, pp. 633–640, 2019.
https://doi.org/10.1007/978-3-030-16692-2_42

when correctly designed, it improves significantly the results of the algorithm as the experiments suggests on two different learning tasks.

2   Background

Recombination or as commonly know as crossover is not new in on-line EER, different authors have proposed implementations with different form of crossover. We briefly review some of that work in the following.

The Probabilistic Gene Transfer Algorithm (PGTA) [2], is commonly cited as the first implementation of a distributed on-line EER algorithm. As the name suggests, recombination is implemented at a gene level. Agents broadcast randomly sampled genes from their genomes and when they receive genes from other agents, they replace the corresponding gene with a probability. The rate at which the agents broadcast their genes is proportional to their fitness[1] and conversely, the rate at which they accept a received gene is inversely proportional to their energy level. This way, selection pressure is introduced in that fit agents transmit their genes to unfit ones.

In the Embodied Evolutionary Algorithm (EEA) [5], crossover is performed between two genomes. Agents select a genome from their local population (received from potential mates) using a binary tournament. This selected genome is then recombined with the current active genome with a probability proportional to the ratio of its fitness and that of the active genome. The newly created genome is the average of both genomes. In this case, crossover is more probable when the selected genomes comes from a fitter agent.

The above mentioned articles implemented the EA on a fixed topology neuro-controller where the competing conventions problem [6] does not rise. It is also worth mentioning that there exists, although fewer, implementation of EER that proposed evolving topology. These implementations adapt the innovation marking introduced in the Neuro-Evolution of Augmenting Topologies algorithm (NEAT) [7], to the distributed case. In this case, innovations appear locally in the population and are not know to all agents who must order them correctly before performing a crossover [8,9].

For a more complete review of the existing work, we recommend the recent review article [1].

3   The $(\mu, 1)$-ON-LINE EEA

The main inspiration of the $(\mu, 1)$-ON-LINE EEA is the original version of mEDEA [10] to which with we add a selection operator that we will describe later (Algorithm 1). The algorithm considers a swarm of λ mobile agents a^j with $j = 1, \ldots, \lambda$ each executing a neuro-controller whose parameters are x^j (the active genome). Each agent maintains a list L^j, initially empty, in which it stores other genomes that it receives from other agents.

---

[1] The authors use a virtual energy level in place of fitness.

At each time step $t < t_{\max}$, an agent executes its active controller and broadcasts its genome within a limited range. In parallel, it listens for genomes originating from other agents, and when a genome is received (a mating event), it is stored in the agent's list L^j (it's local population). This procedure is executed in parallel on all agents during $t_{\max}$ steps, the evaluation period of one generation.

At the end of a generation, the agent selects a genome from its list L^j (the selection step), and replaces its active genome with a mutated copy of the selected one. The list is then emptied and a new generation begins.

In the event where an agent had no mating opportunities and finishes its evaluation period with an empty list $L^j = \emptyset$, it becomes inactive; a state during which the agent is motionless. During this period, the inactive agent continues to listen for incoming genomes from other agents, and once $L^j \neq \emptyset$ the agent becomes active again at the beginning of the next generation.

The number of genomes the agents collects $\mu^j = |L^j|$ $(0 \leq \mu^j \leq \lambda)$ is conditioned by it's mating encounters. Since the communication range is limited, agents that travel long distances will increase their probability of mating. We should note that mating encounters allow agents to spread their active genome and to collect new genetic material. The algorithm stores only one copy of the same genome[2] if it is received more than once.

The $(\mu, 1)$-ON-LINE EEA can be viewed as λ instances of $(|L^j|, 1)$-EA running independently in parallel, where for each instance, $L^j \subseteq \{x^1, \dots, x^\lambda\}$, i.e. each local population is a subset of the set of all active genomes of the generation. Furthermore, the sizes of the individual local populations μ^j are not constant although bounded by λ and depend on the number of mating events (possibly none) the agent had in the previous generation.

3.1   Implementation Details

The original mEDEA [10] was introduced in open-ended environments without specifying a task for the agents to solve. However, when it is applied in a task-driven scenario, we consider an objective or fitness function. In this case selection (line 16 in Algorithm 1) is based on the fitness of the genomes in the individual lists. In this work, the genome is a vector $x \in \mathbb{R}^N$, which represents the weights of the neuro-controller and $f : \mathbb{R}^N \to \mathbb{R}$ is the fitness function. Only the weights undergo evolution (fixed-topology).

In addition to adding a selection operator, there is a significant difference between mEDEA and Algorithm 1 in that we don't consider a listening phase. Agents broadcast if they are active and listen all the time, whereas in mEDEA, agents must be in a specific listening state to record incoming genetic material. This and the added maturation age, described bellow, show that in practice the difference in the results is not significant.

Furthermore, since the EA runs independently on each agent, fitness values are assigned to the individual agents based on their performance with regard to the given task. These values are continuously updated during the agent's lifetime

---

[2] The term "same" is here used in the sense "originating from the same agent".

Algorithm 1. $(\mu, 1)$-ON-LINE EEA

```
 1  for  1 ≤ j ≤ λ in parallel do
 2  |    xʲ ← random()
 3  |_   aʲ is active

 4  repeat
 5  |    for  1 ≤ j ≤ λ in parallel do
 6  |    |    t ← 0, fʲ = 0, Lʲ ← ∅
 7  |    |    while  t < t_max do
 8  |    |    |    t ← t + 1
 9  |    |    |    if aʲ is active then
10  |    |    |    |    exec(xʲ)
11  |    |    |    |    update(fʲ)
12  |    |    |    |    if t > τt_max then
13  |    |    |    |    |_   broadcast(xʲ, fʲ)
14  |    |    |    |_   Lʲ ← Lʲ ∪ listen()
15  |    |    |    if Lʲ ≠ ∅ then
16  |    |    |    |_   xʲ ← mutate(select(Lʲ))
17  |    |    |_   else aʲ is not active
18  until termination condition met
```

(line 11). Each agent stores this value internally and broadcasts it along with its genomes during mating events. Agents on the receiving end store the genome and its fitness values in their lists. Finally, if an agent receives an already seen genome, it updates the genome's fitness as it is the most up to date value. Furthermore, to ensure that genomes are transmitted with accurate fitness values, a *maturation age* is required of the agent before broadcasting [5]. Finally, we should emphasize that fitness evaluation are intrinsically noisy since measurements are performed in varying circumstances.

In the following, we will note $L = \{x_1 \ldots x_\mu\}$ the local population on some agent, and we consider a maximization scenario. Depending on the treatment tested, we apply a different selection scheme. On the one hand, for the case were there is no recombination, the next genome is the winner of a tournament selection. On the other hand, when recombination is considered, the next active genome is the result of the weighted recombination on the local population. After either of these steps, the newly generated genome, which we note $\bar{x}$, is mutated. Mutation is Gaussian with a fixed step size $\sigma \in \mathbb{R}$:

$$x := \bar{x} + \sigma^2 \times \mathcal{N}(1, 0) \tag{1}$$

Selection in $(\mu, 1)$-ON-LINE EEA. We use an adapted k-tournament selection scheme which randomly draws $k > 0$ solutions from the local population of μ parent solutions L and returns the best solution among the parents.

The tournament size k, sets the selection pressure of the procedure, and is usually a parameter of the EA. When $k = 1$, the procedure selects a random solution, and when $k = \mu$, it returns the best solution.

Since the size of the parent populations varies from agent to agent and from one generation to another, the size of the tournament cannot be chosen before hand. In practice we fix k as a function of μ. In it's simpler form this function can be a linear projection such as $k = \lfloor \alpha\mu \rfloor + 1$ with $\alpha \in (0, 1)$, but more sophisticated functions could also be considered. In this case, α is the parameter that tunes the selection pressure.

Weighted Intermediate Recombination in the $(\mu/\mu_{\mathbf{W}}, 1)$-On-line EEA. This recombination scheme is similar to recombination in CMAES [11]. The newly generated genome is defined as:

$$\bar{x} = \sum_{i=1}^{\mu} w_i \, x_{i:\mu}, \qquad \sum_{i=1}^{\mu} w_i = 1, \qquad w_1 \geq w_2 \geq \ldots \geq w_\mu > 0,$$

where $w_{i=1\ldots\mu} \in \mathbb{R}_+$ are the recombination weights and the index $i:\mu$ denotes the i-th ranked individual and $f(x_{1:\mu}) \geq f(x_{2:\mu}) \geq \ldots \geq f(x_{\mu:\mu})$. By assigning different values to w_i it can also be considered as a selection scheme since fitter solutions have greater weights in the sum. In this work, the weights $w_i = \frac{w_i'}{\sum_{k=0}^{\mu} w_k'}$ where $w_i' = \log(\mu + 0.5) - \log i$.

We emphasize that, since population sizes may differ between agents, the weights w_i may also be different from agent to agent due to their normalization (Fig. 1 left). This fact suggests that the EA gives a selective advantage to the fittest genomes that belong in the lists of agents that had fewer mating encounters.

4   Experiments

The experiments were performed on the Roborobo simulator [12] an environment that allows researchers to run experiments on large swarms of agents. In this simulator, agents are e-puck like mobile robots with limited range obstacle sensors and two differential drive wheels. The neuro-controller we consider in this work is a simple feed-forward multi-layered perceptron with one hidden layer that maps sensory inputs (values in $(0, 1)$) to motor outputs (values in $(-1, 1)$). All the parameters of the experiment are summarized in Table 1.

We compare $(\mu/\mu_{\mathbf{W}}, 1)$-On-line EEA and $(\mu, 1)$-On-line EEA on two extensively studied collective evolutionary robotic benchmarks: (1) locomotion and obstacle avoidance, (2) item collection. In the first task, agents must travel the largest distance; the fitness function in this case is the accumulated traveled distance since the beginning of the generation. In the second task the agents must collect the most items; the fitness function is simply the number of items collected since the beginning of the generation. The environment of both tasks are shown on Fig. 1.

In the case of $(\mu, 1)$, we fix the selection pressure to the highest ($\alpha = 1$) to maximize the performance [13]. Furthermore we also compare each algorithm

Table 1. Simulation parameters

t_{max}	300	Arena diam.	400	λ	160
Sensors	24	Sensor range	32	N	135
Effector	2	Com. range	32	g_{max}	300
Topology	$24 \times 5 \times 2$	Nb. items	$\frac{3}{4}\lambda$	σ	0.5
Bias (output)	+1	Agent rad.	3	α	0 or 1
Max tr. vel	2 / tic	Item rad.	10	mat. age	$0.8 \times t_{max}$
Max rot. vel	30 deg/tic	Nb runs	30		

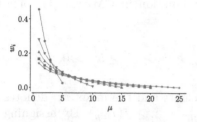

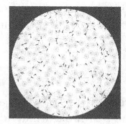

Fig. 1. The recombination weights for different population sizes (left) in $(\mu/\mu_W, 1)$-ON-LINE EEA. An overview of the Roborobo simulator. The locomotion task (middle) and the collection task (right).

with its "naive" instance as a control experiment. In the case of recombination, this is an instance where all weights are equal ($w_i = \frac{1}{\mu}$, $i = 1 \ldots \mu$). Here there is no selection bias towards the fittest solutions. The second "naive" instance performs a tournament with $\alpha = 0$ (selects the next genome randomly from the list). We use the notations $(\mu/\mu_I, 1)$ and $(\mu_{RND}, 1)$ to refer to these "naive" algorithms.

5   Results and Discussion

The results of the experiment are presented in Fig. 2. To compare both algorithms, we are interested in two measures, the fitness and the number of mates the agents inseminated. Both these measures are summarized on the curves as the median (of 30 independent runs) of:

$$f(t) = \frac{1}{\lambda} \sum_{j=1}^{\lambda} f^j(t),$$

where $f^j(t)$ is the fitness of agent j at generation t. We use a similar formulation for the second measure.

On both tasks, $(\mu/\mu_W, 1)$ outperforms $(\mu, 1)$ and the difference is statistically significant. On locomotion for instance, in 95% of the runs, the average traveled

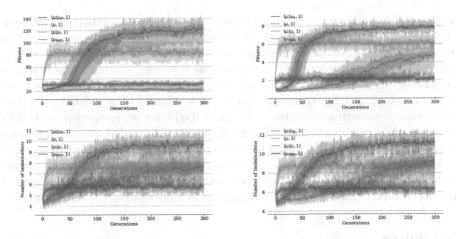

Fig. 2. Fitness and number of mates for the locomotion task (left) and the collection task (right). Curves represent median (solid line), the range between the 25th and the 75th percentile (darker area) and the between the 5th and the 95th percentile (lighter area) of 30 independent runs.

distance by each agent reaches the range $(100, 140)$ compared to $(60, 100)$. On the collection task we have the same trend; agent collect roughly 8 items per generation, two more than in $(\mu, 1)$. The fact the $(\mu/\mu_W, 1)$ performs better than a purely elitist EEA is very interesting, and shows that it is not sufficient to take the best solution and evolve it. There is still room for improvement and better solution can be found by combining traits from multiple solution.

The "naive" instances $(\mu/\mu_I, 1)$ and $(\mu_{\text{RND}}, 1)$ perform very poorly in comparison. Although we note, that in the case of the former, the swarm starts to collect items at around generation 150 almost reaching $(\mu, 1)$ at the end. When we consider the number of insemination, the qualitative results are similar to the exception of $(\mu/\mu_I, 1)$ which seems to evolve agent that learn to successfully spread their genomes.

In the case of $(\mu/\mu_I, 1)$ and $(\mu_{\text{RND}}, 1)$, even though there is no selection pressure induced by the task, the environment exerts a pressure that can reduce the chances of mating for agents, an environmental pressure [10]. The improvements in the fitness values, are due to this. When evolution finds solutions that allow their vehicles to travel longer distances, these solutions spread to other agents and increase their chance of survival. This may explain why $(\mu/\mu_I, 1)$ is able to improve overtime. This is further supported when we see that the improvement happens only in the collection task. The presence of obstacles in locomotion increase the environmental pressure.

6   Conclusions

Weighted intermediate recombination is a crucial component of evolution strategies and there is a large body of work that attest of it importance both theoretically and practically. In this paper we asked the simple question: could this be the case in online EER? We proposed $(\mu/\mu_W, 1)$-ON-LINE EEA and designed a experiment to measure the effect of recombination on simple well studied benchmarks. The results suggests, at least on these tasks, that recombination outperforms a purely mutative strategy. Among the remaining question we would like to pursue, the most natural one is to design a truncation mechanism, akin to what exists in the evolution strategies. Such a procedure takes only the better performing parents into account in the recombination and disregard the rest. The added selection pressure improves convergence in ES which may also be the case in EER. Furthermore, it would be also interesting to compare the algorithms in terms of diversity (genetic and behavioral). Does recombination reduce the diversity? Finally, it is also important to extend the tasks and test the algorithm in more challenging environments.

References

1. Bredeche, N., Haasdijk, E., Prieto, A.: Embodied evolution in collective robotics: a review. Frontiers Robot. AI **5**, 12 (2018)
2. Watson, R., Ficici, S., Pollack, J.: Embodied evolution: distributing an evolutionary algorithm in a population of robots. Robot. Auton. Syst. **39**, 1–18 (2002)
3. Beyer, H.G.: Toward a theory of evolution strategies: on the benefits of sex - the $(\mu/\mu, \lambda)$ theory. Evol. Comput. **3**(1), 81–111 (1995)
4. Arnold, D.V.: Noisy Optimization with Evolution Strategies. Springer, New York (2002). https://doi.org/10.1007/978-1-4615-1105-2
5. Karafotias, G., Haasdijk, E., Eiben, A.E.: An algorithm for distributed on-line, onboard evolutionary robotics. In: Proceedings of GECCO 2011, pp. 171–178. ACM (2011)
6. Schaffer, D.J., Whitley, D., Eshelman, L.J.: Combinations of genetic algorithms and neural networks: a survey of the state of the art. In: Proceedings of COGANN 1992, pp. 1–37 (1992)
7. Stanley, K.O., Miikkulainen, R.: Evolving neural networks through augmenting topologies. Evol. Comput. **10**(2), 99–127 (2002)
8. Silva, F., Urbano, P., Oliveira, S., Christensen, A.L.: odNEAT: an algorithm for distributed online, onboard evolution of robot behaviours. In: Artificial Life. vol. 13, pp. 251–258. MIT Press (2012)
9. Fernández Pèrez, I.n., Boumaza, A., Charpillet, F.: Decentralized innovation marking for neural controllers in embodied evolution. In: Proceedings of GECCO 2015, pp. 161–168. ACM, Madrid (2015)
10. Bredeche, N., Montanier, J.-M.: Environment-driven embodied evolution in a population of autonomous agents. In: Schaefer, R., Cotta, C., KoLodziej, J., Rudolph, G. (eds.) PPSN 2010. LNCS, vol. 6239, pp. 290–299. Springer, Heidelberg (2010). https://doi.org/10.1007/978-3-642-15871-1_30
11. Hansen, N., Ostermeier, A.: Completely derandomized self-adaptation in evolution strategies. Evol. Comput. **9**(2), 159–195 (2001)
12. Bredeche, N., Montanier, J.M., Weel, B., Haasdijk, E.: Roborobo! A fast robot simulator for swarm and collective robotics. CoRR abs/1304.2888 (2013)
13. Fernández Pèrez, I.n., Boumaza, A., Charpillet, F.: Comparison of selection methods in on-line distributed evolutionary robotics. In: Proceedings of ALIFE 2014, pp. 282–289. MIT Press, New York (2014)

Author Index

Printed in the United States
By Bookmasters